Cardiac Regeneration using Stem Cells

Cardiac Regeneration using Stem Cells

Editors

Keiichi Fukuda
Department of Cardiology
Keio University School of Medicine
Tokyo
Japan

Shinsuke Yuasa
Center for Integrated Medical Research
Department of Cardiology
Keio University School of Medicine
Shinanomachi Shinjuku-ku
Tokyo
Japan

CRC Press
Taylor & Francis Group
Boca Raton London New York

CRC Press is an imprint of the
Taylor & Francis Group, an **informa** business
A SCIENCE PUBLISHERS BOOK

CRC Press
Taylor & Francis Group
6000 Broken Sound Parkway NW, Suite 300
Boca Raton, FL 33487-2742

Printed in the United States of America on acid-free paper

International Standard Book Number: 978-1-4665-7839-5 (Hardback)

Library of Congress Cataloging-in-Publication Data

Cardiac regeneration using stem cells / editors, Keiichi Fukuda, Shinsuke Yuasa.
p. ; cm.
Includes bibliographical references and index.
ISBN 978-1-4665-7839-5 (hardcover : alk. paper)
I. Fukuda, Keiichi, editor of compilation. II. Yuasa, Shinsuke, editor of compilation.
[DNLM: 1. Heart--physiology. 2. Regeneration. 3. Pluripotent Stem Cells--transplantation. 4. Stem Cell Transplantation--methods. WG 202]
616.1'2--dc23

2012048691

Visit the Taylor & Francis Web site at
http://www.taylorandfrancis.com

CRC Press Web site at
http://www.crcpress.com

Science Publishers Web site at
http://www.scipub.net

Preface

This book presents the current status and future perspective of cardiac regeneration using pluripotent stem cells. Human induced pluripotent stem (iPS) cells were first reported in 2007, a development that came as a surprise to researchers and patients. iPS cells resemble embryonic stem cells in terms of proliferation ability and multipotency, but can be generated from somatic cells by introducing stem cell specific transcription factors. iPS cells have identical genomic information with donor somatic cells which can be obtained from the patients. Thus iPS cells overcome obstacles such as ethical and immunological problems and thereby are a boon for cardiac regeneration therapy. Since 2007 researchers have felt motivated to carry out cardiac regeneration using pluripotent stem cells. To achieve cardiac regeneration using iPS cells, research workers have to understand many aspects of iPS cells, such as iPS cell generation methods, cardiomyocyte differentiation protocols, cardiomyocyte characterization methods, and tissue engineering and so on. This book is written by many top researchers with brand-new data in these fields and will certainly help readers to get a comprehensive picture of iPS cells in cardiac regeneration.

December 3rd, 2012
Tokyo

Keiichi Fukuda
Shinsuke Yuasa

Contents

CHAPTER 1

Induced Pluripotent Stem Cells

Yoshinori Yoshida[a,*] and Shinya Yamanaka[b,*]

Patients with severe cardiac failure can be treated only by cardiac transplantation. However, the number of transplantable hearts is limited. As an alternative to heart transplantation, cell transplantation can be used to improve cardiac function and cure patients with severe heart failure. As a source of cells for regenerative medicine, ES and iPS cells are now being actively explored in research on cardiac regeneration (Nelson et al. 2010; Yoshida and Yamanaka 2010).

Stem Cells and Pluripotency

Classification of stem cells

Stem cells are classified into two types of cells, somatic stem cells and pluripotent stem cells. Somatic stem cells, which reside in bone marrow, liver, pancreas, and other organs, have the potential for multipotency and self-renewal that enables these cells to regenerate damaged tissues. However, the *in vitro* proliferative capacity of these somatic stem cells is limited. In contrast to somatic stem cells, ES cells and iPS cells have the advantages of both self-renewal and pluripotency, and are therefore expected to provide a potential new source of cells for the treatment of many different degenerative diseases.

Center for iPS Cell, Research and Application, Kyoto University, Sakyo-ku, Kyoto, Japan.
[a]Email: yoshinor@cira.kyoto-u.ac.jp
[b]Email: yamanaka@cira.kyoto-u.ac.jp
*Corresponding authors

Maintenance of pluripotency

Pluripotent stem cells (ES cells and iPS cells) maintain the two characteristics, pluripotency and self-renewal. To maintain these characteristics, several transcriptional factors play important roles. The transcription factors Oct3/4 (POU5F1), SOX2, and Nanog are known to be essential for the formation and maintenance of the epiblasts; embryos null for each of these factors are not able to give rise to ES cells (Avilion et al. 2003; Mitsui et al. 2003; Nichols et al. 1998).

Oct3/4 is expressed only in germ cells, and is reported to be essential for maintaining pluripotency (Nichols et al. 1998; Niwa et al 2000). SRY (sex determining region Y)-box 2, also known as SOX2, is a transcription factor that is essential for maintaining the self-renewal of pluripotent stem cells. SOX-2-deficiency results in embryonic lethality at the post-implantation stage (Avilion et al. 2003). Nanog is also an important factor for maintaining pluripotency. Nanog is expressed in pluripotent mouse and human cell lines and is absent from differentiated cells. In mice, Nanog-deficient inner cell mass failed to generate epiblasts and only produced parietal endoderm-like cells. In addition, Nanog-deficient ES cells lost pluripotency and differentiated into extraembryonic endoderm lineage (Chambers et al. 2003; Mitsui et al. 2003). Transgenic overexpression of Nanog is reported to be sufficient for clonal expansion of ES cells, bypassing the need for Stat3 and maintaining Oct4 levels.

Characterization of pluripotent stem cells

To characterize the pluripotent stem cell lines, an expression analysis using reverse transcription polymerase chain reaction (RT-PCR) for mRNA and immunocytochemistry for proteins can be applied. The analysis of the global expression profiles using a microarray gene expression analysis is also useful. For the evaluation of the pluripotency of these iPS cell lines, demonstration of their differentiation capacity *in vitro* and *in vivo* can be performed. Embryoid body formation can be used to assess the *in vitro* differentiation of iPS cells. To demonstrate of the capacity of stem cell lines to differentiate into specific types of cells, such as neurons or cardiomyocytes, directed differentiation can be performed. The differentiation capacity *in vivo* can be assessed by a teratoma formation assay. The most stringent criterion for mouse ES/iPS cells is their ability to generate germline-competent adult mouse chimeras, showing germline transmission, whereas germline transmission cannot be conducted using human ES/iPS cells. Currently, there is controversy surrounding the criteria that should be applied to define iPS cells (Daley et al. 2009; Ellis et al. 2009).

Differences between human and mouse pluripotent stem cells

Human and mouse pluripotent stem cells are different in several aspects. Mouse ES cells are derived from the inner cell mass of blastocysts. The expression profiles of mouse ES cells resemble those of early epiblasts, whereas the expression patterns of human ES cells are similar to those of late epiblasts. Epiblast stem cells have been generated from epiblasts (Brons et al. 2007; Tesar et al. 2007). However, while these epiblast stem cells show pluripotency *in vitro* and *in vivo*, the mouse ES cells and epiblast stem cells differ in their morphology, responses to signaling pathways, and epigenetic status.

Mouse ES cells require LIF/STAT signaling for self-renewal, and WNT/β-catenin signaling to support pluripotency. Blockade of the FGF/ERK pathway inhibits the differentiation of mouse ES cells, resulting in supporting the maintenance of pluripotency. Mouse ES cells have been reported to be maintained solely by adding a GSK3β inhibitor which promotes WNT/β-catenin signaling, and a MEK inhibitor to block the FGF/ERK pathway (Ying et al. 2008). Human ES cells have similarities to mouse epiblast stem cells, but they are different from mouse ES cells. Human ES cells do not require LIF, but do use the FGF/ERK signal to promote self-renewal.

Mouse female ES cells retain an immature state with an active X-chromosome. In contrast, in human ES cell lines derived from females, the inactivation of the X chromosome is observed (XaXi). The ectopic induction of Oct4, Klf4, and Klf2, combined with LIF, a GSK3β inhibitor, and a MEK inhibitor, is reported to convert the ES cells into a more immature state with an active X-chromosome (XaXa) (Hanna et al. 2010). These converted human ES cells have proliferative properties, gene expression profiles, and a signaling pathway-dependence which share features with mouse ES cells. It was also reported that human ES cells with two active X chromosomes (XaXa) could be generated under hypoxic conditions (5% oxygen) (Lengner et al. 2010).

As noted above, there are marked biological differences between mouse and human pluripotent stem cells. These differences might have a major impact on the effects and the risks of cell therapy using cells derived from pluripotent stem cells from different species.

Advances in iPSC Technology

Generation of ES and iPS cells

Both ES cells and iPS cells have the ability to proliferate robustly while maintaining both their pluripotency and their ability to differentiate.

ES cells are derived from the inner cell mass of mammalian blastocysts. Mouse ES cells were established in 1981 (Evans and Kaufman 1981; Martin 1981), whereas first human ES cells were reported in 1998 (Thomson et al. 1998).

We first established iPS cells in 2006 by the retrovirus-mediated transduction of four transcription factors (c-Myc, Oct3/4, SOX2, and Klf4) into mouse fibroblasts (Takahashi and Yamanaka 2006). In the first report of mouse iPS cell establishment, cells were selected based on their expression of a beta-geo cassette (a fusion of the beta-galactosidase and neomycin resistance genes) driven by the mouse *Fbx15* promoter, and the resultant cells showed the ability to form teratomas with histological evidence of cells differentiating into all three germ layers after subcutaneous injection into nude mice. However, these cells failed to contribute to adult chimeras.

However, when selected by the expression of *Nanog* or *Oct3/4*, which are closely involved in pluripotency, the resultant iPS cells successfully contributed to adult chimeras, while also showing germline transmission (Maherali et al. 2007; Okita et al. 2007; Wernig et al. 2007). Mouse iPSC lines have been reported to show even the ability to autonomously generate full-term mice after injection into tetraploid blastocysts (Kang et al. 2009; Zhao et al. 2009).

Human iPS cells were established in 2007 (Takahashi et al. 2007; Yu et al. 2007) by retrovirus- or lentivirus-mediated transduction of either the same transcription factors (c-Myc, Oct3/4, SOX2, Klf4) or another set of transcriptional factors (Oct3/4, SOX2, Nanog, Lin28) into human fibroblasts. These human iPS cells are similar to human ES cells in terms of their morphology, gene expression, and the epigenetic status of pluripotent cell-specific genes. Furthermore, these cells have been demonstrated to be able to differentiate into cell types of the three germ layers *in vitro* and formed teratomas *in vivo*.

The iPS cell technology provides a new source of cells for regenerative medicine and disease-specific cell research. Additionally, in some species in which ES cells have not yet been established so far, iPS cells may be useful for species preservation or chimera animal research. In the past, iPS cell generation from rats, pigs, rabbits and monkeys (rhesus monkey and marmoset), has been reported (Esteban et al. 2009b; Ezashi et al. 2009; Honda et al. 2010; Li et al. 2009b; Liao et al. 2009; Liu et al. 2008; Tomioka et al. 2010; West et al. 2010; Wu et al. 2010), thereby demonstrating that it appears to be possible to generate cells from various mammals.

Methodological variation of iPS cell generation

In the iPS cells established by retroviral or lentiviral transduction of reprogramming factors, the transgenes randomly inserted into the host genome may cause insertional mutations. Furthermore, a very low level of expression may still remain after the achievement of reprogramming although the expression of the transgenes is silenced after the induction of the pluripotent stem cell state. This persistence of transgenic expression may render the iPS cells refractory to differentiation both *in vitro* and *in vivo*. Furthermore, the aberrant expression of oncogenic transgenes, such as c-Myc, may lead to the development of tumors after the transplantation of iPS-derived cells. Establishment of iPS cells without the integration of transgenes using expression plasmids, including an episomal plasmid system, adenoviruses, and Sendai viruses have been reported (Fusaki et al. 2009; Okita et al. 2008; Stadtfeld et al. 2008b; Yu et al. 2009). Recombinant proteins and modified RNAs were also reported to successfully reprogram somatic cells into iPS cells (Kim et al. 2009a; Warren et al. 2010; Zhou et al. 2009).

Episomal vectors derived from the Epstein-Barr virus have been improved and achieved robust reprogramming with high efficiency (Okita et al. 2011; Wang et al. 2011). On the other hand, minicircle vectors, which are supercoiled DNA molecules that lack a bacterial origin of replication and antibiotic resistance gene, can also reprogram somatic cells into iPS cells (Jia et al. 2010).

Both mouse and human fibroblasts can be reprogrammed using the piggyBac transposon system (Kaji et al. 2009; Woltjen et al. 2009; Yusa et al. 2009). The piggyBac insertions can be removed from the established iPS cells by transposase.

It has been reported that transduction of three transcriptional factors devoid of c-Myc (Oct3/4, Klf4, SOX2) can also reprogram the somatic cells into iPS cells (Nakagawa et al. 2008; Wernig et al. 2008). The use of L-Myc, which has little transformational activity, instead of c-Myc, increases the reprogramming efficiency in human cells and promotes germline transmission, but not tumor formation, in iPSC-derived chimeric mice (Nakagawa et al. 2010).

Mouse neural stem cells, in which the endogenous expression of c-Myc and SOX2 is high, can be reprogrammed by two transcription factors, Oct3/4 and Klf4 or even just one transcription factor (Oct3/4). Human fetal neural stem cells are also reported to be reprogrammed by Oct3/4 alone (Kim et al. 2009b; Kim et al. 2009c; Kim et al. 2008).

Improvement of reprogramming efficiency

The c-Myc and KLF4 transgenes, which are known to causes tumors in mice, may make reprogrammed cells tumorigenic, and reprogramming cells without such oncogenes is a great step toward achievement of safe stem cell therapy. IPS cells can now be established by methods that do not require viral integration, such as via episomal plasmid vectors and Sendai viruses. However, it is still unknown whether transient expression of oncogenes, such as c-Myc, makes iPS cells tumorigenic during the reprogramming process.

It was previously reported that the supplementation of valproic acid (a histone deacetylase inhibitor) increases the reprogramming efficiency (Huangfu et al. 2008a; Huangfu et al. 2008b). It has also been reported that hypoxic cultivation and supplementation of vitamin C enhances the efficiency of iPS cell generation from somatic cells (Esteban et al. 2009a; Yoshida et al. 2009). Lin et al. reported that treatment with a combination of an ALK5 inhibitor (SB431542) and a MEK inhibitor (PD0325901) increased the reprogramming efficiency, and that addition of a small molecule, thiazovivin, which improves the survival of hESCs upon trypsinization, dramatically increased the number of iPS cells after trypsinization (Lin et al. 2009).

The direct delivery of recombinant proteins as reprogramming factors has been reported to reprogram human and mouse somatic cells into iPS cells (Kim et al. 2009a; Zhou et al. 2009). As described above, since some chemical compounds can improve the reprogramming efficiency, iPS cells may therefore be established using only chemical compounds in the near future.

Epigenetic changes in reprogramming

Gene expression is regulated by the modification of chromatin and genomic DNA. Pluripotent stem cells and differentiated somatic cells have different epigenetic status (Bernstein et al. 2006). During the iPS cell reprogramming process, the epigenetic signatures undergo drastic modifications, changing from those of somatic cells into those of pluripotent stem cells.

Methylation of the core histones can contribute to the formation of transcriptionally active and inactive chromatin in response to various signaling pathways, and is a central modification involved in regulating epigenetic transitions in the chromatin. It was reported that the repressive H3K9 methylation, which is mediated by histone methyltransferase G9a, is associated with Oct3/4 inactivation during differentiation in ES cells (Feldman et al. 2006). In accordance with this finding, BIX-01294 (BIX), an inhibitor of the G9a histone methyltransferase, was shown to improve the reprogramming efficiency (Shi et al. 2008).

Huangfu et al. reported that treatment with a DNMT (DNA methyltransferase) inhibitor, 5-Azacytidine, and histone deacetylase (HDAC) inhibitors, such as Trichostatin A and valploic acid, can increase the reprogramming efficiency during iPS cell generation. The mouse embryonic fibroblasts treated with HDAC inhibitors showed upregulation of ES-specific genes and down-regulation of fibroblast-specific genes (Huangfu et al. 2008a; Huangfu et al. 2008b). Furthermore Mali et al. reported that butyrate, which also works as a histone deacetylase inhibitor, can greatly enhance the efficiency of human iPS cell generation by promoting epigenetic remodeling and the expression of pluripotency-associated genes (Mali et al. 2010). These data support the concept that modification of the epigenetic status accelerates the reprogramming process during iPS cell generation.

Cell cycle regulators and reprogramming

The tumor suppressor protein p53 pathway has been reported to protect somatic cells from being reprogrammed into iPS cells, and that the blockade of p53 increases the reprogramming efficiency (Hong et al. 2009; Kawamura et al. 2009; Li et al. 2009a; Marion et al. 2009; Utikal et al. 2009b). Suppression of p16INK4a or p19ARF, which are encoded by Cdkn2a, was also reported to increase the efficiency of iPS cell generation (Banito et al. 2009; Li et al. 2009a; Utikal et al. 2009b). Rem2 enhances the reprogramming efficiency by accelerating the cell cycle and protecting cells from apoptosis via its effects on cyclin D1 expression/localization and its suppression of p53 transcription (Edel et al. 2010).

Mesenchymal-to-epithelial transition and reprogramming

During the reprogramming process, somatic cells, such as fibroblasts, are required to undergo a mesenchymal-to-epithelial transition in order to be reprogrammed to iPS cells. In the mouse, hepatocytes and gastric epithelial cells appear to be more easily reprogrammed and require fewer retroviral integrations than fibroblasts, and in the human, keratinocytes are reported to be reprogrammed more efficiently than fibroblasts (Aasen et al. 2008; Aoi et al. 2008). It is therefore conceivable that epithelial cells *per se* are more amenable to reprogramming without such mesenchymal-to-epithelial transition.

The inhibition of TGF-β signaling has been reported to increase the efficiency of iPS cell generation from mouse embryonic fibroblasts (Ichida et al. 2009; Maherali and Hochedlinger 2009). TGF-β signaling can promote the epithelial-to-mesenchymal transition and blockade of TGF-β signaling may enhance the reprogramming efficiency by supporting the mesenchymal-to-epithelial transition during reprogramming from fibroblasts to iPS cells.

Forced expression of E-cadherin has been reported to increase the efficiency of iPS cell generation (Chen et al. 2010). At the transcriptional level, Sox2/Oct4 suppress the EMT mediator Snail, c-Myc down regulates TGF-β1 and TGF-β receptor 2, Klf4 induces epithelial genes including

E-cadherin, and knock-down of E-cadherin decreases the efficiency of iPS cell generation from mouse embryonic fibroblasts (Li et al. 2010).

Micron cluster 302–367 enhances iPS cell generation by targeting TGF-β receptor 2, promoting increased E-cadherin expression, and finally accelerating the mesenchymal-to-epithelial transition in reprogramming cells (Liao et al. 2011).

Micron in reprogramming

MicroRNAs play important roles in maintaining cell integrity and biological activities in the pluripotent stem cells, as well as other types of cells, through regulation of mRNA stability and translation. ES cells express several types of specific miRNA, such as the miR-290 cluster in mouse cells, the miR-371–373 cluster in human cells, or the miR-302 cluster, which is conserved in both (Houbaviy et al. 2003; Marson et al. 2008; Suh et al. 2004). The miRNAs miR-291-3p, miR-294 and miR-295 increase the efficiency of reprogramming by Oct4, Sox2 and Klf4, but not by these factors plus c-Myc (Judson et al. 2009). Precisely how these miRNAs enhance iPSC reprogramming is unclear, but may be related to their ability to regulate the cell cycle (Judson et al. 2009). MiR-302 is regulated by Oct3/4 and SOX2 in human ES cells, and represses the expression of cyclin D1 (Card et al. 2008). The ectopic expression of the miR-302 cluster has also been reported to alleviate the growth arrest during reprogramming-induced senescence (Banito et al. 2009) and that miRNA cluster 302–367 enhances somatic cell reprogramming by accelerating a mesenchymal-to-epithelial transition (Liao et al. 2011). Anokye-Danso et al. reported that forced expression of the miR302/367 cluster alone rapidly and efficiently reprograms mouse and human somatic cells to an iPSC state without the need for exogenous transcription factors, such as Oct3/4 (Anokye-Danso et al. 2011).

Lin28 is a developmentally regulated RNA binding protein that selectively blocks the processing of miRNA, such as let-7, in embryonic cells (Viswanathan et al. 2008). Lin28 is reported to upregulate the efficiency of human iPS cell generation (Liao et al. 2008; Yu et al. 2009; Yu et al. 2007). However, the precise mechanism of Lin28-mediated enhancement of iPS cell generation is still unknown.

Unresolved Issues

Genetic alterations and reprogramming

Human pluripotent stem cells are a potential source of differentiated cells for regenerative medicine. However, such source cells need to be nontumorigenic. Since genetic alterations are strongly associated with tumorigenesis, the cells used for cell therapy should be free from genetic alterations associated with tumorigenesis. During iPS cell generation, modulating key components of cell cycle, such as Ink4a/ARF or suppression of p53–p21 pathway, improve the reprogramming efficiency. These facts suggest that the reprogramming process might be associated with an inherent risk of genetic instability.

Aneuploidy has been repeatedly detected after prolonged culture of human ES cells (Cowan et al. 2004). The aberrations in copy number are known to correlate with the level of expression of the gene. Human ES cells are considered to acquire chromosomal aberrations that may confer a growth advantage. These adaptive copy number aberrations are frequently found in chromosomes 12, 17, and X (Baker et al. 2007; Draper et al. 2004).

Mayshar et al. analyzed the chromosomal integrity of human iPS and ES cell lines and reported that a substantial number of cell lines harbor chromosomal abbreviations. Some abbreviations resulted from culture adaptation, while others originated from the parent somatic cells. Their analysis also indicated that the other groups of aneuploidy are evident in early passage human iPS cells, suggesting that there is selective pressure during the reprogramming process (Mayshar et al. 2010). Laurent et al. reported that the frequency of subchromosomal copy number variations (CNVs) in human pluripotent stem cells is high compared to non-pluripotent cells. The analysis of affected genes showed that the reprogramming process might be associated with deletions of tumor-suppressor genes, whereas maintenance of the cell lines was associated with duplications of oncogenic genes (Laurent et al. 2011).

Hussein et al. compared the CNVs of different passages of human iPS cells with the original fibroblasts and human ES cells, and showed that more CNVs were present in early-passage human iPS cells than intermediate-passage human iPS cells, fibroblasts or human ES cells. The number of CNVs negatively correlated with the passage number (Hussein et al. 2011). Their findings indicate that CNVs are generated during the reprogramming process, and that most *de novo* CNVs rendered the affected cells at a selective disadvantage during expansion in culture.

Using high-throughput sequencing technology, Gore et al. analyzed the human iPS cell lines generated using different methods. These iPS cell lines contained an average of five protein-coding point mutations in the regions sampled (Gore et al. 2011). Half of these reprogramming-associated mutations were seen in parental fibroblasts at low frequency and the rest may have occurred during the reprogramming process. The selection during iPS cell generation may contribute to the expansion of cells with apparent "mutations".

The karyotyping analysis, CNV analysis using CNV or SNP array, and high-throughput sequencing revealed the risk of genetic alterations, which may result in aberrant behavior of affected cells. Supposedly, the incidence of genetic alteration in iPS cells may be associated with the methods used to generate and maintain the iPS cells. If the reprogramming efficiency could be improved such that colony picking and clonal expansion are unnecessary, the selection pressure which enhances the risk of genetic alterations will be reduced.

Epigenetic status in iPS cells

Regarding their expression profiles, human and mouse iPS cells are similar to ES cells; however there are differences in behavior during differentiation and proliferation (Feng et al. 2010; Hu et al. 2010). These differences cannot be explained solely by the expression profiles of these pluripotent stem cell lines. It was reported that there is a persistent memory of the origin in reprogrammed pluripotent stem cells (Chin et al. 2009). These memories are proposed to result from the differences in DNA methylation (Kim et al. 2010; Polo et al. 2010) and are particularly prominent in the early passages of iPS cells. They are supposed to have an important influence on the cell fate with regard to differentiation, and favor their differentiation along lineages related to the donor cell type.

In a whole-genome DNA methylation analysis of human iPS cells, ES cells, and somatic cells, iPS cells showed significant reprogramming variability. CG-DMRs (differently methylated regions) between iPS cells and ES cells can be classified into a failure to reprogram the somatic cell methylation pattern (somatic epigenetic memory) and iPS cell-specific DMRs, which cannot be seen in original somatic cells and ES cells (Lister et al. 2011). These DMRs, somatic memory and iPS cell-specific DMRs were reported to be transmitted at high frequency through differentiation.

These differences in epigenetic status observed in iPS cells likely have an influence on the behavior of iPS cells. With respect to the acquisition of bona fide pluripotency, the optimal reprogramming of somatic cells to a pluripotent state requires complete reversion of the somatic epigenome into an ES-like state. However, for the purpose of preparing the differentiated cells for cell therapy or disease studies, iPS cells with somatic memories might be useful for favoring the cell fate, thus leading to preferential differentiation into certain types of cells.

Immunogenicity in iPS cells?

Zhao et al. compared the teratoma formation of C57BL/6 (B6) ES cells and iPS cells in B6 mice. They reported that the teratomas formed by B6 iPS cells established by episomal vectors were immunogenic in B6 mice, as indicated by T cell infiltration, whereas teratomas of ES cells derived from B6 mice grew efficiently without any evident immunogenic rejection (Zhao et al. 2011). They also reported that abnormal gene expression in some cells differentiated from iPS cells was detected, and such abnormal expression could induce T cell-dependent immune responses in syngeneic recipients. However, it is yet to be determined whether the reported difference between iPS cells and ES cells is statistically significant. Furthermore, their observation of the immunogenicity of iPS cells was in the case of teratoma formation, and it is still unknown whether such a response would occur in the case of transplantation of differentiated cells, such as neurons or cardiomyocytes.

Heterogeneity and clonal variations in pluripotent stem cell lines

The pluripotent stem cell populations do not comprise a homogenous cellular entity. Stem cells display an inherent heterogeneity at the molecular level, and this heterogeneity affects the fate of differentiation (Chambers et al. 2007; Guo et al. 2010; Stewart et al. 2006). Using single cell PCR methods, Narsinh et al. analyzed the gene expression profiles for an array of 42 genes that characterize the pluripotent and differentiated states of cells, and reported markedly more heterogeneity in the gene expression levels in the hiPSCs (Narsinh et al. 2011). This heterogeneity may affect the behavior of pluripotent stem cells, such as their differentiation and self-renewal. The heterogeneous population of pluripotent cell lines raises the possibility of the existence of differentiation-resistant cells. Even a small population of iPS cells refractory to differentiation could result in immature teratoma formation after transplantation to patients. For the clinical applications, attention should be paid to the heterogeneity of the iPS cells.

iPS cells have been derived from various mouse tissues, including embryonic fibroblasts, adult tail-tip fibroblasts, hepatocytes, gastric epithelial cells, pancreatic cells, melanocytes, adipose-derived cells, neural stem cells, and lymphocytes (Aoi et al. 2008; Eminli et al. 2008; Hanna et al. 2008; Kim et al. 2008; Nakagawa et al. 2008; Okita et al. 2007; Silva et al. 2008; Stadtfeld et al. 2008a; Sugii et al. 2010; Utikal et al. 2009a), and from human skin fibroblasts, keratinocytes, melanocytes, CD34+ hematopoietic stem/progenitor cells, adipose-derived cells, dental pulp cells, and differentiated T cells from the peripheral blood (Aasen et al. 2008; Loh et al. 2009; Loh et al. 2010; Park et al. 2008; Seki et al. 2010; Staerk et al. 2010; Sugii et al. 2010; Takahashi et al. 2007; Tamaoki et al. 2010; Utikal et al. 2009a; Yan et al. 2009; Yu et al. 2007). It is currently unclear whether these iPS cells derived from various tissues have the same characteristics as pluripotent stem cells.

Neurospheres containing neural stem and progenitor cells from mouse iPS cell lines derived from embryonic fibroblasts (MEF-iPS), adult tail-tip fibroblasts (TTF-iPS), hapatocytes, and stomach epithelial cells have been demonstrated to have different teratoma-forming propensities for the secondary neurospheres (Miura et al. 2009). We demonstrated that secondary neurospheres derived from TTF-iPS cells contained significantly higher amounts of undifferentiated cells than those from MEF-iPS cells. The neurospheres from the iPS cells derived from different tissues also varied substantially with regard to their teratoma-forming propensity, which correlated with the persistence of undifferentiated cells. These data suggest that iPS cells from different origins may have different propensities to differentiate, and certain cell types may thus be better for complete reprogramming because of their reduced risk of teratoma formation.

In human ES cells, marked differences have been reported in the differentiation capacity between various cell lines (Osafune et al. 2008). Such differences in the propensity for differentiation and in the persistence of cells refractory to differentiation may exist among the clones derived from different tissues used to generate human iPS cells.

Significant improvements have been rapidly achieved with regard to the efficiency of iPS cell generation and the quality of established iPS cells. However, there are still hurdles to be solved, such as decreasing the genetic alterations and epigenetic aberrations before the technology can advance to clinical applications. Each experiment generates many iPS cell clones. Among these iPS cells, some iPS cells are equivalent to ES cells with full reprogramming with regard to their pluripotency, while other cells are only partially reprogrammed or might be refractory to induced differentiation. Such variations of the derived iPS cells are reprogramming-specific features. In addition, there are many methods that can be used to establish and maintain iPS cells, such as transduction by retroviruses, episomal plasmids, and modified RNAs. It is currently unclear which way is the best to establish and maintain clinically useful iPS cells. Further studies are required to better understand the nature of iPS cells and elucidate the best methods for generating clinically applicable iPS cells.

References

Aasen, T., Raya A., Barrero, M. J., Garetta, E., Consiglio, A., Gonzalez, F., Vassena, R., Bilic, J., Pekarik, V., and Tiscomia, G. 2008. Efficient and rapid generation of induced pluripotent stem cells from human keratinocytes. Nat. Biotechnol. 26, 1276–1284.

Anokye-Danso, F., Trivedi, C. M., Juhr, D., Gupta, M., Cui, Z., Tian, Y., Zhang, Y., Yang, W., Gruber, P. J., Epstein, J. A. et al. 2011. Highly efficient miRNA-mediated reprogramming of mouse and human somatic cells to pluripotency. Cell Stem Cell 8, 376–388.

Aoi, T., Yae, K., Nakagawa, M., Ichisaka, T., Okita, K., Takahashi, K., Chiba, T., and Yamanaka, S. 2008. Generation of pluripotent stem cells from adult mouse liver and stomach cells. Science 321, 699–702.

Avilion, A. A., Nicolis, S. K., Pevny, L. H., Perez, L., Vivian, N., and Lovell-Badge, R. 2003. Multipotent cell lineages in early mouse development depend on SOX2 Function. Genes Dev. 17, 126–140.

Baker, D. E., Harrison, N. J., Maltby, E., Smith, K., Moore, H. D., Shaw, P. J., Heath, P. R., Holden, H., and Andrews, P. W. 2007. Adaptation to culture of human embryonic stem cells and oncogenesis *in vivo*. Nat. Biotechnol. 25, 207–215.

Banito, A., Rashid, S. T., Acosta, J. C., Li, S., Pereira, C. F., Geti, I., Pinho, S., Silva, J. C., Azuara, V., Walsh, M. et al. 2009. Senescence impairs successful reprogramming to pluripotent stem cells. Genes Dev. 23, 2134–2139.

Bernstein, B. E., Mikkelsen, T. S., Xie, X., Kamal, M., Huebert, D. J., Cuff, J., Fry, B., Meissner, A., Wernig, M., Plath, K. et al. 2006. A bivalent chromatin structure marks key developmental genes in embryonic stem cells. Cell 125, 315–326.

Brons, I. G., Smithers, L. E., Trotter, M. W., Rugg-Gunn, P., Sun, B., Chuva de Sousa Lopes, S. M., Howlett, S. K., Clarkson, A., Ahrlund-Richter, L., Pedersen, R.A. et al. 2007. Derivation of pluripotent epiblast stem cells from mammalian embryos. Nature 448, 191–195.

Card, D. A., Hebbar, P. B., Li, L., Trotter, K. W., Komatsu, Y., Mishina, Y., and Archer, T. K. 2008. Oct4/Sox2-regulated miR-302 targets cyclin D1 in human embryonic stem cells. Mol. Cell. Biol. 28, 6426–6438.

Chambers, I., Colby, D., Robertson, M., Nichols, J., Lee, S., Tweedie, S., and Smith, A. 2003. Functional expression cloning of nanog, a pluripotency sustaining factor in embryonic stem cells. Cell 113, 643–655.

Chambers, I., Silva, J., Colby, D., Nichols, J., Nijmeijer, B., Robertson, M., Vrana, J., Jones, K., Grotewold, L., and Smith, A. 2007. Nanog safeguards pluripotency and mediates germline development. Nature 450, 1230–1234.

Chen, T., Yuan, D., Wei, B., Jiang, J., Kang, J., Ling, K., Gu, Y., Li, J., Xiao, L., and Pei, G. 2010. E-cadherin-mediated cell-cell contact is critical for induced pluripotent stem cell generation. Stem Cells 28,1315–1325.

Chin, M. H., Mason, M. J., Xie, W., Volinia, S., Singer, M., Peterson, C., Ambartsumyan, G., Aimiuwu, O., Richter, L., Zhang, J. et al. 2009. Induced pluripotent stem cells and embryonic stem cells are distinguished by gene expression signatures. Cell Stem Cell 5, 111–123.

Cowan, C. A., Klimanskaya, I., McMahon, J., Atienza, J., Witmyer, J., Zucker, J. P., Wang, S., Morton, C. C., McMahon, A. P., Powers, D. et al. 2004. Derivation of embryonic stem-cell lines from human blastocysts. N. Engl. J. Med. 350, 1353–1356.

Daley, G. Q., Lensch, M. W., Jaenisch, R., Meissner, A., Plath, K., and Yamanaka, S. 2009. Broader Implications of Defining Standards for the Pluripotency of iPSCs. Cell Stem Cell 4, 200–201.

Draper, J. S., Smith, K., Gokhale, P., Moore, H. D., Maltby, E., Johnson, J., Meisner, L., Zwaka, T. P., Thomson, J. A., and Andrews, P. W. 2004. Recurrent gain of chromosomes 17q and 12 in cultured human embryonic stem cells. Nat. Biotechnol. 22, 53–54.

Edel, M. J., Menchon, C., Menendez, S., Consiglio, A., Raya, A., and Izpisua Belmonte, J. C. 2010. Rem2 GTPase maintains survival of human embryonic stem cells as well as enhancing reprogramming by regulating p53 and cyclin D1. Genes Dev. 24, 561–573.

Ellis, J., Bruneau, G. B., Keller, G., Lemischka, I. R., Nagy, A., Rossant, J., Srivastava, D., Zandstra, P. W., and Stanford, W. L. 2009. Alternative induced pluripotent stem cell characterization criteria for in vitro applications. Cell Stem Cell 4, 198–199.

Eminli, S., Utikal, J. S., Arnold, K., Jaenisch, R., and Hochedlinger, K. 2008. Reprogramming of neural progenitor cells into iPS cells in the absence of exogenous Sox2 expression. Stem Cells 10, 2467–74

Esteban, M. A., Wang, T., Qin, B., Yang, J., Qin, D., Cai, J., Li, W., Weng, Z., Chen, J., Ni, S. et al. 2009a. Vitamin C enhances the generation of mouse and human induced pluripotent stem cells. Cell Stem Cell 6, 71–9.

Esteban, M. A., Xu, J., Yang, J., Peng, M., Qin, D., Li, W., Jiang, Z., Chen, J., Deng, K., Zhong, M. et al. 2009b. Generation of induced pluripotent stem cell lines from Tibetan miniature pig. J. Biol. Chem 284, 17634–40.

Evans, M. J. and Kaufman, M. H. 1981. Establishment in culture of pluripotential cells from mouse embryos. Nature 292, 154–156.

Ezashi, T., Telugu, B. P., Alexenko, A. P., Sachdev, S., Sinha, S., and Roberts, R. M. 2009. Derivation of induced pluripotent stem cells from pig somatic cells. Proc. Natl. Acad. Sci. USA 106, 10993–10998.

Feldman, N., Gerson, A., Fang, J., Li, E., Zhang, Y., Shinkai, Y., Cedar, H., and Bergman, Y. 2006. G9a-mediated irreversible epigenetic inactivation of Oct-3/4 during early embryogenesis. Nat. Cell Biol. 8, 188–194.

Feng, Q., Lu, S. J., Klimanskaya, I., Gomes, I., Kim, D., Chung, Y., Honig, G. R., Kim, K. S., and Lanza, R. 2010. Hemangioblastic derivatives from human induced pluripotent stem cells exhibit limited expansion and early senescence. Stem Cells 28, 704–12.

Fusaki, N., Ban, H., Nishiyama, A., Saeki, K., and Hasegawa, M. 2009. Efficient induction of transgene-free human pluripotent stem cells using a vector based on Sendai virus, an RNA virus that does not integrate into the host genome. Proc. Jpn. Acad. Se.r B. Phys. Biol. Sci. 85, 348–362.

Gore, A., Li, Z., Fung, H. L., Young, J. E., Agarwal, S., Antosiewicz-Bourget, J., Canto, I., Giorgetti, A., Israel, M. A., Kiskinis, E. et al. 2011. Somatic coding mutations in human induced pluripotent stem cells. Nature 471, 63–67.

Guo, G., Huss, M., Tong, G. Q., Wang, C., Li Sun, L., Clarke, N. D., and Robson, P. 2010. Resolution of cell fate decisions revealed by single-cell gene expression analysis from zygote to blastocyst. Dev. Cell. 18, 675–685.

Hanna, J., Cheng, A. W., Saha, K., Kim, J., Lengner, C. J., Soldner, F., Cassady, J. P., Muffat, J., Carey, B.W., and Jaenisch, R. 2010. Human embryonic stem cells with biological and epigenetic characteristics similar to those of mouse ESCs. Proc. Natl. Acad. Sci. USA 107, 9222–9227.

Hanna, J., Markoulaki, S., Schorderet, P., Carey, B.W., Beard, C., Wernig, M., Creyghton, M. P., Steine, E. J., Cassady, J. P., Foreman, R. et al. 2008. Direct reprogramming of terminally differentiated mature B lymphocytes to pluripotency. Cell 133, 250–264.

Honda, A., Hirose, M., Hatori, M., Matoba, S., Miyoshi, H., Inoue, K., and Ogura, A. 2010. Generation of induced pluripotent stem cells in rabbits: potential experimental models for human regenerative medicine. J. Biol. Chem. 285, 31362–9.

Hong, H., Takahashi, K., Ichisaka, T., Aoi, T., Kanagawa, O., Nakagawa, M., Okita, K., and Yamanaka, S. 2009. Suppression of induced pluripotent stem cell generation by the p53–p21 pathway. Nature 460, 1132–1135.

Houbaviy, H. B., Murray, M. F., and Sharp, P. A. 2003. Embryonic stem cell-specific MicroRNAs. Dev. Cell. 5, 351–358.

Hu, B. Y., Weick, J. P., Yu, J., Ma, L. X., Zhang, X. Q., Thomson, J. A., and Zhang, S. C. 2010. Neural differentiation of human induced pluripotent stem cells follows developmental principles but with variable potency. Proc. Natl Acad. Sci. USA. 107, 4335–4

Huangfu, D., Maehr, R., Guo, W., Eijkelenboom, A., Snitow, M., Chen, A. E., and Melton, D. A. 2008a. Induction of pluripotent stem cells by defined factors is greatly improved by small-molecule compounds. Nat. Biotechnol. 26, 795–797.

Huangfu, D., Osafune, K., Maehr, R., Guo, W., Eijkelenboom, A., Chen, S., Muhlestein, W., and Melton, D. A. 2008b. Induction of pluripotent stem cells from primary human fibroblasts with only Oct4 and Sox2. Nat. Biotechnol. 26, 1269–1275.

Hussein, S. M., Batada, N. N., Vuoristo, S., Ching, R. W., Autio, R., Narva, E., Ng, S., Sourour, M., Hamalainen, R., Olsson, C. et al 2011. Copy number variation and selection during reprogramming to pluripotency. Nature 471, 58–62.

Ichida, J. K., Blanchard, J., Lam, K., Son, E. Y., Chung, J. E., Egli, D., Loh, K. M., Carter, A. C., Di Giorgio, F. P., Koszka, K. et al. 2009. A small-molecule inhibitor of tgf-Beta signaling replaces sox2 in reprogramming by inducing nanog. Cell Stem Cell 5, 491–503.

Jia, F., Wilson, K. D., Sun, N., Gupta, D. M., Huang, M., Li, Z., Panetta, N. J., Chen, Z. Y., Robbins, R. C., Kay, M. A. et al. 2010. A nonviral minicircle vector for deriving human iPS cells. Nat. Methods 7, 197–199.

Judson, R. L., Babiarz, J. E., Venere, M., and Blelloch, R. 2009. Embryonic stem cell-specific microRNAs promote induced pluripotency. Nat. Biotechnol. 27, 459–461.

Kaji, K., Norrby, K., Paca, A., Mileikovsky, M., Mohseni, P., and Woltjen, K. 2009. Virus-free induction of pluripotency and subsequent excision of reprogramming factors. Nature 458, 771–775.

Kang, L., Wang, J., Zhang, Y., Kou, Z., and Gao, S. 2009. iPS cells can support full-term development of tetraploid blastocyst-complemented embryos. Cell Stem Cell 5, 135–8.

Kawamura, T., Suzuki, J., Wang, Y. V., Menendez, S., Morera, L. B., Raya, A., Wahl, G. M., and Belmonte, J. C. 2009.Linking the p53 tumour suppressor pathway to somatic cell reprogramming. Nature 460, 1140–1144.

Kim, D., Kim, C. H., Moon, J. I., Chung, Y. G., Chang, M. Y., Han, B. S., Ko, S., Yang, E., Cha, K. Y., Lanza, R. et al. 2009a. Generation of human induced pluripotent stem cells by direct delivery of reprogramming proteins. Cell Stem Cell 4, 472–476.

Kim, J. B., Greber, B., Arauzo-Bravo, M. J., Meyer, J., Park, K. I., Zaehres, H., and Scholer, H. R. 2009b. Direct reprogramming of human neural stem cells by OCT4. Nature 461, 649–3.

Kim, J. B., Sebastiano, V., Wu, G., Arauzo-Bravo, M. J., Sasse, P., Gentile, L., Ko, K., Ruau, D., Ehrich, M., van den Boom, D. et al. 2009c. Oct4-induced pluripotency in adult neural stem cells. Cell 136, 411–419.

Kim, J. B., Zaehres, H., Wu, G., Gentile, L., Ko, K., Sebastiano, V., Arauzo-Bravo, M. J., Ruau, D., Han, D. W., Zenke, M. et al. 2008. Pluripotent stem cells induced from adult neural stem cells by reprogramming with two factors. Nature 454, 646–650

Kim, K., Doi, A., Wen, B., Ng, K., Zhao, R., Cahan, P., Kim, J., Aryee, M. J., Ji, H., Ehrlich, L. I. et al. 2010. Epigenetic memory in induced pluripotent stem cells. Nature 467, 285–90.

Laurent, L. C., Ulitsky, I., Slavin, I., Tran, H., Schork, A., Morey, R., Lynch, C., Harness, J. V., Lee, S., Barrero, M. J. et al. 2011. Dynamic changes in the copy number of pluripotency and cell proliferation genes in human ESCs and iPSCs during reprogramming and time in culture. Cell Stem Cell 8, 106–118.

Lengner, C. J., Gimelbrant, A. A., Erwin, J. A., Cheng, A. W., Guenther, M. G., Welstead, G. G., Alagappan, R., Frampton, G. M., Xu, P., Muffat, J. et al. 2010. Derivation of pre-X inactivation human embryonic stem cells under physiological oxygen concentrations. Cell 141, 872–883.

Li, H., Collado, M., Villasante, A., Strati, K., Ortega, S., Canamero, M., Blasco, M. A., and Serrano, M. 2009a. The Ink4/Arf locus is a barrier for iPS cell reprogramming. Nature 460, 1136–1139.

Li, R., Liang, J., Ni, S., Zhou, T., Qing, X., Li, H., He, W., Chen, J., Li, F., Zhuang, Q. et al. 2010. A mesenchymal-to-epithelial transition initiates and is required for the nuclear reprogramming of mouse fibroblasts. Cell Stem Cell 7, 51–63.

Li, W., Wei, W., Zhu, S., Zhu, J., Shi, Y., Lin, T., Hao, E., Hayek, A., Deng, H., and Ding, S. 2009b. Generation of rat and human induced pluripotent stem cells by combining genetic reprogramming and chemical inhibitors. Cell Stem Cell 4, 16–19.

Liao, B., Bao, X., Liu, L., Feng, S., Zovoilis, A., Liu, W., Xue, Y., Cai, J., Guo, X., Qin, B. et al. 2011. Microrna cluster 302–367 enhances somatic cell reprogramming by accelerating a mesenchymal-to-epithelial transition. J. Biol. Chem. 286, 17359–64.

Liao, J., Cui, C., Chen, S., Ren, J., Chen, J., Gao, Y., Li, H., Jia, N., Cheng, L., Xiao, H. et al. 2009. Generation of induced pluripotent stem cell lines from adult rat cells. Cell Stem Cell 4, 11–15.

Liao, J., Wu, Z., Wang, Y., Cheng, L., Cui, C., Gao, Y., Chen, T., Rao, L., Chen, S., Jia, N. et al. 2008. Enhanced efficiency of generating induced pluripotent stem iPS cells from human somatic cells by a combination of six transcription factors. Cell Res. 18, 600–603.

Lin, T., Ambasudhan, R., Yuan, X., Li, W., Hilcove, S., Abujarour, R., Lin, X., Hahm, H. S., Hao, E., Hayek, A. et al. 2009. A chemical platform for improved induction of human iPSCs. Nat. Methods 6, 805–808.

Lister, R., Pelizzola, M., Kida, Y. S., Hawkins, R. D., Nery, J. R., Hon, G., Antosiewicz-Bourget, J., O'Malley, R., Castanon, R., Klugman, S. et al. 2011. Hotspots of aberrant epigenomic reprogramming in human induced pluripotent stem cells. Nature 471, 68–73.

Liu, H., Zhu, F., Yong, J., Zhang, P., Hou, P., Li, H., Jiang, W., Cai, J., Liu, M., Cui, K. et al. 2008. Generation of induced pluripotent stem cells from adult rhesus monkey fibroblasts. Cell Stem Cell 3, 587–590.

Loh, Y. H., Agarwal, S., Park, I. H., Urbach, A., Huo, H., Heffner, G. C., Kim, K., Miller, J. D., Ng, K., and Daley, G. Q. 2009. Generation of induced pluripotent stem cells from human blood. Blood 113, 5476–5479.

Loh, Y. H., Hartung, O., Li, H., Guo, C., Sahalie, J. M., Manos, P. D., Urbach, A., Heffner, G. C., Grskovic, M., Vigneault, F. et al. 2010. Reprogramming of T cells from human peripheral blood. Cell Stem Cell 7, 15–19.

Maherali, N. and Hochedlinger, K. 2009. Tgfbeta signal inhibition cooperates in the induction of iPSCs and replaces Sox2 and cMyc. Curr. Biol. 19, 1718–1723.

Maherali, N., Sridharan, R., Xie, W., Utikal, J., Eminli, S., Arnold, K., Stadtfeld, M., Yachechko, R., Tchieu, J., Jaenisch, R. et al. 2007. Directly reprogrammed fibroblasts show global epigenetic remodelling and widespread tissue contribution. Cell Stem Cell 1, 55–70.

Mali, P., Chou, B. K., Yen, J., Ye, Z., Zou, J., Dowey, S., Brodsky, R. A., Ohm, J. E., Yu, W., Baylin, S. B. et al. 2010. Butyrate greatly enhances derivation of human induced pluripotent stem cells by promoting epigenetic remodeling and the expression of pluripotency-associated genes. Stem Cells 28, 713–720.

Marion, R. M., Strati, K., Li, H., Murga, M., Blanco, R., Ortega, S., Fernandez-Capetillo, O., Serrano, M., and Blasco, M. A. 2009. A p53-mediated DNA damage response limits reprogramming to ensure iPS cell genomic integrity. Nature 460, 1149–1153.

Marson, A., Levine, S. S., Cole, M. F., Frampton, G. M., Brambrink, T., Johnstone, S., Guenther, M. G., Johnston, W. K., Wernig, M., Newman, J. et al. 2008. Connecting microRNA genes to the core transcriptional regulatory circuitry of embryonic stem cells. Cell 134, 521–533.

Martin, G. R. 1981. Isolation of a pluripotent cell line from early mouse embryos cultured in medium conditioned by teratocarcinoma stem cells. Proc. Natl. Acad. Sci. USA. 78, 7634–7638.

Mayshar, Y., Ben-David, U., Lavon, N., Biancotti, J. C., Yakir, B., Clark, A. T., Plath, K., Lowry, W. E., and Benvenisty, N. 2010. Identification and classification of chromosomal aberrations in human induced pluripotent stem cells. Cell Stem Cell 7, 521–531.

Mitsui, K., Tokuzawa, Y., Itoh, H., Segawa, K., Murakami, M., Takahashi, K., Maruyama, M., Maeda, M., and Yamanaka, S. 2003. The homeoprotein nanog is required for maintenance of pluripotency in mouse epiblast and ES cells. Cell 113, 631–642.

Miura, K., Okada, Y., Aoi, T., Okada, A., Takahashi, K., Okita, K., Nakagawa, M., Koyanagi, M., Tanabe, K., Ohnuki, M. et al. 2009. Variation in the safety of induced pluripotent stem cell lines. Nat. Biotechnol. 27, 743–5.

Nakagawa, M., Koyanagi, M., Tanabe, K., Takahashi, K., Ichisaka, T., Aoi, T., Okita, K., Mochiduki, Y., Takizawa, N., and Yamanaka, S. 2008. Generation of induced pluripotent stem cells without Myc from mouse and human fibroblasts. Nat. Biotechnol. 26, 101–106.

Nakagawa, M., Takizawa, N., Narita, M., Ichisaka, T., and Yamanaka, S. 2010. Promotion of direct reprogramming by transformation-deficient Myc. Proc. Natl. Acad. Sci. USA. 107, 14152–14157.

Narsinh, K. H., Sun, N., Sanchez-Freire, V., Lee, A. S., Almeida, P., Hu, S., Jan, T., Wilson, K. D., Leong, D., Rosenberg, J. et al. 2011. Single cell transcriptional profiling reveals heterogeneity of human induced pluripotent stem cells. J. Clin. Invest. 121, 1217–1221.

Nelson, T. J., Martinez-Fernandez, A., and Terzic, A. 2010. Induced pluripotent stem cells: developmental biology to regenerative medicine. Nat. Rev. Cardiol. 7, 700–710.

Nichols, J., Zevnik, B., Anastassiadis, K., Niwa, H., Klewe-Nebenius, D., Chambers, I., Scholer, H., and Smith, A. 1998. Formation of pluripotent stem cells in the mammalian embryo depends on the POU transcription factor Oct4. Cell 95, 379–391.

Niwa, H., Miyazaki, J., and Smith, A. G. 2000. Quantitative expression of Oct-3/4 defines differentiation, dedifferentiation or self-renewal of ES cells. Nat. Genet. 24, 372–376.

Okita, K., Ichisaka, T., and Yamanaka, S. 2007. Generation of germ-line competent induced pluripotent stem cells. Nature 448, 313–317.

Okita, K., Matsumura, Y., Sato, Y., Okada, A., Morizane, A., Okamoto, S., Hong, H., Nakagawa, M., Tanabe, K., Tezuka, K. et al. 2011. A more efficient method to generate integration-free human iPS cells. Nat. Methods 8, 409–412.

Okita, K., Nakagawa, M., Hyenjong, H., Ichisaka, T., and Yamanaka, S. 2008. Generation of mouse induced pluripotent stem cells without viral vectors. Science 332, 949–953.

Osafune, K., Caron, L., Borowiak, M., Martinez, R. J., Fitz-Gerald, C. S., Sato, Y., Cowan, C. A., Chien, K. R., and Melton, D. A. 2008. Marked differences in differentiation propensity among human embryonic stem cell lines. Nat. Biotechnol. 26, 313–315.

Park, I. H., Zhao, R., West, J.A., Yabuuchi, A., Huo, H., Ince, T. A., Lerou, P. H., Lensch, M. W., and Daley, G. Q. 2008. Reprogramming of human somatic cells to pluripotency with defined factors. Nature 451, 141–146.

Polo, J. M., Liu, S., Figueroa, M. E., Kulalert, W., Eminli, S., Tan, K. Y., Apostolou, E., Stadtfeld, M., Li, Y., Shioda, T. et al. 2010. Cell type of origin influences the molecular and functional properties of mouse induced pluripotent stem cells. Nat. Biotechnol. 28, 848–855.

Seki, T., Yuasa, S., Oda, M., Egashira, T., Yae, K., Kusumoto, D., Nakata, H., Tohyama, S., Hashimoto, H., Kodaira, M. et al. 2010. Generation of induced pluripotent stem cells from human terminally differentiated circulating T cells. Cell Stem Cell 7, 11–14.

Shi, Y., Do, J. T., Desponts, C., Hahm, H. S., Scholer, H. R., and Ding, S. 2008. A combined chemical and genetic approach for the generation of induced pluripotent stem cells. Cell Stem Cell 2, 525–528.

Silva, J., Barrandon, O., Nichols, J., Kawaguchi, J., Theunissen, T. W., and Smith, A. 2008. Promotion of reprogramming to ground state pluripotency by signal inhibition. PLoS Biol. 6, e253.

Stadtfeld, M., Brennand, K., and Hochedlinger, K. 2008a. Reprogramming of pancreatic Beta cells into induced pluripotent stem cells. Curr. Biol. 18, 890–894.

Stadtfeld, M., Nagaya, M., Utikal, J., Weir, G., and Hochedlinger, K. 2008b. Induced pluripotent stem cells generated without viral integration. Science 322, 945–949.

Staerk, J., Dawlaty, M. M., Gao, Q., Maetzel, D., Hanna, J., Sommer, C. A., Mostoslavsky, G., and Jaenisch, R. 2010. Reprogramming of human peripheral blood cells to induced pluripotent stem cells. Cell Stem Cell 7, 20–24.

Stewart, M. H., Bosse, M., Chadwick, K., Menendez, P., Bendall, S. C., and Bhatia, M. 2006. Clonal isolation of hESCs reveals heterogeneity within the pluripotent stem cell compartment. Nat. Methods 3, 807–815.

Sugii, S., Kida, Y., Kawamura, T., Suzuki, J., Vassena, R., Yin, Y. Q., Lutz, M. K., Berggren, W. T., Izpisua Belmonte, J. C., and Evans, R. M. 2010. Human and mouse adipose-derived cells support feeder-independent induction of pluripotent stem cells. Proc. Natl. Acad. Sci. USA. 107, 3558–3563.

Suh, M. R., Lee, Y., Kim, J. Y., Kim, S. K., Moon, S. H., Lee, J. Y., Cha, K. Y., Chung, H. M., Yoon, H. S., Moon, S. Y. et al. 2004. Human embryonic stem cells express a unique set of microRNAs. Dev. Biol. 270, 488–498.

Takahashi, K., Tanabe, K., Ohnuki, M., Narita, M., Ichisaka, T., Tomoda, K., and Yamanaka, S. 2007. Induction of pluripotent stem cells from adult human fibroblasts by defined factors. Cell 131, 861–872.

Takahashi, K. and Yamanaka, S. 2006. Induction of pluripotent stem cells from mouse embryonic and adult fibroblast cultures by defined factors. Cell 126, 663–676.

Tamaoki, N., Takahashi, K., Tanaka, T., Ichisaka, T., Aoki, H., Takeda-Kawaguchi, T., Iida, K., Kunisada, T., Shibata, T., Yamanaka, S. et al. 2010. Dental pulp cells for induced pluripotent stem cell banking. J. Dent. Res. 89, 773–778.

Tesar, P. J., Chenoweth, J. G., Brook, F.A., Davies, T. J., Evans, E. P., Mack, D. L., Gardner, R. L., and McKay, R. D. 2007. New cell lines from mouse epiblast share defining features with human embryonic stem cells. Nature 448, 196–199.

Thomson, J. A., Itskovitz-Eldor, J., Shapiro, S.S., Waknitz, M. A., Swiergiel, J. J., Marshall, V. S., and Jones, J. M. 1998. Embryonic stem cell lines derived from human blastocysts. Science 282, 1145–1147.

Tomioka, I., Maeda, T., Shimada, H., Kawai, K., Okada, Y., Igarashi, H., Oiwa, R., Iwasaki, T., Aoki, M., Kimura, T. et al. 2010. Generating induced pluripotent stem cells from common marmoset Callithrix jacchus fetal liver cells using defined factors, including Lin28. Genes Cells 15, 959–969.

Utikal, J., Maherali, N., Kulalert, W., and Hochedlinger, K. 2009a. Sox2 is dispensable for the reprogramming of melanocytes and melanoma cells into induced pluripotent stem cells. J. Cell Sci. 122, 3502–3510.

Utikal, J., Polo, J. M., Stadtfeld, M., Maherali, N., Kulalert, W., Walsh, R. M., Khalil, A., Rheinwald, J. G., and Hochedlinger, K. 2009b. Immortalization eliminates a roadblock during cellular reprogramming into iPS cells. Nature 460, 1145–1148.

Viswanathan, S. R., Daley, G. Q., and Gregory, R. I. 2008. Selective blockade of microRNA processing by Lin28. Science 320 , 97–100.

Wang, Y., Chen, J., Hu, J. L., Wei, X. X., Qin, D., Gao, J., Zhang, L., Jiang, J., Li, J. S., Liu, J. et al. 2011. Reprogramming of mouse and human somatic cells by high-performance engineered factors. EMBO Rep. 12, 373–378.

Warren, L., Manos, P. D., Ahfeldt, T., Loh, Y. H., Li, H., Lau, F., Ebina, W., Mandal, P. K., Smith, Z. D., Meissner, A. et al. 2010. Highly efficient reprogramming to pluripotency and directed differentiation of human cells with synthetic modified mRNA. Cell Stem Cell 7, 618–630.

Wernig, M., Meissner, A., Cassady, J. P., and Jaenisch, R. 2008. c-Myc is dispensable for direct reprogramming of mouse fibroblasts. Cell Stem Cell 2, 10–12.

Wernig, M., Meissner, A., Foreman, R., Brambrink, T., Ku, M., Hochedlinger, K., Bernstein, B. E., and Jaenisch, R. 2007. *In vitro* reprogramming of fibroblasts into a pluripotent ES cell-like state. Nature 448, 318–324.

West, F. D., Terlouw, S. L., Kwon, D. J., Mumaw, J. L., Dhara, S. K., Hasneen, K., Dobrinsky, J. R., and Stice, S. L. 010. Porcine induced pluripotent stem cells produce chimeric offspring. Stem Cells Dev. 19, 1211–1220.

Woltjen, K., Michael, I. P., Mohseni, P., Desai, R., Mileikovsky, M., Hamalainen, R., Cowling, R., Wang, W., Liu, P., Gertsenstein, M. et al. 2009. piggyBac transposition reprograms fibroblasts to induced pluripotent stem cells. Nature 458, 766–770.

Wu, Y., Zhang, Y., Mishra, A., Tardif, S. D., and Hornsby, P. J. 2010. Generation of induced pluripotent stem cells from newborn marmoset skin fibroblasts. Stem Cell Res. 4, 180–188.

Yan, X., Qin, H., Qu, C., Tuan, R. S., Shi, S., and Huang, G. T. 2009. iPS cells reprogrammed from mesenchymal-like stem/progenitor cells of dental tissue origin. Stem Cells Dev. 19, 469–480.

Ying, Q. L., Wray, J., Nichols, J., Batlle-Morera, L., Doble, B., Woodgett, J., Cohen, P., and Smith, A. 2008. The ground state of embryonic stem cell self-renewal. Nature 453, 519–523.

Yoshida, Y., Takahashi, K., Okita, K., Ichisaka, T., and Yamanaka, S. 2009. Hypoxia enhances the generation of induced pluripotent stem cells. Cell Stem Cell 5, 237–241.

Yoshida, Y. and Yamanaka, S. 2010. Recent stem cell advances: induced pluripotent stem cells for disease modeling and stem cell-based regeneration. Circulation 122, 80–87.

Yu, J., Hu, K., Smuga-Otto, K., Tian, S., Stewart, R., Slukvin, II, and Thomson, J. A. 2009. Human induced pluripotent stem cells free of vector and transgene sequences. Science 324, 797–801.

Yu, J., Vodyanik, M. A., Smuga-Otto, K., Antosiewicz-Bourget, J., Frane, J. L., Tian, S., Nie, J., Jonsdottir, G. A., Ruotti, V., Stewart, R. et al. 2007. Induced pluripotent stem cell lines derived from human somatic cells. Science 318, 1917–1920.

Yusa, K., Rad, R., Takeda, J., and Bradley, A. 2009. Generation of transgene-free induced pluripotent mouse stem cells by the piggyBac transposon. Nat. Methods 6, 363–9.

Zhao, T., Zhang, Z. N., Rong, Z., and Xu, Y. 2011. Immunogenicity of induced pluripotent stem cells. Nature 474, 212–215.

Zhao, X. Y., Li, W., Lv, Z., Liu, L., Tong, M., Hai, T., Hao, J., Guo, C. L., Ma, Q. W., Wang, L. et al. 2009. iPS cells produce viable mice through tetraploid complementation. Nature 461, 86–90.

Zhou, H., Wu, S., Joo, J. Y., Zhu, S., Han, D. W., Lin, T., Trauger, S., Bien, G., Yao, S., Zhu, Y. et al. 2009. Generation of induced pluripotent stem cells using recombinant proteins. Cell Stem Cell 4, 381–384.

CHAPTER 2

Human iPS Cell Generation Methods for Clinical Usage

Tomohisa Seki[1,2,a] and Shinsuke Yuasa[1,3,b,*]

Advantages of iPS Cell for Clinical Application

In 2006, Yamanaka et al. succeeded in generating pluripotent stem cells from mouse somatic cells (Takahashi and Yamanaka 2006). These cells had the capability of self renewal and multi-differentiation almost equivalent to embryonic stem (ES) cells. This invention was a milestone in regenerative medicine—the epoch making product was named as induced pluripotent stem (iPS) cell. In 2007, they also succeeded in generating iPS cells from human somatic cells (Takahashi et al. 2007b). Because iPS cell can solve an ethical problem and avoid an immune rejection which ES cell contains as a major matter for clinical application, iPS cell is expected to change the future of regenerative medicine. iPS cell offers the possibility of establishing a new approach to therapy for disease which cannot be treated with currently available methods. There are two major approaches to clinical application with iPS cell. First, somatic cells that are derived from iPS cells can be applicable to autogenous transplantation. It is a great advantage in transplantation cell source that iPS cells can be derived from patients' own somatic cells. Secondly, iPS cells that are generated from patients who have genetic disease can make *in vitro* disease models. Specific tissue cells which are differentiated from the patient may be useful as drug screening. It may

[1]Department of Cardiology, Keio University School of Medicine.
[a]Email: danke56@yahoo.co.jp
[b]Email: yuasa@a8.keio.jp
[2]Japan Society for the Promotion of Science.
[3]Center for Integrated Medical Research, Keio University School of Medicine.
*Corresponding author

lead not only to drug discovery and clarification of the disease mechanism, but also to personalized medicine. In this chapter, we discuss the current technology and future problems of human iPS cell generation methods for clinical usage.

In order to artificially generate specific types of human somatic cells, research on ES cell differentiation has been actively undertaken in the field of regenerative medicine. ES cell can be cultured *in vitro* while maintaining its pluripotency and its capability for self-proliferation. Since human ES cells were generated in 1998 (Thomson et al. 1998), they have been expected to become a major cell source of transplantation therapy. However, many problems for clinical application have been left unsolved up until the present. Efficient methods for differentiating ES cells to specific tissue cells and three-dimensioned culture system for construction of organs were not sufficient for a clinical setting. Additionally, ES cells have an ethical problem, in that they are generated from a human zygote and hence there is an immune rejection problem for cell transplantation. In order to apply ES cell to transplantation cell source, many problems have to be solved before clinical application.

iPS cells were first generated from mouse fibroblast by Professor Yamanaka's research group at Kyoto University (Takashi and Yamanaka 2006). They had the capability for self-proliferation and multi-differentiation that was almost equivalent to embryonic stem cells. In 2007, this group and Professor Thomson's group from University of Wisconsin, USA, succeeded in generating iPS cells from human dermal fibroblast (Takahashi et al. 2007b; Yu et al. 2007). Research in cell reprogramming with nuclear transplantation and cell fusion in the past has shown the possibility that there are specific factors that cause cell reprogramming. The researchers hypothesized that some of the factors which involved maintaining pluripotency of ES cell could cause cell reprogramming. Yamanaka et al. picked up 24 factors that contained ES cell specific transcription factors as candidate factors that caused cell reprogramming. They introduced these factors into mouse fibroblast with retrovirus vectors. After requirement study, they found out that mouse fibroblast could be reprogrammed with four factors—Oct3/4, Sox2, Klf4 and c-Myc—to stem cell almost equivalent to ES cell. They also found out that human dermal fibroblast could be reprogrammed with the same four factors. Because the iPS cell can generate from a somatic cell—although generation of human ES cell needs a human zygote—iPS cell can solve the ethical problem of using a human zygote. Additionally, the iPS cell also makes possible, autologous cell implantation. Therefore the iPS cell is expected to be a breakthrough cell source that can solve problems in the ES cell.

Reprogramming Factors for Generating iPS Cells

There are two main areas in which iPS cells can be used. In the first area, iPS cells are utilized as a cell source for transplantation. Here, there are two means of using iPS cells—auto-transplantation and allo-transplantation. In the case of allo-transplantation, the concept of an iPS cell bank is involved. The second method is to use iPS cells as a research tool. Here, iPS cells are generated from patients who have genetic disorders. The purposes of this usage are the pathological analysis of disease and the development of an effective therapy. Currently, there are many methods for generating iPS cells. Many combinations of reprogramming factors, many vehicles to introduce transgenes, and many types of somatic cells that are introduced as reprogramming factors have been reported until today (Gonzalez et al. 2011). Although each of these methods has its own merits and demerits, it has so far been unclear as to which method is the best. The combination of reprogramming factors that was first reported by Yamanaka et al. was composed of four factors: Oct3/4, Sox2, Klf4,and c-Myc (Takahashi and Yamanaka 2006). Shortly after publishing this report, it was reported that c-Myc was dispensable for generating iPS cell, although efficiency of iPS cell generation without c-Myc was lower than that with c-Myc (Wernig et al 2008; Nakagawa et al. 2008). Another report suggested the possibility of upgrading the quality of iPS cells by showing that addition of Tbx3 to the combination of three factors, Oct3/4, Sox2, and Klf4 improves the germ-line competency of mouse iPS cells (Han et al. 2010). Yamanaka's group reported that the replacement of c-Myc by L-Myc which was one of the Myc family members also improved the germ-line competency of mouse iPS cells and the generation efficiency without increasing the risk of oncogenesis (Nakagawa et al. 2010). Furthermore, in 2011, they also reported that the replacement of c-Myc by Glis1 that was screen-detected from 1437 candidate genes significantly improved generation efficiency, repressed incomplete reprogramming, and reduced the risk of oncogenesis (Maekawa et al. 2011). The research for better combination of reprogramming factors has made steady progress. In the near future, further progress in this field is expected.

Various Gene Delivery Systems for Reprogramming

So far, many types of vehicles to introduce reprogramming factors into somatic cells have been used in order to generate iPS cells. It is also a rapidly advancing field of research for suitable transgene delivery vehicles, for clinical application of iPS cell. At first, the delivery of the reprogramming factors into the somatic cells was achieved by using moloney murine leukemia virus (MMLV)-derived retroviruses such as pMXs (Kitamura et al. 2003; Takahashi et al. 2007a; Takahashi et al. 2007b), pLib (Wernig et al. 2007) or pMSCV (Aasen et al. 2008; Hawley et al. 1994). It is known that the transgenes that are delivered with these retrovirus vectors are usually silenced in pluripotent stem cells such as ES cells (Hawley et al. 1994; Jahner et al. 1982). Lentiviral vectors have also been successfully used to generate iPS cells from somatic cells (Blelloch et al. 2007; Yu et al. 2007). They are generally derived from HIV, and usually have higher infection efficiency than MMLV-based retroviruses. Interestingly, transgenes delivered with lentivirus vectors are less effectively repressed in pluripotent stem cells than that delivered with MMLV-derived vectors (Yao et al. 2004). In these iPS cells that are generated with genome integration, transgenes have the potential to break the genes located near the insertion site. These vectors also had the possibility that the delivered transgenes were reactivated after the cell reprogramming. Although these transgenes are silenced through the reprogramming in the high-quality iPS cells as mentioned above, the reactivation of transgenes may drive oncogenesis (Okita et al. 2007). Therefore for the clinical application of iPS cells, the insertion of transgenes into the genome should be avoided. The successful generation of iPS cells using Cre-deletable lentivirus system was also reported (Hanna et al. 2007) and perhaps solved some of these problems, but these viral systems still lack the safety required for transplantation therapy. In order to solve these problems of residual transgenes, many other methods in which iPS cells were generated with temporal expression of transgenes have been reported. To begin with, adenovirus vectors which did not need to insert themselves into host genomes were proven to be able to reprogram somatic cells into iPS cells without residual transgenes (Stadtfeld et al. 2008). A replication-defective adenoviral vector, pHIHG-Ad2 was used in this report. Although this method drew attention as a new method for generating iPS cells without viral integration, it had disadvantages, in that it needed multiple courses of viral infection because of the short transgene duration and the efficiency was significantly lower than that of retrovirus. Additionally, adenoviral vectors have the possibility of integrating into the genome of host cells, although this occurs at extremely low frequencies (Harui et al. 1999). Therefore these disadvantages have to be improved before application for clinical therapy. Besides these reprogramming methods with virus vectors, Sendai virus that replicate in the form of negative-sense single-stranded RNA in the cytoplasm of infected cells were also successfully used to generate iPS cells (Fusaki et al. 2009; Seki et al. 2010). This type of virus vector does not go through a DNA phase nor does it integrate into the host genome. Although a method for erasing residual genomic RNA of Sendai virus vectors from the target cells by using siRNA (Nishimura et al. 2011) and introducing temperature-sensitive mutations (Ban et al. 2011) were also reported, more complete elimination of residual viral component from reprogrammed cells were required for clinical application. As an alternative to integration-defective viruses, reprogramming methods with direct delivery of episomal vectors have also been developed (Jia et al. 2010; Gonzalez et al. 2009; Yu et al. 2009; Okita et al. 2008). But using this method, efficiency of reprogramming was also lower than that of retrovirus and only a low percentage of the iPS cell lines generated are free of plasmid integration.

As another method to avoid the residual transgenes, piggyBac transposon was also successfully used to generate iPS cells (Kaji et al. 2009; Woltjen et al. 2009; Yusa et al. 2009). The piggyBac transposon is a moth-derived DNA transposon (Cary et al. 1989) that is highly active in mammalian cells and mice (Ding et al. 2005). Although iPS cell lines that were integration free were successfully generated with this method, efficiency of reprogramming was also lower than that of retrovirus.

Another way to avoid the introduction of exogenous genetic material into donor cells is to deliver the reprogramming factors as RNAs or proteins. In this field, successful reprogramming by using direct delivery of synthetic mRNAs (Waren et al. 2012) has already been reported. In this report, in order to avoid the endogenous antiviral cell defence response to ssRNA, *in vitro* transcribed RNAs were modified with phosphatase treatment and incorporation of modified ribonucleoside bases substituting 5-methylcytidine for cytidine and pseudouridine for uridine. A recombinant version of B18R protein was also added in the culture medium in order to escape antiviral cell defense. This system allowed for high, dose-dependent levels of protein expression with high cell viability and achieved higher efficiency of iPS generation than that with the original retrovirus system. Although this method was considered attractive in terms of points of high efficiency and absence of genome integration, the high gene dosages of the reprogramming factors resulting from direct mRNA delivery may represent an oncogeneic risk owing to higher expression levels of c-Myc affecting genomic stability. The successful reprogramming with recombinant proteins was also reported (Kim et al. 2009; Zhou et al. 2009). These protein-based strategies were considered attractive for clinical application because of the absence of genome integration, but these methods had problems—they took longer term culture and generation efficiencies were lower than that with the retrovirus system. As just described, there is a variety of delivery systems for reprogramming and each of these methods has advantages and disadvantages. Therefore, further research and discussion are required to identify the most suitable method for clinical application.

The Variety of Donor Cell Types

In order to generate iPS cells from patients in clinical practice, we have to consider the most appropriate type of donor cells. Tissue collection from patients needs a relatively invasive approach. Whether the purpose of generating iPS cells is transplantation therapy or disease research, a less invasive approach is suitable for clinical practice. Originally, iPS cell was generated from mouse fibroblasts (Takahashi and Yamanaka 2006). After that, successful reprogramming of mouse hepatocytes and gastric epithelial cells was reported (Aoi et al. 2008). This report proved that somatic cells other than fibroblast could be reprogrammed. Terminally differentiated somatic cells were also proved to be reprogrammable in reports of successful mouse B lymphocyte (Hanna et al. 2008) and pancreatic beta cell (Stadtfeld et al. 2008a) reprogramming. In studies of human cell reprogramming, since generating iPS cell from human dermal fibroblast was reported (Takahashi et al. 2007b; Yu et al. 2007), many types of human somatic cells have been successfully reprogrammed.

iPS cell generation from plucked hair

In order to obtain human dermal fibroblast, a dermal biopsy that leaves a small scar on a patient's body is needed. Generating iPS cells from human keratinocytes that were derived from plucked human hair is less invasive and more appealing for clinical practice (Aasen and Belmonte 2010; Aasen et al. 2008). Retroviruses encoding the four factors to transduce keratinocytes were used. This showed that reprogramming of human keratinocytes derived from the foreskin biopsy was at least 100-fold more efficient than that of fibroblasts. However, in some cases in this experiment, several hairs were needed to obtain successful cell outgrowth of keratinocytes from a single plucked hair. Although more reliable protocol of primary culture is needed in order to apply this system to clinical practice, this method has an advantage for clinical application in that it does not require an invasive approach to patients.

iPS cell generation from cord blood

As another method of generating iPS cells, cord blood was also used as material of iPS cells (Giorgetti et al. 2009; Haase et al. 2009). Giorgetti et al. used cord blood derived stem cells that were isolated using standard CD133 immunomagnetic selection. They successfully reprogrammed cord blood derived stem cells with retroviral transduction of four (Oct3/4, Sox2, Klf4, c-Myc), three (Oct3/4, Sox2, Klf4), and as few as two (Oct3/4,

Sox2) transcription factors. In their experiments, average efficiency of reprogramming was 0.00625%. Haase et al. also produced iPS cells from human cord blood-derived endothelial cells by means of lentiviral overexpression of Oct3/4, Sox2, Nanog, and Lin28. Their reprogramming efficiency was up to 0.03% which was similar to that previously found after the reprogramming with lentiviral transduction of fibroblasts. Additionally, Takenaka et al. reported that the suppression of p53, that was previously shown to increase the efficiency of mouse and human iPS cell generation (Hong et al. 2009; Kawamura et al. 2009; Li et al. 2009; Marion et al. 2009; Utikal et al. 2009; Zhao et al. 2008), also increases the efficiency of iPS cell generation from CD34 positive cord blood cells (Takenaka et al. 2010). The Sendai virus vectors with temperature-sensitive mutations also successfully generated iPS cells without residual transgenes from CD34 positive cord blood cells (Ban et al. 2011). They introduced four factors—Oct3/4, Sox2, Klf4 and c-Myc—into CD34 positive cord blood cells with Sendai virus vectors which has temperature-sensitive mutations and therefore are easily removed at non-permissive temperatures. Their reprogramming efficiency was up to more than 0.2% which was higher than that previously found after the reprogramming of cord blood cells with lentiviral or retroviral transduction. Cord blood-derived cells are readily available because it does not require mobilization or biopsy and establishment of primary cultures. Additionally, cord blood cells are young cells and minimize the risk of having accumulated genetic mutations. Cord blood also has the advantage of being already banked along with immunological information. To date, more than 400,000 cord blood units are available worldwide in a comprehensive network of cord blood banks, facilitating a rapid and effective search for compatible donors for iPS cell generation (Gluckman and Rocha 2009). Because of these features, cord blood is expected to be an appealing cell source for iPS cell banks.

iPS cell generation from peripheral blood

Another less invasive method of obtaining patients' iPS cells is using blood cells. Peripheral blood is also an appealing cell source that is more easily accessible than skin tissue. In the first report of generating iPS cells from human blood cells (Loh et al. 2009), the authors transduced four factors —Oct3/4, Sox2, Klf4, and c-Myc—into CD34 positive mobilized human peripheral blood cells with retroviral vectors. They obtained iPS cells from CD34 positive cells with an efficiency of 0.01–0.02% that was not higher than that of previous studies. According to their report, mobilized peripheral blood was obtained from the donor who was injected for three days with G-CSF. Although the drug administration was not desirable because it required additional work and could be medically harmful for some patients, this method was very appealing for clinical practice in terms of less invasiveness and easy handling of primary cells. Recently, less invasive methods using peripheral blood have been reported (Brown et al. 2010; Kunisato et al. 2011; Loh et al. 2010; Seki et al. 2010; Staerk et al. 2010). By these methods, authors obtained mononuclear blood cells from donors or frozen samples and introduced four factors into mononuclear blood cells with retrovirus (Brown et al. 2010; Kunisato et al. 2011), lentivirus (Loh et al. 2010; Staerk et al. 2010), or Sendai virus(Seki et al. 2010). According to the report by Loh et al., they reprogrammed fresh peripheral blood mononuclear cells with two rounds of lentiviral infection (Loh et al. 2010). They tested the iPS clones for the presence of functionally rearranged immunoglobulin and T cell receptor genes by using probes specific for IgH, TCR-, and TCR-β2. Among 12 independent clones from three separate individuals, they failed to detect IgH recombination, indicating that none of our lines arose from B lymphocytes. Although they also failed to detect TCR-β2 recombination, six of seven iPS cell lines isolated from a single donor sample exhibited rearrangement of the TCR-locus, indicative of derivation from cells of the T lineage. In their system, they obtained ES cell-like colonies at the low efficiency of 0.0008%–0.001%. Kunisato et al. also successfully generated iPS cells from peripheral mononuclear cells (Kunisato et al. 2011). They introduced retroviruses containing four defined factors into peripheral mononuclear cells in the presence of combination of cytokines. Before retrovirus infection, peripheral mononuclear cells were cultured with human steel factor, thrombopoietin, IL-3, IL-6/IL-6 receptor fusion protein. They also experimented in the condition that they added GM-CSF, M-CSF, and Flt3/L, or Lipopolysaccharide, anti-CD3 antibody, anti-CD28 antibody, and anti-CD40 antibody. Their reprogramming efficiency was 0.00004%–0.00006%. Among four independent clones which they generated from peripheral nuclear cells, they detected TCR rearrangements in one clone and failed to detect IgH recombination. Staerk et al. also successfully reprogrammed peripheral blood (Staerk et al. 2010). They used a doxycycline-inducible lentivirus encoding four factors Oct4, Klf4, Sox2, and c-Myc from a polycistronic expression cassette. They cultured peripheral mononuclear cells in IL-7 or in G-CSF, GM-CSF, IL-6, and IL-3. They obtained the iPS cell colonies more efficiently in the presence of IL-7 expanded and reprogrammed than cells grown in the presence of G-CSF, GM-CSF, IL-3, and IL-6. In their experiments, the majority of iPS cell lines that they generated from mononuclear cells had TCR gene rearrangements which indicated that these cell lines derived from T lymphocytes. Brown et al. and our group (Seki et al. 2010) cultured peripheral mononuclear cells with IL-2 and anti-CD3 antibody which activated and proliferated T lymphocytes (Brown et al. 2010). Four factors—Oct4, Klf4, Sox2, and c-Myc —were introduced into these activated T lymphocytes, with retrovirus in the experiments of Brown et al. or Sendai virus in our group's experiments. The iPS cell lines obtained contained TCR gene rearrangements. Reprogramming efficiencies were approximately 0.01% in experiments with retrovirus and 0.1% in those with Sendai virus. Although the influence of these genome rearrangements on iPS cells needs more experiments, these methods have advantages in that they require less amount of peripheral blood and do not need the pharmacological pre-treatment of patients. In particular, these methods are useful for generating iPS cells from patients who have a specific genetic disease. Sampling of peripheral blood is one of the least invasive procedures performed routinely in clinics. Therefore, these methods can increase the number of patients who are subjected to disease research.

iPS cell generation from dental tissue

As another source of iPS cell, dental tissue has also been studied. Teeth are routinely removed in many clinics and no further procedures are required with respect to the donor. Additionally, some of these dental tissues are treated as biomedical wastes. Yan et al. mobilized dental stem cells from pulp tissue which was obtained from shed primary teeth, extracted permanent teeth and immature permanent teeth. They introduced 4 factors—Oct3/4, Sox2, Nanog, and Lin28—with lentivirus vector or Oct3/4, Sox2, Klf4, and c-Myc with retrovirus vectors (Yan et al. 2010). They successfully reprogrammed three different dental stem cells which were stem cells from exfoliated deciduous teeth, stem cells from apical papilla, and dental pulp stem cells. In the experiments of reprogramming dental pulp stem cells, using double course of retrovirus infection, their reprogramming efficiency was approximately 0.1%. Tamaoki et al. also successfully generated iPS cells from dental pulp cells with retroviral infection of four factors, namely, Oct3/4, Sox2, Klf4, and c-Myc or three factors, Oct3/4, Sox2, and Klf4 (Tamaoki et al. 2010). They also determined the HLA types of 107 dental pulp cell lines and identified two cell lines with homozygous HLA types at all three loci examined, which were estimated to cover approximately 20% of the Japanese population with a perfect match. Oda et al. generated iPS cells from mesenchymal stromal cells derived from human third molars by retroviral transduction of three factors, namely Oct3/4, Sox2, and Klf4 (Oda et al. 2010). In their experiments, reprogramming efficiency was up to more than 0.1% when they introduced reprogramming factors into clonally expanded mesenchymal stromal cells, although the efficiency differed by cell lines they used. In this field, generating iPS cells from oral gingival (Egusa et al. 2010) which is often resected during general dental treatments and oral mucosa fibroblasts (Miyoshi et al. 2010) which can be obtained less invasively were also reported. These methods also have an advantage for clinical application in that they don't need invasive approach to the patients. Moreover, because teeth are routinely removed in many clinics, it allows broad spectrum of HLA types to be easily obtained. Considering the nature of generating iPS cell from dental tissue, these methods have the potential to be a cell source of future iPS cell banks.

Epigenetic Memory in iPS Cells

When we generate iPS cells from patients, we also have to consider the type of donor somatic cells. In some reports, it was pointed out that the epigenetic features of iPS cells differed by type of donor cells. Kim et al. demonstrated that iPS cells retain an epigenetic memory of their tissue of origin although they also demonstrated that it is not apparent when adult somatic cells are reprogrammed to pluripotent stem cells by nuclear transfer (Kim et al. 2010). In experiments using mouse cell lines, they compared the *in vitro* blood-forming potential or osteoblast-forming potential of ES cells derived from naturally fertilized embryos, ES cells derived from nuclei of dermal fibroblasts, iPS cells derived from early bone marrow cells which were Kit positive, Lin negative, and CD45 positive, and iPS cells derived from dermal fibroblasts. In their experiments, blood-derived iPS cells showed higher capability of hematopoietic colony formation than that of fibroblast-derived iPS cells. Adversely, fibroblast-derived iPS cells showed higher capability of differentiation to osteoblasts than that of blood-derived iPS cells. Moreover, analyzing the correlation between methylation and transcriptional silencing in these iPS cells, they indicated the possible presence of differentiation tendency influenced by the epigenetic marks antagonistic to cell lineages distinct from the donor cell type. Polo et al. also reported distinct transcriptional and epigenetic patterns in iPS cells obtained from mouse fibroblasts, hematopoietic and myogenic cells (Polo et al. 2010). They also demonstrated that cellular origin influences the *in vitro* differentiation potentials of iPS cells into embryoid bodies and different hematopoietic cell types, although these differences were almost entirely lost after continuous passage of iPS cells. In human cases, Bar-Nur et al. suggested the same theory about epigenetic influence by using human iPS cells from pancreatic beta cells (Bar-Nur et al. 2011). In their experiments, beta cell derived iPS cells maintained an epigenetic memory at key beta-cell genes and showed preferred differentiation into insulin-producing cells. Therefore, the selection of somatic cell types that influence the differentiation capacities of resultant iPS cells into desired cell lineages may be considered before generating iPS cells from patients. When we generate iPS cells from patients, we have to take account of the epigenetic memory of iPS cells in addition to the invasiveness of cell sampling method and the somatic mutation.

iPS Cell Generation Methods for Clinical Practice

There are two ways of using iPS cells as cell source of transplantation therapies. Auto-transplantation and allo-transplantation of iPS cells each have advantages for clinical application and are expected to be established as novel therapies. In the case of auto-transplantation of iPS cells, iPS cells are generated from the patient's own cells, differentiated to required types of somatic cells and transplanted to the patient. This has then advantage of avoiding an immune rejection and therefore avoiding combination immunosuppressive treatment. On the other hand, allo-transplantation of iPS cells means transplantation of iPS cell-derived cells from an allogenic donor who has foreign HLA. This type of transplantation requires the formulation of strategies to avoid the host immune response to transplanted tissue, for clinical practice. iPS cells that can be generated with less invasive methods have advantage of being suitable for auto-transplantation therapy, although the problem of high cost has to be solved before clinical application. In contrast, cell sources that have been already banked are easy to apply to the iPS cell bank concept and therefore suited to allo-transplantation. Rightfully, for transplantation therapy, in order to avoid driving oncogenesis, we have to choose the most suitable combination of reprogramming factors and gene vehicles. The method that does not leave residual transgenes and does not drive oncogenesis is ideal.

Another possibility of application of iPS cells is for disease modeling. Patient-specific iPS cells that carry disease specific genetic information would be invaluable tools for studying pathogenesis of specific disease and for drug screening. Several reports of generating disease-specific iPS cells have shown the successful modeling of inherited diseases. iPS cells that can be generated with less invasive methods can increase the number of patients willingly consenting to subject themselves to disease research. In order to avoid breaking genes by transgenes, a method that can avoid transgene insertion is suitable for generating disease-specific iPS cells.

Conclusion

There are many combinations of reprogramming factors, many vehicles to introduce transgenes, and many types of somatic cells for generating iPS cell. Each method has advantages and disadvantages and they vary depending on the intended use. Although problems such as avoiding oncogenesis, developing an efficient differentiation method, and developing mass cultivation system have to be solved, we choose a generating method depending on the purpose of using iPS cells.

References

Aasen, T. and Belmonte, J. C. 2010. Isolation and cultivation of human keratinocytes from skin or plucked hair for the generation of induced pluripotent stem cells. Nat. Protoc. 5, 371–382.

Aasen, T., Raya, A., Barrero, M. J., Garreta, E., Consiglio, A., Gonzalez, F., Vassena, R., Bilic, J., Pekarik, V., Tiscornia, G. et al. 2008. Efficient and rapid generation of induced pluripotent stem cells from human keratinocytes. Nat. Biotechnol. 26, 1276–1284.

Aoi, T., Yae, K., Nakagawa, M., Ichisaka, T., Okita, K., Takahashi, K., Chiba, T., and Yamanaka, S. 2008. Generation of pluripotent stem cells from adult mouse liver and stomach cells. Science 321, 699–702.

Ban, H., Nishishita, N., Fusaki, N., Tabata, T., Saeki, K., Shikamura, M., Takada, N., Inoue, M., Hasegawa, M., Kawamata, S. et al. 2011. Efficient generation of transgene-free human induced pluripotent stem cells iPSCs by temperature-sensitive Sendai virus vectors. Proc. Natl. Acad. Sci. USA.

Bar-Nur, O., Russ, H. A., Efrat, S., and Benvenisty, N. 2011. Epigenetic memory and preferential lineage-specific differentiation in induced pluripotent stem cells derived from human pancreatic islet Beta cells. Cell Stem Cell 9, 17–23.

Blelloch, R., Venere, M., Yen, J., and Ramalho-Santos, M. 2007. Generation of induced pluripotent stem cells in the absence of drug selection. Cell Stem Cell 1, 245–247.

Brown, M. E., Rondon, E., Rajesh, D., Mack, A., Lewis, R., Feng, X., Zitur, L. J., Learish, R. D., and Nuwaysir, E. F. 2010. Derivation of induced pluripotent stem cells from human peripheral blood T lymphocytes. PLoS One 5, e11373.

Cary, L. C., Goebel, M., Corsaro, B. G., Wang, H. G., Rosen, E., and Fraser, M. J. 1989. Transposon mutagenesis of baculoviruses: analysis of Trichoplusia ni transposon IFP2 insertions within the FP-locus of nuclear polyhedrosis viruses. Virology 172, 156–169.

Ding, S., Wu, X., Li, G., Han, M., Zhuang, Y., and Xu, T. 2005. Efficient transposition of the piggyBac PB transposon in mammalian cells and mice. Cell 122, 473–483.

Egusa, H., Okita, K., Kayashima, H., Yu, G., Fukuyasu, S., Saeki, M., Matsumoto, T., Yamanaka, S., and Yatani, H. 2010. Gingival fibroblasts as a promising source of induced pluripotent stem cells. PLoS One 5, e12743.

Fusaki, N., Ban, H., Nishiyama, A., Saeki, K., and Hasegawa, M. 2009. Efficient induction of transgene-free human pluripotent stem cells using a vector based on Sendai virus, an RNA virus that does not integrate into the host genome. Proc. Jpn. Acad. Ser. B. Phys. Biol. Sci. 85, 348–362.

Giorgetti, A., Montserrat, N., Aasen, T., Gonzalez, F., Rodriguez-Piza, I., Vassena, R., Raya, A., Boue, S., Barrero, M. J., Corbella, B. A. et al. 2009. Generation of induced pluripotent stem cells from human cord blood using OCT4 and SOX2. Cell Stem Cell 5, 353–357.

Gluckman, E. and Rocha, V. 2009. Cord blood transplantation: state of the art. Haematologica 94, 451–454.

Gonzalez, F., Barragan Monasterio, M., Tiscornia, G., Montserrat Pulido, N., Vassena, R., Batlle Morera, L., Rodriguez Piza, I., and Izpisua Belmonte, J. C. 2009. Generation of mouse-induced pluripotent stem cells by transient expression of a single nonviral polycistronic vector. Proc. Natl. Acad. Sci. USA 106, 8918–8922.

Gonzalez, F., Boue, S., and Izpisua Belmonte, J. C. 2011. Methods for making induced pluripotent stem cells: reprogramming a la carte. Nat. Rev. Genet. 12, 231–242.

Haase, A., Olmer, R., Schwanke, K., Wunderlich, S., Merkert, S., Hess, C., Zweigerdt, R., Gruh, I., Meyer, J., Wagner, S. et al. 2009. Generation of induced pluripotent stem cells from human cord blood. Cell Stem Cell 5, 434–441.

Han, J., Yuan, P., Yang, H., Zhang, J., Soh, B. S., Li, P., Lim, S. L., Cao, S., Tay, J., Orlov, Y. L. et al. 2010. Tbx3 improves the germ-line competency of induced pluripotent stem cells. Nature 463, 1096–1100.

Hanna, J., Markoulaki, S., Schorderet, P., Carey, B. W., Beard, C., Wernig, M., Creyghton, M. P., Steine, E. J., Cassady, J. P., Foreman, R. et al. 2008. Direct reprogramming of terminally differentiated mature B lymphocytes to pluripotency. Cell 133, 250–264.

Hanna, J., Wernig, M., Markoulaki, S., Sun, C. W., Meissner, A., Cassady, J. P., Beard, C., Brambrink, T., Wu, L. C., Townes, T. M. et al. 2007. Treatment of sickle cell anemia mouse model with iPS cells generated from autologous skin. Science 318, 1920–1923.

Harui, A., Suzuki, S., Kochanek, S., and Mitani, K. 1999. Frequency and stability of chromosomal integration of adenovirus vectors. J. Virol. 73, 6141–6146.

Hawley, R. G., Lieu, F. H., Fong, A. Z., and Hawley, T. S. 1994. Versatile retroviral vectors for potential use in gene therapy. Gene. Ther. 1, 136–138.

Hong, H., Takahashi, K., Ichisaka, T., Aoi, T., Kanagawa, O., Nakagawa, M., Okita, K., and Yamanaka, S. 2009. Suppression of induced pluripotent stem cell generation by the p53–p21 pathway. Nature 460, 1132–1135.

Jahner, D., Stuhlmann, H., Stewart, C. L., Harbers, K., Lohler, J., Simon, I., and Jaenisch, R. 1982. De novo methylation and expression of retroviral genomes during mouse embryogenesis. Nature 298, 623–628.

Jia, F., Wilson, K. D., Sun, N., Gupta, D. M., Huang, M., Li, Z., Panetta, N. J., Chen, Z. Y., Robbins, R. C., Kay, M. A. et al. 2010. A nonviral minicircle vector for deriving human iPS cells. Nat. Methods 7, 197–199.

Kaji, K., Norrby, K., Paca, A., Mileikovsky, M., Mohseni, P., and Woltjen, K. 2009. Virus-free induction of pluripotency and subsequent excision of reprogramming factors. Nature 458, 771–775.

Kawamura, T., Suzuki, J., Wang, Y. V., Menendez, S., Morera, L. B., Raya, A., Wahl, G. M., and Belmonte, J. C. 2009. Linking the p53 tumour suppressor pathway to somatic cell reprogramming. Nature 460, 1140–1144.

Kim, D., Kim, C.H., Moon, J. I., Chung, Y. G., Chang, M. Y., Han, B. S., Ko, S., Yang, E., Cha, K. Y., Lanza, R. et al. 2009. Generation of human induced pluripotent stem cells by direct delivery of reprogramming proteins. Cell Stem Cell 4, 472–476.

Kim, K., Doi, A., Wen, B., Ng, K., Zhao, R., Cahan, P., Kim, J., Aryee, M. J., Ji, H., Ehrlich, L. I., et al. 2010. Epigenetic memory in induced pluripotent stem cells. Nature 467, 285–290.

Kitamura, T., Koshino, Y., Shibata, F., Oki, T., Nakajima, H., Nosaka, T., and Kumagai, H. 2003. Retrovirus-mediated gene transfer and expression cloning: powerful tools in functional genomics. Exp. Hematol. 31, 1007–1014.

Kunisato, A., Wakatsuki, M., Shinba, H., Ota, T., Ishida, I., and Nagao, K. 2011. Direct generation of induced pluripotent stem cells from human nonmobilized blood. Stem Cells Dev. 20, 159–168.

Li, H., Collado, M., Villasante, A., Strati, K., Ortega, S., Canamero, M., Blasco, M. A., and Serrano, M. 2009. The Ink4/Arf locus is a barrier for iPS cell reprogramming. Nature 460, 1136–1139.

Loh, Y. H., Agarwal, S., Park, I. H., Urbach, A., Huo, H., Heffner, G. C., Kim, K., Miller, J. D., Ng, K., and Daley, G. Q. 2009. Generation of induced pluripotent stem cells from human blood. Blood 113, 5476–5479.

Loh, Y. H., Hartung, O., Li, H., Guo, C., Sahalie, J. M., Manos, P. D., Urbach, A., Heffner, G. C., Grskovic, M., Vigneault, F. et al. 2010. Reprogramming of T cells from human peripheral blood. Cell Stem Cell 7, 15–19.

Maekawa, M., Yamaguchi, K., Nakamura, T., Shibukawa, R., Kodanaka, I., Ichisaka, T., Kawamura, Y., Mochizuki, H., Goshima, N., and Yamanaka, S. 2011. Direct reprogramming of somatic cells is promoted by maternal transcription factor Glis1. Nature 474, 225–229.

Marion, R. M., Strati, K., Li, H., Murga, M., Blanco, R., Ortega, S., Fernandez-Capetillo, O., Serrano, M., and Blasco, M. A. 2009. A p53-mediated DNA damage response limits reprogramming to ensure iPS cell genomic integrity. Nature 460, 1149–1153.

Miyoshi, K., Tsuji, D., Kudoh, K., Satomura, K., Muto, T., Itoh, K., and Noma, T. 2010. Generation of human induced pluripotent stem cells from oral mucosa. J. Biosci. Bioeng. 110, 345–350.

Nakagawa, M., Koyanagi, M., Tanabe, K., Takahashi, K., Ichisaka, T., Aoi, T., Okita, K., Mochiduki, Y., Takizawa, N., and Yamanaka, S. 2008. Generation of induced pluripotent stem cells without Myc from mouse and human fibroblasts. Nat. Biotechnol. 26, 101–106.

Nakagawa, M., Takizawa, N., Narita, M., Ichisaka, T., and Yamanaka, S. 2010. Promotion of direct reprogramming by transformation-deficient Myc. Proc. Natl. Acad. Sci. USA 107, 14152–14157.

Nishimura, K., Sano, M., Ohtaka, M., Furuta, B., Umemura, Y., Nakajima, Y., Ikehara, Y., Kobayashi, T., Segawa, H., Takayasu, S. et al. 2011. Development of defective and persistent Sendai virus vector: a unique gene delivery/expression system ideal for cell reprogramming. J. Biol. Chem. 286, 4760–4771.

Oda, Y., Yoshimura, Y., Ohnishi, H., Tadokoro, M., Katsube, Y., Sasao, M., Kubo, Y., Hattori, K., Saito, S., Horimoto, K. et al. 2010. Induction of pluripotent stem cells from human third molar mesenchymal stromal cells. J. Biol. Chem. 285, 29270–29278.

Okita, K., Ichisaka, T., and Yamanaka, S. 2007. Generation of germline-competent induced pluripotent stem cells. Nature 448, 313–317.

Okita, K., Nakagawa, M., Hyenjong, H., Ichisaka, T., and Yamanaka, S. 2008. Generation of mouse induced pluripotent stem cells without viral vectors. Science 322, 949–953.

Polo, J. M., Liu, S., Figueroa, M. E., Kulalert, W., Eminli, S., Tan, K. Y., Apostolou, E., Stadtfeld, M., Li, Y., Shioda, T. et al. 2010. Cell type of origin influences the molecular and functional properties of mouse induced pluripotent stem cells. Nat. Biotechnol. 28, 848–855.

Seki, T., Yuasa, S., Oda, M., Egashira, T., Yae, K., Kusumoto, D., Nakata, H., Tohyama, S., Hashimoto, H., Kodaira, M. et al. 2010. Generation of induced pluripotent stem cells from human terminally differentiated circulating T cells. Cell Stem Cell 7, 11–14.

Stadtfeld, M., Brennand, K., and Hochedlinger, K. 2008a. Reprogramming of pancreatic beta cells into induced pluripotent stem cells. Curr. Biol. 18, 890–894.

Stadtfeld, M., Nagaya, M., Utikal, J., Weir, G., and Hochedlinger, K. 2008b. Induced pluripotent stem cells generated without viral integration. Science 322, 945–949.

Staerk, J., Dawlaty, M. M., Gao, Q., Maetzel, D., Hanna, J., Sommer, C.A., Mostoslavsky, G., and Jaenisch, R. 2010. Reprogramming of human peripheral blood cells to induced pluripotent stem cells. Cell Stem Cell 7, 20–24.

Takahashi, K., Okita, K., Nakagawa, M., and Yamanaka, S. 2007a. Induction of pluripotent stem cells from fibroblast cultures. Nat. Protoc. 2, 3081–3089.

Takahashi, K., Tanabe, K., Ohnuki, M., Narita, M., Ichisaka, T., Tomoda, K., and Yamanaka, S. 2007b. Induction of pluripotent stem cells from adult human fibroblasts by defined factors. Cell 131, 861–872.

Takahashi, K. and Yamanaka, S. 2006. Induction of pluripotent stem cells from mouse embryonic and adult fibroblast cultures by defined factors. Cell 126, 663–676.

Takenaka, C., Nishishita, N., Takada, N., Jakt, L. M., and Kawamata, S. 2010. Effective generation of iPS cells from CD34+ cord blood cells by inhibition of p53. Exp. Hematol. 38, 154–162.

Tamaoki, N., Takahashi, K., Tanaka, T., Ichisaka, T., Aoki, H., Takeda-Kawaguchi, T., Lida, K., Kunisada, T., Shibata, T., Yamanaka, S. et al. 2010. Dental pulp cells for induced pluripotent stem cell banking. J. Dent. Res. 89, 773–778.

Thomson, J. A., Itskovitz-Eldor, J., Shapiro, S. S., Waknitz, M. A., Swiergiel, J. J., Marshall, V. S., and Jones, J. M. 1998. Embryonic stem cell lines derived from human blastocysts. Science 282, 1145–1147.

Utikal, J., Polo, J. M., Stadtfeld, M., Maherali, N., Kulalert, W., Walsh, R. M., Khalil, A., Rheinwald, J. G., and Hochedlinger, K. 2009. Immortalization eliminates a roadblock during cellular reprogramming into iPS cells. Nature 460, 1145–1148.

Warren, L., Manos, P. D., Ahfeldt, T., Loh, Y. H., Li, H., Lau, F., Ebina, W., Mandal, P. K., Smith, Z. D., Meissner, A. et al. 2010. Highly efficient reprogramming to pluripotency and directed differentiation of human cells with synthetic modified mRNA. Cell Stem Cell 7, 618–630.

Wernig, M., Meissner, A., Cassady, J. P., and Jaenisch, R. 2008. c-Myc is dispensable for direct reprogramming of mouse fibroblasts. Cell Stem Cell 2, 10–12.

Wernig, M., Meissner, A., Foreman, R., Brambrink, T., Ku, M., Hochedlinger, K., Bernstein, B. E., and Jaenisch, R. 2007. *In vitro* reprogramming of fibroblasts into a pluripotent ES-cell-like state. Nature 448, 318–324.

Woltjen, K., Michael, I. P., Mohseni, P., Desai, R., Mileikovsky, M., Hamalainen, R., Cowling, R., Wang, W., Liu, P., Gertsenstein, M. et al. 2009. piggyBac transposition reprograms fibroblasts to induced pluripotent stem cells. Nature 458, 766–770.

Yao, S., Sukonnik, T., Kean, T., Bharadwaj, R. R., Pasceri, P., and Ellis, J. 2004. Retrovirus silencing, variegation, extinction, and memory are controlled by a dynamic interplay of multiple epigenetic modifications. Mol. Ther. 10, 27–36.

Yan, X., Qin, H., Qu, C., Tuan, R. S., Shi, S., and Huang, G.T. 2010. iPS cells reprogrammed from human mesenchymal-like stem/progenitor cells of dental tissue origin. Stem Cells Dev. 19, 469–480.

Yu, J., Hu, K., Smuga-Otto, K., Tian, S., Stewart, R., Slukvin, II, and Thomson, J. A. 2009. Human induced pluripotent stem cells free of vector and transgene sequences. Science 324, 797–801.

Yu, J., Vodyanik, M. A., Smuga-Otto, K., Antosiewicz-Bourget, J., Frane, J. L., Tian, S., Nie, J., Jonsdottir, G. A., Ruotti, V., Stewart, R. et al. 2007. Induced pluripotent stem cell lines derived from human somatic cells. Science 318, 1917–1920.

Yusa, K., Rad, R., Takeda, J., and Bradley, A. 2009. Generation of transgene-free induced pluripotent mouse stem cells by the piggyBac transposon. Nat. Methods 6, 363–369.

Zhao, Y., Yin, X., Qin, H., Zhu, F., Liu, H., Yang, W., Zhang, Q., Xiang, C., Hou, P., Song, Z. et al. 2008. Two supporting factors greatly improve the efficiency of human iPSC generation. Cell Stem Cell 3, 475–479.

Zhou, H., Wu, S., Joo, J. Y., Zhu, S., Han, D. W., Lin, T., Trauger, S., Bien, G., Yao, S., Zhu, Y. et al. 2009. Generation of induced pluripotent stem cells using recombinant proteins. Cell Stem Cell 4, 381–384.

CHAPTER 3

Reprogramming and Regenerative Medicine: Unique Approach via Protein-derived iPS

Hyun-Jai Cho,[a] Yoo-Wook Kwon[b] and Hyo-Soo Kim[c,*]

Stem cell research after the year 2006 has been epitomized as the era of induced pluripotent stem cells (hereafter called iPS). Stem cells, which can be defined by their ability to renew themselves in a symmetric way and to differentiate into a wide range of specialized cell types, have been the center of cell-based regenerative medicine.

Until recently, stem cell research has been based on a forward concept with the use of *totipotent* zygotes which can give rise to all embryonic and extraembryonic tissues, the *pluripotent* inner cell mass of blastocysts that can give rise to all embryonic tissues with the exclusion of extraembryonic tissues, and *multipotent* adult stem cells that can only give rise to cell types within their lineage. When they reach their terminal differentiation, cells lose their potential to develop to the next stage.

The progress of stem cell research, however, has allowed even fully-differentiated adult cells to be able to dedifferentiate to the pluripotent stage. Reprogramming is simply a technique that involves the dedifferentiation of adult somatic cells to produce donor or patient-specific pluripotent stem cells without using embryos. This change from a forward concept to a reverse concept is groundbreaking and indeed revolutionary.

National Research Laboratory for Cardiovascular Stem Cell Niche, Innovative Research Institute for Cell Therapy, Seoul National University Hospital, Seoul, Korea.
[a]Email: hyunjaicho@snu.ac.kr
[b]Email: ywkwon@snu.ac.kr
[c]Email: hyosoo@snu.ac.kr
*Corresponding author

Nuclear Transfer

The very first approach for reprogramming was attempted through the nuclear transfer method. During the 1950s, Briggs and King developed the technique of somatic cell nuclear transfer (SCNT), or cloning, to replace an oocyte nucleus with a somatic cell nucleus obtained from a donor *in vitro* (Briggs and King 1952; King and Briggs 1955). Cell division was stimulated by chemicals or electricity up to the blastocyst stage; at that time, the inner cell mass is isolated and cultured, resulting in ES cells that are genetically identical to the donor. This and subsequent studies showed that differentiated cells indeed retain the ability to generate their clones (Gurdon 1962; Gurdon et al. 1975). The first mammalian clone, Dolly, was born in 1996 and is thus symbolic of the SCNT technique (Wilmut, Schnieke et al. 1997).

There are numerous advantages to cloning. The resulting ES cells are perfectly matched to the donor's immune system, and no immunosuppressant would be required to prevent rejection. Even though ES cells derived from SCNT have the nuclear genome of the donor cells, mitochondrial DNA inherited by the oocyte could lead to immunogenicity after transplantation. Lanza et al. evaluated the histocompatibility of nuclear transfer-generated tissue; the nucleus of a bovine skin fibroblast was microinjected into an enucleated oocyte (Lanza et al. 2002). Even if the blastocyst was implanted, the purpose was to produce cardiac, renal and skeletal muscle cells, which were then harvested, expanded *in vitro* and seeded onto biodegradable scaffolds. These scaffolds were then implanted into the donor from whom the cells were cloned to determine if cells were histocompatible. The evaluation revealed that T-cell response was not observed in cloned renal cells, suggesting that rejection would not essentially occur in the presence of oocyte-derived mtDNA.

However, SCNT has certain limitations that entail further improvement before its clinical application. First, ethical concerns about the prospect of the resulting embryos of growing into cloned embryos, if implanted into a uterus, should be resolved. Various animal studies have shown that blastocysts derived from SCNT can give rise to a live born infant that is a clone of the donor when implanted into a uterus. But, this reproductive cloning method is banned in most countries for human applications. Second, when the SCNT-derived ES cells are using clinical application, the quality of SCNT-derived ES cells should be assessed carefully. Regrettably, one of the lines derived by SCNT discovered a translocation consisting of an isochromosome comprised of two copies of the long arm of the Y chromosome. It is not known whether chromosomal abnormalities in SCNT-derived ES cells originate from aneuploid embryos or occurred during ES cell isolation and culture. Third, the efficiency of SCNT is very low and the supply of human oocytes is insufficient, since the therapeutic potential of SCNT technique is getting delayed. Taking all this into consideration, SCNT has demanding restrictions that limit its translation into a clinical application.

Somatic Cell Fusion with Embryonic Stem Cell

Another method to produce pluripotent stem cells is through somatic cell fusion with immortal pluripotent cell lines. These cell lines were called embryonal carcinoma cells (ECCs) and obtained from teratocarcinomas or tumors of germ cell origin. In 1976, when Miller and Ruddle fused somatic cells such as thymocytes with ECCs, thymocytes acquired pluripotency (Miller and Ruddle 1976). After that, Tada et al. observed the same results by electrofusion with mouse ES cells (Tada et al. 2001). Transplantation of these cells into nude mice results in the formation of teratomas consisting of various tissues from all three germ layers, substantiating the pluripotency of these cells. A further progress in the cell fusion technique was reported by Cowan et al. to reprogram normal diploid human fibroblasts into human ES cells (Cowan et al. 2005). ES cells fused with the fibroblasts resulted in the hybrids that sustained as predictable, a tetraploid chromosome complement, and displayed morphology, growth rate and surface molecule expression patterns characteristic of human ES cells. Assessment of genome-wide transcriptional changes, allele-specific gene expression and DNA methylation illustrated that the fibroblast genome was reprogrammed to a near embryonic state. Moreover, differentiation of ES-fibroblasts hybrids *in vivo* generated cell types from each germ layer. These findings showed that ES cells have the capacity to reprogram adult somatic cell chromosomes after cell fusion. ES cells may therefore provide a useful substitution for oocytes, for biochemical and genetic studies aimed at understanding how to reprogram differentiated cells to an embryonic state—and thereby increase their growth potential. Ultimately, this technique might lead to an alternative route for creating genetically modified-ES cell lines for use in the study and treatment of various diseases.

However, a significant technical hurdle exists before these hybrid ES-like cells can be used for clinical applications, particularly, the abolition of the ES cell chromosomes either before or after cell fusion. Enucleation of ES cells before fusion may not be practical in circumventing this problem as it has been revealed to eliminate the ability of remaining ES cytoplast to re-activate expression of pluripotency markers in hybrids with somatic cells (Do and Scholer 2004). The selective deletion of a chromosome is possible but may be impracticable for the complete set of chromosomes (Matsumura et al. 2007). Alternatively, ES cells-derived chromosomes carrying the major histocompatibility complex loci could be removed selectively to avoid or at least reduce rejection reactions during treatment. This possibility should be experimentally investigated.

iPSCs by Pluripotency-related Genes

Thirdly—and most strikingly—there comes an innovative approach in the reprogramming of somatic cells into pluripotent stem cells. Takahashi and Yamanaka demonstrated that mouse embryonic fibroblasts (MEFs) and adult mouse fibroblasts can be reprogrammed into pluripotent stem cells (iPS cells) (Takahashi and Yamanaka 2006). They analyzed 24 genes that were considered to be essential for ES cells. From them, they recognized only four genes, namely Oct3/4, Sox2, Klf4 and c-Myc that were essential to give ES-like properties on fibroblasts. MEFs and adult fibroblasts were co-transduced with retroviral vectors, each carrying Oct3/4, Sox2, Klf4 and c-Myc. The consequential iPS cells possessed the immortal growth characteristics of self-renewing ES cells, expressed genes specific for ES cells and generated embryoid bodies (EBs) *in vitro* and teratomas *in vivo*. When the iPS cells were injected into mouse blastocysts, they contributed to a variety of different cell types representing their growth potential. Dissimilar to ES cells, chimeras of iPS cells did not effect in germline transmission. Gene expression profiles of the iPS cells demonstrated that they possessed a diverse gene expression marker compared to ES cells. The epigenetic state of the iPS cells was someplace between their somatic origins and fully reprogrammed ES cells. These results were signifying the incomplete reprogramming.

Wernig et al. reported that these results were considerably improved (Wernig et al. 2007). They infected fibroblast with retroviral vectors and have examined the activation degree of endogenous Oct4 or Nanog genes. This study demonstrated that DNA methylation, gene expression profiles, and chromatic state of the reprogrammed cells were very similar to those of ES cells. Teratomas induced by these cells have differentiated cell types representing all three germ layers. Most prominently, the reprogrammed cells from the experiment were able to form viable chimeras and give to the germ-line-transmittable ES cells, suggesting that these iPS cells were completely reprogrammed.

Moreover, Takahashi et al. and Yu et al. demonstrated the reprogramming of human cells using transduction of the four pluripotent genes (Takahashi et al. 2007; Yu et al. 2007). They demonstrated that retrovirus-mediated transfection of Oct3/4, Sox2, Klf4 and c-Myc produces human iPS cells, which are comparable to human ES cells in terms of morphology, proliferation, gene expression, surface markers, and teratoma formation. In dissimilarity, Thomson's research group demonstrated that retroviral transduction of

Oct4, Sox2, Nanog, and Lin28 could also produce pluripotent stem cells without introducing any oncogenes. These two studies depicted that human iPS cells were analogous to human ES cells. However, it is disquieting that these iPS cells contain three to six retroviral integrations which may increase the risk of tumorigenicity. Okita et al. reported tumor formation in chimeric mice generated from Nanog-iPS cells and observed that 20% of the progeny developed tumors due to the retroviral expression of c-Myc (Okita et al. 2007). Another approach would be to use a brief expression method, since both Meissner et al. and Okita et al. demonstrated strong silencing of the viral controlled transcripts in iPS cells (Meissner et al. 2007). Their results depict that exogenous defined factors are only required for the induction, not for the maintenance of pluripotency.

Recently Kim et al. did a meaningful achievement in somatic cell reprogramming (Kim et al. 2008). They reported that adult mouse neural stem cells (NSCs) express higher endogenous levels of Sox2 and c-Myc compared to ES cells and then exogenous Oct4 together with either Klf4 or c-Myc is sufficient to generate iPS cells from NSCs. These two factor iPS cells are analogous to ES cells at the molecular level, contribute to development and form chimeras. Very recently, the same group achieved another progress (Kim et al. 2009). They demonstrated that exogenous expression of the germline specific transcription factor Oct4 is sufficient to induce pluripotent stem cells from adult mouse NS cells. This Oct4 induced pluripotent stem cells are similar to ES cells *in vitro* and *in vivo*. These cells can be efficiently differentiated into neural cells, cardiomyocytes and germ cells *in vitro*, but they are also able to form teratoma and germline transmission *in vivo*. This study suggests that Oct4 is required and sufficient to directly reprogram NS cells to pluripotency. According to these results, for the induction of pluripotency, the number of reprogramming factors can be reduced when using somatic cells that endogenously express appropriate levels of complementing factors.

At present, researchers are trying to figure out how to prevent iPS cell-related tumor formation; it is a major concern in clinical applications. Nakagawa et al. generated iPS cells without c-Myc from mouse and human fibroblasts, because c-Myc retrovirus was suspected to increase tumorigenicity in germline transmitted-progeny mice (Nakagawa et al. 2008). This study showed significantly less non-iPS background cells and the iPS cells generated were consistently of high efficiency. Mice generated from c-Myc-free iPS cells did not show tumors during the experimental period. In addition, this study generated human iPS cells from adult dermal fibroblasts without c-Myc.

Another disquieting factor in retrovirus or lentivirus-mediated iPS cells is that viral integration into the host genome may increase the risk of tumorigenicity. Hence, Stadtfeld et al. generated mouse iPS cells by using non-integrating adenoviruses transiently expressing Oct4, Sox2, Klf4 and c-Myc (Stadtfeld et al. 2008). These adenovirus-mediated iPS cells illustrated DNA demethylation characteristics of ES cells, expressed endogenous pluripotency-related genes, formed teratomas and contributed to multiple tissues including the germ cells in chimeric mice. This result bestows strong evidence that mutagenesis with genomic integration is not required for *in vitro* reprogramming. The technique of non-genomic integrating reprogramming may give a better method for induction and studying patient-specific stem cells and for comparing the characteristics between ES cells and iPS cells. Besides, Okita et al. also generated mouse iPS cells without viral vectors (Okita et al. 2008). They demonstrated that repeated transfection of a single plasmid containing the cDNAs of Oct3/4, Sox2, and Klf4, jointly with a c-Myc expression plasmid, into MEFs resulted in iPS cells without evidence of plasmid integration, which produced well differentiated teratomas and contributed to adult chimeras. The generation of virus-free iPS cells, although from embryonic fibroblasts, concentrated on critical safety concern for potential use of iPS cells in regenerative medicine.

iPSCs by Protein Transfer

Lastly, reprogramming using various types of proteins, also works effectively. Recent advances in integration-free reprogramming methods, using non-integrating adenoviruses (Stadtfeld et al. 2008), transient transfection of plasmids (Okita et al. 2008), and RNA delivery (Warrenet al. 2010) as well, have to produce iPS cells with the use of genetic materials. However, with the reprogramming proteins, either recombinant form or direct extracts from cells, iPS cells can be produced without any genetic modification. So far, there have been three different approaches for generating protein-iPS: by delivering purified recombinant proteins (Zhouet al. 2009), by extracts from genetically engineered HEK293 cells (Kim et al. 2009), or by whole extracts from ESCs (Cho et al. 2010). Each of the three methods is covered consecutively below.

1. Recombinant cell-penetrating reprogramming proteins

The idea of avoiding the use of exogenous genetic materials and introducing reprogramming proteins directly into cells was first proposed by Ding and his colleagues (Zhou et al. 2009). But there was a hurdle to deliver macromolecules such as proteins across the cell membrane (Belting et al. 2005). The solution was glimpsed from findings by Frankel and Pabo (Frankel et al. 1988; Frankel and Pabo 1988). In 1988, Frankel et al. found that the human immunodeficiency virus transactivator of transcription (hereafter called HIV-TAT) protein can overcome this hurdle with a short basic segment residing at amino acids 48–60 that allows this protein to penetrate the cell membrane. These peptides contain a high proportion of basic amino acids, like arginine or lysine, and are known as CPPs (Ziegler et al. 2005; El-Sayed et al. 2009).

Protocols

To design cell-penetrating recombinant proteins, Ding designed and fused a poly-arginine (i.e., 11R) protein transduction domain to the C terminus of four reprogramming factors: Oct4, Sox2, Klf4 and c-Myc (Zhou et al. 2009). These proteins were expressed in *E. coli* in inclusion bodies and then were solubilized, refolded, and further purified. The identities and purity of proteins were confirmed by mass spectrometry and western blot. The ability of proteins to penetrate cell membranes was also tested by treating mouse embryonic fibroblast (MEF) cells with the recombinant proteins. They found that the purified 11R-tagged recombinant transcription factors readily entered cells at concentrations of 0.5–8ug/ml within 6 hours and could translocate into the nucleus. In addition, they found that the transduced proteins appeared to be stable inside cells for up to 48 hours. Because reprogramming through the iPSC mechanism typically requires sustained activity of reprogramming proteins for 7–10 days, they treated the cells with proteins in four cycles. In each cycle the fibroblasts (seeded at the density of $5X10^4$ cells/well in a six-well plate) were first treated overnight with the recombinant proteins at 8ug/ml in the mESC growth media supplemented with or without 1mM valproic acid (VPA), a HDAC inhibitor that is known to improve reprogramming efficiency (Huangfu et al. 2008). After that, cells were transferred to the same media without the recombinant reprogramming proteins and VPA, and cultured for an additional 36 hours before the next cycle of the treatment. After completing four repeated protein transductions of reprogramming proteins, the treated cells were transferred onto irradiated MEF feeder cells and simply kept in mESC growth media until colonies emerged around day 30–35. Three colonies per $5X10^4$ cells were obtained when they were transduced with four proteins and treated with VPA, and one colony per $5X10^4$ cells was obtained when they were transduced with only three proteins (except c-Myc) and treated with VPA. However, not any colony was obtained by transducing the three or four reprogramming proteins without VPA only for the same period of time.

In vitro and in vivo characteristics

After over 30 passages, the generated murine protein-iPS cells were morphologically indistinguishable from classic mESCs. They also expressed typical pluripotency markers, including ALP, Oct4, Nanog, Sox2, and stage-specific embryonic antigen (SSEA)-1. To examine developmental potential of protein-iPS cells, standard *in vitro* differentiation using embryoid bodies (EBs) or monolayer chemically defined stepwise differentiation, and *in vivo* chimerism assays were performed. Protein-iPS cells could efficiently form EBs and differentiate into three germ cell layers and most importantly, such cells could successfully incorporate into the inner cell mass of a blastocyst following aggregation with an eight-cell embryo and led to high-level chimerism with apparent germline contribution *in vivo*. These *in vitro* and *in vivo* characterizations confirm that the purified cell-penetrating recombinant reprogramming proteins are sufficient to reprogram MEFs to become protein iPS cells molecularly, morphologically, and functionally similar to mES cells.

2. Proteins from genetically engineered HEK293 cell line

Protocols

Another way to deliver recombinant proteins into target cells has been suggested by Kim via generating stable HEK293 cell lines that could express each of the four human reprogramming factors (Oct4, Sox2, Klf4 and c-Myc) fused with 9R and the myc tag (Kim et al. 2009). High expression of these proteins was confirmed in HEK293 cell lines by western blot analysis. When human newborn fibroblasts (hereafter called HNFs) were treated with cell extracts from the HEK293 cells, efficient intracellular translocation of each recombinant protein was observed within eight hours. In a series of experiments, $5X10^5$ HNFs were treated with combined total extracts of four HEK293 cell lines for 16 hours. After washing, cells were incubated for six days in ES media 1 and then transferred onto mouse embryonic feeders (MEFs). The transferred cells were incubated with ES media 2 for up to four weeks. After successive failure to achieve pluripotent cells with treating combined total extracts only once, they tested repeated protein treatment cycles could yield protein-iPS cells. Using this approach, after three or four rounds of treatment, several colonies with iPSC-like morphology but AP negative were observed; for further cycles, the number of colonies which were AP positive significantly increased. AP-positive colonies with iPSC-like morphology were handpicked and transferred onto MEFs in the presence of ES media 2 and ES media 3 for seven days each. Five iPSC-like colonies were established, and two of them were maintained and characterized. Overall, the establishment of these colonies took about eight weeks—approximately double that seen with viral transduction (Takahashi et al. 2007; Yu et al. 2007; Park et al. 2008).

In vitro and in vivo characteristics

In order to determine if the protein-iPS cells have hESC-like properties, expression of markers was examined including AP, Oct4, Nanog, tumor-rejection antigen (TRA) 1-60, stage-specific embryonic antigen (SSEA)-3, and SSEA-4. The expression patterns of pluripotency markers were indistinguishable from hESCs, suggesting that appropriate reprogramming had occurred in protein-iPS cells. In addition, bisulfate sequencing analyses and global gene expression also indicated the successful reprogramming of protein-iPS cells. When the protein-iPSCs were allowed to form embryoid bodies (EBs) by suspension culture, they readily differentiated into cells of all three germ layers. Moreover, teratoma formation was observed after transplantation of protein-iPSCs under the kidney capsule of nude mice for 6-8 weeks. These teratomas contained tissues from all three germ layers. These data mentioned above collectively confirmed that protein-iPS cells exhibit pluripotency both *in vitro* and *in vivo.*

3. Proteins from ES cell extracts

Studies using cell extracts or proteins have shown a modest effect on reprogramming into specific lineages (Hakelien et al. 2002; Qin et al. 2005). Hakelien et al. reported that 293T cells, when lymphocytes extract transferred, expressed lymphocyte markers. The same group demonstrated that dedifferentiation and reprogramming of 293T cells after treated with extracts obtained from pluripotent cells such as embryonal carcinoma cells (ECCs) and ES cells (Taranger et al. 2005). Upregulation of a number of pluripotency genes and downregulation of somatic genes such as lamin A, were subsequently detected up to four weeks following treatment. Also, these cells showed the ability to differentiate to mesoderm and ectoderm lineages. The phenotypic alterations in these cells were the result of epigenetic modifications of the chromatin mediated by chromatin remodeling factors, histone acetylation and expression of specific genes and protein synthesis.

Recently Bru et al. attempted the same approach into human cells (Bru et al. 2008). Pluripotency-related genes expressed when human somatic 293T cells were permeabilized and incubated in extracts of mouse ES cells. Expression of all four genes (Oct3/4, Sox2, Klf4 and c-Myc) was induced over 1 to 8 hours. Gene expression was associated with loss of repressive histone

H3 modifications and increased recruitment of RNA polymerase II at the promoters. Lamin A/C, which is typically found only in differentiated cells, was also removed from the nuclei. When 293T cells were turned to culture after exposure to ES cell extract, the expression of pluripotency-related genes continued to rise over the following 48 hours of culture, suggesting that long term reprogramming of gene expression had been induced. This provides a methodology for studying the dedifferentiation of somatic cells that can potentially lead to an efficient way of reprogramming somatic cells to a pluripotent state without genetically altering them. Moreover, Rajasingh et al. also reported that, on exposure to mouse ES cells extracts by reversible permeabilization, NIH3T3 cells underwent dedifferentiation and subsequent redifferentiation by stimulus into multiple lineage cell types (Rajasingh et al. 2008). However, genome wide expression profiling revealed significant differences between NIH3T3 and ES extract-treated NIH3T3 cells. Epigenetically, ES cell extracts induced CpG demethylation of Oct4 promoter, hyperacetylation of histone 3 and 4, and decreased lysine 9 (K-9) dimethylation of histone 3 in NIH3T3 cells. This study provides evidence for the generation of functional multipotent stem-like cells from terminally somatic cells, which would represent a major step in the use of regenerative medicine.

These studies so far have shown a modest effect on reprogramming into specific lineages or dedifferentiation into the pluripotent state. The reprogrammed multipotent cells produced by transferring extracts of embryonic carcinoma or ES cell did not reach the pluripotent state in terms of the transcriptional state, *in vivo* 3 germ layer differentiation capacity, and developmental potential. Cho reported that a single transfer of ES cell-derived extract proteins into primarily cultured adult somatic cells could achieve full reprogramming up to the pluripotent state (Cho et al. 2010). This result can provide an alternative and safe strategy for the reprogramming of somatic cells that can be used to facilitate protein-iPS-based cell therapy.

Protocols

As a target cell of reprogramming, primary cardiac fibroblasts (hereafter called cFBs) from C57 mouse were cultured and tested for the efficacy of streptolysin O-mediated reversible permeabilization using fluorescent dextran (Fig. 1). C57 background ES cell-derived extract proteins were introduced into cFB by reversible permeabilization. Starting with $1X10^6$ cFBs, approximately 5 to 10 colonies were observed from day 4 to 7 after induction. On day 10, colonies were disaggregated into a single cell and reseeded onto supporting feeder cell layers. Ten to fifteen days later (days 20–25 from induction), numerous secondary colonies, which were similar to the ES cell colonies in morphology, were observed on the feeder layers.

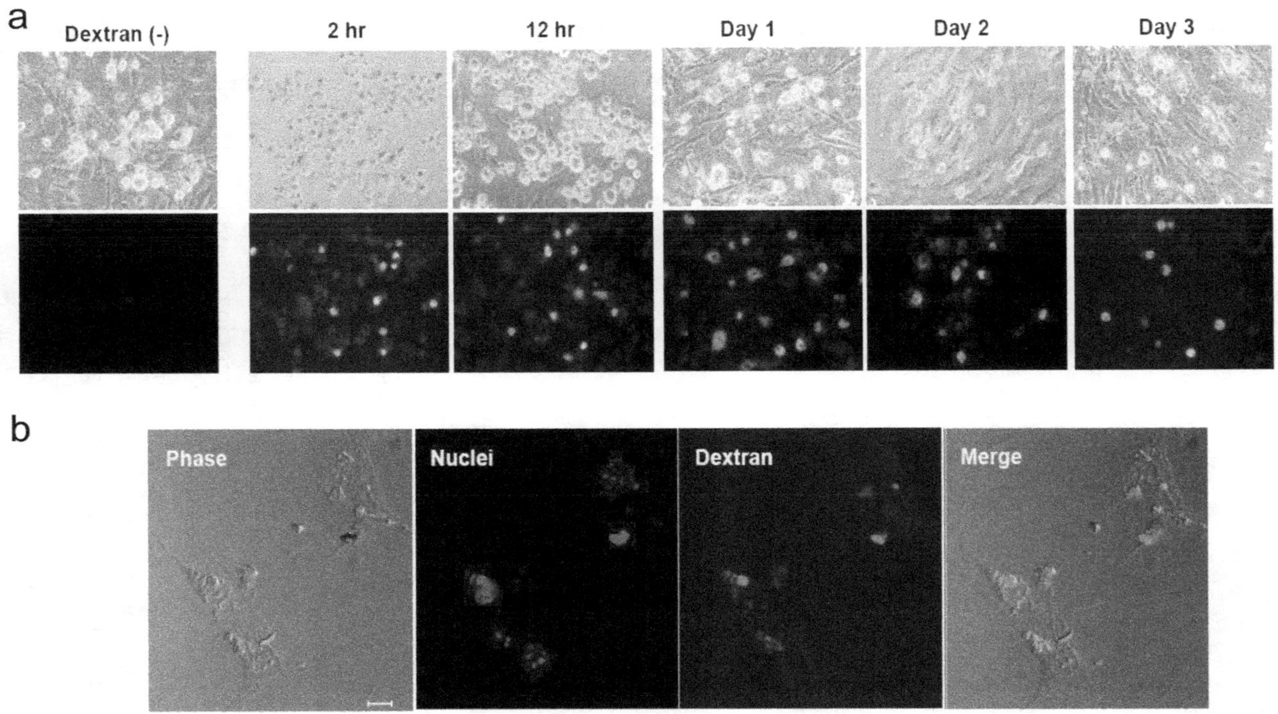

Figure 1. Generation of protein-iPS cells from cardiac fibroblasts by protein delivery. (a) Fibroblasts were reversibly permeabilized, incubated with 70 kDa Texas Red–conjugated dextran and resealed. Serial fluorescent images. (b) Confocal images at day 7.

Color image of this figure appears in the color plate section at the end of the book.

The ES-like cells (colonies) were expanded with subculture every three to five days using the standard ES cell culture protocol. These cells were named cFB-protein-iPS cells.

In vitro characteristics of the protein-iPS

After 45 to 55 days of culture (passages 5–7), several assays to compare the cFB-protein-iPS with ES cell, that is the donor of extract proteins, were carried out. Both cFB-protein-iPS and ES colonies exhibited comparably strong ALP activity, whereas cFB showed very weak ALP activity. Immunohistochemistry revealed that the cFB-protein-iPS cells expressed pluripotency markers including SSEA1 and Oct4. The RT-PCR showed that the cFB-protein-iPS expressed typical pluripotent ES cell genes, such as Nanog, Oct4, Sox2, and E-Ras at consistent levels compared with the ES cells, whereas the cFB did not express these genes. Klf4 and c-Myc expression in the cFB-protein-iPS was also comparable with the ES cell. To confirm the expression levels of the pluripotency genes, real-time PCR was performed to compare the relative expression levels of Oct4 and Nanog between cFB-protein-iPS cells and fibroblasts. There is no significant difference in the expression levels of Oct4 and Nanog between ES and c-FB-protein-iPS cells. The expression levels in ES and protein-iPS cells are at least 10,000-fold higher than those in fibroblasts.

Next, a microarray analysis was performed to assess the global gene expression profiles. Scatterplots demonstrated that the cFB-protein-iPS was very different from the cFB, and cFB-protein-iPS had an ES-like global gene expression pattern. Hierarchical clustering also showed a similarity between the cFB-protein-iPS and ES cell, but not between the cFB-protein-iPS and the cFB. Interestingly, compared with cFB, CD44 was down-regulated and Dnmt3a, Dnmt3b, Dnmt31, histone methyl-transferase 1 (HMT1), leukemia inhibitory factor receptor (LIF-Rc), undifferentiated embryonic cell transcription factor 1 (Utf1), and T-cell lymphoma breakpoint 1 (Tcl1) were markedly up-regulated in the cFB-protein-iPS. Analyzed in terms of biologic processes and molecular functions, up-regulated genes in ES cells were categorized to metabolic and cellular processes and cell cycle-related genes, and up-regulated genes in cFB-protein-iPS cells were categorized to anatomical structure development and metabolic process-related genes. Collectively, there is not any meaningful difference related to reprogramming process between ES and cFB-protein-iPS cells.

In addition to the global gene expression level, epigenetic modifications are critical for reprogramming and dedifferentiation up to the pluripotent state (Freberg et al. 2007; Meissner et al. 2008). Therefore, DNA at the methylation level was evaluated. Bisulfite genomic sequencing analysis showed that the promoter regions of Oct4 and Nanog were largely unmethylated in both cFB-protein-iPS and ES cell, but not in cFB. The histone modification status was also assessed. The ChIP assays of the Oct4 and Nanog promoters showed that cFB-protein-iPS had histone H3 lysine 4 (H3K4) tri-methylation, whereas the cFB had histone H3 lysine 27 (H3K27) tri-methylation. Furthermore, cFB-protein-iPS showed increased acetylation of histone H3 of Oct4 and Nanog promoters. Taken together, patterns of DNA methylation and histone modifications in cFB-protein-iPS were different from those of cFB and very similar to those of the ES cell, implying that cFB-protein-iPS underwent epigenetic reprogramming.

In vitro and in vivo differentiation potentials of protein-iPS

To determine the differentiation potential of the cFB-protein-iPS, EB-based spontaneous differentiation protocol was used. After 7 days of suspension, the cFB-protein-iPS formed cell aggregates that looked like typical EBs. The ES cell-derived aggregates served as a positive control. To further assess the differentiation potential, the aggregated cells (EBs) were plated onto a 0.1% gelatin-coated dish. Beginning at 4 to 7 days of differentiation, some of the EBs contracted spontaneously and expressed cardiac troponin T and alpha-sarcomeric actinin, indicating the properties of cardiomyocytes. On days 7 and 14, the cells were harvested, and the expression of 3 germ layer marker genes and proteins was evaluated. These cells expressed ectoderm (glial fibrillary acidic protein), mesoderm (alpha-smooth muscle actin), and endoderm (alpha-fetoprotein) lineage markers.

Next, to investigate the *in vivo* pluripotency and differentiation potential, cFB-protein-iPS cells were injected into NOD/SCID mice. The cFB-protein-iPS gave rise to subcutaneous mass on days 14 to 21, and well-demarcated tumors were harvested on day 28. The tumors contained the development of a well-differentiated teratoma (Fig. 2). These results suggest that the cFB-protein-iPS cell is comparable with the ES cell with regard to *in vitro* and *in vivo* pluripotency.

Consolidation: Protein-based Reprogramming is Neither by the Contamination of Donor ES Cells nor by DNAs/RNAs from Donor ES Cells

The reprogramming of somatic cells into pluripotent stem cells has been generally considered a slow stochastic process, which requires continuous ectopic expression of the defined transcription factors in a certain period and also requires multiple cell divisions (Chan et al. 2009; Hanna et al. 2009). In contrast, Cho observed that the transfer of ES cell-derived proteins reprogrammed somatic cells in the relatively short time window (Cho et al.

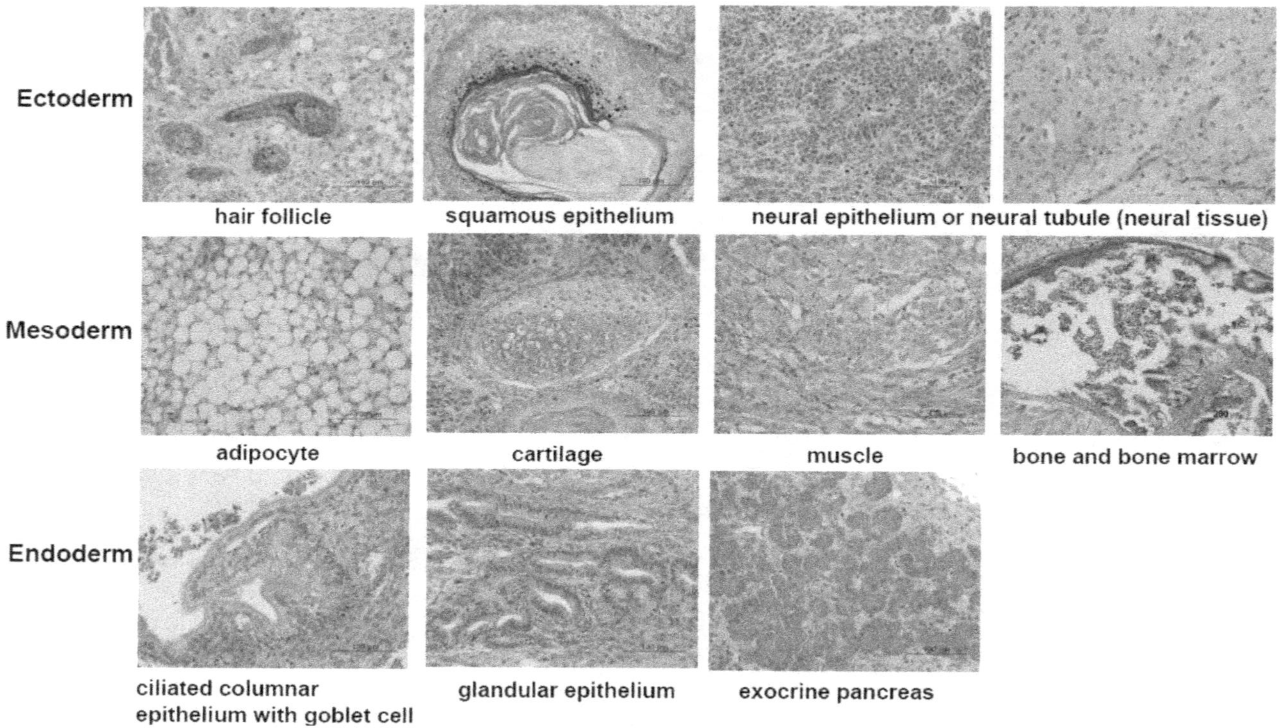

Figure 2. *In vivo* differentiation potentials of the protein-iPS cells. Hematoxylin and eosin (H&E) staining of a tumor from cFB-protein-iPS cells showing a well-differentiated teratoma.

Color image of this figure appears in the color plate section at the end of the book.

2010). Therefore, it is necessary to exclude the possibility of experimental artifacts to consolidate its efficacy.

First, to rule out the possibility of ES cell contamination during the preparation of ES-derived extract proteins, iPS cells were generated using adult skin fibroblasts of a different genetic strain from protein-donor ES cells. The donor of ES-derived extract proteins was a C57 background ES cell, and the recipient of ES proteins (a target cell of reprogramming) was a FVB background skin fibroblast. FVB strain has been widely used for the generation of transgenic animals but has been considered a resistant or nonpermissive strain for ES cell derivation.

By the same protein transfer method, ES-like colonies from skin fibroblasts (hereafter called sFBs) were acquired. The origin of the sFB-protein-iPS cells was confirmed by multiple microsatellite markers of genomic DNA and exact size measurement. sFB-protein-iPS cell was a FVB background. Further characterization steps were done: the expression of pluripotency-related genes and proteins, global gene expression patterns, DNA methylation status, and *in vitro* differentiation potential of FVB-sFB-protein-iPS cells. All of them were similar to the authentic ES cells, indicating that sFB-protein-iPS cells are fully reprogrammed and comparable with the ES cell in terms of pluripotency.

Next, to demonstrate that ES extract-mediated reprogramming is dependent upon proteins, DNA or RNA contamination in the ES-derived extract was first checked. To further exclude the possibility of contamination by genetic material or RNA, the extract was treated with DNase or RNase before transfer. In a series of experiments, Oct4 promoter-driven GFP expression fibroblast was used as a reporter. To verify proteins are an effector, the extract was inactivated by heat. Data showed that DNA or RNA from ES cell has no effect on the reprogramming and that ES cell-derived proteins are wholly responsible (Fig. 3).

In this study, Cho demonstrate that the delivery of ES cell-derived proteins enables the reprogramming of adult fibroblasts, converting them into pluripotent stem cells without the forced expression of certain genetic factors (Cho et al. 2010). The results are different in several aspects, compared with the studies previously conducted using a similar protocol (Fig. 4) (Taranger et al. 2005; Rajasingh et al. 2008).

First, full reprogramming of adult fibroblast up to the pluripotent state and redifferentiation into three germ layers *in vitro* and *in vivo* were achieved. *In vivo* developmental potential of protein-iPS cells was also demonstrated. Second, instead of immortalized cell lines or fetal cells used in the previous studies, primarily cultured adult fibroblasts, which obviously possess a limited lifespan were used; this might have resulted in the natural selection of pluripotent stem cells during the reprogramming. Third, after

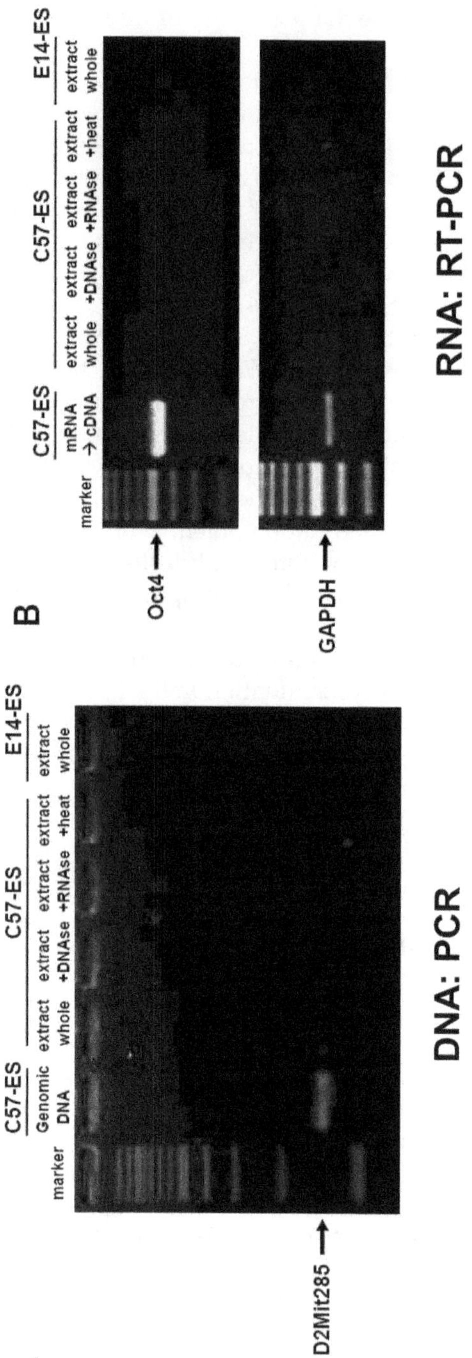
A
marker
C57-ES
Genomic DNA
extract whole
extract +DNAse
extract +RNAse
extract +heat
E14-ES
extract whole
D2Mit285
DNA: PCR
B
marker
C57-ES
mRNA → cDNA
C57-ES
extract whole
extract +DNAse
extract +RNAse
extract +heat
E14-ES
extract whole
Oct4
GAPDH
RNA: RT-PCR

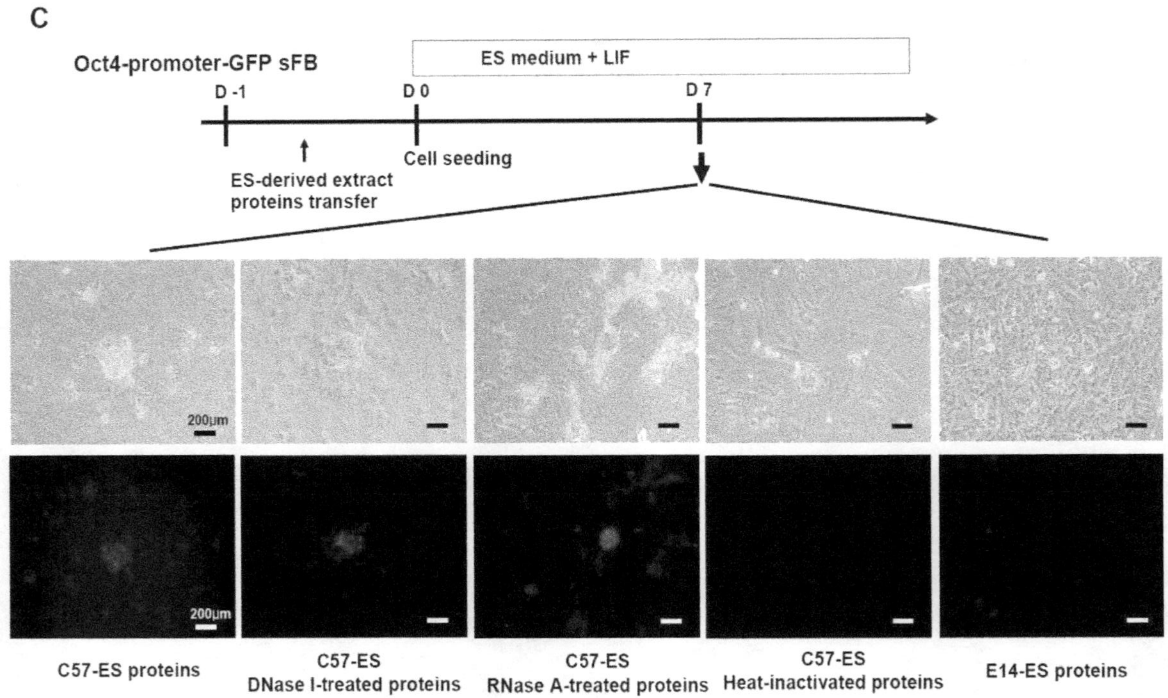

Figure 3. Effector of reprogramming. Protein-based reprogramming is not mediated by DNAs/RNAs from donor ES cell. (A-B) The absence of contaminating DNA or RNA in the extract proteins was confirmed by PCR and RT-PCR. (C) Colony formation of Oct4-promoter–driven GFP reporter cells (skin fibroblast) at seven days after ES derived extract proteins transfer.

Color image of this figure appears in the color plate section at the end of the book.

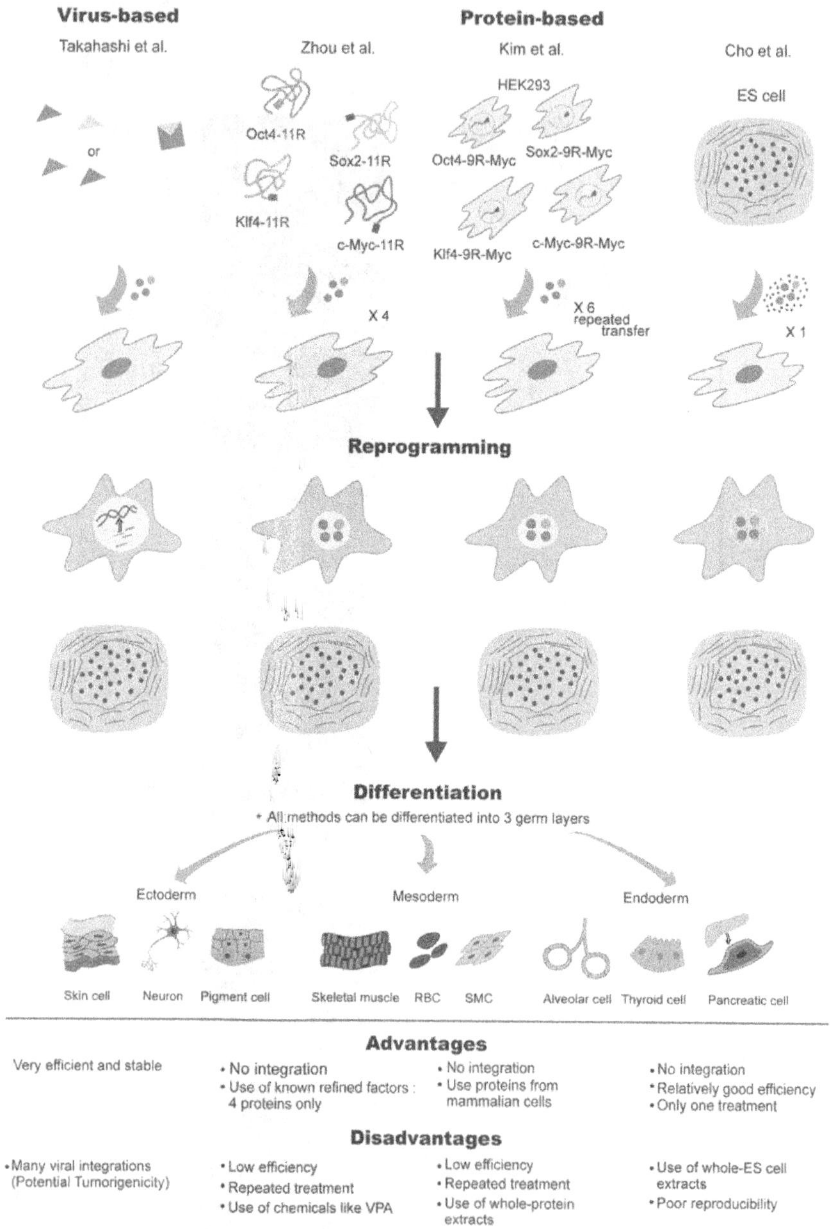

Figure 4. Methods of genetic and non-genetic derived iPS. A flow diagram summarizing the methods to generate iPS cells, with their advantages and disadvantages shown below.

Color image of this figure appears in the color plate section at the end of the book.

the initial colony formation, the colonies were subcultured and maintained on the feeder cell layers. Fourth, C57 strain ES cells were used as a protein donor and transferred to both C57 background adult fibroblasts, which are considered permissive strain for ES cell derivation, and FVB background fibroblasts, which are considered a resistant or non-permissive strain. Collectively, protein-based reprogramming in both strains was successful. However, when other genetic background cells, such as the 129 strain of ES cells (E14), were tried as a protein donor, C57 adult fibroblasts failed to achieve pluripotency. To further investigate the cause, proteomic analysis of the 129 and C57 strain of ES cells was done (Jin et al. 2011). Using iTRAQ and mRNA microarray, the experiment notably showed that reprogramming-competent C57-mES cells highly expressed proteins that regulate protein synthesis and metabolism, when compared to reprogramming-incompetent 129-mES cells, suggesting that there is a threshold that protein synthetic machinery must exceed to initiate reprogramming (Fig. 5).

Currently however, little is known about the molecular mechanisms underlying reprogramming process by cellular proteins. Early colony formation (four to seven days after protein transfer) and Oct4 expression are unique features of this approach, compared with reprogramming sequences used in previous studies based on 4-factor virus/plasmid/protein methods. Accordingly, to further assess the role of four factors like Oct4 in this protein-based reprogramming process, the concentration of Oct4 in the ES-derived protein extracts was measured. Oct4 accounted only for 0.00014% of total extract proteins. Usually 200uL extract proteins were transferred for 10^6 cells, and thus, the absolute amount of Oct4 available for reprogramming was 9.25ng. A recent study reported that repeated transfer of 8000ng/mL Oct4 protein with valproic acid could induce pluripotent stem cells. Compared with that study, substantially lower concentration of Oct4 protein could also generate pluripotent stem cells, which suggests that Oct4 may not have been an essential functional effector molecule in this study. Taken together, ES cell-derived proteins may have a combination of many factors like cytoplasms of meiotic oocyte and mitotic zygote, which can induce pluripotency of somatic cells.

Cell fusion studies indicated that the ES cell nucleus is needed to reprogram the somatic cell nucleus (Do and Scholer 2004; Cowan et al. 2005). Accordingly, a fractionation experiment was performed. Dividing ES proteins into cytosolic and nuclear fraction, neither cytosolic nor nuclear fraction was able to induce reprogramming of adult fibroblasts, whereas whole cellular proteins reproducibly generated protein-iPS cells. Regarding the molecular sizes of effector proteins, extract proteins larger than 30kDa induced reprogramming, but proteins larger than 100kDa did not, suggesting that a certain group of ES cell-derived proteins between 30 to 100 kDa could

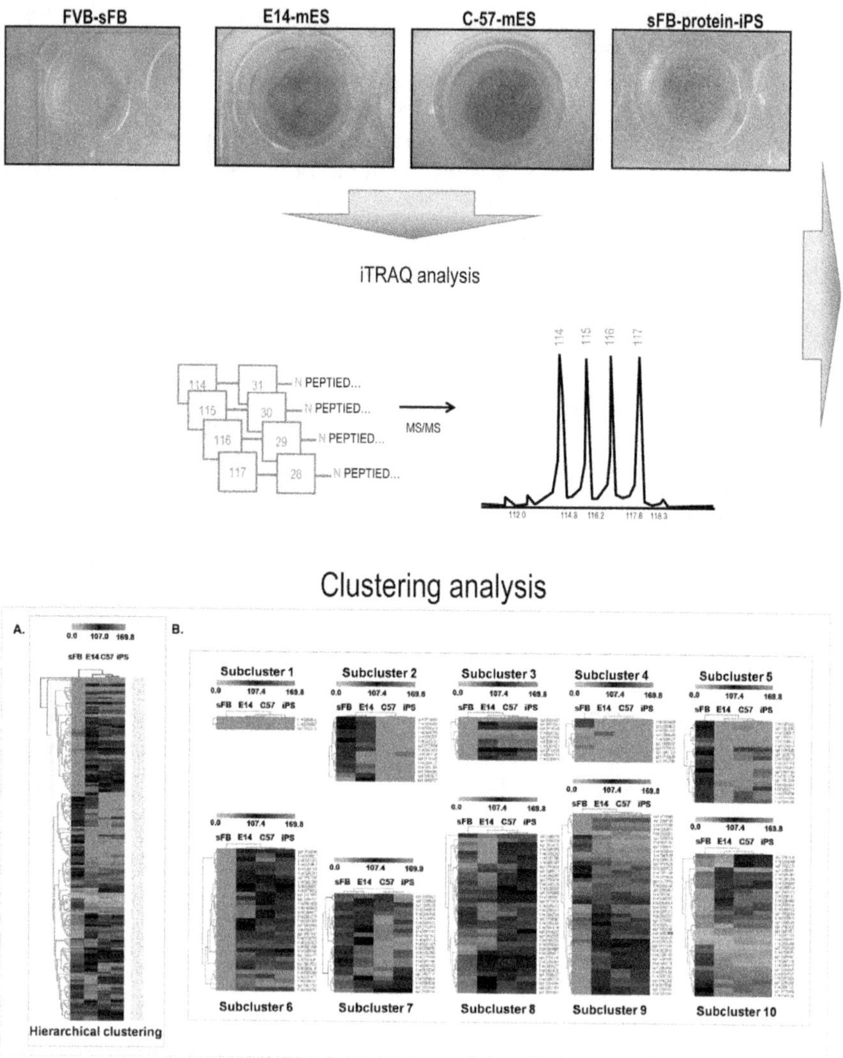

Figure 5. Analysis of differential proteomes of induced pluripotent stem cells by protein-based reprogramming of fibroblasts.

Color image of this figure appears in the color plate section at the end of the book.

be responsible for protein-based iPS generation (Fig. 6). Further studies that reveal the characteristics of specific effector proteins would provide insight and understanding of the mechanism of the protein-based reprogramming process. As one promising example, BAF complex was proposed (Singhal et al. 2010). It is said that ATP-dependent BAF chromatin-remodeling complex increases reprogramming efficiency by facilitating enhanced Oct4 binding to target promoters during reprogramming.

In summary, a single transfer of proteins, rather than using fetus or newborn-origin cells, can induce reprogramming of adult fibroblasts to reach the pluripotent state. This protein-based reprogramming is relatively simple and reproducible and does not requires neither repeated transfer nor prolonged exposure to materials, or a combinatorial approach involving proteins and chemicals. These results provide a safe and effective alternative strategy for reprogramming adult somatic cells and suggest that the described technique could be further developed to provide tailored or patient-specific cell therapy. The protein based iPS cells, however, could reopen ethical issues, since the protein extracts were derived from ES cells. Meanwhile, the second iPSCs that are derived from proteins of iPS cells instead of ES cells could possibly be the ethical substitutes of the first iPSCs. As stem cell research progresses, a new technique is emerging and seeming to blossom. Indisputably, iPSCs have not only changed in their *gene* level but have also changed our current *meme*.

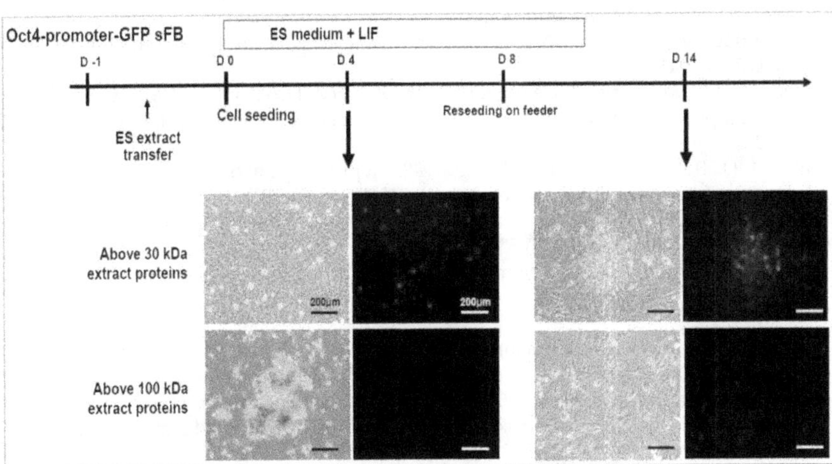

Figure 6. The size of speculated effector of reprogramming. Extract proteins above 30 kDa induced reprogramming, whereas proteins above 100kDa did not.

Color image of this figure appears in the color plate section at the end of the book.

Acknowledgment

This work is supported by grants from the National Research Foundation funded by the Korea Government (MEST) (2010-0020257) and from the Innovative Research Institute for Cell Therapy, Republic of Korea. Dr Hyo-Soo Kim is also a professor of Molecular Medicine and Biopharmaceutical Sciences, Seoul National University, sponsored by World Class University Program from the Ministry of Education, Science, and Technology of Korea.

References

Belting, M., Sandgren, S., Wittrup, A. et al. 2005. Nuclear delivery of macromolecules, barriers and carriers. Adv. Drug. Deliv. Rev. 57(4), 505–527.

Briggs, R. and King, T. J. 1952. Transplantation of living nuclei from blastula cells into enucleated frogs' eggs. Proc. Natl. Acad. Sci. USA. 38(5), 455–463.

Bru, T., Clarke, C., Sang, H.M. et al. 2008. Rapid induction of pluripotency genes after exposure of human somatic cells to mouse ES cell extracts. Exp. Cell Res. 314(14), 2634–2642.

Chan, E. M., Ratanasirintrawoot, S., Park, I. H. et al. 2009. Live cell imaging distinguishes bona fide human iPS cells from partially reprogrammed cells. Nat. Biotechnol. 27(11), 1033–1037.

Cho, H. J., Lee, C. S., Kwon, Y. W. et al. 2010. Induction of pluripotent stem cells from adult somatic cells by protein-based reprogramming without genetic manipulation. Blood 116(3), 386–395.

Cowan, J., Atienza, C. A., Melton, D. A. et al. 2005. Nuclear reprogramming of somatic cells after fusion with human embryonic stem cells. Science 309(5739), 1369–1373.

Do, J. T. and Scholer, H. R. 2004. Nuclei of embryonic stem cells reprogram somatic cells. Stem Cells 22(6), 941–949.

El-Sayed, A., Futaki, S., and Harashima, H. 2009. Delivery of macromolecules using arginine-rich cell-penetrating peptides, ways to overcome endosomal entrapment. AAPS J. 11(1), 13–22.

Frankel, A. D., Bredt, D. S., Pavo, C. O. et al. 1988. Tat protein from human immunodeficiency virus forms a metal-linked dimer. Science 240(4848), 70–73.

Frankel, A. D. and Pabo, C. O. 1988. Cellular uptake of the tat protein from human immunodeficiency virus. Cell 55(6), 1189–1193.

Freberg, C. T., Dahl, J. A., Timoskainen, S. et al. 2007. Epigenetic reprogramming of OCT4 and NANOG regulatory regions by embryonal carcinoma cell extract. Mol. Biol. Cell 18(5), 1543–1553.

Gurdon, J. B. 1962. The developmental capacity of nuclei taken from intestinal epithelium cells of feeding tadpoles. J. Embryol. Exp. Morphol. 10, 622–640.

Gurdon, J. B., Laskey, R. A. Reeves, O. R. et al. 1975. The developmental capacity of nuclei transplanted from keratinized skin cells of adult frogs. J. Embryo.l Exp. Morphol. 34(1), 93–112.

Hakelien, A. M., Landsverk, H. B., Robl, J. M. et al. 2002. Reprogramming fibroblasts to express T-cell functions using cell extracts. Nat. Biotechnol. 20(5), 460–466.

Hanna, J., Saha, K., Pando, B. et al. 2009. Direct cell reprogramming is a stochastic process amenable to acceleration. Nature 462(7273), 595–601.

Huangfu, D., Osafune, K., Maehr, R. et al. 2008. Induction of pluripotent stem cells from primary human fibroblasts with only Oct4 and Sox2. Nat. Biotechnol. 26(11), 1269–1275.

Jin, J., Kwon, Y. W., Paek, J. S. et al. 2011. Analysis of differential proteomes of induced pluripotent stem cells by protein-based reprogramming of fibroblasts. J. Proteome. Res. 10(3), 977–989.

Kim, D., Kim, C. H., Moon, J. I. et al. 2009. Generation of human induced pluripotent stem cells by direct delivery of reprogramming proteins. Cell Stem Cell 4(6), 472–476.

Kim, J. B., Sebastiano, V. et al. 2009. Oct4-induced pluripotency in adult neural stem cells. Cell 136(3), 411–419.

Kim, J. B., Zaehres, H., Wu, G. et al. 2008. Pluripotent stem cells induced from adult neural stem cells by reprogramming with two factors. Nature 454(7204), 646–650.

King, T. J. and Briggs, R. 1955. Changes in the Nuclei of Differentiating Gastrula Cells, as Demonstrated by Nuclear Transplantation. Proc. Natl. Acad. Sci. USA 41(5), 321–325.

Lanza, R. P., Chung, H. Y., Yoo, J. J. et al. 2002. Generation of histocompatible tissues using nuclear transplantation. Nat. Biotechnol. 20(7), 689–696.

Matsumura, H., Tada, M., Otsuji, T. et al. 2007. Targeted chromosome elimination from ES-somatic hybrid cells. Nat Methods 4(1), 23–25.

Meissner, A., Mikkelsen, T. S. et al. 2008. Genome-scale DNA methylation maps of pluripotent and differentiated cells. Nature 454(7205), 766–770.

Meissner, A., Wernig, M., Gu, H. et al. 2007. Direct reprogramming of genetically unmodified fibroblasts into pluripotent stem cells. Nat. Biotechnol. 25(10), 1177–1181.

Miller, R. A. and Ruddle F. H. 1976. Pluripotent teratocarcinoma-thymus somatic cell hybrids. Cell 9(1), 45–55.

Nakagawa, M., Koyanagi, M., Tanabe, K. et al. 2008. Generation of induced pluripotent stem cells without Myc from mouse and human fibroblasts. Nat. Biotechnol. 26(1), 101–106.

Okita, K., Ichisaka, T. et al. 2007. Generation of germline-competent induced pluripotent stem cells. Nature 448(7151), 313–317.

Okita, K., Nakagawa, M., Yamanaka, S. et al. 2008. Generation of mouse induced pluripotent stem cells without viral vectors. Science 322(5903), 949–953.

Park, I. H., Zhao, R., West, J. A. et al. 2008. Reprogramming of human somatic cells to pluripotency with defined factors. Nature 451(7175), 141–146.

Qin, M., Tai, G., Polak J. M. et al. 2005. Cell extract-derived differentiation of embryonic stem cells. Stem Cells 23(6), 712–718.

Rajasingh, J., Lambers, E., Hamada, H. et al. 2008. Cell-free embryonic stem cell extract-mediated derivation of multipotent stem cells from NIH3T3 fibroblasts for functional and anatomical ischemic tissue repair. Circ. Res. 102(11), e107–117.

Singhal, N., Graumann, J., Wu, G. et al. 2010. Chromatin-remodeling components of the BAF complex facilitate reprogramming. Cell 141(6), 943–955.

Stadtfeld, M., Nagaya, M. et al. 2008. Induced pluripotent stem cells generated without viral integration. Science 322(5903), 945–949.

Tada, M., Takahama, Y., Utikal, J. et al. 2001. Nuclear reprogramming of somatic cells by *in vitro* hybridization with ES cells. Curr. Biol. 11(19), 1553–1558.

Takahashi, K., Tanabe, K., Ohnuki, M. et al. 2007. Induction of pluripotent stem cells from adult human fibroblasts by defined factors. Cell 131(5), 861–872.

Takahashi, K. and Yamanaka, S. 2006. Induction of pluripotent stem cells from mouse embryonic and adult fibroblast cultures by defined factors. Cell 126(4), 663–676.

Taranger, C. K., Noer, A., Sørensen, A. L. et al. 2005. Induction of dedifferentiation, genomewide transcriptional programming, and epigenetic reprogramming by extracts of carcinoma and embryonic stem cells. Mol. Biol. Cell. 16(12), 5719–5735.

Warren, L., Manos, P. D., Ahfeldt, T. et al. 2010. Highly efficient reprogramming to pluripotency and directed differentiation of human cells with synthetic modified mRNA. Cell Stem Cell 7(5), 618–630.

Wernig, M., Meissner, A., Foreman, R. et al. 2007. *In vitro* reprogramming of fibroblasts into a pluripotent ES-cell-like state. Nature 448(7151), 318–324.

Wilmut, I., Schnieke, A. E., McWhir, J. et al. 1997. Viable offspring derived from fetal and adult mammalian cells. Nature 385(6619), 810–813.

Yu, J., Vodyanik, M. A., Smuga-Otto, K. et al. 2007. Induced pluripotent stem cell lines derived from human somatic cells. Science 318(5858), 1917–1920.

Zhou, H., Wu, S., Joo, J. Y. et al. 2009. Generation of induced pluripotent stem cells using recombinant proteins. Cell Stem Cell 4(5), 381–384.

Ziegler, A., Nervi, P., Dürrenberger, M. et al. 2005. The cationic cell-penetrating peptide CPP(TAT) derived from the HIV-1 protein TAT is rapidly transported into living fibroblasts, optical, biophysical, and metabolic evidence. Biochemistry 44(1), 138–148.

CHAPTER 4

Lessons from Heart Development to Regeneration

Hiroyuki Yamagishi

Introduction

The vertebrate heart is an organ with four chambers, constituted by a multilayered striated myocardium, and two great arteries. The primitive heart tube is the first organ to function as a circulatory system by beating at embryonic Day (E) 9 in mice and at Day 22 in humans to meet the requirement to deliver nutrients throughout the developing embryo. From the beating tube, the shape of the heart changes to create a rightward looping of the tube, followed by the steps to form a four-chambered structure as a result of the orchestration of multiple cardiac progenitor cells that are regulated by numerous signaling molecules and transcription factors with hemodynamic changes. Maintaining the pumping function, with concomitant shape changes during development, may make the cardiovascular system relatively susceptible to a variety of mistakes. It is known that congenital heart defects are the most among human congenital anomalies that are estimated at approximately 1% of newborns. Immediately after delivery, usually at 20 days in mice and around 270 days after conception in humans, the cardiovascular system switches the course of blood circulation for an oxygen source from the placenta to the lung, which involves closure of the ductus arteriosus, ductus venosus and foramen ovale. The morphogenesis of the heart, thus, seems to be a natural art brought about by evolution in order to obtain a higher function for maintaining a higher organism.

Department of Pediatrics, Division of Pediatric Cardiology, Keio University School of Medicine, 35 Shinanomachi, Shinjuku-ku, Tokyo 160-8582, Japan.
Email: hyamag@sc.itc.keio.ac.jp

Multiple Developmental Origins of Cardiac Progenitors

Decades of descriptive embryology and cell lineage tracings have improved our understanding of the development of the heart, including developmental origins of cardiac progenitors of each component of the heart (Fig. 1) (Vincent and Buckingham 2010; Srivastava 2006; Kodo and Yamagishi 2011). In the first step of cardiac development, a specific subset of mesodermal cells, generated during embryonic gastrulation migrate anterolaterally to form a crescent shape on approximately E7.5 in the mouse embryo, corresponding roughly to Week 2 of human gestation. By E8.0 in mice, or Week 3 in humans, a pair of endocardial tubes appears in the cardiac crescent and fuses along the ventral midline to form the primitive heart tube, consisting of an interior layer of endocardial cells and an exterior layer of myocardial cells, separated by the extracellular matrix, or "cardiac jelly", that mediates reciprocal signaling between the two layers. The crescent shaped pool of cardiogenic progenitors that are derived from the lateral plate mesoderm is referred to as the "first heart field" (FHF).

In addition to the FHF, there is a second source of cardiac progenitors. In the mouse embryo, this source lies medially to the cardiac crescent and then lies behind the primitive heart tube, extending into the mesodermal layer of the pharyngeal arches (Fig. 1). This second source is referred to as the "second heart field" (SHF) (Vincent and Buckingham 2010; Srivastava 2006; Kodo and Yamagishi 2011). The heart tube derived from the FHF may predominantly provide a scaffold upon which cells from the SHF migrate into both arterial and venous poles of the heart tube, where they subsequently construct the requisite cardiac components. Upon rightward looping of the heart tube, SHF-derived progenitor cells cross the pharyngeal mesoderm into the arterial and venous portions, populating a large portion of the outflow tract, future right ventricle, and atria. The addition of the SHF-derived myocardium to the outflow tract results in its elongation that is necessary to allow the outflow tract to rotate for building a correct alignment of the pulmonary and aortic trunks with their respective ventricles. In the arterial pole of the heart, the SHF-derived progenitors give rise to both myocardial and smooth muscle lineages. Finally, progenitors derived from the FHF contribute exclusively to the left ventricle and all other parts of the heart except the outflow tract, whereas those from the SHF, which segregates rather later, contribute exclusively to the outflow tract and all other parts of the heart, except the left ventricle. In murine embryos, the entire outflow tract myocardium and most of right ventricular myocardium are formed from progenitors derived from the anterior part of the SHF, whereas in chick embryos, the myocardial contribution of the SHF cells appears to be limited to a part of the outflow tract (Waldo et al. 2001).

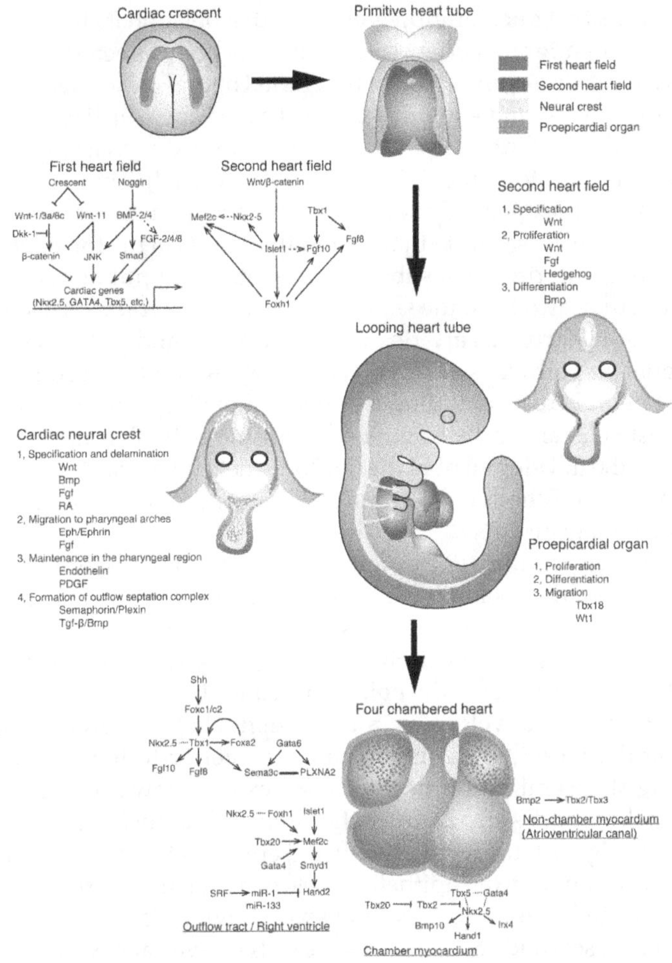

Figure 1. Developmental origins of the heart. Frontal views of embryos at cardiac crescent and primitive heart tube stage, and a lateral view of embryo at looping heart stage with transverse sections, and the four-chambered heart are shown. In the cardiac crescent, the first heart field (FHF) cells form a crescent shape in the anterior region of embryo with the second heart field (SHF) cells medial to it. In the primitive heart tube, the SHF cells lie dorsal to the heart tube in the splanchnic mesoderm of pharyngeal region. In the looping heart tube, the SHF cells migrate into the anterior and posterior ends of the tube (blue arrows). Subsequently, cardiac neural crest cells also migrate into the outflow tract from the dorsal neural tube (yellow arrows) to remodel the outflow tract and the aortic arch system. In the four-chambered heart, the FHF cells form the left ventricle and atria (red) whereas the SHF cells form the outflow tract, the right ventricle, and part of atria (blue). Cardiac neural crest cells contribute to the outflow tract cushions (yellow) that eventually form the aorto-pulmonary septum. Epicardium is derived from proepicardial organ (green). Signaling and transcriptional network for the development of the cardiac progenitors is shown (See detail in the text).

Color image of this figure appears in the color plate section at the end of the book.

Neural crest cells are multipotential cells that delaminate from the dorsal neural tube and migrate widely throughout the body as ectoderm-derived mesenchymal cells. A sub-region of the cranial neural crest cells originating between the otocyst and somite 3 has been called the "cardiac neural crest cells" that migrates into the third, fourth, and sixth pharyngeal arches and the cardiac outflow tract (Hutson and Kirby 2007). They contribute to remodeling of the six pairs of bilaterally symmetric pharyngeal arch arteries connecting to the dorsal aorta that eventually results in the formation of the thoracic aorta, and proximal subclavian, carotid, and pulmonary arteries. In addition, they also form the septum of the outflow tract that divides the embryonic single truncus arteriosus into the aortic and pulmonary trunk. Many signaling pathways are involved in the migration and condensation of cardiac neural crest, including reciprocal signaling between the cardiac neural crest cells and cells originated from the SHF, that are eventually essential for the development of the outflow tract and the aortic arch system (Srivastava 2006; Scholl and Kirby 2009; Vincent and Buckingham 2010; Kodo and Yamagishi 2011).

The proepicardium (PE) is another lineage of cardiac progenitors developing from the coelomic mesothelium which overlays the liver bud. The expression of PE-specific genes is induced in naïve mesothelial cells in response to a localized liver-derived signal (Srivastava 2006; Vincent and Buckingham 2010). Cells of the epicardium and the coronary vessels arise from the PE, which develops as multiple epithelial villi protruding from the pericardial mesothelium immediately posterior to the sinoatrium of the looping stage embryonic heart. The PE extends towards the primitive heart, attaches, and spreads over the myocardial surface, forming the epicardium (Fig. 1). During PE growth and epicardial formation, some PE/epicardial cells undergo an epithelial–mesenchymal transformation (EMT) and are localized to the subepicardial space, giving rise to the precursors of coronary vessels and connective tissue cells (Olivey and Svensson 2010). As for the development of the coronary vasculature, smooth muscle cells arise from the epicardium, and endothelial cells and capillaries are derived from the venous plexus, at the sinus venosus, which invades the heart after formation of the epicardium.

Islet1 in Cardiac Progenitors

Insight into the molecular mechanism underlying the development of cardiac progenitors would provide new therapeutic approaches for cardiac repair. In the past 10 years, numerous transcription factors that function during cardiac development have been identified. LIM-homeodomain transcription factor, Islet1, has been defined as a marker for a cardiac progenitor lineage that is capable of differentiating into all three major cell types of the heart, namely cardiomyocytes, smooth muscle cells, and endothelial cells (Laugwitz et al. 2005; Moretti et al. 2006). Thereafter, it has been shown that a diverse set of human fetal Islet1-positive cardiovascular progenitors give rise to the cardiomyocyte, smooth muscle and endothelial cell lineages, and that purified Islet1-positive primordial progenitors were capable of self-renewal and expansion before differentiation into the three major cell types in the heart (Bu et al. 2009). Islet1 knockout mice showed developmental defects of the right ventricle and outflow tract (Cai et al. 2003). During development, the expression of Islet1 is first detectable in the area comprising the dorsal-medial aspect of the cardiac crescent, and then behind the forming heart tube, expanding the splanchnic mesoderm of the pharyngeal region. From these observations, the expression domain of Islet1 is now commonly defined as the SHF. Expression of Islet1 is extinguished as progenitor cells express markers of cardiac differentiation, although Islet1 is necessary for SHF-derived progenitor cells to populate the heart. Cardiac progenitors in the SHF appear to be regulated by complex positive and negative signaling networks, including members of the Wnt, fibroblast growth factor (Fgf), bone morphogenetic protein (Bmp), hedgehog (Hh), etc. Such signals may be autocrine, but often arise from the adjacent endoderm or other cell lineages.

Wnt Signaling for Cardiac Progenitors

Canonical Wnt signaling promotes progenitor cell proliferation, in addition to playing earlier roles in mesodermal development and negative modulation of cardiac specification (Cohen et al. 2007). After the heart tube has formed, canonical Wnt signaling is maintained in the SHF. SHF cells may remain in an undifferentiated progenitor state until they are incorporated into the heart because they are close to inhibitory Wnt signals that emanate from the midline. Canonical Wnts (Wnt1, Wnt3a) from the dorsal neural tube are important for the maintenance of proliferating progenitors within the SHF and for the inhibition of differentiation. Canonical Wnt β-catenin pathway is also required for neural crest induction (De Calisto et al. 2005).

β-catenin is required for SHF progenitors. The genetic ablation of β-catenin results in pharyngeal arch defects and a single outflow tract, probably as a result of a reduction in proliferation of SHF-derived progenitors (Cohen et al. 2007; Lin et al. 2007; Kwon et al. 2007). The expression of Islet1, Fgf10, and Shh are reduced, suggesting an upstream role of canonical Wnt signaling in the SHF, and in fact, β-catenin can directly activate transcription of Islet1 and Fgf10 promoters. Loss of Wnt signaling reduces the number of Islet1-positive cells, leading to defects of the outflow tract and right ventricle, whereas excess Wnt signaling expands the Islet1-positive population. Conditional expression of stabilized β-catenin results in the expansion of SHF progenitors. Transcriptome analysis of these Islet1-positive cells shows an up-regulation of Fgfs, which promote proliferation, when canonical Wnt signaling is increased (Kwon et al. 2009). Of Wnt ligands potentially involved in autocrine signaling within the SHF, Wnt2 is expressed in the posterior region and directly regulates proliferation of venous pole progenitors in concert with β-catenin (Tian et al. 2010).

In addition to its role in promoting proliferation in the SHF, there is an evidence to suggest that canonical Wnt signaling also prevents the onset of differentiation (Kwon et al. 2009). Conditional expression of stabilized β-catenin in Islet1-positive cells also shows downregulation of the gene encoding myocardin that promotes myocardial and smooth muscle differentiation. On the other hand, β-catenin induces expression of both Bmp4 and non-canonical Wnt11 in the SHF, suggesting that early canonical Wnt signaling also sets the stage for differentiation. Although Islet1 had been mainly associated with proliferation, it may, at least when expressed at a high level, promote differentiation and can directly activate a regulatory element of the myocardin promoter. Differences in the levels of Islet1 in determining its effects on differentiation or proliferation are likely to be associated with the level of stabilized β-catenin in cardiac progenitor cells.

Non-canonical Wnt signaling is specifically required for the outflow tract development (Pandur et al. 2002). *Wnt11* mutant embryos have a short outflow tract and develop alignment and septation defects likely mediated by the loss of transforming growth factor (Tgf) β2 signaling. Tgfβ2 signaling is a critical regulator of the neural crest and endocardial cushion development during the outflow tract septation (Zhou et al. 2007). *Wnt5a* is expressed in the SHF. *Wnt5a* mutant embryos have a common outflow tract and abnormal neural crest invasion, although the expression domains of SHF genes *Fgf8*, *Fgf10*, and *Tbx1* are unaltered, suggesting a role for noncanonical Wnt signaling from the SHF to neural crest cells (Schleiffarth et al. 2007).

Fgf Signaling for Cardiac Progenitors

Fgf signaling promotes proliferation of the SHF. Fgf8 plays an important role in specifying the early cardiac mesoderm and it is likely to be the major Fgf ligand driving the SHF development (Alsan et al. 2002). Mice with *Fgf8* hypomorphic alleles develop outflow tract defects reminiscent of those in the mouse mutant for T-box transcription factor, Tbx1 (Abu-Issa et al. 2002). Regulatory elements on the *Fgf8* locus are directly activated by Tbx1, suggesting that Tbx1 may positively regulate proliferation of cardiac progenitor cells in the SHF through transcriptional regulation of the Fgf8-Fgfr1 signaling pathway (Hu et al. 2004). Tbx1 itself is regulated positively by Shh signaling (Garg et al. 2001; Yamagishi et al. 2003), and negatively by retinoic acid signaling, and thus, plays an important role in integrating different signaling inputs during SHF development.

Fgf8 is produced by the mesoderm, endoderm, and ectoderm of pharyngeal arches. Among those, mesodermal Fgf8 expression is clearly important for the development of the SHF (Abu-Issa and Kirby 2007). Fgf10 is also expressed in the SHF. Although mice with mutation of Fgf10 do not show defects in SHF development, mesodermal Fgf8/Fgf10 compound mutants display increasingly severe defects in the outflow tract and pharyngeal arch arteries, suggesting that proliferation of progenitor cells in the SHF that contribute to the outflow tract myocardium is sensitive to the dosage of Fgf signaling (Watanabe et al. 2010).

Conditional ablation of both *Fgfr1* and *Fgfr2*, genes encoding Fgf receptors, specifically in the SHF, but not in endothelial or neural crest cells, shows a reduction in proliferation of progenitor cells within the SHF, suggesting that autocrine Fgf signaling from the SHF-derived myocardium are likely to drive proliferation and deployment of SHF cells (Park et al. 2008). Fgf signaling, probably through its effects on proliferation, increases the extent of the heart field and the consequent numbers of cardiomyocytes that eventually form the outflow tract.

There has been some evidence suggesting the role of Fgf signaling for migration of neural crest cells. Fgf8 and Fgf2 have chemotactic activity for neural crest cells (Kubota and Ito 2000; Sato et al. 2011). Interestingly, cranial neural crests express a variety of ephrinB2 ligand and the Eph receptors during their migration (Smith et al. 1997). Their migratory pathways are clearly bordered by non-neural crest cells expressing ephrinB1 ligand and the EphB2 receptor. Endothelin and Platelet-derived growth factor (PDGF) play a role in maintenance and/or proliferation of the cardiac neural crest cells after they migrate into pharyngeal arches or the outflow tract (Yanagisawa et al. 1998; Richarte et al. 2007). Endothelin-A receptor (ETA) is expressed by cardiac neural crest in the pharyngeal arches and the outflow tract, while endothelin-1 (ET1) and endothelin converting enzyme-1 (ECE) are expressed by pharyngeal arch epithelia. Knockout of ET1, ETA or ECE results in similar abnormal patterning of the aortic arch system, although their initial patterning of pharyngeal arch arteries develops normally. Both of the PDGF receptor α and β subtypes are expressed in cardiac neural crests. Conditional double knockout of PDGFα/β in the cardiac neural crest also results in failure of septation of the outflow tract and abnormal pattering of the aortic arch.

Fgf signaling is also thought to be involved in the interaction between cardiac progenitors derived from the SHF and neural crest. In mesodermal *Fgf8* mutant mice, Bmp/Tgfβ signaling, that normally provides a relay that influences the neural crest, is downregulated in the SHF (Park et al. 2008;

Nie et al. 2008). In chick embryos, neural crest ablation results in an increase in Fgf8 that perturbs the development of SHF-derived progenitors, and the defect can be rescued when the level of Fgf8 is reduced (Hutson et al. 2006). An increase in Fgf signaling might increase the cardiac progenitor pool at the level of differentiation.

In addition to Fgf8 and Fgf10, Fgf15 is also implicated in the outflow tract development. Fgf15 is expressed in the pharyngeal arch mesoderm (arch 3) and in the SHF. *Fgf15* mutants show outflow tract defects, probably at least partly as a result of abnormal development of the neural crest (Vincentz et al. 2005).

Hh Signaling for Cardiac Progenitors

Hh signaling is also often associated with proliferation. Sonic hedgehog (Shh), from the endoderm, the notochord, and the floorplate affects arterial pole formation, probably through an effect on the SHF proliferation, and is required for the formation of the atrial septum at the venous pole. *Shh* mutant embryos display defects in the SHF deployment and survival, resulting in a common outflow tract. The conditional deletion of *Shh* in mice has revealed that the pharyngeal endoderm is the source of a ligand required for the SHF and outflow tract development (Goddeeris et al. 2008). In these conditional mutant embryos, extensive cell death in the SHF results in elongation defects of the outflow tract.

Shh expression is regulated by Islet1 in the pharyngeal endoderm (Lin et al. 2006). At the arterial pole, Shh from the endoderm affects the pharyngeal arch mesoderm with additional effects on the maintenance and deployment of the neural crest (Dyer and Kirby 2009). In the absence of *Shh*, the pharyngeal vasculature, as well as outflow tract development is affected. Shh maintains the expression of Tbx1 in the SHF, particularly implicated in the formation of the pulmonary trunk myocardium (Yamagishi et al. 2003; Théveniau-Ruissy et al. 2008). In chick embryos, Shh signaling is clearly important for maintaining progenitor cell proliferation in the critical time frame which precedes addition of cells to the heart tube (Dyer and Kirby 2009). Manipulation of Shh signaling in chicks affects migration of SHF-derived progenitor cells into the outflow tract, suggesting that it is also important for SHF recruitment. Moreover, the absence of Shh signaling in the posterior SHF of mouse embryos results in specific venous pole defects. The source of the signal is the underlying pulmonary endoderm, where Islet1 is also present and is required for Shh expression (Théveniau-Ruissy et al. 2008). Taken together, Hh signal reception is necessary in the SHF and in the pharyngeal endoderm itself to regulate a secondary signal that controls the survival and deployment of progenitor cells derived from the SHF. It is also required for the survival of neural crest cells in the pharyngeal region.

Shh signaling in the pharyngeal mesoderm is required not only for the outflow tract development but also for the atrioventricular septation (Goddeeris et al. 2008). Development of the dorsal mesocardial protrusion (DMP) derived from the SHF depends on the Shh signaling. The DMP eventually gives rise to part of the atrial septum extending toward the atrioventricular cushions. Shh may play a role in the specification of atrial septal progenitors and in the recruitment of cells from this part of the SHF.

Bmp Signaling for Cardiac Progenitors

Bmp signaling is required for cardiac mesoderm specification and differentiation (Schultheiss et al. 1997). SHF progenitors lie medially to the differentiating cardiac crescent, at a distance from sources of Bmp emanating from the lateral mesoderm, consistent with their maintenance in an undifferentiated state. Mouse mutants with *Bmpr1a,* that encodes the Bmp receptor type 1 in the early cardiac mesoderm, results in a failure to form a differentiating heart tube, although Islet1-positive cardiac progenitor cells are present (Klaus et al. 2007). Later conditional deletion of *Bmpr1a* in Islet1-expressing cells results in an abnormal right ventricle and outflow tract with increased numbers of Islet1-positive cells, suggesting a differentiation defect as Islet1 is normally downregulated in cardiomyocytes. This is accompanied by a reduction in Tbx20 that is required to repress Islet1 in the outflow tract (Yang et al. 2006).

Bmp signaling is also required for cardiac neural crest induction. Bmp4 from adjacent ectoderm appears to play a major role. At the time of induction, Bmp4 is homogenously expressed in the dorsal neural tube whereas the Bmp inhibitor noggin is expressed in a cranial-coudal gradient, with high expression in the presumptive cardiac neural crest in order to control the level of Bmp activity to precise concentration for induction (Sela-Donenfeld and Kalcheim 1999). A gradient of Bmp activity determined by noggin specifies the expression of downstream transcription factor, Msx genes, in the neural fold which in turn upregulates early neural crest specific markers, such as Snail1/2 and FoxD3 (Tribulo et al. 2003). For the delamination from the neural tube, neural crest cells synchronously undergo epithelial-mesenchymal transition in the S phase of the cell cycle. Delamination is disturbed by inhibition of the cell cycle transition from G1 to S (Burstyn-Cohen and Kalcheim 2002). Bmp4 upregulates the expression of Slug/Snail2 through Smad1 and concomitantly downregulates N-cadherin by stimulating its cleavage via a metalloproteinase, ADAM10.

During the outflow tract development, Bmp4 is essential for myocardium formation as well as the survival of the neural crest (Nie et al. 2008). Outflow tract elongation is reduced in embryos lacking both *Bmp4* and

Bmp7 (Liu et al. 2004). Blocking of Bmp signaling inhibits differentiation of the SHF-derived cardiac progenitors. Bmp2, and also Bmp4, are proposed to be candidate molecules for inducing differentiation of SHF progenitors at the arterial pole of the heart tube, and may play an additional role in the recruitment of these cells (Uchimura et al. 2009; Waldo et al. 2001). Bmp signaling downregulates Islet1 expression as SHF-derived progenitor cells enter the outflow tract.

Bmp4, together with other genes for the Bmp/Tgfβ family members, are targets of Fgf signaling in the SHF (Park et al. 2008). Wnt/β-catenin signaling, which plays a role in proliferation and maintenance of an undifferentiated state in the SHF, also up-regulates Bmp4 (Ai et al. 2007). The expression levels of these signal transductions are probably critical. Different signaling thresholds may be required for an effect on proliferation or differentiation of cardiac progenitors. In this context, a balance between Fgfs and Bmps should be important. The importance of regulating the level of Bmp/Tgfβ signaling within the SHF is demonstrated by the role of Nkx2-5 that represses Bmp2 expression (Prall et al. 2007). In the absence of Nkx2-5, an over specification of cardiomyocytes and a reduction in proliferating SHF progenitors leads to the outflow tract truncation. Smad1 is supposed to mediate the effect of Bmp2 on the SHF. In mesodermal Smad1 mutant mice, SHF proliferation and the outflow tract length are increased. The Nkx2-5 phenotype is alleviated on loss of Smad1 alleles. These findings demonstrate a negative feedback loop involving Nkx2-5, Bmp2, and Smad1 that controls the balance between progenitor proliferation and differentiation in the SHF. Bmp/Tgfβ signaling can also be regulated by Tbx1 through direct interaction with Smad1 (Fulcoli et al. 2009). Development of the SHF involves Bmp driven differentiation at the distal end of the heart tube and Fgf driven proliferation in the pharyngeal mesoderm.

The TGFβ superfamily, including BMP and TGFβ ligands, also affects the differentiation or maintenance of cardiac neural crest. In general terms, TGFβs bind to the TGFβ type II receptor (TGFβRII) and TGFβ type I receptor (ALK5) activating Smad2 and 3, while BMPs bind to the BMP type II receptor and type I receptors ALK2, 3, or 6, activating Smad1, 5 and 8 (Chen and Lechleider 2004). Alk2 is required for maintenance of cardiac neural crest in pharyngeal arches and the outflow tract (Stottmann et al. 2004). Canonical TGFβ or BMP signaling results in phosphorylation of different Smads, but these phosphorylated Smads all bind the universally used co-Smad4. The Smad4 complex is then translocated to the nucleus where it alters transcription. Conditional deletion of Smad4 in cardiac neural crest using Wnt1-cre leads to failure of the outflow tract septation (Jia et al. 2007).

GATA Transcription Factors in Cardiac Progenitors

Members of zinc-finger transcription factors, termed GATA factors, have essential roles in the differentiation and survival of many cell types, including cardiomyocytes (Brewer and Pizzey 2006). Focusing on the heart, transcripts of GATA4, 5, and 6 are all present in the precardiac mesoderm at E7–7.5, and are thought to be some of the earliest genes reported to be expressed in cardiomyocyte precursors in murine embryos. As for GATA4, early cardiac expression at E8.0 is confined to the posterior end of the heart tube. During the looping stage at E8.5–9.0, its expression extends throughout the heart tube (Nemer and Nemer 2003). The expression pattern of GATA6 has been reported to be almost identical to that of GATA4 within the developing heart up to E9.5. By E11.5, GATA4 and GATA6 are detected in the myocardium and endocardium of the atria and ventricles, and within the proximal outflow tract. Both proteins are detected in the endocardial cushions of the atrioventricular canal and the outflow tract at E10.5–11.5. Cardiac expression of GATA4 and 6 persists in the adult in the endocardium and myocardium.

The process of early embryonic development involved in cardiogenesis is well conserved across species. In flies, cardiac precursor cells are specified from paired progenitor fields and move into a linear tubular structure like that in vertebrates. In Drosophila, GATA binding factor Pannier is expressed within the dorsal, heart-forming mesoderm of embryos and is required for cardiogenesis (Sorrentino et al. 2005). In vertebrate embryos, cardiac GATA factors are co-expressed within the cardiac mesoderm from the very early stages in its specification (Nemer and Nemer 2003). Targeted disruption of GATA 4, 5, and 6 in the mouse has revealed phenotypes consistent with their individual expression patterns. Mice null for GATA4 die between E8.0–9.0 as a result of defects in heart morphogenesis and ventral closure of the foregut (Molkentin et al. 1997). GATA4 null mice show cardia bifida resulting from the failure of ventral fusion of the lateral aspects of the embryo. This defect of heart formation in GATA4 null mice is likely a secondary effect associated with an intrinsic defect in the endoderm that underlies the splanchnic mesoderm containing the cardiac field. GATA6 null mice die during early embryogenesis (E5.5–7.5) as a result of defects in visceral endoderm function and subsequent extraembryonic development that are consistent with the expression pattern of GATA6 in the embryonic primitive endoderm (Morrisey et al. 1998). Expression of GATA4 and 6 is dependent on each other, given the observation that GATA6 is up-regulated in GATA4 null mice and that GATA6 null embryos show down-regulation of GATA4. Although the mechanism underlying this regulation has not been elucidated, it is suggested that GATA4 negatively regulates GATA6 gene expression, whereas GATA6 positively regulates GATA4 gene expression. Conditional inactivation of GATA6, specifically in the cardiac NCC, causes persistent truncus arteriosus, a severe congenital heart defect characterized by an embryonic single outflow tract without septation, suggesting an essential role of GATA6 during development of the outflow tract (Lepore et al. 2006). Mice compound heterozygous for GATA4 and GATA6 null alleles die *in utero* and exhibit a spectrum of congenital heart defects, involving the atrioventricular septum and outflow tract (Xin et al. 2006). GATA6 may function in concert with GATA4 to direct tissue-specific gene expression essential for formation of the mammalian heart.

HAND Transcription Factors in Cardiac Progenitors

Hand1/eHand and *Hand2/dHand* belong to a family of basic-helix-loop-helix (bHLH) transcription factors and are indispensable for cardiogenesis, especially in ventricular development. *Hand* genes share common bHLH regions for DNA binding which are almost identical between *Hand1* and *Hand2*. The *Hand2* proteins share approximately 95% homology of the amino acid sequences among human, mouse, chicken, frog, and zebra fish, whereas the *Hand1* proteins are less conserved across species (Srivastava et al. 1995). *Hand* genes direct a unique expression pattern during cardiac ventricular development. In chicks, *Hand1* and *Hand2* are expressed in both developing ventricles. In contrast, in mouse embryos, *Hand1* is predominantly expressed in the left ventricle whereas *Hand2* is expressed in the right ventricle in a complementary fashion (Srivastava et al. 1997). Intriguingly, genetic analysis of the developing heart *in situs inversus* mice showed that *Hand1* and *Hand2* are not expressed in the left-sided ventricle and right-sided ventricle, but in the morphological left ventricle and right ventricle, respectively (Thomas et al. 1998). *Hand1* and *Hand2* are also expressed in the NCC in partially overlapping patterns. *Hand2* is more broadly expressed in the cardiac outflow tract than *Hand1*. *Hand* genes are downregulated after E13.5 during murine cardiac development and also after birth. In humans, *Hand* genes are expressed in both ventricles of adult hearts and *Hand1* is reported to be downregulated in cardiomyopathies (Natarajan et al. 2001). Hence, the molecular mechanism that restricts the expression of *Hand* genes in the developing heart may be downregulated in the adult heart (Firulli 2003). Among the transcriptional factors related to developing cardiac ventricles other than *Hand* genes, *Nkx2-5* and *Gata4* are expressed throughout the ventricles during early development. *Mef2c* is predominantly expressed in the developing right ventricle, while *Tbx5* is predominantly expressed in the left ventricle (Srivastava 2006). *Nkx2-5* null mice died early *in utero* and showed a downregulation of *Hand1* expression (Tanaka et al. 1999).

In contrast, GATA4 cardiac-specific knockout mice showed hypoplasia of the developing right ventricle and a downregulation of *Hand2* expression (Zeisberg et al. 2005). The right ventricle-specific enhancer of *Hand2* contains conserved GATA binding elements suggesting that the expression of Hand2 is under the control of GATA4 in right ventricular formation (McFadden et al. 2000). The expression of *Hand1* is, on the other hand, under the control of *Nkx2-5* in left ventricular formation, although the *cis*-regulatory element of *Hand1* in the developing heart remains to be identified.

Genetic manipulation of *Hand* genes revealed that both *Hand* genes are required for the development of cardiac ventricles. *Hand2* null mice die around E10.5 due to cardiac failure resulting from hypoplasia of the right ventricle and pharyngeal arch artery, suggesting that *Hand2* is essential for the formation of the right ventricle (Srivastava et al. 1997). Specific deletion of *Hand2* in the SHF using *Isl1*-Cre demonstrated a phenocopy of *Hand2* null embryos and excessive apoptosis was observed in SHF derived progenitor cells, suggesting that the function of *Hand2* in SHF cells is essential for their survival and the formation of the right ventricle (Tsuchihashi et al. 2011). Specific deletion of *Hand2* in a neural crest lineage using *Wnt1*-Cre lines also resulted in defects of the outflow tract, suggesting a requirement of *Hand2* in the neural crest during development of the heart (Morikawa et al. 2008). *Hand1* knockout mice showed an early embryonic demise before cardiac looping around E8.0 due to placental defects. Cardiac specific knockout of *Hand1* resulted in defects of the interventricular septum and atrioventricular valve and death after birth (McFadden et al. 2004). These cardiac phenotypes were much milder than *Hand2* null mice. Crossing *Hand1* cardiac-specific knockout mice with *Hand2* knockout mice enabled the analysis of the effect of the gradual reduction of *Hand* genes during heart development. Cardiac-specific *Hand1* null; *Hand2* heterozygous embryos showed two immature ventricles with a failure of ventricular ballooning. The ventricular cardiomyocyte marker, *Mlc2v*, was expressed in these immature ventricles, indicating that ventricular cell fate was specified. Both *Hand1* null; *Hand2* null embryos showed more severe cardiac phenotypes, showing only a single primitive ventricle. These results suggest that *Hand* genes are required for the ballooning process and *Hand2* may genetically compensate *Hand1* function in a dose dependent fashion during cardiac ventricular development. Either *Nkx2-5* or *Hand2* null embryos showed a single developing ventricle. Both *Nkx2-5* and *Hand2* knockout embryos showed only single cardiac chambers with an atrium-like appearance. This immature chamber directed atrial cardiomyocyte-specific marker, *Mlc2a*, suggesting that a genetic interaction between *Nkx2-5* and *Hand2* is essential for cardiac ventricular development. The overexpression of *Hand1* or *Hand2* throughout developing ventricles using a *Mlc2v* promoter resulted in a single ventricular chamber with abnormal development of the interventricular septum (Yamagishi et al. 2001; Togi et al. 2006). Ballooning of the developing left and right ventricles robustly expressing *Hand1* and *Hand2*, respectively, may form the interventricular groove between these ventricles and, as a result, the interventricular septum may emerge in a passive manner by the mechanical force of blood flow into each ventricle.

T-box Transcription Factors in Cardiac Progenitor

Members of the T-box gene family (Tbx) are essential for normal development, and mutations in some human *TBX* genes cause congenital malformations. During development of the heart, T-box genes play important roles in early cardiac lineage determination, chamber specification, valvuloseptal development, and diversification of the specialized conduction system (Greulich et al. 2011). The vertebrate genome has at least 18 different T-box genes in five sub-families with diverse regulatory functions in development and disease. Of the six T-box genes essential for cardiac development, Tbx1 and Tbx18 ensure elongation of the heart tube at the arterial and venous pole, respectively. Tbx1 acts in the pharyngeal mesoderm to maintain proliferation of mesenchymal precursor cells for formation of a myocardialized and septated outflow tract. Tbx18 is expressed in the sinus venosus region and is required for myocardialization of the caval veins and the sinoatrial node. Tbx5 and Tbx20 function in the early heart tube and independently activate the chamber myocardial gene program, whereas Tbx2 and Tbx3 locally repress this program and promote valvuloseptal formation and development of the cardiac conduction system.

Tbx1 is expressed in the SHF, but not in the cardiac NCC (Garg et al. 2001; Yamagishi et al. 2003). Pharyngeal expression of Tbx1 is maintained by Shh. Shh is produced in the pharyngeal endoderm, and in the adjacent mesoderm induces the expression of forkhead class transcription factors, FoxA2, FoxC1, and FoxC2 that directly act on the Tbx1 promoter. *Tbx1*-Cre transgenic mice using the Fox binding element of the Tbx1 promoter demonstrated that Tbx1 plays an important role for the SHF that eventually gives rise to the right ventricle and the outflow tract (Maeda et al. 2006). Although little is known about the molecular mechanisms underlying the interaction between the SHF and cardiac neural crests, Tbx1 may play a key role in its signal transduction (see previous section). Formation of pharyngeal arch arteries requires the pharyngeal surface ectoderm as a signaling center from which Gbx2, in response to Tbx1 from the SHF, triggers essential directions for the migration of the cardiac neural crests to the caudal pharyngeal arches (Calmont et al. 2009).

Tbx18 is expressed in the PE, in the epicardium, and both in mesenchymal progenitors and the differentiated myocardium of the sinus venosus region (Kraus et al. 2001). This region of the developing heart is a particularly complex area since it contributes to the outer tissue layer of the heart, the cardiac pacemaker, as well as to the myocardium of the sinus venosus. Genetic lineage analysis showed that Tbx18 expressing cells form the sinus horn and the sinoatrial node myocardium, but do not contribute to the atrial and pulmonary vein myocardium that instead are descendents of FHF and SHF progenitors. *Tbx18* null mutant mice die perinatally and display a range of rib and vertebrae defects, along with specific defects of the systemic venous return (Wiese et al. 2009), suggesting that Tbx18 may play a role in proliferation, differentiation, and selective recruitment of precursor cells of a distinct posterior pool during formation of the systemic venous return.

Tbx2 and Tbx3 are two closely related T-box transcription factor genes that are expressed exclusively in the non-chamber myocardium of the atrioventricular canal and the outflow tract during cardiac development, suggesting a functional role in setting up this early non-myocardial area (Christoffels et al. 2004). Given the biochemical similarity and the overlapping expression domains, both genes are likely to act redundantly in the atrioventricular canal, and show unique requirements in the other expression domains. Tbx2, *in vitro*, interacts with Nkx2–5 and Gata4 and binds to consensus T-half sites on the Nppa promoter, resulting in the effective inhibition of Tbx5-mediated transactivation (Habets et al. 2002). Tbx2 and Tbx3 can repress target genes, they do not merely behave as competitors that occupy binding sites of the strong activator, Tbx5 or Tbx20. Since chamber myocardial patterning appears to be achieved by localized expression of the transcriptional repressors Tbx2 and Tbx3, the cellular and molecular mechanisms that activate and restrict these factors in the linear and looping heart are critical.

Tbx5 plays an important role for myogenesis in the FHF. The early embryonic expression of *Tbx5* in the cardiac primordia and posterior region of the heart tube in mice is consistent with functions in early atrial cardiomyocyte development (Bruneau et al. 1999). Biochemical analyses suggest that Tbx5 acts as a transcriptional activator that is associated with different co-factors to achieve differential activation of genes essential for the chamber myocardium. Analysis of the murine Nppa promoter revealed two T-half sites that are bound by Tbx5, and a site for the NK-type homeobox transcription factor Nkx2–5, in close proximity. Nkx2-5 is an evolutionary conserved cardiac transcription factor that is expressed in the FHF and SHF, and is also required for chamber formation together with Tbx5 and Hands (Bruneau et al. 2001). *Tbx5* null mutant mice result in an embryonic lethality at E9.5 as a result of severe hypoplasia of posterior cardiac structures originating from the FHF.

Tbx20 is expressed in the FHF and SHF throughout cardiac development. Tbx20 activates Nmycl and suppresses Tbx2 expression to promote proliferation and specification of the atrial and ventricular chamber myocardium. Tbx20 also indirectly promotes Tbx2 expression through Bmp2. The spatiotemporal expression of Tbx2 in the atrioventricular canal appears to be balanced between these two opposing signals mediated by Tbx20 (Cai et al. 2011). Mice lacking Tbx20 fail to form atrioventricular canal constriction and severely disturb the EMT for the atrioventricular cushion.

Cardiac Transcription Factors Implicated in Congenital Heart Diseases

Most congenital heart diseases involve specific developmental processes and usually result from a failure of specific morphogenetic events. Such processes in cardiogenesis frequently involve transcriptional regulatory circuits and, thus, it is unsurprising to find transcriptional regulators as genetic causes in the etiology of several congenital heart diseases.

TBX1 is a major genetic determinant of the 22q11.2 deletion syndrome in humans that is highly associated with outflow tract defects, such as persistent truncus arteriosus and tetralogy of Fallot (Yamagishi 2002; Yamagishi and Srivastava 2003; Yagi et al. 2003). *Tbx1* null mutant mice phenocopy the human phenotype of the 22q11.2 deletion syndrome (Jerome and Papaioannou 2001). Mice with *Tbx1* hypomorphic alleles display cardiovascular defects, but no cleft palate, a phenotype milder than *Tbx1* null mutant mice (Hu et al. 2004). A gradual Tbx1 dosage reduction from 100 to 2% using two hypomorphic mouse models demonstrated a variety of outflow tract defects including persistent truncus arteriosus, tetralogy of Fallot, and double outlet right ventricle (Zhang and Baldini 2008).

In humans, mutations of TBX3 gene are associated with ulnar mammary syndrome (Bamshad et al. 1997). Mutations of TBX20 gene have been reported to be associated with diverse cardiac defects including ventricular septal defects, aberrant valvulogenesis, tetralogy of Fallot, and cardiomyopathy (Liu et al. 2008). A number of different mutations in the TBX5 gene are identified in patients with Holt-Oram syndrome, a disease associated with severe limb and congenital heart diseases including septal defects, as well as conduction disturbance (Huang et al. 2002a). Heterozygous Tbx5 mutant mice remarkably resemble the phenotype of Holt-Oram syndrome, confirming the dose-sensitive role of Tbx5 to induce and maintain the posterior chamber and septal myocardium (Bruneau et al. 1999; Bruneau et al. 2001).

Abnormal expression of GATA4, 5, or 6 might be expected to be associated with specific human congenital heart diseases, based on their essential roles in cardiac development. It was reported that GATA4 mutations caused familial cardiac septal defects (Garg et al. 2003) whereras NKX2-5 mutations caused familial atrial septal defect with atrioventricular block (Schott et al. 1998). Thereafter, a wide range of mutations has been identified in patients with various congenital heart diseases and, in some cases, specific genotype-phenotype correlations were observed.

Recently, we screened for mutations in cardiac transcription factors in patients with selected non-syndromic persistent truncus arteriosus, and identified two different GATA6 mutations in two probands that were not found in 182 unrelated controls with no congenital heart disease (Kodo et al. 2009). Our subsequent biological analyses revealed that SEMA3C and PLXNA2 were directly regulated by GATA6, and both GATA6 mutant proteins failed to transactivate these genes. SEMA3C and PLXNA2 mediate a signal transduction essential for neurovascular guidance as a ligand and a receptor, respectively, during development. Transgenic analysis further suggested that the expression of SEMA3C and PLXNA2 in the outflow tract was dependent on GATA transcription factors during heart development. Together, our data implicate mutations in GATA6 as novel genetic causes of a type of congenital heart disease involving outflow tract development, as a result of the disruption of the direct regulation of semaphorin-plexin signaling.

Two other reports (Lin et al. 2010; Maitra et al. 2010) came out on mutations in GATA6 after our study that are associated with congenital heart diseases including atrioventricular septal defects, tetralogy of Fallot, and atrial septal defects. It is of note that the GATA6 mutations identified in humans are predominantly associated with outflow tract defects, whereas GATA4 mutations are commonly associated with septal defects, although there are some phenotypic overlaps (Kodo and Yamagishi, unpublished observations) likely to result from a redundant role of both genes. Although the clinical phenotype of individuals with GATA6 mutations commonly involves the outflow tract, it is distinct from that of the 22q11.2 deletion (TBX1 haploinsufficiency) syndrome, the most common genetic cause of outflow tract defects, rather manifesting as non-syndromic congenital heart diseases. Although no direct molecular link has been demonstrated between GATA6 and TBX1 until recently, a report (Théveniau-Ruissy et al. 2008) and our preliminary data suggest that the expression of Sema3c in the outflow tract and pharyngeal region is altered in mouse embryos deficient for *Tbx1* (Kodo and Yamagishi, unpublished observations), suggesting that GATA6 may share, at least in part, a common molecular pathway with TBX1 during outflow tract development.

Recent genetic screening of patients with non-familial hypoplastic left heart syndrome revealed a common frameshift mutation of the loop region of the bHLH domain in *Hand1* (Reamon-Buettner et al. 2008). The truncated *Hand1* protein failed to promote transcription *in vitro*, suggesting that this heterozygous *Hand1* mutation may result in a loss in the function of the *Hand1* protein essential for heart development, and has been implicated in hypoplastic left heart syndrome in humans. A heterozygous deletion of *Hand2* was detected in terminal deletions of chromosome 4q in humans. $4q^-$ syndrome is commonly associated with right ventricular outflow tract defects, such as pulmonary valve stenosis, although cytogenetic analysis failed to confirm an association between the *Hand2* haploinsufficiency and cardiac phenotype in this syndrome (Huang et al. 2002b).

MicroRNAs in Cardiac Progenitors

Eukaryotic genomes transcribe up to 90 % of the genomic DNA, but only 1%–2% of these transcripts encode for proteins and the vast majority are transcribed as protein non-coding microRNAs (Kaikkonen et al. 2011). miR-1 and miR-133a are microRNAs that expressed specifically in cardiac and skeletal muscles during development and in the adult human. They arise from two precursors, coded in two highly conserved chromosomal loci. miR-1-1 and miR-133a-2 are closely linked on mouse chromosome 2 or human chromosome 20. miR-1-2 and miR-133a-1 are closely linked on mouse chromosome 18 or human chromosome 18 (Rao et al. 2006). During early embryonic stages, miR-1 and miR-133 function in concert to promote mesoderm induction, while suppressing differentiation into the ectodermal or endodermal lineages (Ivey et al. 2008). However, miR-1 and miR-133 have antagonistic effects for subsequent differentiation with miR-1 promoting differentiation toward a cardiac fate, whereas miR-133 inhibits differentiation into cardiac muscle (Ivey et al. 2008). The cardiac transcription of miR-1/miR-133 bicistronic microRNAs is directly regulated by Mef2 and SRF (Zhao et al. 2005). Both miR-1 and miR-133 are co-expressed in cardiac and skeletal muscle throughout mouse development, and *Hand2* was found as the target of miR-1. Homozygous deletion of miR-1-2 in mice resulted in the mortality of up to 50% of offsprings in late embryonic-postnatal stages with ventricular septal defects (Zhao et al. 2007). Mice lacking either miR-133a-1 or miR-133a-2 are normal, whereas deletion of both miRNAs causes lethal defects in approximately half of double-mutant embryos or neonates (Liu et al. 2008). Deletion of both miR-133a genes resulted in the aberrant expression of smooth muscle genes in the heart, and these abnormalities can be attributed, at least in part, to elevated expression of SRF and cyclin D2, which are targets for repression by miR-133a. There has been increasing evidence for regulatory roles of microRNAs in cardiac progenitors, making microRNAs one of the targets of future regenerative medicine.

Conclusion

Developmental biology of the cardiovascular system has progressed over the last few decades. We can now recognize the heart as one of the well-understood tissues at the molecular, physiologic, and anatomic levels. Recombinant technology has been instrumental in revealing the factors governing developmental process of the cardiovascular system. Complex genetic networks for signaling pathways and transcriptional regulators have been documented for various types of cardiac progenitors as reviewed above. Advances in human genetic tools have also increased understanding of the importance of developmental pathways of the cardiovascular system implicated in human congenital heart defects. A deeper understanding of the interactions between various signaling and transcriptional networks and their ultimate downstream targets will be necessary to identify potential approaches in parents at risk. Towards the disease-related developmental biology, exciting future technologies may generate disease-specific embryonic stem cell lines or induced pluripotent stem cells for mechanistic studies of disease etiology and development of patient-specific stem cells as therapeutics. Although issues such as stem cell expansion, delivery, incorporation, electrical coupling, and safety remain to be addressed, it may now become possible to guide stem cells or cardiac progenitors, based on our knowledge of early developmental pathways, ready for future regenerative medicine.

Acknowledgements

The author expresses thanks to Dr. Chihiro Yamagishi for preparation of Figs. This work was supported by Funds for Promoting Science and Technology, Ministry of Education, Culture, Sports, Science, and Technology, Japan.

References

Abu-Issa, R. and Kirby, M. L. 2007. Heart field: from mesoderm to heart tube. Ann. Rev. Cell. Dev. Biol. 23, 45–68.

Abu-Issa, R., Smyth, G., Smoak, I., Yamamura, K., and Meyers, E. N. 2002. Fgf8 is required for pharyngeal arch and cardiovascular development in the mouse. Development 129, 4613–4625.

Ai, D., Fu, X., Wang, J., Lu, M. F., Chen, L., Baldini, A. et al. 2007. Canonical wnt signaling functions in second heart field to promote right ventricular growth. Proc. Natl. Acad. Sci. USA 104, 9319–9324.

Alsan, B. H. and Schultheiss, T. M. 2002. Regulation of avian cardiogenesis by Fgf8 signaling. Development 129, 1935–1943

Bamshad, M., Lin, R. C., Law, D. J., Watkins, W. C., Krakowiak, P. A., Moore, M. E. et al. 1997. Mutations in human TBX3 alter limb, apocrine and genital development in ulnar-mammary syndrome. Nat. Genet. 16, 311–315.

Brewer, A. and Pizzey, J. 2006. GATA factors in vertebrate heart development and disease. Expert Rev. Mol. Med. 8, 1–20.

Bruneau, B. G., Logan, M., Davis, N., Levi, T., Tabin, C. J., Seidman, J. G. et al. 1999. Chamber-specific cardiac expression of Tbx5 and heart defects in Holt-Oram syndrome. Dev. Biol. 211, 100–108.

Bruneau, B. G., Nemer, G., Schmitt, J. P., Charron, F., Robitaille, L., Caron, S. et al. 2001. A murine model of Holt-Oram syndrome defines roles of the T-box transcription factor Tbx5 in cardiogenesis and disease. Cell 106, 709–721.

Bu, L., Jiang, X., Martin-Puig, S., Caron, L., Zhu, S., Shao, Y., Roberts, D. J., Huang, P. L., Domian, I. J., and Chien, K. R. 2009. Human ISL1 heart progenitors generate diverse multipotent cardiovascular cell lineages. Nature 460, 113–117.

Burstyn-Cohen, T. and Kalcheim, C. 2002. Association between the cell cycle and neural crest delamination through specific regulation of G1/S transition. Dev. Cell 3, 383–395.

Cai, C. L., Liang, X., Shi, Y., Chu, P. H., Pfaff, S. L., Chen, J. et al. 2003. Isl1 identifies a cardiac progenitor population that proliferates prior to differentiation and contributes a majority of cells to the heart. Dev. Cell 5, 877–889.

Cai, X., Nomura-Kitabayashi, A., Cai, W., Yan, J., Christoffels, V. M., and Cai, C.L. 2011. Myocardial Tbx20 regulates early atrioventricular canal formation and endocardial epithelial-mesenchymal transition via Bmp2. Dev. Biol. 360, 381–390.

Calmont, A., Ivins, S., Van Bueren, K. L., Papangeli, I., Kyriakopoulou, V., Andrews, W. D. et al. 2009. Tbx1 controls cardiac neural crest cell migration during arch artery development by regulating Gbx2 expression in the pharyngeal ectoderm. Development 136, 3173–83.

Chen, S. and Lechleider, R. J. 2004. Transforming growth factor-beta-induced differentiation of smooth muscle from a neural crest stem cell line. Circ. Res. 94, 1195–1202.

Christoffels, V. M., Hoogaars, W. M., Tessari, A., Clout, D. E., Moorman, A. F., and Campione, M. 2004. T-box transcription factor Tbx2 represses differentiation and formation of the cardiac chambers. Dev. Dyn. 229, 763–770

Cohen, E. D., Wang, Z., Lepore, J. J., Lu, M. M., Taketo, M. M., Epstein, D. J. et al. 2007 Wnt/beta-catenin signaling promotes expansion of Isl-1-positive cardiac progenitor cells through regulation of FGF signaling. J. Clin. Invest. 117, 1794 –1804.

Dyer, L. and Kirby, M. 2009. Sonic hedgehog maintains proliferation in secondary heart field progenitors and is required for normal arterial pole formation. Dev. Biol. 330, 305–317.

De Calisto, J., Araya, C., Marchant, L., Riaz, C. F. and Mayor, R. 2005. Essential role of non-canonical Wnt signalling in neural crest migration. Development 132, 2587–2597.

Firulli, A. B. 2003. A HANDful of questions: The molecular biology of the heart and neural crest derivatives (HAND)-subclass of basic helix-loop-helix transcription factors. Gene 312, 27–40.

Fulcoli, F. G., Huynh, T., Scambler, P. J. and Baldini, A. 2009. Tbx1 regulates the BMP smad1 pathway in a transcription independent manner. PLoS ONE 4, e6049.

Garg, V., Yamagishi, C., Hu, T., Kathiriya, I. S., Yamagishi, H., and Srivastava, D. 2001. Tbx1, a DiGeorge syndrome candidate gene, is regulated by sonic hedgehog during pharyngeal arch development. Dev. Biol. 235, 62–73.

Garg, V., Kathiriya, I. S., Barnes, R., Schluterman, M. K., King, I. N., Butler, C. A. et al. 2003. GATA4 mutations cause human congenital heart defects and reveal an interaction with TBX5. Nature 424, 443–447.

Goddeeris, M. M., Rho, S., Petiet, A., Davenport, C. L., Johnson, G. A., Meyers, E. N., and Klingensmith, J. 2008. Intracardiac septation requires hedgehog-dependent cellular contributions from outside the heart. Development 135, 1887–1895.

Greulich, F., Rudat, C., and Kispert, A. 2011. Mechanisms of T-box gene function in the developing heart. Cardiovasc. Res. 91, 212–22.

Habets, P. E., Moorman, A. F., Clout, D. E., van Roon, M. A., Lingbeek, M., van Lohuizen, M. et al. 2002. Cooperative action of Tbx2 and Nkx2.5 inhibits ANF expression in the atrioventricular canal: implications for cardiac chamber formation. Genes Dev. 16, 1234–1246.

Hu, T., Yamagishi, H., Maeda, J., McAnally, J., Yamagishi, C., and Srivastava, D. 2004. Tbx1 regulates fibroblast growth factors in the anterior heart field through a reinforcing auto regulatory loop involving forkhead transcription factors. Development 131, 5491–5502.

Huang, T., Lock, J. E., Marshall, A. C., Basson, C., Seidman, J. G., and Seidman, C. E. 2002a. Causes of clinical diversity in human TBX5 mutations. Cold Spring Harb. Symp. Quant. Biol. 67, 115–120.

Huang, T., Lin, A. E., Cox, G. F., Golden, W. L., Feldman, G. L., Ute, M. et al. 2002b. Cardiac phenotype in chromosome 4q-syndrome with and without a deletion of the dHAND gene. Genet. Med. 4, 464–467.

Hutson, M. R., Zhang, P., Stadt, H. A., Sato, A. K., Li, Y. X., Burch, J., Creazzo, T. L., and Kirby, M. L. 2006. Cardiac arterial pole alignment is sensitive to FGF8 signaling in the pharynx. Dev. Biol. 295: 486–497.

Hutson, M. R. and Kirby, M. L. 2007. Model systems for the study of heart development and disease. Cardiac neural crest and conotruncal malformations. Semin. Cell Dev. Biol. 18, 101–110.

Ivey, K. N., Muth, A., Arnold, J., King, F. W., Yeh, R. F., Fish, J. E. et al. 2008. MicroRNA regulation of cell lineages in mouse and human embryonic stem cells. Cell Stem Cell 2, 219–229.

Jerome, L.A. and Papaioannou, V. E. 2001. DiGeorge syndrome phenotype in mice mutant for the T-box gene, Tbx1. Nat. Genet. 27, 286–291.

Jia, Q., McDill, B. W., Li, S. Z., Deng, C., Chang, C. P., and Chen, F. 2007. Smad signaling in the neural crest regulates cardiac outflow tract remodeling through cell autonomous and non-cell autonomous effects. Dev. Biol. 311, 172–184.

Kaikkonen, M. U., Lam, M. T. Y., and Glass, C. K. 2011. Non-coding RNAs as regulators of gene expression and epigenetics. Cardiovasc. Res. 90, 430–440.

Klaus, A., Saga, Y., Taketo, M. M., Tzahor, E., and Birchmeier, W. 2007. Distinct roles of wnt/beta-catenin and bmp signaling during early cardiogenesis. Proc. Natl. Acad. Sci. USA 104, 18531–18536.

Kodo, K., Nishizawa, T., Furutani, M., Arai, S., Yamamura, E., Joo, K. et al. 2009. GATA6 mutations cause human cardiac outflow tract defects by disrupting semaphorin-plexin signaling. Proc. Natl. Acad. Sci. USA 106, 13933–13938.

Kodo, K. and Yamagishi, H. 2011. A decade of advances in the molecular embryology and genetics underlying congenital heart defects. Circ. J. 75, 2296–2304.

Kraus, F., Haenig, B., and Kispert, A. 2001. Cloning and expression analysis of the mouse T-box gene Tbx18. Mech. Dev. 100, 83–86.

Kubota, Y. and Ito, K. 2000. Chemotactic migration of mesencephalic neural crest cells in the mouse. Developmental Dynamics 217, 170–179.

Kwon, C., Arnold, J., Hsiao, E. C., Taketo, M. M., Conklin, B. R., and Srivastava, D. 2007. Canonical wnt signaling is a positive regulator of mammalian cardiac progenitors. Proc. Natl. Acad. Sci. USA 104, 10894–10899.

Kwon, C., Qian, L., Cheng, P., Nigam, V., Arnold, J., and Srivastava, D. 2009. A regulatory pathway involving notch1/beta-catenin/isl1 determines cardiac progenitor cell fate. Nat. Cell Biol. 11, 951–957.

Laugwitz, K. L., Moretti, A., Lam, J., Gruber, P., Chen, Y. H., Woodard, S. et al. 2005. Postnatal isl1+ cardioblasts enter fully differentiated cardiomyocyte lineages. Nature 433, 647–653.

Lepore, J. J., Mericko, P. A., Cheng, L., Lu, M. M., Morrisey, E. E., and Parmacek, M. S. 2006. GATA-6 regulates semaphorin 3C and is required in cardiac neural crest for cardiovascular morphogenesis. J. Clin. Invest. 116, 929–939.

Lin, L., Bu, L., Cai, C.L., Zhang, X., and Evans, S. 2006. Isl1 is upstream of sonic hedgehog in a pathway required for cardiac morphogenesis. Dev. Biol. 295, 756 –763.

Lin, L., Cui, L., Zhou, W., Dufort, D., Zhang, X., Cai, C. et al. 2007. Beta-catenin directly regulates Islet1 expression in cardiovascular progenitors and is required for multiple aspects of cardiogenesis. Proc. Natl. Acad. Sci. USA 104, 9313–9318.

Lin, X., Huo, Z., Liu, X., Zhang, Y., Li, L., Zhao, H., Yan, B., Liu, Y., Yang, Y., and Chen, Y. H. 2010. A novel GATA6 mutation in patients with tetralogy of Fallot or atrial septal defect. J. Hum. Genet. 55, 662–667.

Liu, W., Selever, J., Wang, D., Lu, M.F., Moses, K. A., Schwartz, R. J., and Martin, J. F. 2004. Bmp4 signaling is required for outflow-tract septation and branchial-arch artery remodeling. Proc. Natl. Acad. Sci. USA 101, 4489–4494.

Liu, C., Shen, A., Li, X., Jiao, W., Zhang, X., and Li, Z. 2008. T-box transcription factor TBX20 mutations in Chinese patients with congenital heart disease. Eur. J. Med. Genet. 51, 580–587.

Liu, N., Bezprozvannaya, S., Williams, A. H., Qi, X., Richardson, J. A., Bassel-Duby, R., and Olson, E. N. 2008. microRNA-133a regulates cardiomyocyte proliferation and suppresses smooth muscle gene expression in the heart. Genes Dev. 22, 3242–3254.

Maeda, J., Yamagishi, H., McAnally, J., Yamagishi, C., and Srivastava, D. 2006. Tbx1 is regulated by forkhead proteins in the secondary heart field. Dev. Dyn. 235, 701–710.

Maitra, M., Koenig, S. N., Srivastava, D., and Garg, V. 2010. Identification of GATA6 sequence variants in patients with congenital heart defects. Pediatr. Res, 68, 281–285.

McFadden, D. G., Charite, J., Richardson, J. A., Srivastava, D., Firulli, A. B., and Olson, E. N. 2000. A GATA-dependent right ventricular enhancer controls dHAND transcription in the developing heart. Development 127, 5331–5341.

McFadden, D. G., Barbosa, A. C., Richardson, J. A., Schneider, M. D., Srivastava, D., and Olson, E. N. 2004. The Hand1 and Hand2 transcription factors regulate expansion of the embryonic cardiac ventricles in a gene dosage-dependent manner. Development 132, 189–201.

Molkentin, J. D., Lin, Q., Duncan, S. A., and Olson, E. N. 1997. Requirement of the transcription factor GATA4 for heart tube formation and ventral morphogenesis. Genes Dev. 11, 1061–1072.

Moretti, A., Caron, L., Nakano, A., Lam, J. T., Bernhausen, A., Chen, Y., Qyang, Y., Bu, L., Sasaki, M., Martin-Puig, S., Sun, Y., Evans, S. M., Laugwitz, K. L., and Chien, K. R. 2006. Multipotent embryonic isl1+ progenitor cells lead to cardiac, smooth muscle, and endothelial cell diversification. Cell 127, 1151–1165.

Morikawa, Y. and Cserjesi, P. 2008. Cardiac neural crest expression of Hand2 regulates outflow and second heart field development. Circ. Res. 103, 1422–1429.

Morrisey, E. E., Tang, Z., Sigrist, K., Lu, M. M., Jiang, F., Ip, H. S., and Parmacek, M. S. 1998. GATA6 regulates HNF4 and is required for differentiation of visceral endoderm in the mouse embryo. Genes Dev. 12, 3579–90.

Natarajan, A., Yamagishi, H., Ahmad, F., Li, D., Roberts, R., Matsuoka, R., Hill, S. et al. 2001. Human eHAND, but not dHAND, is down-regulated in cardiomyopathies. J. Mol. Cell. Cardiol. 33, 1607–1614.

Nemer, G. and Nemer, M. 2003. Transcriptional activation of BMP-4 and regulation of mammalian organogenesis by GATA-4 and -6. Dev. Biol. 254, 131–148.

Nie, X., Deng, C. X., Wang, Q., and Jiao, K. 2008. Disruption of smad4 in neural crest cells leads to mid-gestation death with pharyngeal arch, craniofacial and cardiac defects. Dev. Biol. 316, 417–430.

Olivey, H. E. and Svensson, E. C. 2010. Epicardial-myocardial signaling directing coronary vasculogenesis. Circ. Res. 106, 818–832.

Pandur, P., Lasche, M., Eisenberg, L. M., and Kuhl, M. 2002. Wnt-11 activation of a noncanonical Wnt signalling pathway is required for cardiogenesis. Nature 418, 636–641.

Papaioannou, V. E. 2001. T-box genes in development: from hydra to humans. Int. Rev. Cytol. 207, 1–70.

Park, E., Watanabe, Y., Smyth, G., Miyagawa-Tomita, S., Meyers, E., Klingensmith, J., Camenisch, T., Buckingham, M., and Moon, A. 2008. An FGF autocrine loop initiated in second heart field mesoderm regulates morphogenesis at the arterial pole of the heart. Development 135, 3599–3610.

Prall, O. W. J., Menon, M. K., Solloway, M. J., Watanabe, Y., Zaffran, S., Bajolle, F. et al. 2007. An Nkx2-5/bmp2/smad1 negative feedback loop controls heart progenitor specification and proliferation. Cell 128, 947–959.

Rao, P. K., Kumar, R. M., Farkhondeh, M., Baskerville, S., and Lodish, H. F. 2006. Myogenic factors that regulate expression of muscle-specific microRNAs. Proc. Natl. Acad. Sci. USA 103, 8721–8726.

Rao, P. K., Toyama, Y., Chiang, H. R., Gupta, S., Bauer, M., Medvid, R. et al. 2009. Loss of cardiac microRNA-mediated regulation leads to dilated cardiomyopathy and heart failure. Circ. Res. 105, 585–594.

Reamon-Buettner, S. M., Ciribilli, Y., Inga, A., and Borlak, J. 2008. A loss-of-function mutation in the binding domain of HAND1 predicts hypoplasia of the human hearts. Hum. Mol. Genet. 17, 1397–1405.

Richarte, A. M., Mead, H. B., and Tallquist, M. D. 2007. Cooperation between the PDGF receptors in cardiac neural crest cell migration. Dev. Biol. 306, 785–796.

Sato, A., Scholl, A. M., Kuhn, E. B., Stadt, H. A., Decker, J. R., Pegram, K., Hutson, M. R., and Kirby, M. L. 2011. FGF8 signaling is chemotactic for cardiac neural crest cells. Dev. Biol. 354, 18–30.

Schleiffarth, J. R., Person, A. D., Martinsen, B. J., Sukovich, D. J., Neumann, A., Baker, C. V. et al. 2007. Wnt5a is required for cardiac outflow tract septation in mice. Pediatr. Res. 61, 386–391.

Scholl, A. M. and Kirby, M. L. 2009. Signals controlling neural crest contributions to the heart. Wiley Interdiscip. Rev. Syst. Biol. Med. 1, 220–227.

Schott, J. J., Benson, D. W., Basson, C. T., Pease, W., Silberbach, G. M., Moak, J. P., Maron, B. J., Seidman, C. E., and Seidman, J. G. 1998. Congenital heart disease caused by mutations in the transcription factor NKX2-5. Science 281, 108–111.

Schultheiss, T. M., Burch, J. B., and Lassar, A. B. 1997. A role for bone morphogenetic proteins in the induction of cardiac myogenesis. Genes Dev., 11, 451–462.

Sela-Donenfeld, D. and Kalcheim, C. 1999. Regulation of the onset of neural crest migration by coordinated activity of BMP4 and Noggin in the dorsal neural tube. Development 126, 4749–4762.

Smith, A. et al. 1997. The EphA4 and EphB1 receptor tyrosine kinases and ephrin-B2 ligand regulate targeted migration of branchial neural crest cells. Curr. Biol. 7, 561–570.

Sorrentino, R. P., Gajewski, K. M., and Schulz, R. A. 2005. GATA factors in Drosophila heart and blood cell development. Semin. Cell Dev. Biol. 16, 107–116.

Srivastava, D., Cserjesi, P., and Olson, E. N. 1995. A subclass of bHLH proteins required for cardiac morphogenesis. Science 270, 1995–1999.

Srivastava, D., Thomas, T., Lin, Q., Kirby, M. L., Brown, D., and Olson, E. N. 1997. Regulation of cardiac mesodermal and neural crest development by the bHLH transcription factor, dHAND. Nat. Genet. 16, 154–160.

Srivastava, D. 2006. Making or breaking the heart: From lineage determination to morphogenesis. Cell 126, 1037–1048.

Stottmann, R. W., Stottmann, R. W., Choi, M., Mishina, Y., Meyers, E. N., and Klingensmith, J. 2004. BMP receptor IA is required in mammalian neural crest cells for development of the cardiac outflow tract and ventricular myocardium. Development 131, 2205–2218.

Tanaka, M., Chen, Z., Bartunkova, S., Yamasaki, N., and Izumo, S. 1999. The cardiac homeobox gene Csx/Nkx2.5 lies genetically upstream of multiple genes essential for heart development. Development 126, 1269–1280.

Théveniau-Ruissy, M., Dandonneau, M., Mesbah, K., Ghez, O., Mattei, M. G., Miquerol, L., and Kelly, R. G. 2008. The del22q11.2 candidate gene tbx1 controls regional outflow tract identity and coronary artery patterning. Circ. Res. 103, 142–148.

Thomas, T., Yamagishi, H., Overbeek, P. A., Olson, E. N., and Srivastava, D. 1998. The bHLH factors, dHAND and eHAND, specify pulmonary and systemic cardiac ventricles independent of left-right sidedness. Dev. Biol. 196, 228–23.

Tian, Y., Yuan, L., Goss, A. M., Wang, T., Yang, J., Lepore, J. J. et al. 2010. Characterization and in vivo pharmacological rescue of a wnt2-gata6 pathway required for cardiac inflow tract development. Dev. Cell 18, 275–287.

Togi, K., Yoshida, Y., Matsumae, H., Nakashima, Y., Kita, T., and Tanaka, M. 2006. Essential role of Hand2 in interventricular septum formation and trabeculation during cardiac development. Biochem. Biophys. Res. Commun. 343, 144–151.

Tribulo, C. et al. 2003. Regulation of Msx genes by a Bmp gradient is essential for neural crest specification. Development 130, 6441–6452.

Tsuchihashi, T., Maeda, J., Shin, C. H., Ivey, K. N., Black, B. L., Olson, E. N. et al. 2011. Hand2 function in second heart field progenitors is essential for cardiogenesis. Dev. Biol. 351, 62–69.

Uchimura, T., Komatsu, Y., Tanaka, M., McCann, K., and Mishina, Y. 2009. Bmp2 and bmp4 genetically interact to support multiple aspects of mouse development including functional heart development. Genesis 47, 374–384.

Vincent, S. D. and Buckingham, M. E. 2010. How to make a heart: The origin and regulation of cardiac progenitor cells. Curr. Top. Dev. Biol. 90, 1–41.

Vincentz, J. W., McWhirter, J. R., Murre, C., Baldini, A., and Furuta, Y. 2005. Fgf15 is required for proper morphogenesis of the mouse cardiac outflow tract. Genesis 41, 192–201.

Waldo, K. L., Kumiski, D. H., Wallis, K. T., Stadt, H. A., Hutson, M. R., Platt, D. H. et al. 2001. Conotruncal myocardium arises from a secondary heart field. Development 128, 3179–3188.

Waldo, K. L., Kumiski, D. H., Wallis, K. T., Stadt, H. A., Hutson, M. R., Platt, D. H. et al. 2001. Conotruncal myocardium arises from a secondary heart field. Development 128, 3179–3188.

Watanabe, Y., Miyagawa-Tomita, S., Vincent, S. D., Kelly, R. G., Moon, A. M., and Buckingham, M. E. 2010. Role of mesodermal FGF8 and FGF10 overlaps in the development of the arterial pole of the heart and pharyngeal arch arteries. Circ. Res. 106, 495–503.

Wiese, C., Grieskamp, T., Airik, R., Mommersteeg, M. T., Gardiwal, A., de Gier-de Vries, C. et al. 2009. Formation of the sinus node head and differentiation of sinus node myocardium are independently regulated by Tbx18 and Tbx3. Circ. Res. 104, 388–397.

Xin, M., Davis, C. A., Molkentin, J. D., Lien, C. L., Duncan, S. A., Richardson, J. A. et al. 2006. A threshold of GATA4 and GATA6 expression is required for cardiovascular development. Proc. Natl. Acad. Sci. USA 103, 11189–11194.

Yagi, H., Furutani, Y., Hamada, H., Sasaki, T., Asakawa, S., Minoshima, S. et al. 2003. Role of TBX1 in human del22q11.2 syndrome. Lancet 362, 1366–1373.

Yamagishi, H., Yamagishi, C., Nakagawa, O., Harvey, R. P., Olson, E. N., and Srivastava, D. 2001. The combinatorial activities of Nkx2.5 and dHAND are essential for cardiac ventricle formation. Dev. Biol. 239, 190–203.

Yamagishi, H. 2002. The 22q11.2 deletion syndrome. Keio J. Med. 51, 77–88.

Yamagishi, H., Maeda, J., Hu, T., McAnally, J., Conway, S. J., Kume, T. et al. 2003. Tbx1 is regulated by tissue-specific forkhead proteins through a common sonic hedgehog-responsive enhancer. Genes Dev. 17, 269–281.

Yamagishi, H. and Srivastava, D. 2003. Unraveling the genetic and developmental mysteries of 22q11 deletion syndrome. Trends Mol. Med. 9, 383–389.

Yanagisawa, H. et al. 1998. Role of endothelin-1/endothelin-A receptor-mediated signaling pathway in the aortic arch patterning in mice. J. Clin. Invest 102, 22–33.

Yang, L., Cai, C. L., Lin, L., Qyang, Y., Chung, C., Monteiro, R. M. et al. 2006. Isl1cre reveals a common bmp pathway in heart and limb development. Development 133, 1575–1585.

Zhao, Y., Samal, E., and Srivastava, D. 2005. Serum response factor regulates a muscle-specific mircroRNA that targets Hand2 during cardiogenesis. Nature 436, 214–220.

Zhao, Y., Ransom, J. F., Li, A., Vedantham, V., von Drehle, M., Muth, A.N. et al. 2007. Dysregulation of cardiogenesis, cardiac conduction, and cell cycle in mice lacking miRNA-1-2. Cell 129, 303–317.

Zeisberg, E. M., Ma, Q., Juraszek, A. L., Moses, K., Schwartz, R. J., Izumo, S., and Pu, W. T. 2005. Morphogenesis of the right ventricle requires myocardial expression of Gata4. J. Clin. Invest 115, 1522–1531.

Zhang, Z. and Baldini, A. 2008. *In vivo* response to high-resolution variation of Tbx1 mRNA dosage. Hum. Mol. Genet. 17, 150–157.

Zhou, W., Lin, L., Majumdar, A., Li, X., Zhang, X., Liu, W. et al. 2007. Modulation of morphogenesis by noncanonical Wnt signaling requires ATF/CREB family-mediated transcriptional activation of TGFbeta2. Nat. Genet. 39, 1225–1234.

CHAPTER 5

Directed and Systematic Differentiation of Cardiovascular Cells from Mouse and Human Pluripotent Stem Cells

Hideki Uosaki[1,2,3,a] and Jun K. Yamashita[1,2,b,*]

Introduction

The discovery of human embryonic stem cells and induced pluripotent stem cells (ESCs/iPSCs) offers potential for the application and usage of differentiated cells in novel and vast research fields such as regenerative medicine, developmental biology, cell physiology, disease model, and drug discovery (Thomson et al. 1998; Takahashi et al. 2007). For these applications, it is important to induce differentiation of specific cell types and to purify them. Cardiomyocytes derived from ESCs are easily detected because they start to beat soon after differentiation. However, no efficient and robust induction method was available until recently. In these last few years, directed differentiation methods for cardiac lineages have been

[1]Center for iPS Cell Research and Application, Kyoto University, 53, Shogoin Kawahara-cho, Sakyo-ku, Kyoto, Japan, 606-8507.
[2]Institute for Frontier Medical Sciences, Kyoto University, 53, Shogoin Kawahara-cho, Sakyo-ku, Kyoto, Japan, 606-8507.
[3]Current address: Division of Cardiology, The Johns Hopkins Medical Institutions, 720 Rutland Ave, Ross 954, Baltimore MD 21205.
[a]Email: huosaki1@jhmi.edu
[b]Email: juny@frontier.kyoto-u.ac.jp
*Corresponding author

rapidly developed and improved. In this chapter, we describe the directed and systematic differentiation of cardiovascular cell lineage from mouse and human ESCs/iPSCs and briefly summarize the cell surface makers for the purification of cardiomyocytes.

Cardiovascular Development

To understand cardiovascular cell differentiation, it would be helpful to understand cardiovascular development *in vivo* (Fig. 1A). The circulatory system is one of the first functional organs that arise from the mesoderm. The heart begins to beat as early as embryonic day (ED) 8.25 and 22 in mice and humans, respectively (Srivastava 2006). At the onset of gastrulation (ED 6.5 and 14 in mice and humans, respectively), the primitive streak forms at the most proximal site on the posterior midline of the epiblast. The cells in the primitive streak (PS) express *Brachyury/T* (Wilkinson et al. 1990). PS cells give rise to both mesoderm and definitive endoderm cells. A subset of mesoderm cells differentiate and migrate into the anterior laterally to form lateral plate mesoderm (LPM) marked by *Mesp1, Flk1/KDR and PdgfRα* at ED 6.5–7.0 in mice (Kataoka et al. 1997; Saga et al. 1999). *Mesp1* and *Mesp2* double null mice fail to form LPM and *Mesp1* regulates the expressions of various cardiovascular progenitor specific genes. Thus, Mesp1 is considered as one of the master regulator genes of cardiovascular progenitors (Saga et al. 2000; Bondue et al. 2008; Lindsley et al. 2008). LPM further migrate to the anterior and form the cardiac crescent at ED 7.5–7.75 in mice (Harvey 2002). Cardiac progenitor cells are found in the cardiac crescent and consist of two distinct populations. *Nkx2.5* marks the first heart field in the cardiac crescent, which is the left ventricular primordial, and *Isl1* marks the second heart field found in anterior of the first heart field and contributes to right ventricle and outflow tract (Schwartz and Olson 1999; Cai et al. 2003). During mesoderm to cardiac progenitor development, various soluble molecules (Bone Morphogenic Protein (BMP), Wnt, Noggin, Dkk1, Cerberus and fibroblast growth factors (FGF)) secreted from surrounding endoderm and ectoderm regulate cell differentiation (Nakajima et al. 2009; Harvey, 2002). As early heart development proceeds, the cardiac crescent fuses at the midline to form a single, beating linear heart tube. Subsequently, the heart tube loops rightward and forms a four-chamber heart.

Endothelial cells also arise from LPM cells and require *Etv2/ER71* expression and VEGF (vascular endothelial growth factor) signaling for their proper development (de Val et al. 2008; Lee et al. 2008; Coultas et al. 2005). Recent studies demonstrated that hematopoietic cells arise from aortic endothelium (Bertrand et al. 2010; Boisset et al. 2010; Kissa and Herbomel 2010).

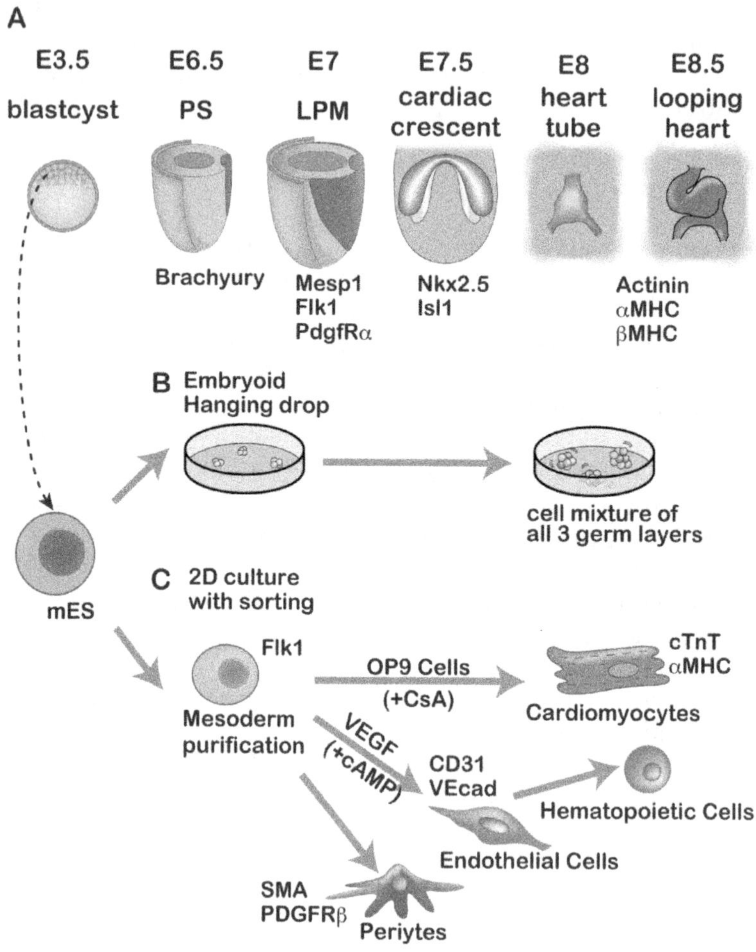

Figure 1. Heart development *in vivo* and cardiovascular differentiation *in vitro*. (A) Heart development in mouse. At gastrulation stage, mesendodem cells express Brachyury. Part of mesoderm cells migrate lateral anterior to form lateral plate mesoderm expressing *Mesp1*, Flk1 and PdgfRα. These cells further migrate to form cardiac crescent expressing *Nkx2.5* and/or *Isl1*, then fuse midline to form heart tube and start beating. Linear heart tube loops to form four-chambered heart. Abbreviations; PS: primitive streak, LPM: lateral plate mesoderm. (B-C) Embryonic stem cell (ESC) differentiation. ESCs are generated from inter cell mass (ICM) of blastocysts. (B) Conventional differentiation method is called hanging drop or embryoid body methods, in which ESC differentiation is induced as small aggregates. These aggregates spontaneously differentiate into cells containing all three germ layers. (C) Two-dimensional culture and sequential cell sorting allow dissecting differentiation processes in detail. As the first step, Flk1-positive mesoderm cells are sorted by FACS or MACS. Cardiomyocytes are induced by culturing Flk1-positive cells on OP9 feeder cells with supplementation of cyclosporin A. Endothelial cells and mural cells are induced on collagen IV or gelatin-coated dishes with VEGF treatment (modified from Uosaki and Yamashita 2011a).

Color image of this figure appears in the color plate section at the end of the book.

Overview of ESC/iPSC Differentiation Methods

For cardiovascular differentiation from ESCs/iPSCs, various methods have been reported. The most conventional method is called hanging drop or embryoid body (EB) method (Fig. 1B), in which ESCs/iPSCs are aggregated in droplets or suspension and spontaneously differentiate into all three germ layers (meso-, endo-, and ectoderm). Some EBs start to beat when cardiomyocytes appear in EBs. During differentiation, lineage markers should appear sequentially in similar order as *in vivo*. Although the EB method is convenient for inducing differentiation, it is hard to dissect cellular and molecular mechanisms during differentiation, because of difficulty in dissecting differentiation processes and observing differentiating cells at the cellular level. To overcome these disadvantages, Yamashita and his colleagues have developed induction method for the systematic, stepwise differentiation of cardiovascular lineage based on cultures in monolayer and purification using Flk1 (also designated as vascular endothelial cell growth factor receptor type 2; VEGFR2 or kinase insert domain receptor; KDR) as a cell surface marker for mesoderm cells (Fig. 1C) (Yamashita et al. 2000; 2005; Narazaki et al. 2008). Most cardiovascular cell types—cardiomyocytes, endothelial cells (arterial, venous, and lymphatic), mural cells (pericytes and vascular smooth muscle cells), and heamatopoietic cells—are induced in appropriate condition from Flk1-positive mesoderm cells. Furthermore, a single Flk1-positive cell gives rise to cardiomyocyte, endothelial cells, and pericyte in this system.

Based on the knowledge of heart development and cell surface markers, novel differentiation methods, called directed differentiation, have been established recently to induce differentiation toward cardiomyocytes from human ESC in a chemically defined medium with some growth factors (Laflamme et al. 2007; Yang et al. 2008; Kattman et al. 2011). We applied and modified one of the directed differentiation methods for human iPSC differentiation and succeeded in identifying one marker specifically expressed on cardiomyocytes and efficiently purifying human iPS cell-derived cardiomyocytes (Uosaki et al. 2011b).

Mouse Embryonic Stem and Induced Pluripotent Stem Cell Differentiation

Mesodermal differentiation

As cardiomyocytes and endothelial cells come from mesoderm cells, the initial step in differentiation is mesoderm induction. Currently, *Brachyury*, *Mesp1*, *Flk1* and *PdgfRα* are widely used to determine mesoderm cells (Fehling et al. 2003; Bondue et al. 2011; Yamashita et al. 2000; Kattman et al. 2011). Among them, Flk1 and PdgfRα are cell surface markers; double positive cells mark cardiac mesoderm expressing a transcriptional factor, *Mesp1* (Bondue et al. 2011). Thus, these two markers are useful in various ESC/iPSCs rather than transcriptional factors in purifying mesoderm cells. Recently, novel cell surface markers (such as N-CAM/CD56 and N-cadherin/CD326) were also identified as early mesodermal markers (Evseenko et al. 2010; Honda et al. 2006). These markers are also expressed on neural lineages and therefore have not been used widely.

In our differentiation system, Flk1 expressing cells start to appear around 3–3.5 days and peak at 4.5 days after differentiation in the monolayer method of mouse ESC/iPSC differentiation. Time-course of appearance is similar between ESCs and iPSCs (Narazaki et al. 2008). The differentiation sequence was described in detail: Flk1+/Bry-GFP+ (green fluorescent protein knocked in to Brachyury locus) at day 3.25 in EBs give rise to primitive blood and endothelial cells and Flk1+ cells at day 4.25 differentiated from Flk1-/Bry-GFP+ at day 3.25 give rise to cardiomyocytes and endothelial cells (Kattman et al. 2006).

Several growth factors and soluble factors such as Activin A, BMP4, ascorbic acid, bFGF and Wnt positively regulate mesodermal differentiation in a temporal manner. Especially, Activin A, BMP4, and bFGF are used in several directed differentiations of cardiomyocytes from human ESC/iPSC (see below).

Cardiac differentiation

To induce cardiomyocyte differentiation from mesoderm cells, several approaches have been intensively established (Yamashita et al. 2005; Kattman et al. 2006). All these methods require high cell density, re-aggregation or feeder cells, indicating that cell–cell contact might mediate cardiac differentiation.

Wnt and BMP signalings have stage-specific effects on cardiac differentiation (Naito et al. 2006; Ueno et al. 2007; Kattman et al. 2011). Soluble antagonists such as Dkk1, Crescent and Noggin expressed in anterior visceral endoderm (AVE) positively regulate cardiac differentiation from mesoderm cells (Foley and Mercola 2005; Yuasa et al. 2005).

A feeder cell line, OP9, which was established from the bone marrow of op/op monocyte colony stimulating factor-deficient mice, supports cardiac differentiation from Flk1-positive mesoderm cells. Even a single Flk1+ cell gives rise to both cardiomyocytes and endothelial cells on OP9 cells (Yamashita et al. 2005). Self-beating cardiomyocytes appear at day 3–4 after Flk1 cells are cultured on OP9 feeder cells. Flk1+/CxCR4+/VE-cadherin- (FCV) cells at day 2 are a late stage cardiac progenitor cell population and mostly express Nkx2.5 and/or Isl1 (Yamashita et al. 2005; Yan et al. 2009).

Combined with OP9 method and chemical biology, an immunosuppressant Cyclosporin A has been identified to induce cardiac progenitor and cardiomyocyte differentiation potently (Yan et al. 2009). Cyclosporin A induced FCV cells up to 10–20 times and induced cardiomyocytes up to 10 times from Flk1+ cells. Interestingly, Cyclosporin A reciprocally suppressed endothelial cell differentiation.

Vascular differentiation

As endothelial cells and mural cells come from mesoderm cells *in vivo*, Flk1-positive cells give rise to endothelial cells and mural cells when cells are treated with VEGF (Yamashita et al. 2000). Additional to VEGF, cAMP (cyclic adenosine mono phosphate) signaling increases endothelial number and arterial endothelial cell differentiation (Yurugi-Kobayashi et al. 2006). In the downstreams of cAMP signals, two distinct pathways—PI3K (Phosphoinositide 3-kinase)-Notch/β-catenin and PKA (protein kinase A)-Flk1/NRP1 (Neuropilin 1)—are involved in arterial specification and endothelial cell differentiation, respectively (Yamamizu et al. 2009, 2010). cAMP-PI3K activate Notch and β-catenin. Activated Notch intracellular domain and β-catenin translocate to nucleus and form a complex with RBPjκ to activate target arterial genes such as *EphrinB2*, *Cxcr4*, *NRP1*, and *Dll4* (Yamamizu et al. 2010). cAMP-PKA regulates the VEGF competency of Flk1-positive vascular progenitor cells through modulation of Flk1 and NRP1 expression. Upregulated Flk1 and NRP1 form a receptor complex for VEGF165 and increase VEGF signal in progenitor cells thereby inducing endothelial cell differentiation (Yamamizu et al. 2009). Flk1 positive cells also give rise to lymphatic endothelial cells expressing LIVE1 and Prox1 on OP9 feeder cells (Kono et al. 2006).

Other cardiovascular cell differentiation

Haematopoietic cells can arise from endothelial cells differentiated from Flk1-positive cells on OP9 feeder cells *in vitro* (Eilken et al. 2009; Lancrin et al. 2009). These results were confirmed *in vivo* recently (Bertrand et al. 2010; Boisset et al. 2010; Kissa and Herbomel 2010).

Human Embryonic Stem Cells and Induced Pluripotent Stem Cells

Directed differentiation

Currently there are two well-established directed differentiation methods towards cardiovascular cells from human ESCs (Fig. 2A) (Laflamme

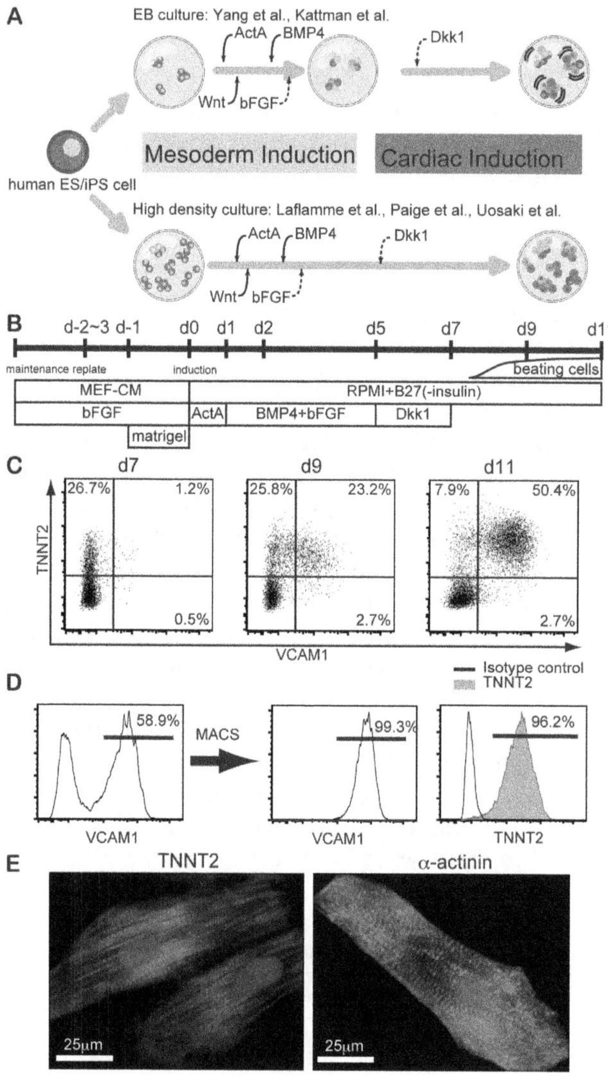

Figure 2. Directed differentiation and cardiomyocyte specific cell surface marker. (A) Schemes of two established directed differentiation methods—EB culture and high density culture methods. In both of them, cells are treated with Activin A and BMP4 to induce differentiation of mesoderm. Wnt3a, bFGF, or Dkk1 are occasionally added to enhance differentiation to mesoderm or cardiomyocytes, respectively. (B) Differentiation scheme of modified directed differentiation. (C) TNNT2 (cTnT)-positive cells start to appear at day 7 after Activin A treatment. Approximately 80% of TNNT2-positive cells become VCAM1-positive cells at day 11 after induction. (D) Approximately 95% of cells were positive for cTnT after purification of VCAM1-positive cells by MACS. (E) Purified cells showed clear sarcomeric structure (modified from Uosaki et al. PLoS One 2011).

Color image of this figure appears in the color plate section at the end of the book.

et al. 2007; Yang et al. 2008). Based on these methods, several modified differentiation methods have been also established and applied to human iPSCs (Paige et al. 2010; Melkoumian et al. 2010; Kattman et al. 2011; Uosaki et al. 2011b).

2. Cardiac differentiation

Keller and his colleagues established a cardiac differentiation method based on EB method under physiological hypoxic condition. In this method, Activin A, BMP4, and bFGF were used to induce KDR1mid, cKIT- mesoderm cells (Yang et al. 2008). They have further extensively improved upon this method and demonstrated that the concentration of Activin A and BMP4 is critical to induce cardiac mesoderm cells expressing KDR1 and PDGFRα (Kattman et al. 2011). Purified KDRmid/cKIT- or KDR1/PDGFRα cells give rise to cardiomyocytes in high efficiency when cells were cultured in high density or re-aggregate. As KDR1/PDGFRα cells appear efficiently in this method, cells can differentiate to cardiomyocytes even with no cell purification processes. Zambidis and his colleagues established an EB-based cardiac differentiation system using single cell culture for human ESCs/ iPSCs and forced aggregation (Burridge et al. 2011). Polyvinyl alcohol and hypoxia eliminated interline variability among many human ESCs/ iPSCs.

Murry and his colleagues also established a cardiac differentiation method under the condition of high cell density, and sequential administration of Activin A and BMP4 (Laflamme et al. 2007). Early surge of Wnt3a expression is critical for mesoderm and cardiac differentiation from human ESCs (Paige et al. 2010). Because this method is based on high cell density but not EBs, it is highly scalable and easy to handle compared to EB-based method. Kamp and his colleague improved this method with prior administration of Matrigel and achieved higher efficiency (Zhang et al. 2012). We have also improved upon this method and applied it to human iPSCs (Uosaki et al. 2011b). bFGF treatment during mesoderm differentiation and Dkk1 treatment during cardiac differentiation increase cardiomyocyte appearance to 30–70% (Fig. 2B-C).

3. Purification of cardiomyocytes

The establishment of high-efficient cardiac differentiation methods provides us with the opportunity to use human cardiomyocytes for drug discovery, toxicology and disease-specific physiology, and even for clinical application. For further application of human cardiomyocytes derived from ESCs/ iPSCs, the purification methods of cardiomyocytes should be established.

The requirements for the purification method in clinical application would be (1) easy access, (2) high purity, (3) scalability (speed), and (4) non-genetic factors.

Currently there are no appropriate methods covering all these four requirements. Mercola and his colleague established a lentiviral vector-based drug resistance selection of cardiomyocytes derived from human ESCs (Kita-Matsuo et al. 2009). This method provides an easy, high-purity, and scalable method but it requires gene modification. Fukuda and his colleague reported that a mitochondorial dye—TMRM (tetramethylrhodamine methyl ester perchlorate) specifically labels mouse and human ESC/iPSC derived cardiomyocytes (Hattori et al. 2010). After TMRM sorting using FACS (fluorescent activated cell sorting), NKX2.5-positive cells were 99.9% in purified cells. The requirement of the FACS process in this method lengthens the time taken to obtain plenty of cells. ALCAM1 is one of the cell surface markers expressed on cardiomyocytes (Hirata et al. 2006; Rust et al. 2009). Using anti-ALCAM antibody and MACS (magnetic activated cell sorting), αMHC-positive cardiomyocytes can be enriched to approximately 60% from 10%. Recently, we have identified a novel cardiomyocyte specific marker, VCAM1 (vascular cell adhesion molecule 1) (Uosaki et al. 2011b). Using this marker, we successfully purified human iPSC-derived cardiomyocytes by MACS. Cardiac troponin T-positive cells reached were 95% of sorted cells (Fig. 2C-E).

4. Endothelial cell differentiation

Mesoderm cells derived from human ESC/iPSCs differentiate into endothelial cells and mural cells under the presence of VEGF, similar to mouse ESC/iPSCs (Sone et al. 2007; Taura et al. 2009). Both the human ESC/iPSC differentiation methods above efficiently induce mesoderm cells (Yang et al. 2008; unpublished data), and administration of VEGF to mesoderm cells induces endothelial cell differentiation.

Closing Remarks

Induction and purification methods for various cardiovascular cells have been rapidly improved in recent years. Currently, cardiomyocytes, endothelial cells, and mural cells can be efficiently induced and purified from both mouse and human ESC/iPSCs, which are a critical technological basis for drug discovery, safety tests, disease modeling, and cell therapy. Utilizing these technologies, studies using ESC/iPSCs would move to the next stage that actually provides fruitful benefits to patients through development of novel therapeutic strategies.

References

Bertrand, J. Y., Chi, N. C., Santoso, B., Teng, S., Stainier, D. Y. R., and Traver, D. 2010. Haematopoietic stem cells derive directly from aortic endothelium during development. Nature 464, 108–111.

Boisset, J.-C., van Cappellen, W., Andrieu-Soler, C., Galjart, N., Dzierzak, E., and Robin, C. 2010. *In vivo* imaging of haematopoietic cells emerging from the mouse aortic endothelium. Nature 464, 116–120.

Bondue, A., Tännler, S., Chiapparo, G., Chabab, S., Ramialison, M., Paulissen, C., Beck, B., Harvey, R., and Blanpain, C. 2011. Defining the earliest step of cardiovascular progenitor specification during embryonic stem cell differentiation. J. Cell Biol. 192, 751–765.

Bondue, A., Lapouge, G., Paulissen, C., Semeraro, C., Iacovino, M., Kyba, M., and Blanpain, C. 2008. Mesp1 acts as a master regulator of multipotent cardiovascular progenitor specification. Cell Stem Cell 3, 69–84.

Burridge, P. W., Thompson, S., Millrod, M. A., Weinberg, S., Yuan, X., Peters, A., Mahairaki, V., Koliatsos, V. E., Tung, L., and Zambidis, E. T. 2011. A universal system for highly efficient cardiac differentiation of human induced pluripotent stem cells that eliminates interline variability. PLoS ONE 6, e18293.

Cai, C.-L., Liang, X., Shi, Y., Chu, P.-H., Pfaff, S. L., Chen, J., and Evans, S. 2003. Isl1 identifies a cardiac progenitor population that proliferates prior to differentiation and contributes a majority of cells to the heart. Dev. Cell 5, 877–889.

Coultas, L., Chawengsaksophak, K., and Rossant, J. 2005. Endothelial cells and VEGF in vascular development. Nature 438, 937–945.

De Val, S., Chi, N. C., Meadows, S. M., Minovitsky, S., Anderson, J. P., Harris, I. S., Ehlers, M. L., Agarwal, P., Visel, A., Xu, S. M., Pennacchio, L. A., Dubchak, I., Krieg, P. A., Stainier, D. Y., and Black, B. L. 2008. Combinatorial regulation of endothelial gene expression by ets and forkhead transcription factors. Cell 135, 1053–1064.

Eilken, H. M., Nishikawa, S. I., and Schroeder, T. 2009. Continuous single-cell imaging of blood generation from haemogenic endothelium. Nature 457, 896–900.

Evseenko, D., Zhu, Y., Schenke-Layland, K., Kuo, J., Latour, B., Ge, S., Scholes, J., Dravid, G., Li, X., MacLellan, W. R., and Crooks, G. M. 2010. Mapping the first stages of mesoderm commitment during differentiation of human embryonic stem cells. Proc. Natl. Acad. Sci. USA 107, 13742–13747.

Fehling, H. J., Lacaud, G., Kubo, A., Kennedy, M., Robertson, S., Keller, G., and Kouskoff, V. 2003. Tracking mesoderm induction and its specification to the hemangioblast during embryonic stem cell differentiation. Development 130, 4217–4227.

Foley, A. C., and Mercola, M. 2005. Heart induction by Wnt antagonists depends on the homeodomain transcription factor Hex. Genes Dev. 19, 387–396.

Harvey, R. P. 2002. Patterning the vertebrate heart. Nat. Rev. Genet. 3, 544–556.

Hattori, F., Chen, H., Yamashita, H., Tohyama, S., Satoh, Y.S., Yuasa, S., Li, W., Yamakawa, H., Tanaka, T., Onitsuka, T., Shimoji, K., Ohno, Y., Egashira, T., Kaneda, R., Murata, M., Hidaka, K., Morisaki, T., Sasaki, E., Suzuki, T., Sano, M., Makino, S., Oikawa, S., and Fukuda, K. 2010. Nongenetic method for purifying stem cell-derived cardiomyocytes. Nat. Methods 7, 61–66.

Hirata, H., Murakami, Y., Miyamoto, Y., Tosaka, M., Inoue, K., Nagahashi, A., Jakt, L. M., Asahara, T., Iwata, H., Sawa, Y., and Kawamata, S. 2006. ALCAM (CD166) is a surface marker for early murine cardiomyocytes. Cells Tissues Organs 184, 172–180.

Honda, M., Kurisaki, A., Ohnuma, K., Okochi, H., Hamazaki, T. S., and Asashima, M. 2006. N-cadherin is a useful marker for the progenitor of cardiomyocytes differentiated from mouse ES cells in serum-free condition. Biochem. Biophys. Res. Commun. 351, 877–882.

Kataoka, H., Takakura, N., Nishikawa, S., Tsuchida, K., Kodama, H., Kunisada, T., Risau, W., Kita, T., and Nishikawa, S. I. 1997. Expressions of PDGF receptor alpha, c-Kit and Flk1 genes clustering in mouse chromosome 5 define distinct subsets of nascent mesodermal cells. Dev. Growth Differ. 39, 729–740.

Kattman, S. J., Huber, T. L., and Keller, G. M. 2006. Multipotent flk-1+ cardiovascular progenitor cells give rise to the cardiomyocyte, endothelial, and vascular smooth muscle lineages. Dev. Cell 11, 723–732.

Kattman, S. J., Witty, A. D., Gagliardi, M., Dubois, N. C., Niapour, M., Hotta, A., Ellis, J., and Keller, G. 2011. Stage-specific optimization of Activin/Nodal and BMP signaling promotes cardiac differentiation of mouse and human pluripotent stem cell lines. Cell Stem Cell 8, 228–240.

Kissa, K. and Herbomel, P. 2010. Blood stem cells emerge from aortic endothelium by a novel type of cell transition. Nature 464, 112–115.

Kita-Matsuo, H., Barcia, M., Prigozhina, N., Salomonis, N., Wei, K., Jacot, J. G., Nelson, B., Spiering, S., Haverslag, R., Kim, C., Talantova, M., Bajpai, R., Calzolari, D., Terskikh, A., McCulloch, A. D., Price, J. H., Conklin, B. R., Chen, H. S., and Mercola, M. 2009. Lentiviral vectors and protocols for creation of stable hESC lines for fluorescent tracking and drug resistance selection of cardiomyocytes. PLoS ONE 4, e5046.

Kono, T., Kubo, H., Shimazu, C., Ueda, Y., Takahashi, M., Yanagi, K., Fujita, N., Tsuruo, T., Wada, H., and Yamashita, J. K. 2006. Differentiation of lymphatic endothelial cells from embryonic stem cells on OP9 stromal cells. Arterioscler. Thromb. Vasc. Biol. 26, 2070.

Laflamme, M. A., Chen, K. Y., Naumova, A. V., Muskheli, V., Fugate, J. A., Dupras, S. K., Reinecke, H., Xu, C., Hassanipour, M., Police, S., O'Sullivan, C., Collins, L., Chen, Y., Minami, E., Gill, E. A., Ueno, S., Yuan, C., Gold, J., and Murry, C. E. 2007. Cardiomyocytes derived from human embryonic stem cells in pro-survival factors enhance function of infarcted rat hearts. Nat. Biotechnol. 25, 1015–1024.

Lancrin, C., Sroczynska, P., Stephenson, C., Allen, T., Kouskoff, V., and Lacaud, G. 2009. The haemangioblast generates haematopoietic cells through a haemogenic endothelium stage. Nature 457, 892–895.

Lee, D., Park, C., Lee, H., Lugus J. J., Kim, S. H., Arentson, E., Chung, Y. S., Gomez, G., Kyba, M., Lin, S., Janknecht, R., Lim, D. S., and Choi, K. 2008. ER71 acts downstream of BMP, Notch, and Wnt signaling in blood and vessel progenitor specification. Cell Stem Cell 2, 497–507.

Lindsley, R. C., Gill, J. G., Murphy, T. L., Langer, E. M., Cai, M., Mashayekhi, M., Wang, W., Niwa, N., Nerbonne, J. M., Kyba, M., and Murphy, K. M. 2008. Mesp1 coordinately regulates cardiovascular fate restriction and epithelial-mesenchymal transition in differentiating ESCs. Cell Stem Cell 3, 55–68.

Melkoumian, Z., Weber, J.L., Weber, D. M., Fadeev, A. G., Zhou, Y., Dolley-Sonneville, P., Yang, J., Qiu, L., Priest, C. A., Shogbon, C., Martin, A. W., Nelson, J., West, P., Beltzer, J. P., Pal, S., and Brandenberger, R. 2010. Synthetic peptide-acrylate surfaces for long-term self-renewal and cardiomyocyte differentiation of human embryonic stem cells. Nat. Biotechnol. 28, 606–610.

Naito, A. T., Shiojima, I., Akazawa, H., Hidaka, K., Morisaki, T., Kikuchi, A., and Komuro, I. 2006. Developmental stage-specific biphasic roles of Wnt/beta-catenin signaling in cardiomyogenesis and hematopoiesis. Proc. Natl. Acad. Sci. USA 103, 19812–19817.

Nakajima, Y., Sakabe, M., Matsui, H., Sakata, H., Yanagawa, N., and Yamagishi, T. 2009. Heart development before beating. Anat. Sci. Int. 84, 67–76.

Narazaki, G., Uosaki, H., Teranishi, M., Okita, K., Kim, B., Matsuoka, S., Yamanaka, S., and Yamashita, J. K. 2008. Directed and systematic differentiation of cardiovascular cells from mouse induced pluripotent stem cells. Circulation 118, 498–506.

Paige, S. L., Osugi, T., Afanasiev, O. K., Pabon, L., Reinecke, H., and Murry, C. E. 2010. Endogenous Wnt/beta-catenin signaling is required for cardiac differentiation in human embryonic stem cells. PLoS ONE 5, e11134.

Rust, W., Balakrishnan, T., and Zweigerdt, R. 2009. Cardiomyocyte enrichment from human embryonic stem cell cultures by selection of ALCAM surface expression. Regenerative Medicine 4, 225–237.

Saga, Y., Kitajima, S., and Miyagawa-Tomita, S. 2000. Mesp1 expression is the earliest sign of cardiovascular development. Trends Cardiovasc. Med. 10, 345–352.

Saga, Y., Miyagawa-Tomita, S., Takagi, A., Kitajima, S., Miyazaki, J. I., and Inoue, T. 1999. MesP1 is expressed in the heart precursor cells and required for the formation of a single heart tube. Development 126, 3437–3447.

Schwartz, R. J. and Olson, E. N. 1999. Building the heart piece by piece: modularity of cis-elements regulating Nkx2-5 transcription. Development 126, 4187–4192.

Sone, M., Itoh, H., Yamahara, K., Yamashita, J. K., Yurugi-Kobayashi, T., Nonoguchi, A., Suzuki, Y., Chao, T. H., Sawada, N., Fukunaga, Y., Miyashita, K., Park, K., Oyamada, N., Sawada, N., Taura, D., Tamura, N., Kondo, Y., Nito S., Suemori, H., Nakatsuji, N., Nishikawa, S., and Nakao, K. 2007. Pathway for differentiation of human embryonic stem cells to vascular cell components and their potential for vascular regeneration. Arterioscler. Thromb. Vasc. Biol. 27, 2127–2134.

Srivastava, D. 2006. Making or breaking the heart: from lineage determination to morphogenesis. Cell 126, 1037–1048.

Takahashi, K., Tanabe, K., Ohnuki, M., Narita, M., Ichisaka, T., Tomoda, K., and Yamanaka, S. 2007. Induction of pluripotent stem cells from adult human fibroblasts by defined factors. Cell 131, 861–872.

Taura, D., Sone, M., Homma, K., Oyamada, N., Takahashi, K., Tamura, N., Yamanaka, S., and Nakao, K. 2009. Induction and isolation of vascular cells from human-induced pluripotent stem cells. Arterioscler. Thromb. Vasc. Biol. 29, 1100–1103.

Thomson, J. A., Itskovitz-Eldor, J., Shapiro, S. S., Waknitz, M. A., Swiergiel, J. J., Marshall, V. S., and Jones, J. M. 1998. Embryonic stem cell lines derived from human blastocysts. Science 282, 1145–1147.

Ueno, S., Weidinger, G., Osugi, T., Kohn, A. D., Golob, J. L., Pabon, L., Reinecke, H., Moon, R. T., and Murry, C. E. 2007. Biphasic role for Wnt/beta-catenin signaling in cardiac specification in zebra fish and embryonic stem cells. Proc. Natl. Acad. Sci. USA 104, 9685–9690.

Uosaki, H. and Yamashita, J. K. 2011a. Chemicals Regulating Cardiomyocyte Differentiation. pp. 471–486. *In:* Craig Atwood [Eds.]. 2011. Embryonic stem cells: The hormonal regulation of pluripotency and embryogenesis. InTech. Available from: http://www.intechopen.com/articles/show/title/chemicals-regulating-cardiomyocyte-differentiation.

Uosaki, H., Fukushima, H., Takeuchi, A., Matsuoka, S., Nakatsuji, N., Yamanaka, S., and Yamashita, J. K. 2011b. Efficient and scalable purification of cardiomyocytes from human embryonic and induced pluripotent stem cells by VCAM1 Surface Expression. PLoS ONE 6, e23657.

Wilkinson, D. G., Bhatt, S., and Herrmann, B. G. 1990. Expression pattern of the mouse T gene and its role in mesoderm formation. Nature 343, 657–659.

Yamamizu, K., Kawasaki, K., Katayama, S., Watabe, T., and Yamashita, J. K. 2009. Enhancement of vascular progenitor potential by protein kinase A through dual induction of Flk-1 and Neuropilin-1. Blood 114, 3707–3716.

Yamamizu, K., Matsunaga, T., Uosaki, H., Fukushima, H., Katayama, S., Hiraoka-Kanie, M., Mitani, K., and Yamashita, J. K. 2010. Convergence of Notch and beta-catenin signaling induces arterial fate in vascular progenitors. J. Cell. Biol. 189, 325–338.

Yamashita, J., Itoh, H., Hirashima, M., Ogawa, M., Nishikawa, S., Yurugi, T., Naito, M., Nakao, K., and Nishikawa, S. 2000. Flk1-positive cells derived from embryonic stem cells serve as vascular progenitors. Nature 408, 92–96.

Yamashita, J. K., Takano, M., Hiraoka-Kanie, M., Shimazu, C., Peishi, Y., Yanagi, K., Nakano, A., Inoue, E., Kita, F., and Nishikawa, S.-I. 2005. Prospective identification of cardiac progenitors by a novel single cell-based cardiomyocyte induction. FASEB J. 19, 1534–1536.

Yan, P., Nagasawa, A., Uosaki, H., Sugimoto, A., Yamamizu, K., Teranishi, M., Matsuda, H., Matsuoka, S., Ikeda, T., Komeda, M. et al. 2009. Cyclosporin-A potently induces highly cardiogenic progenitors from embryonic stem cells. Biochem. Biophys. Res. Commun. 379, 115–120.

Yang, L., Soonpaa, M. H., Adler, E. D., Roepke, T. K., Kattman, S. J., Kennedy, M., Henckaerts, E., Bonham, K., Abbott, G. W., Linden, R. M., Field, L. J., and Keller, G. M. 2008. Human cardiovascular progenitor cells develop from a KDR+ embryonic-stem-cell-derived population. Nature 453, 524–528.

Yuasa, S., Itabashi, Y., Koshimizu, U., Tanaka, T., Sugimura, K., Kinoshita, M., Hattori, F., Fukami, S.-I., Shimazaki, T., Ogawa, S. et al. 2005. Transient inhibition of BMP signaling by Noggin induces cardiomyocyte differentiation of mouse embryonic stem cells. Nat. Biotechnol. 23, 607–611.

Yurugi-Kobayashi, T., Itoh, H., Schroeder, T., Nakano, A., Narazaki G., Kita, F., Yanagi, K., Hiraoka-Kanie, M., Inoue, E., Ara, T., Nagasawa ,T., Just, U., Nakao, K., Nishikawa, S., and Yamashita, J. K. 2006. Adrenomedullin/cyclic AMP pathway induces Notch activation and differentiation of arterial endothelial cells from vascular progenitors. Arterioscler. Thromb. Vasc. Biol. 26, 1977–1984.

Zhang, J., Klos, M., Wilson, G.F., Herman, A.M., Lian, X., Raval, K.K., Barron, M.R., Hou, L., Soerens, A.G., Yu, J., Palecek, S.P., Lyons, G.E., Thomson, J.A., Herron, T.J., Jalife, J., and Kamp, T.J. 2012. Extracellular matrix promotes highly efficient cardiac differentiation of human pluripotent stem cells: the matrix sandwich method. Circ. Res. 111: 1125–36.

CHAPTER 6

Wnt Signaling in Cardiogenesis

Atsuhiko T. Naito,[1] Seitaro Nomura,[2] Ichiro Shiojima[3] and Issei Komuro[1,4,*]

Introduction

Wnts constitute a large family of secreted proteins that activate evolutionarily conserved intracellular signaling cascade, and are hence called the Wnt signaling (Clevers 2006; Logan and Nusse 2004). Wnt signaling plays critical roles during embryonic development through regulating gene transcription, proliferation, polarity, and migration of cells (Angers and Moon 2009; Clevers 2006; Logan and Nusse 2004). The role of Wnt signaling in heart development was first described in 1995 using *Drosophila* (Wu et al. 1995) and it is currently well recognized that Wnt signaling is one of the most important signaling cascades that affect heart development through multiple actions in distinct phases (Gessert and Kuhl 2010). Wnt signaling may also play a role in the adult heart, through regulating the proliferation/differentiation of the so called 'cardiac stem cells' (Cohen et al. 2008). In this chapter, we first provide an introduction to the basics of Wnt signal transduction. We then comprehensively review the multiple effects of Wnt signaling during heart development *in vivo* and during cardiomyocyte differentiation from pluripotent stem cells *in vitro*.

[1]Department of Cardiovascular Medicine, Osaka University Graduate School of Medicine, Osaka 565-0871, Japan; Email: atnaito0508@gmail.com
[2]Department of Cardiovascular Medicine, Chiba University Graduate School of Medicine, Chiba 260-8670, Japan; Email: senomura-cib@umin.ac.jp
[3]Department of Medicine II, Kansai Medical University, Osaka 573-1191, Japan. Email: shiojima@hirakata.kmu.ac.jp
[4]Department of Cardiovascular Medicine, Graduate School of Medicine, The University of Tokyo; Email: komuro-tky@umin.ac.jp
*Corresponding author

Wnt Signal Transduction

In late 1970s, *wingless* was first identified as a *Drosophila* mutant which lacks wings (Sharma and Chopra 1976) and later isolated as a gene that regulates segmental pattern of the larva (Nusslein-Volhard and Wieschaus 1980). Mouse *int1* gene was identified independently as a preferential integration site of the Mouse Mammary Tumor Virus that induces mouse breast cancer (Nusse and Varmus 1982). In 1987, Nusse and colleagues identified that these genes were homologous and renamed the gene *int1* to '*Wnt1*' taking the letters from *wingless* and *int1* (Rijsewijk et al. 1987). In mammals, 19 Wnt ligands are identified and three different pathways are activated by these Wnt ligands: the canonical Wnt/β-catenin pathway, the Wnt/planar cell polarity (PCP) pathway and the Wnt/Ca^{2+} pathway. The latter two are together called non-canonical Wnt pathways. Among these, canonical Wnt/β-catenin pathway is the best characterized pathway and has been extensively studied in the field of cancer biology and developmental biology. The non-canonical Wnt pathways are less characterized, but recent findings suggest that they also play important roles during development and in various diseases through activation of their specific downstream signaling cascade or their function to inhibit canonical Wnt/β-catenin signaling. In this first section, we will overview the details about the canonical Wnt/β-catenin signaling and non-canonical Wnt/PCP signaling.

Canonical Wnt signaling

Molecular players that transduce canonical Wnt signaling were originally identified through searching for the *Drosophila* mutants that show similar (*porcupine* (van den Heuvel et al. 1993), *dishevelled* (Klingensmith et al. 1994), *armadillo* (Peifer and Wieschaus 1990)) or opposite (*shaggy/zeste-white 3* (Siegfried et al. 1992)) phenotypes of *wingless* mutants. Epistatic experiments using these fly mutants outlined the hierarchy of the molecules (Noordermeer et al. 1994; Siegfried et al. 1994). In parallel, researches on tumor suppressor gene APC (Groden et al. 1991; Kinzler et al. 1991) have developed into identification of β-catenin as a binding partner of APC (Rubinfeld et al. 1993; Su et al. 1993) and Tcf/Lef as an effector gene of β-catenin (Behrens et al. 1996; Korinek et al. 1997; Morin et al. 1997). Combination of these two totally different approaches revealed many players of canonical Wnt signaling and their hierarchy within (Fig. 1). In vertebrates, canonical Wnt/β-catenin signaling is initiated by formation of ternary complex composed of Wnt proteins, the Frizzled (Fz) family of seven-transmembrane proteins, and the single transmembrane protein low-density-lipoprotein receptor-related protein 5/6 (LRP5/6) (Angers and Moon 2009; MacDonald et al. 2009). Formation of this ternary complex

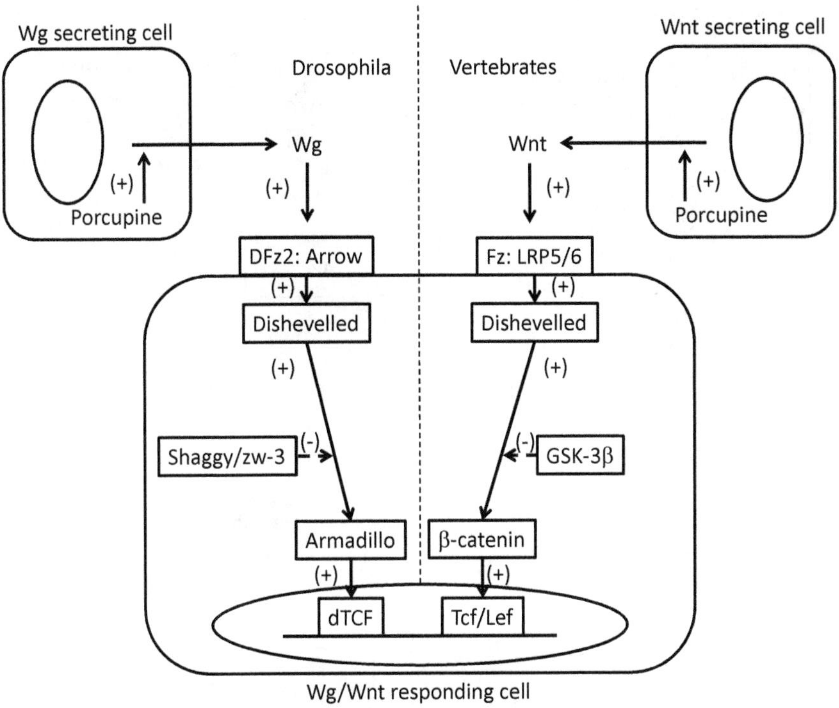

Figure 1. Molecular Hierarchy of Wg/Wnt signaling.

initiates a series of events including phosphorylation of the cytoplasmic tail of LRP5/6 (Davidson et al. 2005; Zeng et al. 2005), aggregation of the membrane and intracellular components of the canonical Wnt signaling, and inactivation of the β-catenin destruction complex (Bilic et al. 2007; Zeng et al. 2008). β-catenin destruction complex is composed of multiple proteins such as tumor suppressor APC (Kishida et al. 1998), casein kinase (-1α, -1ε, and -1δ) (Price 2006), glycogen synthase kinase 3 (-α and -β) (Rubinfeld et al. 1996), and a scaffold protein Axin (Ikeda et al. 1998; Sakanaka et al. 1998). In the absence of Wnt stimulation, cytosolic β-catenin is usually trapped and phosphorylated by two kinases in the destruction complex. Phosphorylation marks β-catenin to become recognized by E3 ubiquitin complex protein β-TrCP (Kitagawa et al. 1999; Liu et al. 1999) and induces degradation via ubiquitin-proteasome pathway (Aberle et al. 1997). However, under the stimulation with Wnt protein, β-catenin destruction complex can no longer phosphorylate β-catenin and unphosphorylated β-catenin accumulates in the cytosol (Fig. 2). The mechanism by which receptor activation inhibits β-catenin phosphorylation remains elusive but recruitment of Axin to the cytoplasmic tail of LRP5/6 and/or the Fz-Dishevelled complex has been

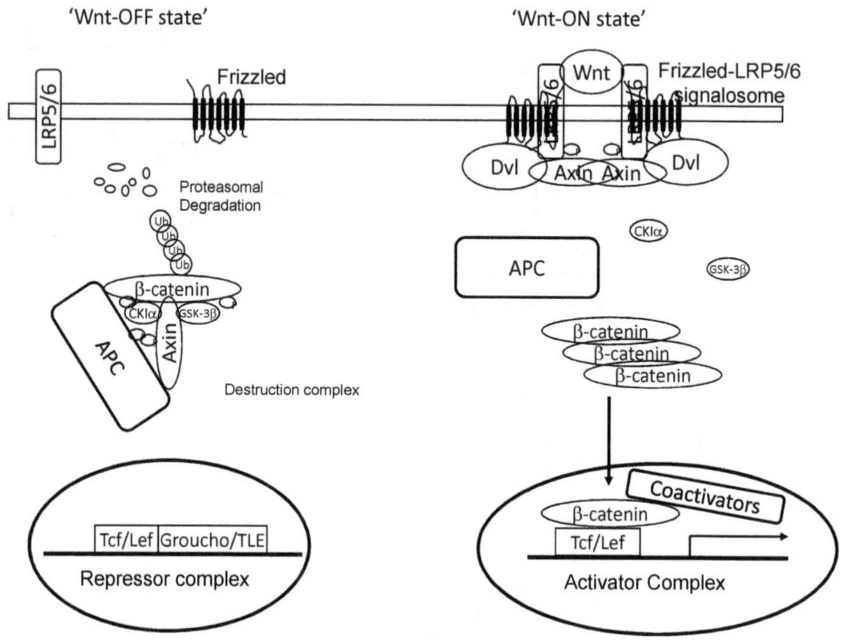

Figure 2. The Canonical Wnt Signaling.

recognized as a key event that releases β-catenin from the destruction complex (Bilic et al. 2007; Davidson et al. 2005; Zeng et al. 2008; Zeng et al. 2005) (Fig. 2).

After stabilization, β-catenin translocates to the nucleus and alters the gene expression by binding and activating transcription factor Tcf/Lef. In the absence of β-catenin, Tcf/Lef forms a complex with Groucho proteins and binds to their target gene acting as a transcriptional repressor (Cavallo et al. 1998). β-catenin binds to Tcf/Lef through its armadillo repeat domain and converts Tcf/Lef to a transcriptional activator by dislocating Groucho protein (Daniels and Weis 2005). The C-terminus of β-catenin is also critical for target gene transcription and has been recognized as a transactivation domain (van de Wetering et al. 1997). Later studies revealed that this domain works as a transactivator by binding with many transcriptional co-activators, such as CBP/p300 histone acetyltransferase, MLL1/2 histone methyltransferases, and Brg1, a component of SWI/SNF family of ATPases, and altering the epigenetic state of the target genes (Mosimann et al. 2009) (Fig. 2).

It is noteworthy that β-catenin/Tcf/Lef complex can in some cases repress the transcription of certain genes through competitive repression of other signaling cascade or transcription factors (Kahler and Westendorf 2003; Piepenburg et al. 2000), recruitment of co-repressors instead of co-activators (Olson et al. 2006; Theisen et al. 2007), and binding with repressive Tcf/Lef binding element (Blauwkamp et al. 2008). Nuclear β-catenin also binds to other nuclear receptors (such as estrogen receptor (ERα) and glucocorticoid receptor (GR)) and transcription factors (such as Foxo and HIF1α) to regulate the transcription of their target genes (MacDonald et al. 2009). Collectively, the output of canonical Wnt signaling activation is not limited to transcriptional activation of Tcf/Lef target genes but also includes transcriptional repression or interaction with other signaling cascades.

Non-canonical Wnt signaling

The non-canonical Wnt pathway includes Wnt/PCP pathway and Wnt/Ca2+ pathway (Fig. 3). In this section, we will focus on the Wnt/PCP pathway because of its role in heart development and cardiomyocyte differentiation. For comprehensive overviews on Wnt/Ca2+ signaling, the reader is referred to the excellent review by Kohn and Moon (Kohn and Moon 2005). Wnt/PCP pathway is less understood, while the presence of the pathway has been recognized as old as the canonical pathway (Vinson and

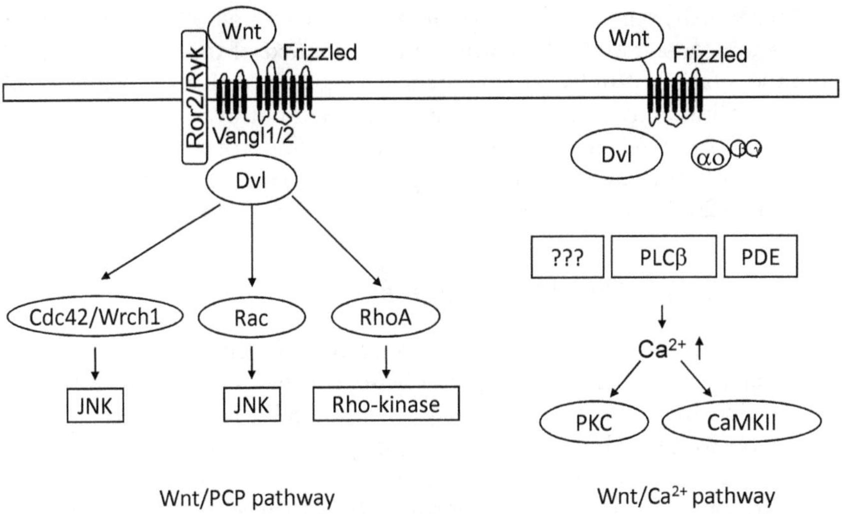

Figure 3. The Non-Canonical Wnt Signaling.

Adler 1987). The major role of Wnt/PCP pathway is to regulate cell polarity and cell movement by organizing the cytoskeleton (Simons and Mlodzik 2008; Zallen 2007). Planar cell polarity can be most visibly understood in *Drosophila*, whose hairs in the wing cells, and the bristles on the thorax are beautifully arranged along the planar cell polarity (Simons and Mlodzik 2008). In mammals, Wnt/PCP pathway is critical for convergent extension movements during gastrulation. Wnt/PCP pathway also regulates proper alignment of the cilia in the respiratory tract and in the organ of Corti during development (Barrow 2006; Tada et al. 2002). Gene knockout studies have revealed that PCP pathway is also important for proper formation of the tubular organs such as lung, kidney, digestive tissues, and blood vessels (Karner et al. 2006). The receptors which transduce Wnt/PCP pathway are not fully understood, but Frizzled, four-pass transmembrane protein Vang-like (Vangl) 1 and 2, and single pass transmembrane receptor-type tyrosine kinase Ror2 (Gao et al. 2011; Green et al. 2008; Oishi et al. 2003) have been reported to play a role in Wnt/PCP signal transduction. Serial activation of Rho family of GTPases, Rho kinase, and jun N-terminal kinase (JNK) is the essential intracellular event that regulates cellular cytoskeleton or cellular movement and s planar cell polarity phenotypes (Angers and Moon 2009) (Fig. 3).

It is noteworthy that individual Wnt ligands are not linked to binding with specific Frizzled receptors or activation of either canonical or non-canonical Wnt signaling pathways (van Amerongen and Nusse 2009). In other words, particular Wnt proteins may activate both canonical and non-canonical signaling depending on the context of the target cells. For example, Wnt1 or Wnt3A, which are considered to be activators of canonical Wnt signaling, may activate non-canonical Wnt pathway, and vice versa; Wnt5A or Wnt11, which are considered to be activators of non-canonical Wnt signaling, can activate canonical Wnt pathway in certain situations (Mikels and Nusse 2006; Takada et al. 2005; van Amerongen and Nusse 2009).

Wnt Signaling in Heart Development

The heart is one of the first functional organs that develop during embryogenesis. Development of the heart can be divided into different phases. Specification of the cardiac progenitor cells, formation of a linear heart tube, followed by cardiac looping, chamber formation, and septation of the heart. Cardiac progenitor cells are specified from the mesodermal cells and form the bilateral heart-forming fields within the anterior lateral mesoderm. There are two temporally distinct cardiogenic cell pools within the heart forming field. The cell pool, which becomes cardiogenic initially and forms a linear heart tube, is called the first heart field. The second heart field is located in the most medial portion of the heart-forming field, become cardiogenic afterward and added at either the arterial or venous pole to form the outflow tract and right ventricle of the heart.

Timely expression of various transcription factors is essential for proper heart development. These transcription factors are also used as markers that label the cells with specific developmental stages, and recently as a driver of *Cre* gene expression to induce cell-lineage specific gene recombination. *Mesp1* gene is the earliest marker gene that labels the cells of cardiac lineage. Double knockout mice for *Mesp1* and *Mesp2*, which function redundantly for Mesp1, lack any mesodermal tissue including the heart (Kitajima et al. 2000). Thus, *Mesp1-Cre* mice are used to induce gene recombination in the cells of cardiac progenitor cells of both first heart field and second heart field (Saga et al. 1999). *Nkx2-5* is a transcription factor expressed slightly after *Mesp1* and *Mesp2*, but more specifically in the cells that are committed to differentiate into cardiomyocytes (Komuro and Izumo 1993). Although linear heart tube is observed, cardiac looping was not initiated in mice lacking *Nkx2-5* (Lyons et al. 1995), and later experiments revealed that *Nkx2-5* knockout mice lack the component of left ventricle and atrium of the heart (Tanaka et al. 1999). As *Nkx2-5* gene expression is observed in all cells which are committed to cardiomyocytes, *Nkx2-5-Cre* mice are used to induce gene recombination in cardiac progenitor cells and cardiomyocytes which belong to both first heart field and second heart field (Stanley et al. 2002). *Mef2c* is a MADS-box transcription factor and its cardiac expression is observed shortly after *Nkx2-5*. In mice lacking *Mef2c* gene, the linear heart tube does not undergo cardiac looping and the future right ventricle is missing (Lin et al. 1997). *Mef2c* gene expression is initially observed in the entire heart-forming fields, but its expression is later restricted to the outflow tract and right ventricle. These tissues are derivatives of so called 'anterior heart field (AHF)', which overlaps very well with second heart field. A *Mef2c*-AHF-*Cre* mice are used to induce gene recombination in these fields (Verzi et al. 2005). *Isl1* encodes a transcription factor that contains LIM homeodomain. *Isl1* gene expression in the heart shows a complementary pattern with the marker for differentiated cardiomyocyte marker, *Mlc2a*, suggesting that it is expressed exclusively in progenitor cells (Cai et al. 2003). The cardiac phenotype of *Isl1*-knockout mice is similar to *Mef2c*-knockout mice in that they do not undergo cardiac looping and lack outflow tract and right ventricle (Cai et al. 2003). Cell lineage analysis using *Isl1-Cre* mice revealed that *Isl1*-expressing cells contribute to outflow tract and right ventricle, suggesting that *Isl1-Cre* mice can be used to induce gene recombination in the second heart field (Cai et al. 2003). However, we should be aware that incomplete recombination of *Cre*-dependent reporter genes may influence the results, which is the important limitation of the *Cre*-based fate-mapping experiments (Vooijs et al. 2001). In fact, reassessment using other reported strains revealed that *Isl1-Cre* drives recombination in both the first heart field and second heart field (Ma et al. 2008). This report suggests that the difference between first and second heart field is not distinguished by the expression of specific genes but by the timing when they differentiate, in other words, when they lose the expression of *Isl1* gene.

Canonical Wnt signaling in heart formation

The canonical Wnt pathway is triggered by Wingless (Wg) in flies. *Drosophila* Wg is required for the induction of cardiac mesoderm, because elimination of Wg function for a short time period after gastrulation using temperature sensitive *wg* mutant flies results in selective loss of heart precursor cells without significant changes in formation of the body wall or visceral muscle cells (Wu et al. 1995). Genetic epistasis experiments revealed that this effect of Wg on heart development is mediated by *dishevelled*, *armadillo*, and *dTCF*, suggesting that canonical pathway is required for heart formation in flies (Park et al. 1998; Park et al. 1996; Wu et al. 1995). However, regulation of heart development by Wnt signaling has been once considered to be opposite in vertebrates. In chick embryos, Wnt inhibitors are expressed in the anterior part of the embryo and induce heart formation by suppressing canonical Wnt activity in the anterior part of the embryo (Marvin et al. 2001). At a later stage of development, Wnt ligands are secreted from neural tube and block heart formation in the paraxial part of the mesoderm (Tzahor and Lassar 2001). Likewise, in *Xenopus* embryos, ectopic expression of Wnt antagonists induces ectopic heart formation in the non-cardiogenic ventral marginal zone mesoderm, while forced expression of *Wnt3A* gene or *Wnt8* gene in the four-cell stage embryos blocks heart formation (Schneider and Mercola 2001). In addition, conditional deletion of β-catenin in the endodermal cells in the forming foregut using *Cytokeratin 19-Cre* mice led to multiple heart formation (Lickert et al. 2002), collectively suggesting that active canonical Wnt signaling blocks heart formation in vertebrates.

In year 2007, six groups independently reported the role of canonical Wnt signaling in heart formation using various Cre strains and conditional loss-of-function (LOF) and gain-of-function (GOF) β-catenin floxed mice (Grigoryan et al. 2008) (Table 1). Birchmeier and colleagues used *Mesp1-Cre* mice (Saga et al. 1999) to conditionally mutate β-catenin in cardiac progenitor cells that give rise to both first heart field and second heart field (Klaus et al. 2007). They observed defective cardiac looping and shortened outflow tract in β-catenin LOF mutants whereas formation of linear heart tube was perturbed in β-catenin GOF mutants. Gene expression analysis using *in situ* hybridization showed the marked reduction of *Isl1* gene expression in the heart of β-catenin LOF mutants whereas *Isl1* positivity and *Isl1*-positive area was enhanced and broadened in the early stage embryo of β-catenin GOF mutants. Interestingly, proliferation of the cells in the first heart field was

Table 1. Role of the Canonical Wnt Signaling in Heart Development; Summary of the Genetically Engineered Mice.

	Cre-Mice	Floxed in CPC/CM	Floxed in FHF/SHF	Phenotype GOF	Phenotype LOF	Proliferation	Isl1 expression
Birchmeier	Mesp1-Cre	(Pre-)CPC	FHF+SHF	Cardia Bifida	No looping Short OFT	Up GOF: FHF cells Down LOF: SHF cells	GOF: Increased LOF: Reduced
Srivastava	Isl1-Cre	CPC	SHF	Large RV	Small RV	Up GOF: SHF cells	LOF: No change
	Nkx2-5-Cre (Ventricular)	CM	FHF+SHF	Not described	Reduced ventricular mass (RV>LV)	Up GOF: FHF cells Down LOF: SHF cells	Not described
Evans	Isl1-Cre	CPC	SHF		Small, thin RV Dilated OFT	Down LOF: SHF cells	LOF: Reduced
Chien	Isl1-Cre	CPC	SHF		Not described	Down LOF: SHF cells	Not described
	Mef2c-AHF-Cre	CM	SHF	Small RV Dilated OFT Short OFT		Up GOF: SHF cells	GOF: Increased
Morrissey	SM22a-Cre	CM	FHF+SHF	Large RV	Small, thin RV	Up GOF: FHF cells Down LOF: SHF cells	GOF: Increased LOF: Reduced
Martin	Mef2c-AHF-Cre	CM	SHF	Small, thick RV Thick LV	Small, thin RV	Down LOF: SHF cells	Not described

increased in β-catenin GOF mutants, whereas the proliferation of the cells in the second heart field was reduced in β-catenin LOF mutants. These results suggest the two roles of canonical Wnt signaling in heart development. First, canonical Wnt signaling contributes to the formation of second heart field derivatives by positively regulating *Isl1* gene expression and *Isl1*-positive cell proliferation. Second, activation of canonical Wnt signaling induces the proliferation of the cells from the first heart field, but does not affect the cardiac (first heart field)-specific gene expression in these cells.

Srivastava and colleagues used two Cre strains: *Isl1-Cre* mice to induce recombination in cardiac progenitor cells and modified *Nkx2-5-Cre* mice, in which *Cre* gene expression is under the control of a ventricular specific enhancer of *Nkx2-5*, to induce recombination in more differentiated cells (Kwon et al. 2007). Conditional β-catenin LOF mutation in *Isl1*-positive cells resulted in defective right ventricle formation without affecting *Isl1* expression, whereas conditional β-catenin GOF mutation resulted in enlarged right ventricle with enhanced proliferation and expansion of the second heart field pool. On the other hand, conditional β-catenin LOF mutation in differentiated cardiomyocytes suppressed the proliferation of ventricular cardiomyocytes resulting in reduction of ventricular size, especially right ventricular size, and thinning of the ventricular wall. In clear contrast, β-catenin GOF mutation promoted the proliferation of ventricular cardiomyocytes and enhanced the expression of *Mef2c* gene. These results suggest that canonical Wnt signaling promotes the proliferation of both cardiac progenitor cells and differentiated cardiomyocytes during heart development.

Evans and colleagues also used *Isl1-Cre* mice (Lin et al. 2007). They observed a small right ventricle, dilated outflow tract, and thin ventricular wall in β-catenin LOF mutants, which is roughly the same with the phenotype reported by Srivastava's group. Evans's group and Chien's group (Qyang et al. 2007) also observed that proliferation of cells in the outflow tract is decreased in the heart of β-catenin LOF mutant mice. However, in contrast to Srivastava's group, Evans's group observed that the expression level of *Isl1* was significantly decreased in β-catenin LOF mutants and identified the Tcf/Lef binding site in the promoter region of *Isl1* gene, concluding that *Isl1* gene is the target of canonical Wnt signaling. Srivastava's group even reported that *Isl1* gene expression is repressed by canonical Wnt signaling (Kwon et al. 2009). In our *in vitro* differentiation system, *Isl1* gene expression is up-regulated after Wnt stimulation, favoring the result from Evans's group. However, ChIP-seq analysis showed that the binding of β-catenin was not observed in the Isl1 promoter region but in the third intronic region where Tcf/Lef binding site was not observed, suggesting that canonical Wnt signaling regulates *Isl1* transcription through more complicated mechanism (A.T.N. and S.N., unpublished observations).

Morrissey and colleagues used *SM22α-Cre* mice (Cohen et al. 2007). This strain induces gene recombination in smooth muscle and cardiac muscle cells, but the recombination is observed only from embryonic day 9.5 (Lepore et al. 2005). Therefore, the recombination is supposed to occur in differentiated cardiomyocytes, rather than in cardiac progenitor cells, which are derived from both the first heart field and the second heart field (Lepore et al. 2005). They observed small and hypoplastic right ventricle in β-catenin LOF mutant heart whereas the diameter of the right ventricle was increased in β-catenin GOF mutants. They also observed that the number and the proliferation of *Isl1*-positive cells were decreased and increased in β-catenin LOF and GOF mutants, respectively. This phenotype is apparently confusing because gene recombination is supposed to have occurred in differentiated cardiomyocytes rather than in *Isl1*-positive cardiac progenitor cells. As a possible answer, they showed that the expression of mitogenic fibroblast growth factors (FGFs) are positively regulated by canonical Wnt signaling, suggesting the unique mechanism by which the proliferation of cardiac progenitor cells is regulated by a paracrine factor from differentiated cardiomyocytes. It is also interesting that only the right ventricle is affected after β-catenin mutation, which is consistent with the phenotype of *Nkx2-5-Cre* mice from Srivastava's group, suggesting that the proliferation of the cells derived from the second heart field are more affected by canonical Wnt signaling.

Martin and colleagues used *Mef2c*-AHF-*Cre* mice (Verzi et al. 2005) to induce gene recombination exclusively in AHF (Ai et al. 2007). They observed a small, thin right ventricle and shortened outflow tract in β-catenin LOF mutants whereas they observed also small, but thickened, hypercellular right ventricle in β-catenin GOF mutants. It is intriguing to see that the left ventricle is also hypoplastic in LOF mutants and hyperplastic in GOF mutants whereas the gene recombination was observed exclusively in right ventricle, again suggesting the presence of paracrine mechanisms, also proposed by Birchmeier's group (via BMP4) (Klaus et al. 2007).

These studies collectively indicate that the major role of canonical Wnt signaling during early heart development is to induce proliferation of cardiac cells, and that the cells from second heart field are more affected by canonical Wnt signaling (Fig. 4). Further experiments and debates are required to answer whether regulation of *Isl1* gene expression by canonical Wnt signaling mediates the proliferative phenotype. It is also important to determine whether the effect of canonical Wnt signaling on cell proliferation is cell autonomous or non-cell autonomous.

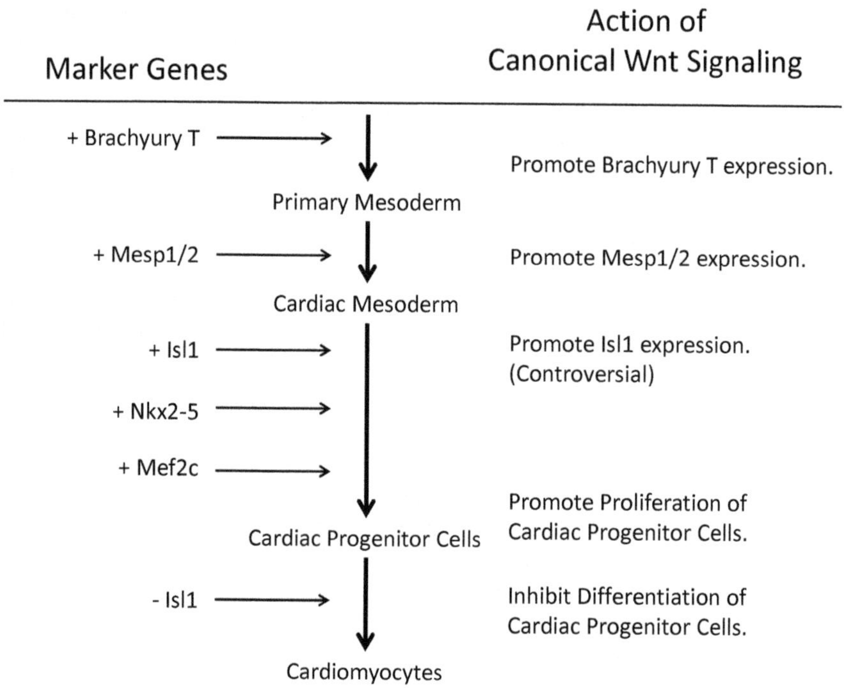

Figure 4. Role of Canonical Wnt Signaling in Cardiomyocyte Differentiation.

Non-canonical Wnt signaling in heart formation

The non-canonical Wnt signaling, or Wnt/PCP signaling, plays an essential role in embryonic morphogenesis, particularly in regulation of convergent extension movements during gastrulation (Barrow 2006; Tada et al. 2002). In 2002, Kuhl and colleagues reported that Wnt11 and Wnt/PCP pathway is required and sufficient for heart induction in *Xenopus*. Introduction of morpholino oligonucleotide against *XWnt11* (later renamed to *Wnt11b* because of its low homology to Wnt11 of other vertebrates) or dominant negative *Wnt11b* strongly inhibited the expression of cardiac markers in the presumptive cardiogenic region and blocked heart formation (Pandur et al. 2002). Conversely, overexpression of *Wnt11b* in pluripotent animal cap or non-cardiogenic ventral marginal zone induced cardiac gene expression and converted these tissues into contracting tissue. These results strongly suggest that *Wnt11b* is one of the essential factors for induction of the heart. However, later study pointed out that *Wnt11b* is not expressed in the heart of the developing *Xenopus* embryo (Ku and Melton 1993), arguing against the role of endogenous *Wnt11b* in heart formation (Garriock et al. 2005). Rather, Krieg and colleagues cloned *XWnt11-R* gene (later renamed to *Wnt11*), which is the orthologue of mammalian and chick *Wnt11* gene, and showed that *Xenopus Wnt11* is expressed in the developing heart. Morpholino oligonucleotide against *Wnt11* delayed the formation of the heart tube and induced cardia bifida in some embryos but never affected heart induction. They also injected *Wnt11* to non-cardiogenic ventral marginal zone and observed the expression of cardiac marker genes. However, they concluded that the cardiogenic activity of *Wnt11*, and *Wnt11b* as well, is an experimental artifact of the overexpression assay because the amount of RNA required for cardiac induction is extraordinarily high. These results collectively suggest the critical role of Wnt/PCP signaling in heart induction and in heart morphogenesis, but whether endogenous Wnt ligands are involved in heart induction through activation of the Wnt/PCP pathway still remains elusive.

Gene knockout studies in mice also supports that the role of Wnt/PCP signaling is in proper formation of the outflow tract rather than in heart induction. Various outflow tract defects were observed in Wnt5a knockout mice (Schleiffarth et al. 2007; Yamaguchi et al. 1999a) and in Wnt11 knockout mice (Zhou et al. 2007). Interestingly, the similar outflow tract defects were also observed in *Vangl2* mutant mice (*Loop-tail*) (Henderson et al. 2001; Phillips et al. 2005) and in *Vangl2*/*Vangl1* double heterozygotic mice (Torban et al. 2008) further suggesting the critical role of Wnt/PCP pathway in morphogenesis of the mammalian heart.

Wnt Signaling in Cardiomyocyte Differentiation

Beside the genetic studies using frog, chick, or mouse embryos, signaling molecules that regulate cardiomyocyte differentiation have also been extensively studied using *in vitro* differentiation system of P19 embryonic carcinoma cells and embryonic stem (ES) cells.

Canonical Wnt signaling in cardiomyocyte differentiation

After the initial apparently contradicting results regarding the role of canonical Wnt signaling in vertebrate and invertebrate embryos (Lickert et al. 2002; Lyons et al. 1995; Marvin et al. 2001; Schneider and Mercola 2001; Tzahor and Lassar 2001), we, among others, tried to solve this problem using differentiation system *in vitro*. Efficiency of cardiomyocyte differentiation was significantly decreased when canonical Wnt signaling was blocked during dimethylsulfoxide (DMSO)-induced P19 cell differentiation, whereas activation of canonical Wnt signaling initiated cardiomyocyte differentiation by itself and activation of canonical Wnt signaling increased the efficiency of cardiomyocyte differentiation of P19 cells (Naito et al. 2005; Nakamura et al. 2003). Furthermore, we identified that canonical Wnt signaling possesses biphasic roles during *in vitro* cardiomyocyte differentiation of ES cells. Cardiomyocyte differentiation from embryonic stem cells was promoted by canonical Wnt signaling when it was activated at the initial stage, whereas inhibition of canonical Wnt signaling at this stage blocked the expression of cardiac transcription factors and cardiomyocyte differentiation. On the contrary, activation of canonical Wnt signaling at the later stage of differentiation inhibited the appearance of spontaneously contracting cardiomyocytes whereas inhibition of canonical Wnt signaling rather enhanced the differentiation into contracting cardiomyocytes (Naito et al. 2006). This observation was later underscored by reports from several other groups with the addition of a beautiful *in vivo* study using transgenic zebrafish embryo that harbors heat-shock inducible *Wnt8* gene. The *Nkx2-5*-positive heart-forming region became larger when *Wnt8* was induced during the pre-gastrula stages, whereas *Wnt8* gene induction during the gastrulation rather reduced the size of the heart-forming region (Ueno et al. 2007).

Mechanistically, canonical Wnt signaling may affect cardiomyocyte differentiation by regulating the expression of several transcription factors that is essential for induction of cardiac progenitor cells from undifferentiated mesodermal cells. *Brachyury T* (*T*) is a T-box transcription factor that is a direct target of canonical Wnt signaling and expressed in all mesodermal cells (Yamaguchi et al. 1999b). *Mesp1* is the earliest marker that labels the cardiac cell-lineage and also requires canonical Wnt signaling for its expression during differentiation of ES cells (Lindsley et al. 2006). *Isl1* also labels undifferentiated cardiac progenitor cells both *in vivo* and *in vitro* (Laugwitz et al. 2005; Moretti et al. 2006) and its expression is positively regulated by canonical Wnt signaling (Lin et al. 2007). In contrast, the mechanism or the direct target gene which is responsible for the repressive effect of canonical Wnt signaling on cardiomyocyte differentiation still remains elusive. Because the expression of *Isl1* is observed exclusively in cardiac progenitor cells but lost in differentiated cardiomyocytes, one might speculate that canonical Wnt signal blocks the differentiation of cardiac progenitor cells by retaining the expression of *Isl1* in cardiac progenitor cells.

Non-canonical Wnt signaling in cardiomyocyte differentiation

As described above, observations from the genetically modified mice clearly show that Wnt/PCP pathway is involved in the morphogenesis of the heart, but not in induction of the endogenous heart nor in differentiation of the endogenous cardiomyocytes (Schleiffarth et al. 2007; Takeuchi et al. 2000;

Yamaguchi et al. 1999a; Zhou et al. 2007). Nonetheless, Wnt11 is a powerful stimulator of cardiomyocyte differentiation because overexpression of *Wnt11* gene or stimulation with Wnt11-conditioned medium successfully induced cardiomyocyte differentiation in many systems including frog explant culture, P19 cells (Pandur et al. 2002), ES cells (Terami et al. 2004), circulating progenitor cells (Koyanagi et al. 2005), and bone marrow mononuclear cells (Flaherty et al. 2008). Even so, it still remains elusive whether this phenotype is solely due to the effect of Wnt11 on activating non-canonical Wnt/PCP pathway, because such strong cardiogenic activity is not reported on other Wnt proteins that preferentially activate Wnt/PCP pathway, such as Wnt5a.

Conclusion

In this chapter, we have comprehensively reviewed the previous reports regarding the role of canonical and non-canonical Wnt signaling in heart development and/or cardiomyocyte differentiation. In future studies, we should classify the multiple role of Wnt signaling—whether it is cell-autonomous or non-cell-autonomous. We should also investigate the temporal and spatial expression patterns of endogenous Wnt ligands (as well as atypical activators or inhibitors of Wnt signaling, such as Norrin (Xu et al. 2004), R-Spondin1 (Wei et al. 2007), IGFBP-4 (Zhu et al. 2008) and C1q (Naito et al. in press) to elucidate which ligand induces the observation from gene mutation studies or *in vitro* studies. Given that cytoplasmic and nuclear β-catenin might function more than we knew previously (Tcf/Lef-dependent transcription), we should re-examine the output of canonical Wnt stimulation or β-catenin stabilization more comprehensively, using new technologies including the next-generation sequencer.

We especially tried to organize the role of canonical and non-canonical Wnt signaling from the viewpoint of whether it is performed *in vivo* (heart development) or *in vitro* (cardiomyocyte differentiation). Intelligent readers now recognize that many controversies have been solved only after gathering the knowledge from both *in vivo* and *in vitro* studies. Although *in vitro* differentiation system is a brilliant model to study complicated developmental processes, including those of the cardiovascular system, we should always be aware that the *in vitro* differentiation system is biased in its nature and should confirm that the observations are also physiologically relevant *in vivo*. This holds true for the 'fashionable' science or medicine based on induced pluripotent stem cell technology. Because of their unlimited possibilities, we should always be aware of the limitations of these man-made cells and should perform experiments and interpret their results with unbiased vision.

References

Aberle, H., Bauer, A., Stappert, J., Kispert, A., and Kemler, R. 1997. beta-catenin is a target for the ubiquitin-proteasome pathway. EMBO J. 16, 3797–3804.

Ai, D., Fu, X., Wang, J., Lu, M. F., Chen, L., Baldini, A., Klein, W. H., and Martin, J. F. 2007. Canonical Wnt signaling functions in second heart field to promote right ventricular growth. Proc. Natl. Acad. Sci. USA 104, 9319–9324.

Angers, S. and Moon, R. T. 2009. Proximal events in Wnt signal transduction. Nat. Rev. Mol. Cell. Biol. 10, 468–477.

Barrow, J. R. 2006. Wnt/PCP signaling: A veritable polar star in establishing patterns of polarity in embryonic tissues. Dev. Biol.Semin. Cell Dev. Biol. 17, 185–193.

Behrens, J., von Kries, J. P., Kuhl, M., Bruhn, L., Wedlich, D., Grosschedl, R., and Birchmeier, W. 1996. Functional interaction of beta-catenin with the transcription factor LEF-1. Nature 382, 638–642.

Bilic, J., Huang, Y. L., Davidson, G., Zimmermann, T., Cruciat, C. M., Bienz, M., and Niehrs, C. 2007. Wnt induces LRP6 signalosomes and promotes dishevelled-dependent LRP6 phosphorylation. Science 316, 1619–1622.

Blauwkamp, T. A., Chang, M. V., and Cadigan, K. M. 2008. Novel TCF-binding sites specify transcriptional repression by Wnt signalling. EMBO J. 27, 1436–1446.

Cai, C. L., Liang, X., Shi, Y., Chu, P. H., Pfaff, S. L., Chen, J., and Evans, S. 2003. Isl1 identifies a cardiac progenitor population that proliferates prior to differentiation and contributes a majority of cells to the heart. Dev. Cell 5, 877–889.

Cavallo, R. A., Cox, R. T., Moline, M. M., Roose, J., Polevoy, G. A., Clevers, H., Peifer, M., and Bejsovec, A. 1998. Drosophila Tcf and Groucho interact to repress Wingless signalling activity. Nature 395, 604–608.

Clevers, H. 2006. Wnt/beta-catenin signaling in development and disease. Cell 127, 469–480.

Cohen, E. D., Tian, Y., and Morrisey, E. E. 2008. Wnt signaling, an essential regulator of cardiovascular differentiation, morphogenesis and progenitor self-renewal. Development 135, 789–798.

Cohen, E. D., Wang, Z., Lepore, J. J., Lu, M. M., Taketo, M. M., Epstein, D. J., and Morrisey, E. E. 2007. Wnt/beta-catenin signaling promotes expansion of Isl-1-positive cardiac progenitor cells through regulation of FGF signaling. J. Clin. Invest. 117, 1794–1804.

Daniels, D. L. and Weis, W. I. 2005. Beta-catenin directly displaces Groucho/TLE repressors from Tcf/Lef in Wnt-mediated transcription activation. Nat. Struct. Mol. Biol. 12, 364–371.

Davidson, G., Wu, W., Shen, J., Bilic, J., Fenger, U., Stannek, P., Glinka, A., and Niehrs, C. 2005. Casein kinase 1 gamma couples Wnt receptor activation to cytoplasmic signal transduction. Nature 438, 867–872.

Flaherty, M. P., Abdel-Latif, A., Li, Q., Hunt, G., Ranjan, S., Ou, Q., Tang, X. L., Johnson, R. K., Bolli, R., and Dawn, B. 2008. Non-canonical Wnt11 signaling is sufficient to induce cardiomyogenic differentiation in unfractionated bone marrow mononuclear cells. Circulation 117, 2241–2252.

Gao, B., Song, H., Bishop, K., Elliot, G., Garrett, L., English, M. A., Andre, P., Robinson, J., Sood, R., Minami, Y., Economides, A. N., and Yang, Y. 2011. Wnt signaling gradients establish planar cell polarity by inducing Vangl2 phosphorylation through Ror2. Dev. Cell 20, 163–176.

Garriock, R. J., D'Agostino, S. L., Pilcher, K. C., and Krieg, P. A. 2005. Wnt11-R, a protein closely related to mammalian Wnt11, is required for heart morphogenesis in Xenopus. Dev. Biol. 279, 179–192.

Gessert, S. and Kuhl, M. 2010. The multiple phases and faces of wnt signaling during cardiac differentiation and development. Circ. Res. 107, 186–199.

Green, J. L., Kuntz, S. G., and Sternberg, P. W. 2008. Ror receptor tyrosine kinases: orphans no more. Trends Cell. Biol. 18, 536–544.

Grigoryan, T., Wend, P., Klaus, A., and Birchmeier, W. 2008. Deciphering the function of canonical Wnt signals in development and disease: conditional loss- and gain-of-function mutations of beta-catenin in mice. Genes Dev. 22, 2308–2341.

Groden, J., Thliveris, A., Samowitz, W., Carlson, M., Gelbert, L., Albertsen, H., Joslyn, G., Stevens, J., Spirio, L., Robertson, M. et al. 1991. Identification and characterization of the familial adenomatous polyposis coli gene. Cell 66, 589–600.

Henderson, D. J., Conway, S. J., Greene, N. D., Gerrelli, D., Murdoch, J. N., Anderson, R. H., and Copp, A. J. 2001. Cardiovascular defects associated with abnormalities in midline development in the Loop-tail mouse mutant. Circ. Res. 89, 6–12.

Ikeda, S., Kishida, S., Yamamoto, H., Murai, H., Koyama, S., and Kikuchi, A. 1998. Axin, a negative regulator of the Wnt signaling pathway, forms a complex with GSK-3beta and beta-catenin and promotes GSK-3beta-dependent phosphorylation of beta-catenin. EMBO J. 17, 1371–1384.

Kahler, R. A. and Westendorf, J. J. 2003. Lymphoid enhancer factor-1 and beta-catenin inhibit Runx2-dependent transcriptional activation of the osteocalcin promoter. J. Biol. Chem. 278, 11937–11944.

Karner, C., Wharton, K. A., Jr., and Carroll, T. J. 2006. Planar cell polarity and vertebrate organogenesis. Dev. Biol.Semin. Cell Dev. Biol. 17, 194–203.

Kinzler, K. W., Nilbert, M. C., Su, L. K., Vogelstein, B., Bryan, T. M., Levy, D.B., Smith, K.J., Preisinger, A. C., Hedge, P., McKechnie, D. et al. 1991. Identification of FAP locus genes from chromosome 5q21. Science 253, 661–665.

Kishida, S., Yamamoto, H., Ikeda, S., Kishida, M., Sakamoto, I., Koyama, S., and Kikuchi, A. 1998. Axin, a negative regulator of the wnt signaling pathway, directly interacts with adenomatous polyposis coli and regulates the stabilization of beta-catenin. J. Biol. Chem. 273, 10823–10826.

Kitagawa, M., Hatakeyama, S., Shirane, M., Matsumoto, M., Ishida, N., Hattori, K., Nakamichi, I., Kikuchi, A., and Nakayama, K. 1999. An F-box protein, FWD1, mediates ubiquitin-dependent proteolysis of beta-catenin. EMBO J. 18, 2401–2410.

Kitajima, S., Takagi, A., Inoue, T., and Saga, Y. 2000. MesP1 and MesP2 are essential for the development of cardiac mesoderm. Development 127, 3215–3226.

Klaus, A., Saga, Y., Taketo, M. M., Tzahor, E., and Birchmeier, W. 2007. Distinct roles of Wnt/beta-catenin and Bmp signaling during early cardiogenesis. Proc. Natl. Acad. Sci. USA 104, 18531–18536.

Klingensmith, J., Nusse, R., and Perrimon, N. 1994. The Drosophila segment polarity gene disheveled encodes a novel protein required for response to the wingless signal. Genes Dev. 8, 118–130.

Kohn, A. D. and Moon, R. T. 2005. Wnt and calcium signaling: beta-catenin-independent pathways. Cell Calcium 38, 439–446.

Komuro, I. and Izumo, S. 1993. Csx: a murine homeobox-containing gene specifically expressed in the developing heart. Proc. Natl. Acad. Sci. USA 90, 8145–8149.

Korinek, V., Barker, N., Morin, P. J., van Wichen, D., de Weger, R., Kinzler, K. W., Vogelstein, B., and Clevers, H. 1997. Constitutive transcriptional activation by a beta-catenin-Tcf complex in APC-/- colon carcinoma. Science 275, 1784–1787.

Koyanagi, M., Haendeler, J., Badorff, C., Brandes, R. P., Hoffmann, J., Pandur, P., Zeiher, A. M., Kuhl, M., and Dimmeler, S. 2005. Non-canonical Wnt signaling enhances differentiation of human circulating progenitor cells to cardiomyogenic cells. J. Biol. Chem. 280, 16838–16842.

Ku, M. and Melton, D. A. 1993. Xwnt-11: a maternally expressed Xenopus wnt gene. Development 119, 1161–1173.

Kwon, C., Arnold, J., Hsiao, E. C., Taketo, M. M., Conklin, B. R., and Srivastava, D. 2007. Canonical Wnt signaling is a positive regulator of mammalian cardiac progenitors. Proc. Natl. Acad. Sci. USA 104, 10894–10899.

Kwon, C., Qian, L., Cheng, P., Nigam, V., Arnold, J., and Srivastava, D. 2009. A regulatory pathway involving Notch1/beta-catenin/Isl1 determines cardiac progenitor cell fate. Nat. Cell Biol. 11, 951–957.

Laugwitz, K. L., Moretti, A., Lam, J., Gruber, P., Chen, Y., Woodard, S., Lin, L. Z., Cai, C. L., Lu, M. M., Reth, M., Platoshyn, O., Yuan, J. X., Evans, S., and Chien, K. R. 2005. Post-natal isl1+ cardioblasts enter fully differentiated cardiomyocyte lineages. Nature 433, 647–653.

Lepore, J. J., Cheng, L., Min Lu, M., Mericko, P. A., Morrisey, E. E., and Parmacek, M. S. 2005. High-efficiency somatic mutagenesis in smooth muscle cells and cardiac myocytes in SM22alpha-Cre transgenic mice. Genesis 41, 179–184.

Lickert, H., Kutsch, S., Kanzler, B., Tamai, Y., Taketo, M. M., and Kemler, R. 2002. Formation of multiple hearts in mice following deletion of beta-catenin in the embryonic endoderm. Dev. Cell 3, 171–181.

Lin, L., Cui, L., Zhou, W., Dufort, D., Zhang, X., Cai, C. L., Bu, L., Yang, L., Martin, J., Kemler, R., Rosenfeld, M. G., Chen, J., and Evans, S. M. 2007. Beta-catenin directly regulates Islet1 expression in cardiovascular progenitors and is required for multiple aspects of cardiogenesis. Proc. Natl. Acad. Sci. USA 104, 9313–9318.

Lin, Q., Schwarz, J., Bucana, C., and Olson, E. N. 1997. Control of mouse cardiac morphogenesis and myogenesis by transcription factor MEF2C. Science 276, 1404–1407.

Lindsley, R. C., Gill, J. G., Kyba, M., Murphy, T. L., and Murphy, K. M. 2006. Canonical Wnt signaling is required for development of embryonic stem cell-derived mesoderm. Development 133, 3787–3796.

Liu, C., Kato, Y., Zhang, Z., Do, V. M., Yankner, B. A., and He, X. 1999. beta-Trcp couples beta-catenin phosphorylation-degradation and regulates Xenopus axis formation. Proc. Natl. Acad. Sci. USA 96, 6273–6278.

Logan, C. Y. and Nusse, R. 2004. The Wnt signaling pathway in development and disease. Annu. Rev. Cell. Dev. Biol. 20, 781–810.

Lyons, I., Parsons, L. M., Hartley, L., Li, R., Andrews, J. E., Robb, L., and Harvey, R. P. 1995. Myogenic and morphogenetic defects in the heart tubes of murine embryos lacking the homeo box gene Nkx2-5. Genes Dev. 9, 1654–1666.

Ma, Q., Zhou, B., and Pu, W. T. 2008. Reassessment of Isl1 and Nkx2-5 cardiac fate maps using a Gata4-based reporter of Cre activity. Dev. Biol. 323, 98–104.

MacDonald, B. T., Tamai, K., and He, X. 2009. Wnt/beta-catenin signaling: components, mechanisms, and diseases. Dev. Cell 17, 9–26.

Marvin, M. J., Di Rocco, G., Gardiner, A., Bush, S. M., and Lassar, A. B. 2001. Inhibition of Wnt activity induces heart formation from posterior mesoderm. Genes Dev. 15, 316–327.

Mikels, A. J. and Nusse, R. 2006. Purified Wnt5a protein activates or inhibits beta-catenin-TCF signaling depending on receptor context. PLoS Biol. 4, e115.

Moretti, A., Caron, L., Nakano, A., Lam, J. T., Bernshausen, A., Chen, Y., Qyang, Y., Bu, L., Sasaki, M., Martin-Puig, S., Sun, Y., Evans, S. M., Laugwitz, K. L., and Chien, K. R. 2006. Multipotent embryonic isl1+ progenitor cells lead to cardiac, smooth muscle, and endothelial cell diversification. Cell 127, 1151–1165.

Morin, P. J., Sparks, A. B., Korinek, V., Barker, N., Clevers, H., Vogelstein, B., and Kinzler, K. W. 1997. Activation of beta-catenin-Tcf signaling in colon cancer by mutations in beta-catenin or APC. Science 275, 1787–1790.

Mosimann, C., Hausmann, G., and Basler, K. 2009. Beta-catenin hits chromatin: regulation of Wnt target gene activation. Nat. Rev. Mol. Cell. Biol. 10, 276–286.

Naito, A. T., Akazawa, H., Takano, H., Minamino, T., Nagai, T., Aburatani, H., and Komuro, I. 2005. Phosphatidylinositol 3-kinase-Akt pathway plays a critical role in early cardiomyogenesis by regulating canonical Wnt signaling. Circ. Res. 97, 144–151.

Naito, A. T., Shiojima, I., Akazawa, H., Hidaka, K., Morisaki, T., Kikuchi, A., and Komuro, I. 2006. Developmental stage-specific biphasic roles of Wnt/beta-catenin signaling in cardiomyogenesis and hematopoiesis. Proc. Natl. Acad. Sci. USA 103, 19812–19817.

Nakamura, T., Sano, M., Songyang, Z., and Schneider, M. D. 2003. A Wnt- and beta -catenin-dependent pathway for mammalian cardiac myogenesis. Proc. Natl. Acad. Sci. USA 100, 5834–5839.

Noordermeer, J., Klingensmith, J., Perrimon, N., and Nusse, R. 1994. Dishevelled and armadillo act in the wingless signaling pathway in Drosophila. Nature 367, 80–83.

Nusse, R. and Varmus, H. E. 1982. Many tumors induced by the mouse mammary tumor virus contain a provirus integrated in the same region of the host genome. Cell 31, 99–109.

Nusslein-Volhard, C. and Wieschaus, E. 1980. Mutations affecting segment number and polarity in Drosophila. Nature 287, 795–801.

Oishi, I., Suzuki, H., Onishi, N., Takada, R., Kani, S., Ohkawara, B., Koshida, I., Suzuki, K., Yamada, G., Schwabe, G. C., Mundlos, S., Shibuya, H., Takada, S., and Minami, Y. 2003. The receptor tyrosine kinase Ror2 is involved in non-canonical Wnt5a/JNK signalling pathway. Genes Cells 8, 645–654.

Olson, L. E., Tollkuhn, J., Scafoglio, C., Krones, A., Zhang, J., Ohgi, K. A., Wu, W., Taketo, M. M., Kemler, R., Grosschedl, R., Rose, D., Li, X., and Rosenfeld, M. G. 2006. Homeodomain-mediated beta-catenin-dependent switching events dictate cell-lineage determination. Cell 125, 593–605.

Pandur, P., Lasche, M., Eisenberg, L. M., and Kuhl, M. 2002. Wnt-11 activation of a non-canonical Wnt signalling pathway is required for cardiogenesis. Nature 418, 636–641.

Park, M., Venkatesh, T. V., and Bodmer, R. 1998. Dual role for the zeste-white3/shaggy-encoded kinase in mesoderm and heart development of Drosophila. Dev. Genet. 22, 201–211.

Park, M., Wu, X., Golden, K., Axelrod, J. D., and Bodmer, R. 1996. The wingless signaling pathway is directly involved in Drosophila heart development. Dev. Biol.177, 104–116.

Peifer, M. and Wieschaus, E. 1990. The segment polarity gene armadillo encodes a functionally modular protein that is the Drosophila homolog of human plakoglobin. Cell 63, 1167–1176.

Phillips, H. M., Murdoch, J. N., Chaudhry, B., Copp, A. J., and Henderson, D. J. 2005. Vangl2 acts via RhoA signaling to regulate polarized cell movements during development of the proximal outflow tract. Circ. Res. 96, 292–299.

Piepenburg, O., Vorbruggen, G., and Jackle, H. 2000. Drosophila segment borders result from unilateral repression of hedgehog activity by wingless signaling. Mol. Cell. 6, 203–209.

Price, M. A. 2006. CKI, there's more than one: casein kinase I family members in Wnt and Hedgehog signaling. Genes Dev. 20, 399–410.

Qyang, Y., Martin-Puig, S., Chiravuri, M., Chen, S., Xu, H., Bu, L., Jiang, X., Lin, L., Granger, A., Moretti, A., Caron, L., Wu, X., Clarke, J., Taketo, M. M., Laugwitz, K. L., Moon, R. T., Gruber, P., Evans, S. M., Ding, S., and Chien, K. R. 2007. The renewal and differentiation of Isl1+ cardiovascular progenitors are controlled by a Wnt/beta-catenin pathway. Cell Stem Cell 1, 165–179.

Rijsewijk, F., Schuermann, M., Wagenaar, E., Parren, P., Weigel, D., and Nusse, R. 1987. The Drosophila homolog of the mouse mammary oncogene int-1 is identical to the segment polarity gene wingless. Cell 50, 649–657.

Rubinfeld, B., Albert, I., Porfiri, E., Fiol, C., Munemitsu, S., and Polakis, P. 1996. Binding of GSK3beta to the APC-beta-catenin complex and regulation of complex assembly. Science 272, 1023–1026.

Rubinfeld, B., Souza, B., Albert, I., Muller, O., Chamberlain, S. H., Masiarz, F. R., Munemitsu, S., and Polakis, P. 1993. Association of the APC gene product with beta-catenin. Science 262, 1731–1734.

Saga, Y., Miyagawa-Tomita, S., Takagi, A., Kitajima, S., Miyazaki, J., and Inoue, T. 1999. MesP1 is expressed in the heart precursor cells and required for the formation of a single heart tube. Development 126, 3437–3447.

Sakanaka, C., Weiss, J. B., and Williams, L. T. 1998. Bridging of beta-catenin and glycogen synthase kinase-3beta by axin and inhibition of beta-catenin-mediated transcription. Proc. Natl. Acad. Sci. USA 95, 3020–3023.

Schleiffarth, J. R., Person, A. D., Martinsen, B. J., Sukovich, D. J., Neumann, A., Baker, C. V., Lohr, J. L., Cornfield, D. N., Ekker, S. C., and Petryk, A. 2007. Wnt5a is required for cardiac outflow tract septation in mice. Pediatr. Res. 61, 386–391.

Schneider, V. A. and Mercola, M. 2001. Wnt antagonism initiates cardiogenesis in Xenopus laevis. Genes Dev. 15, 304–315.

Sharma, R. P. and Chopra, V. L. 1976. Effect of the Wingless (wg1) mutation on wing and haltere development in Drosophila melanogaster. Dev. Biol. 48, 461–465.

Siegfried, E., Chou, T. B., and Perrimon, N. 1992. wingless signaling acts through zeste-white 3, the Drosophila homolog of glycogen synthase kinase-3, to regulate engrailed and establish cell fate. Cell 71, 1167–1179.

Siegfried, E., Wilder, E. L., and Perrimon, N. 1994. Components of wingless signalling in Drosophila. Nature 367, 76–80.

Simons, M. and Mlodzik, M. 2008. Planar cell polarity signaling: from fly development to human disease. Annu. Rev. Genet. 42, 517–540.

Stanley, E. G., Biben, C., Elefanty, A., Barnett, L., Koentgen, F., Robb, L., and Harvey, R. P. 2002. Efficient Cre-mediated deletion in cardiac progenitor cells conferred by a 3′UTR-ires-Cre allele of the homeobox gene Nkx2-5. Int. J. Dev. Biol. 46, 431–439.

Su, L. K., Vogelstein, B., and Kinzler, K. W. 1993. Association of the APC tumor suppressor protein with catenins. Science 262, 1734–1737.

Tada, M., Concha, M. L., and Heisenberg, C. P. 2002. Non-canonical Wnt signalling and regulation of gastrulation movements. Dev. Biol.Semin. Cell Dev. Biol. 13, 251–260.

Takada, R., Hijikata, H., Kondoh, H., and Takada, S. 2005. Analysis of combinatorial effects of Wnts and Frizzleds on beta-catenin/armadillo stabilization and Dishevelled phosphorylation. Genes Cells 10, 919–928.

Takeuchi, S., Takeda, K., Oishi, I., Nomi, M., Ikeya, M., Itoh, K., Tamura, S., Ueda, T., Hatta, T., Otani, H., Terashima, T., Takada, S., Yamamura, H., Akira, S., and Minami, Y. 2000. Mouse Ror2 receptor tyrosine kinase is required for the heart development and limb formation. Genes Cells 5, 71–78.

Tanaka, M., Chen, Z., Bartunkova, S., Yamasaki, N., and Izumo, S. 1999. The cardiac homeobox gene Csx/Nkx2.5 lies genetically upstream of multiple genes essential for heart development. Development 126, 1269–1280.

Terami, H., Hidaka, K., Katsumata, T., Iio, A., and Morisaki, T. 2004. Wnt11 facilitates embryonic stem cell differentiation to Nkx2.5-positive cardiomyocytes. Biochem. Biophys. Res. Commun. 325, 968–975.

Theisen, H., Syed, A., Nguyen, B. T., Lukacsovich, T., Purcell, J., Srivastava, G. P., Iron, D., Gaudenz, K., Nie, Q., Wan, F. Y., Waterman, M. L., and Marsh, J. L. 2007. Wingless directly represses DPP morphogen expression via an armadillo/TCF/Brinker complex. PLoS One 2, e142.

Torban, E., Patenaude, A. M., Leclerc, S., Rakowiecki, S., Gauthier, S., Andelfinger, G., Epstein, D. J., and Gros, P. 2008. Genetic interaction between members of the Vangl family causes neural tube defects in mice. Proc. Natl. Acad. Sci. USA 105, 3449–3454.

Tzahor, E. and Lassar, A. B. 2001. Wnt signals from the neural tube block ectopic cardiogenesis. Genes Dev. 15, 255–260.

Ueno, S., Weidinger, G., Osugi, T., Kohn, A. D., Golob, J. L., Pabon, L., Reinecke, H., Moon, R. T., and Murry, C. E. 2007. Biphasic role for Wnt/beta-catenin signaling in cardiac specification in zebrafish and embryonic stem cells. Proc. Natl. Acad. Sci. USA 104, 9685–9690.

van Amerongen, R. and Nusse, R. 2009. Towards an integrated view of Wnt signaling in development. Development 136, 3205–3214.

van de Wetering, M., Cavallo, R., Dooijes, D., van Beest, M., van Es, J., Loureiro, J., Ypma, A., Hursh, D., Jones, T., Bejsovec, A., Peifer, M., Mortin, M., and Clevers, H. 1997. Armadillo coactivates transcription driven by the product of the Drosophila segment polarity gene dTCF. Cell 88, 789–799.

van den Heuvel, M., Harryman-Samos, C., Klingensmith, J., Perrimon, N., and Nusse, R. 1993. Mutations in the segment polarity genes wingless and porcupine impair secretion of the wingless protein. EMBO J. 12, 5293–5302.

Verzi, M. P., McCulley, D. J., De Val, S., Dodou, E., and Black, B. L. 2005. The right ventricle, outflow tract, and ventricular septum comprise a restricted expression domain within the secondary/anterior heart field. Dev. Biol. 287, 134–145.

Vinson, C. R. and Adler, P. N. 1987. Directional non-cell autonomy and the transmission of polarity information by the frizzled gene of Drosophila. Nature 329, 549–551.

Vooijs, M., Jonkers, J., and Berns, A. 2001. A highly efficient ligand-regulated Cre recombinase mouse line shows that LoxP recombination is position dependent. EMBO Rep. 2, 292–297.

Wei, Q., Yokota, C., Semenov, M. V., Doble, B., Woodgett, J., and He, X. 2007. R-spondin1 is a high affinity ligand for LRP6 and induces LRP6 phosphorylation and beta-catenin signaling. J. Biol. Chem. 282, 15903–15911.

Wu, X., Golden, K., and Bodmer, R. 1995. Heart development in Drosophila requires the segment polarity gene wingless. Dev. Biol. 169, 619–628.

Xu, Q., Wang, Y., Dabdoub, A., Smallwood, P. M., Williams, J., Woods, C., Kelley, M. W., Jiang, L., Tasman, W., Zhang, K., and Nathans, J. 2004. Vascular development in the retina and inner ear: control by Norrin and Frizzled-4, a high-affinity ligand-receptor pair. Cell 116, 883–895.

Yamaguchi, T. P., Bradley, A., McMahon, A. P., and Jones, S. 1999a. A Wnt5a pathway underlies outgrowth of multiple structures in the vertebrate embryo. Development 126, 1211–1223.

Yamaguchi, T. P., Takada, S., Yoshikawa, Y., Wu, N., and McMahon, A. P. 1999b. T (Brachyury) is a direct target of Wnt3a during paraxial mesoderm specification. Genes Dev. 13, 3185–3190.

Zallen, J. A. 2007. Planar polarity and tissue morphogenesis. Cell 129, 1051–1063.

Zeng, X., Huang, H., Tamai, K., Zhang, X., Harada, Y., Yokota, C., Almeida, K., Wang, J., Doble, B., Woodgett, J., Wynshaw-Boris, A., Hsieh, J. C., and He, X. 2008. Initiation of Wnt signaling, control of Wnt coreceptor Lrp6 phosphorylation/activation via frizzled, dishevelled and axin functions. Development 135, 367–375.

Zeng, X., Tamai, K., Doble, B., Li, S., Huang, H., Habas, R., Okamura, H., Woodgett, J., and He, X. 2005. A dual-kinase mechanism for Wnt co-receptor phosphorylation and activation. Nature 438, 873–877.

Zhou, W., Lin, L., Majumdar, A., Li, X., Zhang, X., Liu, W., Etheridge, L., Shi, Y., Martin, J., Van de Ven, W., Kaartinen, V., Wynshaw-Boris, A., McMahon, A. P., Rosenfeld, M. G., and Evans, S. M. 2007. Modulation of morphogenesis by noncanonical Wnt signaling requires ATF/CREB family-mediated transcriptional activation of TGFbeta2. Nat. Genet. 39, 1225–1234.

Zhu, W., Shiojima, I., Ito, Y., Li, Z., Ikeda, H., Yoshida, M., Naito, A. T., Nishi, J., Ueno, H., Umezawa, A., Minamino, T., Nagai, T., Kikuchi, A., Asashima, M., and Komuro, I. 2008. IGFBP-4 is an inhibitor of canonical Wnt signalling required for cardiogenesis. Nature 454, 345–349.

CHAPTER 7

The Control of Proliferation of Pluripotent Stem Cell-derived Cardiomyocytes

Shinsuke Yuasa

Introduction

Although recent evidences indicate that the adult human heart has a limited capacity for regeneration in physiological and pathological conditions (Becker et al. 1974; Fukuda and Yuasa 2006; Yuasa et al. 2004; Beltrami et al. 2001; Bergmann et al. 2009), the mechanisms of regeneration remain unknown and it continues to be difficult to control cardiac regeneration in diseased human heart (Yuasa and Fukuda 2008a,b). For cardiac regenerative medicine using cell replacement therapy to become a viable option, which of the stem cells is most suitable (i.e., embryonic stem (ES) or somatic stem cells) needs to be determined at first. Thus far, there is no clear consensus as to which is better, because each has its own advantages and disadvantages. Somatic stem cells can be obtained from adult tissues with less ethical problems; autologous somatic stem cells don't have immunological problems after transplantation. But its proliferative capacity and differentiation ability is limiting compared to those of ES cells. Although ES cells are promising pluripotent cells with a strong proliferative capacity and multi-potency, there are ethical considerations that constrain the use (and destruction) of early human embryos to establish new human ES

Cardiology Division, Department of Internal Medicine; Center for Integrated Medical Research, Keio University School of Medicine
Email: yuasa@a8.keio.jp

cells; furthermore, the ES cells do not display the autologous genotype of the patients in whom they are to be used, which suggests that we have to take immunosuppressive drugs after cell-transplantation (Evans and Kaufman 1981). In an attempt to overcome these problems and yet maintain pluripotent stem cell characteristics, many studies had investigated various techniques, such as cell fusion and somatic nuclear transplantation; however, as yet, none has progressed to successful clinical application (Cowan et al. 2005; Wilmut et al. 1997).

In 2006, the first report of the establishment of murine induced pluripotent stem (iPS) cells was announced to the great surprise of the scientific community (Takahashi and Yamanaka 2006). Next year, in 2007, human iPS cells were generated from human adult skin fibroblasts using the virus mediated gene transfer of *Oct3/4*, *Sox2*, *c-Myc*, and *Klf4* (Yu et al. 2007; Takahashi et al. 2007). The fact is that generating human iPS cells that exhibit pluripotency and have the ability to differentiate into any type of cell in the human body has created an alternative to the use of human ES cells, which are hampered by ethical problems and their allogenic genetic background (Takahashi and Yamanaka 2006; Yu et al. 2007; Takahashi et al. 2007). The morphology, growth characteristics, and pluripotency of iPS cells are similar to those of ES cells. Moreover, the germ line competency of murine iPS cells has been also demonstrated (Okita et al. 2007). Of the transcription factors introduced to generate iPS cells, there is a risk of tumorigenicity caused by the reactivation of the oncogene c-Myc. However, it has already been proven that iPS cells can be generated from mouse and human fibroblasts without c-Myc; although these cells produce less or no tumors, the efficiency of iPS cell generation is extremely compromised. Transgene integration is a risk of host gene mutation and tumor formation, but recent innovation showed that iPS cells can be generated by transient transgene expression without transgene integration into the host genome (Nakagawa et al. 2008). These successful basic studies shed light on the future of regenerative medicine using human iPS cells.

Differentiation of iPS/ES Cells into Cardiomyocytes

Because of the similarities between ES and iPS cells, the differentiation system used for ES cells can be applied to iPS cells. The first report of the establishment of a mouse ES cell line was reported by Evans and Kaufman in 1981 (Evans and Kaufman 1981; Manis 2007; Murry and Keller 2008); ES cells were used as a model of early development as well as to generate genetically modified mice (Evans and Kaufman 1981; Manis 2007; Murry and Keller 2008). In the mid-1990s, ES cell research slowly moved to the development of stem cell-based cell transplantation therapy in various organs using *in vivo* animal models (Klug et al. 1996; Kolossov et al. 1998; Wobus et al. 1997).

The differentiation of mouse ES cells into cardiomyocytes *in vitro* was first demonstrated in 1985 without precise characteristics (Doetschman et al. 1985). Research into the use of ES cells in cardiac regenerative medicine was initiated after the mid-1990s. In 1996, Klug et al. reported that stable transfection of ES cells with the aminoglycoside phosphotransferase gene under the control of the α-cardiac myosin heavy chain promoter succeeded in purifying cardiomyocytes after differentiation by drug selection *in vitro* (Klug et al. 1996). Those studies shed light on the use of ES cells in cardiac regenerative medicine. The development of more selective and efficient methods of differentiating ES cells into cardiomyocytes progressed slowly but steadily after the late 1990s (Hattan et al. 2005). In 1998, Thomson et al. firstly reported the establishment of human ES cells from human early embryos. Although ethical issues remained, human ES cells attracted significant attention for their potential in regeneration therapy.

There have been many methods reported for the differentiation of ES cells into cardiomyocytes. The differentiation of ES cells *in vitro* mimics normal embryonic development *in vivo*, thereby providing essential information on developmental processes, including heart development. So it is generally accepted that the humoral factors that are essential for cardiomyogenesis *in vivo* will stimulate ES cells to differentiate into cardiomyocytes *in vitro*. In fact, there are generally two different strategies used to achieve the differentiation of pluripotent stem cells into cardiomyocytes, namely the embryoid body (EB) formation system and the fluorescence-activated cell sorting (FACS)-based system. Conventionally, the EB formation system is used only for spontaneous ES cell differentiation. In this system, ES cells are cultured in floating or hanging drops to form cell aggregates and differentiate in a manner partially similar to that seen during normal early embryonic development. Differentiated EBs contain several types of differentiated cells, such as cardiomyocytes, hematopoietic cells, neural cells etc. In this system, physiological factors that promote cardiomyocyte differentiation may increase the population of cardiomyocytes obtained from pluripotent stem cells. Thus far, many factors involved in early heart development have been tried in the EB formation culture system. Particular advantages of the EB formation system are that it is technically easy and suited for large-scale culture. Drawbacks include the fact that the EBs contain many types of cells, so the mechanism controlling differentiation is often difficult to understand. Also, the dissociation of EBs and the selection of cardiomyocytes is technically difficult. The underlying concept of the FACS-based system depends on the collection of cardiac progenitor cells and/or mature cardiac myocytes. The FACS system is relatively sophisticated and can provide some clues as to the mechanisms involved in cardiomyocyte differentiation at the single cell level. However, in terms of its practical application in human cardiac regenerative medicine, it is difficult to obtain large numbers of cardiac myocytes using the FACS system and there is also a risk of damage to the sorted cells caused by the laser emission and/or the process of single-cell sorting. Thus, considerable technical advances are required for future cardiac regenerative medicine.

EB Formation-based Cardiomyocyte Differentiation from ES/iPS cells

Information obtained in genetically modified mice displaying cardiac abnormalities has provided key information on the essential factors in heart development and cardiomyocyte differentiation. Many attempts have been made to utilize this information to increase the efficiency of ES/iPS cell differentiation into cardiomyocytes. For example, early research showed that cardiac anomalies occur in mice lacking the receptor for retinoic acid (RA), a vitamin A derivative, or when vitamin A is deficient during embryonic development, suggesting vitamin A or RA is essential for cardiac differentiation and development (Dyson et al. 1995; Kastner et al. 1994; Osmond et al. 1991). Furthermore, RA induces the differentiation of cardiomyocytes from embryonal carcinoma (EC) cells *in vitro* in a time- and concentration-dependent manner (Edwards et al. 1983). Based on these findings, in 1997 Wobus et al. succeeded in increasing the efficiency of cardiomyocyte induction by exposing ES cells to RA under strictly controlled conditions with respect to concentration and timing (Wobus et al. 1997).

The differentiation of ES cells into any cell lineage depends, in part, on the regulatory mechanisms underlying normal early development. Although several signaling proteins, including bone morphogenetic proteins (BMPs) (Gaussin et al. 2002; Schlange et al. 2000; Winnier et al. 1995; Zhang and Bradley 1996), Wnts (Foley and Mercola 2005; Marvin et al. 2001; Schneider and Mercola 2001), Notch (Timmerman et al. 2004), and fibroblast growth factors (FGFs) (Mima et al. 1995), are involved in heart development, little was known as to the precise temporal and spatial regulatory signals that mediate the differentiation of ES cells into cardiomyocytes. In mouse embryos, cardiac progenitor cells appear around embryonic day (E) 7.0 and the cardiac crescent is formed by E7.5, indicating that the growth factors expressed in these regions or in surrounding areas at the relevant developmental stage may be important for efficient cardiomyocyte induction from ES/iPS cells. Indeed, in the past decade may studies have investigated the effects of BMPs, BMP inhibitors, Wnt, Wnt inhibitors, and Notch on the induction of cardiomyocyte from ES cells at specific developmental stages. Pivotal roles were demonstrated for these factors in cardiomyocyte differentiation from ES cells using the EB formation system and cardiac myocyte differentiation systems (Behfar et al. 2002; Chen et al. 2008; Klaus et al. 2007; Lin et al. 2007; Naito et al. 2006; Ueno et al. 2007; Yang et al. 2008; Yuasa et al. 2005). However, there is no single growth factor or cytokine that acts constantly throughout the entire process of organ induction during the development of multiple organ systems, which suggests that we should pay particular attention to fine spatial and temporal regulation. Alternatively, chemical compound screening is underway to discover compounds that can promote cardiac differentiation (Sadek et al. 2008; Takahashi et al. 2003).

Cell Sorting-based Cardiomyocyte Differentiation from ES/iPS Cells

Cardiomyocytes are derived from cardiovascular progenitor cells, mesodermal progenitor cells, and pluripotent stem cells. Several marker genes have been reported for each stage, and cell surface markers are particularly useful for cell sorting without genetic manipulation. Those marker-based cardiac progenitor cell selections enhance cardiac differentiation and increased the purity of differentiated cardiomyocytes from pluripotent stem cells. Markers that can be used for cardiomyocyte progenitor sorting are detailed below.

1. Investigations into the temporal expression of the primitive streak (PS) marker *brachyury*, which is also a mesodermal marker, have demonstrated the sequential allocation of mesodermal cells to the hemangioblast and cardiac fates during embryonic development (Kinder et al. 1999). On the basis of these results, cell-sorting for *Brachyury*-positive ES cells is likely to increase the population of cells with cardiac differentiation potential (Kattman et al. 2006).
2. Flk-1 (vascular endothelial growth factor receptor-2; also known as kinase insert domain protein receptor (Kdr)) is known as a lateral plate marker of gastrulation stage embryo. An Flk-1-positive cell population appears to develop as cells exit the PS and begin to migrate to form the cardiac crescent (Ema et al. 2006; Kataoka et al. 1997). These observations support the notion that the myocardial and endothelial lineages develop through a common Flk-1-positive progenitor from ES/iPS cells and that Flk-1-positive cells may be cardiac progenitor cells (Nelson et al. 2008; Yamashita et al. 2005; Yang et al. 2008).
3. Stage-specific embryonic antigen 1 (SSEA-1) can be used as an index of the differentiation of human ES cells as well as of the human blastocyst. Because SSEA-1 is one of the earliest markers of human ES cell differentiation and loss of pluripotency, combined FACS selection

for this in addition to other markers enables the isolation of an early population of cardiovascular progenitor cells (Blin et al. 2010; Leschik et al. 2008).

4. The cellular prion protein (PrP) is highly expressed in cardiac crescent of mouse early embryo. And PrP cells selection from EBs serves as an effective surface marker for isolating nascent cardiomyocytes as well as cardiomyogenic progenitors (Hidaka et al. 2010).

The Control of Cardiomyocyte Proliferation

Cardiomyocyte development is a multistep process that includes initial mesodermal induction, the emergence of the cardiomyoblast, cardiomyoblast proliferation, and cardiomyocyte maturation (Srivastava 2006). To improve the efficiency of cardiomyocyte acquisition, we focused on another step of the differentiation pathway, namely controlling cardiomyocyte proliferation during development. Cardiomyocyte proliferation is one of the most important physiological steps in heart development, and is regulated by several growth factors and cytokines (Chen et al. 2004; Kajstura et al. 1994; Kastner et al. 1997; Lavine et al. 2005; Olson and Schneider 2003; Sucov et al. 1994). Several studies about cardiac development show that Insulin-like growth factor-1 (IGF-1), Insulin-like growth factor-2 (IGF-2) and its receptor (Kajstura et al. 1994; Li et al. 2011), as well as several fibroblast growth factors and their receptors (Lavine et al. 2005) regulate myocardial proliferation during mid-gestational heart development. Retinoic acid produced by the epicardium is also a critical regulator of cardiac growth, and $RXR\alpha^{-/-}$ mice die during embryogenesis from a failure in the proliferative expansion of ventricular cardiomyocytes (Sucov et al. 1994; Kastner et al. 1997). BMP10 is specifically expressed in the ventricular trabecular myocardium from E9.0–13.5 and $BMP10^{-/-}$ mice have a cardiac growth defect (Chen et al. 2004). These factors synergistically regulate cardiomyocyte proliferation under normal physiological conditions. Therefore there were many attempts to use these factors to achieve cardiomyocyte proliferation during ES cell differentiation, but they failed to establish an efficient cardiomyocyte proliferation protocol. We thought these factors would synergistically regulate cardiomyocyte proliferation with other factors. However, the requirement for multiple factors has hampered basic research into regenerative medical approaches to controlling cardiomyocyte proliferation. Previously, we reported cardiomyocyte differentiation methods from ES cells by influencing cardiac commitment of mesodermal progenitor cells. In order to establish a more efficient cardiomyocyte differentiation system, we used microarray analysis to screen humoral factors by comparing cardiomyocyte rich embryoid bodies and control embryoid bodies (Yuasa et al. 2010). Then, we searched for humoral factors and receptors that were upregulated in the cardiomyocyte differentiation group and found that the expression level of *csf3r*, which encodes the granulocyte colony-stimulating factor receptor (G-CSFR), was significantly higher in cardiomyocyte rich embryoid bodies. G-CSF is a very famous hematopoietic cytokine (Demetri and Griffin 1991; Welte et al. 1996), and its role in hematopoiesis and stem cell mobilization has been investigated in detail (Chen et al. 2006; Morrison et al. 1997). However, little is known about the role of G-CSF in heart development. To further explore the role of G-CSF signaling in cardiomyocyte development and its involvement in the differentiation of ES cells into cardiomyocytes, we investigated the expression pattern of G-CSFR *in vivo* and the effect of G-CSF during cardiomyogenesis *in vitro* and *in vivo*.

We showed that murine embryonic cardiomyocytes expressed G-CSFR at approximately E9.5, and intra-uterine G-CSF administration strongly promoted the proliferation of embryonic cardiomyocytes. Moreover, G-CSFR gene targeting resulted in fetal death in approximately 50% of samples due to the thinning of atrioventricular walls. Based on these *in vivo* findings, we applied G-CSF to ES cell-derived cardiomyocyte differentiation protocol. Stimulation with extrinsic G-CSF significantly increased the number of ES cell-derived cardiomyocytes. In contrast, neutralizing intrinsic G-CSF with a specific antibody or mitotic inhibitory agents decreased the number of ES cell-derived cardiomyocytes. The effect of G-CSF is also demonstrated in primate ES and human iPS (hiPS) cells. These results indicated that G-CSF directly regulates cardiomyocyte proliferation *in vitro* and *in vivo* (Shimoji et al. 2010). Thus, cardiomyocyte proliferation promoting factor treatment can boost cardiomyocyte yield from ES cells, possibly in combination with other cardiomyocyte differentiation protocols (Fig. 1).

Application of iPS Cells in Cardiac Regenerative Medicine

Cardiovascular diseases are important targets for regenerative medicine because they are associated with high morbidity and mortality. Most pathological processes that initiate irreversible heart dysfunction, such as myocardial infarction and cardiomyopathies, either result from or are exacerbated by a loss of heart cells. Because the human adult heart lacks the capacity for self-repair, the recovery of heart dysfunction requires the replacement of damaged cells by transplantation with large quantities of healthy cardiomyocytes. To this end, human stem cells may be the source cells for transplantation because of their strong proliferative capacity and ability to differentiate reliably. Recent studies have shown that human iPS cells can differentiate into cardiomyocyte-like cells that are similar in terms of their gene expression profiles and physiologic properties to native

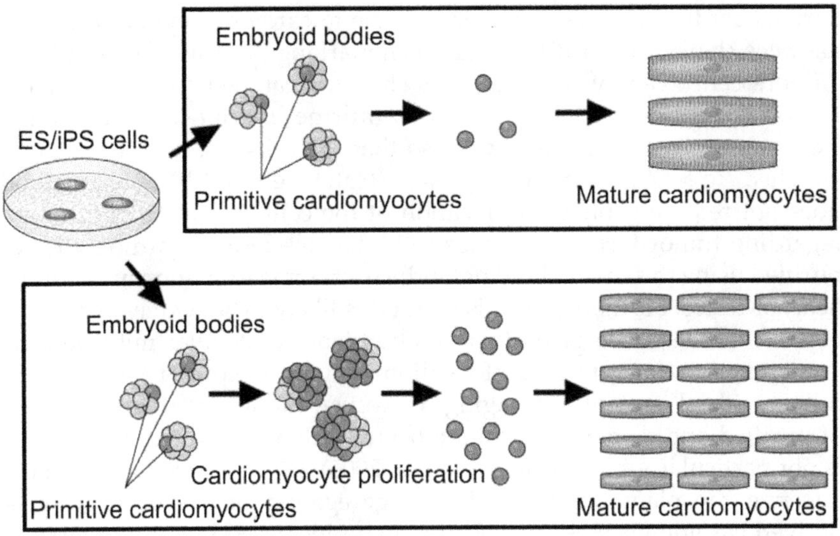

Figure 1. The effects of factors on cardiomyocyte proliferation. Embryonic stem (ES)/induced pluripotent stem (iPS) cells can differentiate into many cell types, including cardiomyocytes. Conventionally, primitive cardiomyocytes emerge spontaneously and differentiate into mature cardiomyocytes (upper panel). Cardiomyocyte proliferation promoting factors can promote the proliferation of primitive cardiomyocytes, boosting the yield of mature cardiomyocytes (lower panel).

Color image of this figure appears in the color plate section at the end of the book.

cardiomyocytes and ES cell-derived cardiomyocytes (Tanaka et al. 2009; Yokoo et al. 2009; Zhang et al. 2009). These studies highlight the potential of human iPS cells in cardiovascular regenerative medicine (Murata et al. 2010).

The first application of iPS cell technology for cardiac regenerative medicine is likely to be the transplantation of iPS cell-derived cardiomyocytes into diseased hearts to restore pump function. A considerable advantage of using human iPS cells is the possibility of creating isogenic cardiomyocytes that are genetically equivalent to the cells in the transplant recipient, thus avoiding immune rejection, which is likely to be seen with allogenic transplants (i.e., ES cell-derived cardiomyocytes may be rejected by the recipient's immune system). However, transplantation of stem cell-derived cardiomyocytes into animal models has raised many questions that need to be addressed before clinical transplantation into humans. Most studies have used a mixture of human ES cell derivatives containing a significant proportion of cardiomyocytes in addition to other cell types (Caspi et al. 2008; Laflamme et al. 2007; van Laake et al. 2008). Because the transplantation of undifferentiated stem cells may lead to the formation of teratomas, it is necessary to establish purification strategies for stem cell-derived cardiomyocytes. A major obstacle to cardiomyocyte purification has been the lack of cardiac-specific cell marking systems. We found that a fluorescent dye that labels the mitochondria can be used to selectively mark rat cardiomyocytes, as well as murine, marmoset, and human stem cell-derived cardiomyocytes, so that the cells can be subsequently enriched (to >99% purity) using FACS (Hattori et al. 2010). This method does not require genetic modification of the cells and reduces concerns regarding tumor formation. Indeed, the ES cell-derived cardiomyocytes purified using this method did not induce teratoma formation in either the heart or testes. Furthermore, this method is likely to be widely applicable beyond these species given that an abundance of cellular mitochondria is a common characteristic of cardiomyocytes, despite a reduction in mitochondrial content with aging. However, the reported survival rate of purified and dispersed cardiomyocytes differentiated from stem cells is not particularly high (Kolossov et al. 2006). We also developed a more effective transplantation method by reaggregating purified ES or iPS cell-derived cardiomyocytes that resulted in an increased long-term survival of the purified cardiomyocytes. Overall, these methods may increase the success rate for using stem cell-derived cardiomyocytes in basic and clinical applications.

Another cell transplantation therapy is likely to be a biological pacemaker created from stem cell-derived cardiomyocytes. It is estimated that currently three million people have an implantable pacemaker to control cardiac rhythm disturbances, including sick sinus syndrome and atrioventricular block (Wood and Ellenbogen 2002). Although these devices treat patients successfully, there are shortcomings associated with their use, including cost, patient discomfort, cosmetic problems, and increased susceptibility to infection. These issues have led to recent interest in the creation of a biological pacemaker. Xue et al. have demonstrated the pacemaking ability of transplanted human ES cell-derived cardiomyocytes in a guinea pig preparation (Xue et al. 2005) and Kehat et al. have demonstrated that EBs injected into a swine model of complete atrioventricular block could function as an ectopic pacemaker (Kehat et al. 2004). These two studies confirmed the capacity of transplanted stem cell-derived cardiomyocytes to couple with host myocardium to function as an ectopic pacemaker, alluding to the potential of a biological pacemaker for clinical application. However, the follow-up period after transplantation in both studies was relatively short (<3 weeks) despite the use of EBs, which contain non-cardiomyocytes or undifferentiated cells. A longer follow-up period is necessary to confirm that this technique is safe without any risk of tumor formation.

Conclusion

Although the adult human heart may have some potential for innate regeneration, it remains impractical to control innate cardiac regeneration for therapeutic purposes in the diseased heart. Many clinical trials have been undertaken to investigate cardiac regeneration using adult stem cells and/or cytokines, and both clinician and patient expectations have been high. However, the results have been disappointing so far. We are now in the position to use newly identified stem cells, namely iPS cells, which show considerable potential. The realization of cardiac regeneration depends on the outcome of many basic experimental investigations and the subsequent application of these results to the clinical setting. Careful clinical trials may result in gradual, yet significant, advances this field, and should be performed in tandem with further intensive investigations with the aim of realizing cardiac regeneration therapy in the clinical setting.

References

Becker, R. O., Chapin, S., and Sherry, R. 1974. Regeneration of the ventricular myocardium in amphibians. Nature 248, 145–147.

Behfar, A., Zingman, L. V., Hodgson, D. M., Rauzier, J.-M., Kane, G. C., Terzic, A., and Puceat, M. 2002. Stem cell differentiation requires a paracrine pathway in the heart. FASEB J. 16, 1558–1566.

Beltrami, A. P., Urbanek, K., Kajstura, J., Yan, S.-M., Finato, N., Bussani, R., Nadal-Ginard, B., Silvestri, F., Leri, A., Beltrami, C. A. et al. 2001. Evidence that human cardiac myocytes divide after myocardial infarction. N. Engl. J. Med. 344, 1750–1757.

Bergmann, O., Bhardwaj, R. D., Bernard, S., Zdunek, S., Barnabe-Heider, F., Walsh, S., Zupicich, J., Alkass, K., Buchholz, B. A., Druid, H. et al. 2009. Evidence for cardiomyocyte renewal in humans. Science 324, 98–102.

Blin, G., Nury, D., Stefanovic, S., Neri, T., Guillevic, O., Brinon, B., Bellamy, V., Rücker-Martin, C., Barbry, P., Bel, A. et al. 2010. A purified population of multipotent cardiovascular progenitors derived from primate pluripotent stem cells engrafts in postmyocardial infarcted nonhuman primates. The Journal of Clinical Investigation 120, 1125–1139.

Caspi, O., Itzhaki, I., Arbel, G., Kehat, I., Gepstien, A., Huber, I., Satin, J., and Gepstein, L. 2008. *In vitro* electrophysiological drug testing using human embryonic stem cell derived cardiomyocytes. Stem Cells Dev.

Chen, C., Huang, X., Atakilit, A., Zhu, Q.-S., Corey, S.J., and Sheppard, D. 2006. The integrin [alpha]9[beta]1 contributes to granulopoiesis by enhancing granulocyte colony-stimulating factor receptor signaling. Immunity 25, 895–906.

Chen, H., Shi, S., Acosta, L., Li, W., Lu, J., Bao, S., Chen, Z., Yang, Z., Schneider, M. D., and Chien, K. R. 2004. BMP10 is essential for maintaining cardiac growth during murine cardiogenesis. Development 131, 2219–2231.

Chen, V. C., Stull, R., Joo, D., Cheng, X., and Keller, G. 2008. Notch signaling respecifies the hemangioblast to a cardiac fate. Nat. Biotech. 26, 1169–1178.

Cowan, C. A., Atienza, J., Melton, D. A., Eggan, K. 2005. Nuclear reprogramming of somatic cells after fusion with human embryonic stem cells. Aug 26; Science 309(5739): 1369–73.

Demetri, G. D. and Griffin, J. D. 1991. Granulocyte colony-stimulating factor and its receptor. Blood 78, 2791–2808.

Doetschman, T. C., Eistetter, H., Katz, M., Schmidt, W., and Kemler, R. 1985. The *in vitro* development of blastocyst-derived embryonic stem cell lines: formation of visceral yolk sac, blood islands and myocardium. J. Embryol. Exp. Morphol. 87, 27–45.

Dyson, E., Sucov, H. M., Kubalak, S. W., Schmid-Schonbein, G. W., DeLano, F. A., Evans, D. M., Ross, J., Jr., and Chien, K. R. 1995. Atrial-like phenotype is associated with embryonic ventricular failure in retinoid X receptor {alpha} -/- mice. Proceedings of the National Academy of Sciences 92, 7386–7390.

Edwards, M. K., Harris, J. F., and McBurney, M. W. 1983. Induced muscle differentiation in an embryonal carcinoma cell line. Mol. Cell. Biol. 3, 2280–2286.

Ema, M., Takahashi, S., and Rossant, J. 2006. Deletion of the selection cassette, but not cis-acting elements, in targeted Flk1-lacZ allele reveals Flk1 expression in multipotent mesodermal progenitors. Blood 107, 111–117.

Evans, M. J. and Kaufman, M. H. 1981. Establishment in culture of pluripotential cells from mouse embryos. Nature 292, 154–156.

Foley, A. C. and Mercola, M. 2005. Heart induction by Wnt antagonists depends on the homeodomain transcription factor Hex. Genes Dev. 19, 387–396.

Fukuda, K. and Yuasa, S. 2006. Stem Cells as a Source of Regenerative Cardiomyocytes. Circ. Res. 98, 1002–1013.

Gaussin, V., Van de Putte, T., Mishina, Y., Hanks, M. C., Zwijsen, A., Huylebroeck, D., Behringer, R. R., and Schneider, M. D. 2002. Endocardial cushion and myocardial defects after cardiac myocyte-specific conditional deletion of the bone morphogenetic protein receptor ALK3. Proceedings of the National Academy of Sciences 99, 2878–2883.

Hattan, N., Kawaguchi, H., Ando, K., Kuwabara, E., Fujita, J., Murata, M., Suematsu, M., Mori, H., and Fukuda, K. 2005. Purified cardiomyocytes from bone marrow mesenchymal stem cells produce stable intracardiac grafts in mice. Cardiovascular Research 65, 334–344.

Hattori, F., Chen, H., Yamashita, H., Tohyama, S., Satoh, Y.-S., Yuasa, S., Li, W., Yamakawa, H., Tanaka, T., and Onitsuka, T. 2010. Nongenetic method for purifying stem cell-derived cardiomyocytes. Nat. Meth. 7, 61–66.

Hidaka, K., Shirai, M., Lee, J.-K., Wakayama, T., Kodama, I., Schneider, M. D., and Morisaki, T. 2010. The cellular prion protein identifies bipotential cardiomyogenic progenitors. Circulation Research 106, 111–119.

Kajstura, J., Cheng, W., Reiss, K., and Anversa, P. 1994. The IGF-1-IGF-1 receptor system modulates myocyte proliferation but not myocyte cellular hypertrophy *in vitro*. Experimental Cell Research 215, 273–283.

Kastner, P., Grondona, J. M., Mark, M., Gansmuller, A., LeMeur, M., Decimo, D., Vonesch, J.-L., Dolle, P., and Chambon, P. 1994. Genetic analysis of RXR[alpha] developmental function: Convergence of RXR and RAR signaling pathways in heart and eye morphogenesis. Cell 78, 987–1003.

Kastner, P., Messaddeq, N., Mark, M., Wendling, O., Grondona, J. M., Ward, S., Ghyselinck, N., and Chambon, P. 1997. Vitamin A deficiency and mutations of RXRalpha, RXRbeta and RARalpha lead to early differentiation of embryonic ventricular cardiomyocytes. Development 124, 4749–4758.

Kataoka, H., Takakura, N., Nishikawa, S., Tsuchida, K., Kodama, H., Kunisada, T., Risau, W., Kita, T., and Nishikawa, S. I. 1997. Expressions of PDGF receptor alpha, c-Kit and Flk1 genes clustering in mouse chromosome 5 define distinct subsets of nascent mesodermal cells. Dev. Growth Differ. 39, 729–740.

Kattman, S. J., Huber, T. L., and Keller, Gordon, M. 2006. Multipotent Flk-1+ Cardiovascular progenitor cells give rise to the cardiomyocyte, endothelial, and vascular smooth muscle lineages. Developmental Cell 11, 723–732.

Kehat, I., Khimovich, L., Caspi, O., Gepstein, A., Shofti, R., Arbel, G., Huber, I., Satin, J., Itskovitz-Eldor, J., and Gepstein, L. 2004. Electromechanical integration of cardiomyocytes derived from human embryonic stem cells. Nat. Biotechnol. 22, 1282–1289.

Kinder, S. J., Tsang, T. E., Quinlan, G. A., Hadjantonakis, A. K., Nagy, A., and Tam, P. P. 1999. The orderly allocation of mesodermal cells to the extraembryonic structures and the anteroposterior axis during gastrulation of the mouse embryo. Development 126, 4691–4701.

Klaus, A., Saga, Y., Taketo, M. M., Tzahor, E., and Birchmeier, W. 2007. Distinct roles of Wnt/beta-catenin and Bmp signaling during early cardiogenesis. Proceedings of the National Academy of Sciences 104, 18531–18536.

Klug, M. G., Soonpaa, M. H., Koh, G. Y., and Field, L. J. 1996. Genetically selected cardiomyocytes from differentiating embronic stem cells form stable intracardiac grafts. J. Clin. Invest. 98, 216–224.

Kolossov, E., Bostani, T., Roell, W., Breitbach, M., Pillekamp, F., Nygren, J. M., Sasse, P., Rubenchik, O., Fries, J. W., and Wenzel, D. 2006. Engraftment of engineered ES cell-derived cardiomyocytes but not BM cells restores contractile function to the infarcted myocardium. J. Exp. Med. 203, 2315–2327.

Kolossov, E., Fleischmann, B. K., Liu, Q., Bloch, W., Viatchenko-Karpinski, S., Manzke, O., Ji, G. J., Bohlen, H., Addicks, K., and Hescheler, J. 1998. Functional characteristics of ES cell-derived cardiac precursor cells identified by tissue-specific expression of the green fluorescent Protein. J. Cell Biol. 143, 2045–2056.

Laflamme, M. A., Chen, K. Y., Naumova, A. V., Muskheli, V., Fugate, J. A., Dupras, S. K., Reinecke, H., Xu, C., Hassanipour, M., and Police, S. 2007. Cardiomyocytes derived from human embryonic stem cells in pro-survival factors enhance function of infarcted rat hearts. Nat. Biotechnol. 25, 1015–1024.

Lavine, K. J., Yu, K., White, A. C., Zhang, X., Smith, C., Partanen, J., and Ornitz, D. M. 2005. Endocardial and epicardial derived FGF signals regulate myocardial proliferation and differentiation *in vivo*. Developmental Cell 8, 85–95.

Leschik, J., Stefanovic, S., Brinon, B., and Puceat, M. 2008. Cardiac commitment of primate embryonic stem cells. Nat. Protocols 3, 1381–1387.

Li, P., Cavallero, S., Gu, Y., Chen, T. H. P., Hughes, J., Hassan, A. B., Brüning, J. C., Pashmforoush, M., and Sucov, H. M. 2011. IGF signaling directs ventricular cardiomyocyte proliferation during embryonic heart development. Development 138, 1795–1805.

Lin, L., Cui, L., Zhou, W., Dufort, D., Zhang, X., Cai, C.-L., Bu, L., Yang, L., Martin, J., Kemler, R. 2007. beta-Catenin directly regulates Islet1 expression in cardiovascular progenitors and is required for multiple aspects of cardiogenesis. Proceedings of the National Academy of Sciences 104, 9313–9318.

Manis, J. P. 2007. Knock out, knock in, knock down—genetically manipulated mice and the Nobel Prize. N. Engl. J. Med. 357, 2426–2429.

Marvin, M. J., Di Rocco, G., Gardiner, A., Bush, S. M., and Lassar, A. B. 2001. Inhibition of Wnt activity induces heart formation from posterior mesoderm. Genes Dev. 15, 316–327.

Mima, T., Ueno, H., Fischman, D. A., Williams, L. T., and Mikawa, T. 1995. Fibroblast growth factor receptor is required for *in vivo* cardiac myocyte proliferation at early embryonic stages of heart development. Proceedings of the National Academy of Sciences 92, 467–471.

Morrison, S. J., Wright, D. E., and Weissman, I. L. 1997. Cyclophosphamide/granulocyte colony-stimulating factor induces hematopoietic stem cells to proliferate prior to mobilization. Proceedings of the National Academy of Sciences 94, 1908–1913.

Murata, M., Tohyama, S., and Fukuda, K. 2010. Impacts of recent advances in cardiovascular regenerative medicine on clinical therapies and drug discovery. Pharmacology & Therapeutics 126, 109–118.

Murry, C. E. and Keller, G. 2008. Differentiation of embryonic stem cells to clinically relevant populations: lessons from embryonic development. Cell 132, 661–680.

Naito, A. T., Shiojima, I., Akazawa, H., Hidaka, K., Morisaki, T., Kikuchi, A., and Komuro, I. 2006. Developmental stage-specific biphasic roles of Wnt/beta-catenin signaling in cardiomyogenesis and hematopoiesis. Proceedings of the National Academy of Sciences 103, 19812–19817.

Nakagawa, M., Koyanagi, M., Tanabe, K., Takahashi, K., Ichisaka, T., Aoi, T., Okita, K., Mochiduki, Y., Takizawa, N., and Yamanaka, S. 2008. Generation of induced pluripotent stem cells without Myc from mouse and human fibroblasts. Nat. Biotech. 26, 101–106.

Nelson, T. J., Faustino, R. S., Chiriac, A., Crespo-Diaz, R., Behfar, A., and Terzic, A. 2008. CXCR4+/FLK-1+ biomarkers select a cardiopoietic lineage from embryonic stem cells. Stem Cells 26, 1464–1473.

Okita, K., Ichisaka, T., and Yamanaka, S. 2007. Generation of germline-competent induced pluripotent stem cells. Nature 448, 313–317.

Olson, E. N. and Schneider, M. D. 2003. Sizing up the heart: development redux in disease. Genes Dev 17, 1937–1956.

Osmond, M. K., Butler, A. J., Voon, F. C., and Bellairs, R. 1991. The effects of retinoic acid on heart formation in the early chick embryo. Development 113, 1405–1417.

Sadek, H., Hannack, B., Choe, E., Wang, J., Latif, S., Garry, M. G., Garry, D. J., Longgood, J., Frantz, D. E., and Olson, E. N. 2008. Cardiogenic small molecules that enhance myocardial repair by stem cells. Proceedings of the National Academy of Sciences 105, 6063–6068.

Schlange, T., Andree, B., Arnold, H.-H., and Brand, T. 2000. BMP2 is required for early heart development during a distinct time period. Mechanisms of Development 91, 259–270.

Schneider, V. A. and Mercola, M. 2001. Wnt antagonism initiates cardiogenesis in Xenopus laevis. Genes Dev. 15, 304–315.

Shimoji, K., Yuasa, S., Onizuka, T., Hattori, F., Tanaka, T., Hara, M., Ohno, Y., Chen, H., Egasgira, T., and Seki, T. 2010. G-CSF Promotes the proliferation of developing cardiomyocytes *in vivo* and in derivation from ESCs and iPSCs. Cell Stem Cell 6, 227–237.

Srivastava, D. 2006. Making or breaking the heart: from lineage determination to morphogenesis. Cell 126, 1037–1048.

Sucov, H. M., Dyson, E., Gumeringer, C. L., Price, J., Chien, K. R., and Evans, R. M. 1994. RXR alpha mutant mice establish a genetic basis for vitamin A signaling in heart morphogenesis. Genes Dev. 8, 1007–1018.

Takahashi, T., Lord, B., Schulze, P. C., Fryer, R. M., Sarang, S. S., Gullans, S. R., and Lee, R. T. 2003. Ascorbic acid enhances differentiation of embryonic stem cells into cardiac myocytes. Circulation 107, 1912–1916.

Takahashi, K. and Yamanaka, S. 2006. Induction of pluripotent stem cells from mouse embryonic and adult fibroblast cultures by defined factors. Aug 25; Cell 126(4): 663–76.

Takahashi, K., Tanabe, K., Ohnuki, M., Narita, M., Ichisaka, T., Tomoda, K., and Yamanaka, S. 2007. Induction of pluripotent stem cells from adult human fibroblasts by defined factors. Nov 30; Cell 131(5): 861–72.

Tanaka, T., Tohyama, S., Murata, M., Nomura, F., Kaneko, T., Chen, H., Hattori, F., Egashira, T., Seki, T., and Ohno, Y. 2009. *In vitro* pharmacologic testing using human induced pluripotent stem cell-derived cardiomyocytes. Biochemical and Biophysical Research Communications 385, 497–502.

Timmerman, L. A., Grego-Bessa, J., Raya, A., Bertran, E., Perez-Pomares, J. M., Diez, J., Aranda, S., Palomo, S., McCormick, F., and Izpisua-Belmonte, J. C. 2004. Notch promotes epithelial-mesenchymal transition during cardiac development and oncogenic transformation. Genes Dev. 18, 99–115.

Ueno, S., Weidinger, G., Osugi, T., Kohn, A. D., Golob, J. L., Pabon, L., Reinecke, H., Moon, R. T., and Murry, C. E. 2007. From the Cover: Biphasic role for Wnt/beta-catenin signaling in cardiac specification in zebrafish and embryonic stem cells. Proceedings of the National Academy of Sciences 104, 9685–9690.

van Laake, L. W., Passier, R., Doevendans, P. A., and Mummery, C. L. 2008. Human embryonic stem cell-derived cardiomyocytes and cardiac repair in rodents. Circ. Res. 102, 1008–1010.

Welte, K., Gabrilove, J., Bronchud, M. H., Platzer, E., and Morstyn, G. 1996. Filgrastim r-metHuG-CSF: the first 10 years. Blood 88, 1907–1929.

Wilmut, I., Schnieke, A. E., McWhir J., Kind, A. J., and Campbell, K. H. 1997. Viable offspring derived from fetal and adult mammalian cells. Feb 27; Nature 385(6619): 810–3.

Winnier, G., Blessing, M., Labosky, P. A., and Hogan, B. L. 1995. Bone morphogenetic protein-4 is required for mesoderm formation and patterning in the mouse. Genes Dev. 9, 2105–2116.

Wobus, A. M., Kaomei, G., Shan, J., Wellner, M.-C., Rohwedel, J., Guanju, J., Fleischmann, B., Katus, H. A., Hescheler, J., and Franz, W.-M. 1997. Retinoic acid accelerates embryonic stem cell-derived cardiac differentiation and enhances development of ventricular cardiomyocytes. Journal of Molecular and Cellular Cardiology 29, 1525–1539.

Wood, M. A. and Ellenbogen, K. A. 2002. Cardiology patient pages. Cardiac pacemakers from the patient's perspective. Circulation 105, 2136–2138.

Xue, T., Cho, H. C., Akar, F. G., Tsang, S. Y., Jones, S. P., Marban, E., Tomaselli, G. F., and Li, R. A. 2005. Functional integration of electrically active cardiac derivatives from genetically engineered human embryonic stem cells with quiescent recipient ventricular cardiomyocytes: insights into the development of cell-based pacemakers. Circulation 111, 11–20.

Yamashita, J. K., Takano, M., Hiraoka-Kanie, M., Shimazu, C., Peishi, Y., Yanagi, K., Nakano, A., Inoue, E., Kita, F., and Nishikawa, S.-I. 2005. Prospective identification of cardiac progenitors by a novel single cell-based cardiomyocyte induction. FASEB J. 04-3540fje.

Yang, L., Soonpaa, M. H., Adler, E. D., Roepke, T. K., Kattman, S. J., Kennedy, M., Henckaerts, E., Bonham, K., Abbott, G. W., and Linden, R. M. 2008. Human cardiovascular progenitor cells develop from a KDR+ embryonic-stem-cell-derived population. Nature 453, 524–528.

Yokoo, N., Baba, S., Kaichi, S., Niwa, A., Mima, T., Doi, H., Yamanaka, S., Nakahata, T., and Heike, T. 2009. The effects of cardioactive drugs on cardiomyocytes derived from human induced pluripotent stem cells. Biochem. Biophys. Res. Commun. 387, 482–488.

Yu, J., Vodyanik, M. A., Smuga-Otto, K., Antosiewicz-Bourget, J., Frane, J. L., Tian, S., Nie, J., Jonsdottir, G. A., Ruotti, V., Stewart, R., Slukvin, I. I., and Thomson, J. A. 2007. Induced pluripotent stem cell lines derived from human somatic cells. Dec 21; Science 318(5858): 1917–20.

Yuasa, S. and Fukuda, K. 2008a. Cardiac regenerative medicine. Circulation Journal 72, A49–A55.

Yuasa, S. and Fukuda, K. 2008b. Recent advances in cardiovascular regenerative medicine: the induced pluripotent stem cell era. Expert Review of Cardiovascular Therapy 6, 803–810.

Yuasa, S., Fukuda, K., Tomita, Y., Fujita, J., Ieda, M., Tahara, S., Itabashi, Y., Yagi, T., Kawaguchi, H., and Hisaka, Y. 2004. Cardiomyocytes undergo cells division following myocardial infarction is a spatially and temporally restricted event in rats. Molecular and Cellular Biochemistry 259, 177–181.

Yuasa, S., Itabashi, Y., Koshimizu, U., Tanaka, T., Sugimura, K., Kinoshita, M., Hattori, F., Fukami, S., Shimazaki, T., and Ogawa, S. 2005. Transient inhibition of BMP signaling by Noggin induces cardiomyocyte differentiation of mouse embryonic stem cells. Nat. Biotechnol. 23, 607–611.

Yuasa, S., Onizuka, T., Shimoji, K., Ohno, Y., Kageyama, T., Yoon, S. H., Egashira, T., Seki, T., Hashimoto, H., and Nishiyama, T. 2010. Zac1 is an essential transcription factor for cardiac morphogenesis. Circ. Res. 106, 1083–1091.

Zhang, H. and Bradley, A. 1996. Mice deficient for BMP2 are nonviable and have defects in amnion/chorion and cardiac development. Development 122, 2977–2986.

Zhang, J., Wilson, G. F., Soerens, A. G., Koonce, C. H., Yu, J., Palecek, S. P., Thomson, J. A., and Kamp, T. J. 2009. Functional cardiomyocytes derived from human induced pluripotent stem cells. Circ. Res. 104, e30–41.

CHAPTER 8

How to Purify Cardiomyocytes for Research and Therapeutic Purposes

Fumiyuki Hattori

Introduction

In every cell-based assay system, the homogeneity of the cells is one of the crucial factors affecting the reproducibility of the results. Cardiac research has been hampered for some time by limited cell availability: only neonatal rat cardiomyocytes can be obtained relatively easily, which are then enriched by density gradient centrifugation. One of our methods, the mitochondrial method, can be used to obtain relatively high yields of purified cardiomyocytes from embryos, neonates, and adult hearts, as well as from pluripotent stem cell (PSC) s, irrespective of species (Hattori et al. 2010).

Human PSCs are thought to be an unlimited source of cardiomyocytes for heart regenerative therapies. Numerous studies have tried to induce directional differentiation of mouse, monkey, and human embryonic stem cell (ESC)s into pure cardiomyocytes (Mummery et al. 2003; Nemir et al. 2006; Yuasa et al. 2005), but without success. In another study, the transplantation of undifferentiated ESCs into the myocardium resulted in the formation of teratomas (Kolossov et al. 2006), highlighting the importance of separating undifferentiated cells from therapeutic cells. Thus, purifying cardiomyocytes is one way in which to minimize the risk of tumorigenesis.

Department of Cardiology, Keio University School of Medicine, 35 Shinanomachi, Shinjuku-ku, Tokyo 160-8582, Japan; and, Asubio Pharma Co., Ltd. 6-4-3, Minatojima Minami-cho, Chuo-ku, Kobe City 650-0047, Japan.
Email: hattori.fumiyuki.ef@asubio.co.jp

In this chapter, various methods for the enrichment of cardiomyocytes are discussed in terms of their application in basic research, pharmaceutical testing, and therapeutics.

Genetic Methods for Purifying Cardiomyocytes

Genetic purification of ESC-derived cardiomyocytes was first reported by Klug et al. in 1996, who established mouse ESC lines by permanent gene transfection of the *aminoglycoside phosphotransferase* (*neo*) gene driven by the *α-myosin heavy chain* promoter and obtained 99.6% pure mESC-derived cardiomyocytes (Klug et al. 1996). Thereafter, several studies tested various combinations of cardiomyocyte-specific promoters and reporters to obtain pure ESC-derived cardiomyocyte cell populations. For example, Muller et al. stably transfected the *enhanced green fluorescent protein* (*EGFP*)-tagged *myosin light chain-2v* promoter into mESC to mark a few of the differentiated beating cells (Muller et al. 2000). They then added the *cytomegalovirus enhancer* sequence upstream of the *myosin light chain-2v* promoter, which resulted in an increase in the number of EGFP-positive cells. Enzymatic digestion of the transgenic embryoid bodies (EB), followed by Percoll gradient centrifugation and fluorescence-activated cell sorting (FACS) resulted in a 97% pure population of cardiomyocytes. In another study, Hidaka et al. used *EGFP* knock-in to the *Nkx2.5* locus of mESCs, followed by enrichment procedures, to produce a 97% pure cardiomyocyte population (Hidaka et al. 2003). Using a different approach, Fijnvandraat et al. stably transfected the *Na^+/Ca^{2+} exchanger* (*Ncx*) *1* promoter coupled with *neo* to acquire stable, non-ectopic, and non-mosaic expression (Fijnvandraat et al. 2003). The *Ncx1* promoter was then integrated into the *hypoxanthine phosphoribosyl transferase* gene, which is ubiquitously expressed in all tissues at all stages of development. However, Fijnvandraat et al. did not provide any data on the purity of the cardiomyocytes obtained using this approach. In 2004, Gassanov et al. reported a technique using stable transfection of the *EGFP*-tagged *atrial natriuretic peptide* promoter that yielded 91% pure cardiomyocytes (Gassanov et al. 2004). Anderson et al. first enriched human ESC (hESC)-derived cardiomyocytes using genetic modifications and then designed a human *myosin heavy chain* promoter driving a bicistronic reporters: *green fluorescent protein* (*GFP*) and *puromycin-N*-acetyltransferase to obtain a 91.5% pure cardiomyocyte population (Anderson et al. 2007). All these strategies, in addition to the non-genetic methods for purifying cardiomyocytes described in the following section, are summarized in Table 1.

Table 1.

	Methods	Cell sources	Purity	Notes	Ref.
Genetic	MHC promoter /neo	mouse ES cells	99.6%		Klug et al.
	CMV enhancer+MLC2v promoter / EGFP	mouse ES cells	97.0%	Pre-Percoll enrichment was done.	Muller et al.
	Nkx2.5 knock-in / EGFP	mouse ES cells	97.0%	FACS was used	Hidaka et al.
	Na+/Ca2+ exchanger 1 promoter / neo	mouse ES cells	-	Targeted integration into the hypoxanthine phosphoribosyl transferase	Fijnvandraat et al.
	atrial natriuretic peptide promoter / EGFP	mouse ES cells	91.0%	FACS was used	Gassanov et al.
	MHC promoter / GFP-IRES-puro	human ES cells	91.5%	FACS was used	Anderson et al.
Nongenetic	Density Gradient Centrifugation (DGC)	neonatal rat	99.0%		Pretlow et al.
		mouse ES cells	-		Doevendans et al.
		human ES cells	60.0%		Xu et al.
		human ES cells	82.6%	directed cardiac differentiation method was used	Laflamme et al.
	Mitochondria specific staining	neonatal rat	100.0%	FACS was used	Hattori et al.
		whole rat embryo	99.0%		
		mouse ES cells	99.5%		
		mouse iPS cells	99.0%		
		marmoset ES cells	99.0%		
		human ES cells	99.0%		
		human iPS cells	99.3%		
	Glucose withdrawal and Lactate administration	neonatal rat	-		Tohyama et al.
		whole rat embryo	-		
		mouse ES cells	-		
		marmoset ES cells	-		
		human ES cells	98.3%		
		human iPS cells	98.6%		
	Cell surface marker of flk-1	mouse ES cells	-	multipotential cardiogenic progenitors	Yan et al.
	Cell surface marker of prion protein	mouse ES cells	-	bipotential cardiogenic progenitors	Hidaka et al.
	Cell surface marker of SSEA-1	human ES cells	-	multipotential cardiogenic progenitors	Blin et al.

Non-genetic Methods for Purifying Cardiomyocytes

Density gradient centrifugation-based methods

Cardiomyocyte enrichment by density gradient centrifugation (DGC) was first reported by Pretlow et al. and has remained the golden standard for the enrichment of primary cardiomyocytes (Pretlow et al. 1972). In 2000, Doevendans et al. applied DGC to mESCs-derived cardiomyocytes and noted that one limitation of this method is the heterogeneity of the cell population isolated (Doevendans et al. 2000). Xu et al. also applied DGC to hESC-derived cardiomyocytes, using partially digested cardiomyocytes containing embryoid bodies, including single cells and small clumps, and obtained a cardiomyocyte-enriched fraction (containing ~60% cardiomyocytes) (Xu et al. 2006). Finally, Laflamme et al. and Xu et al. combined a direct differentiation method for cardiomyocytes and DGC to obtain an enriched cardiomyocyte population (82.6% purity) (Laflamme et al. 2007; Xu et al. 2006).

Mitochondrial method

Background

Of all the organs in the body, the heart consumes the most energy. Adult cardiomyocytes contain abundant mitochondria, which can occupy up to 40% of a cardiomyocyte's volume (Knaapen et al. 1997) (Fig. 1). In addition, the heart is the first functional organ to appear in the developing embryo. From ultrastructural studies, the surface area of mitochondrial cristae increases threefold from embryonic day (E) 11 to E17, and the mitochondrial volume per cell volume increases markedly (3.5-fold) after birth (Knaapen et al. 1997). Thus, we hypothesized that mitochondria-selective fluorescent dyes could be used to label cardiomyocytes (Hattori et al. 2010) (Fig. 2).

Cell sorting of MitoTracker Red-stained neonatal heart cells by FACS resulted in the separation of three cell populations (Hattori et al. 2010) (Fig. 3a, b). The two populations with the highest (Fraction 1) and intermediate (Fraction 2) fluorescence intensity were sorted and cultured separately. All cells in Fraction 1 showed rhythmic beating. Immunofluorescent staining for α-actinin revealed that Fraction 1 consisted of cardiomyocytes only, whereas very few cardiomyocytes were identified in Fraction 2 (Fig. 3c, d). Fraction 3 consisted of hematocytes and dead cells.

Fluorescent dyes that are used biologically can be divided into two groups: nernstian or non-nernstian dyes. The former group, which includes tetramethylrhodamine methyl ester (TMRM), can enter and exit the mitochondrial matrix freely depending on the mitochondrial membrane

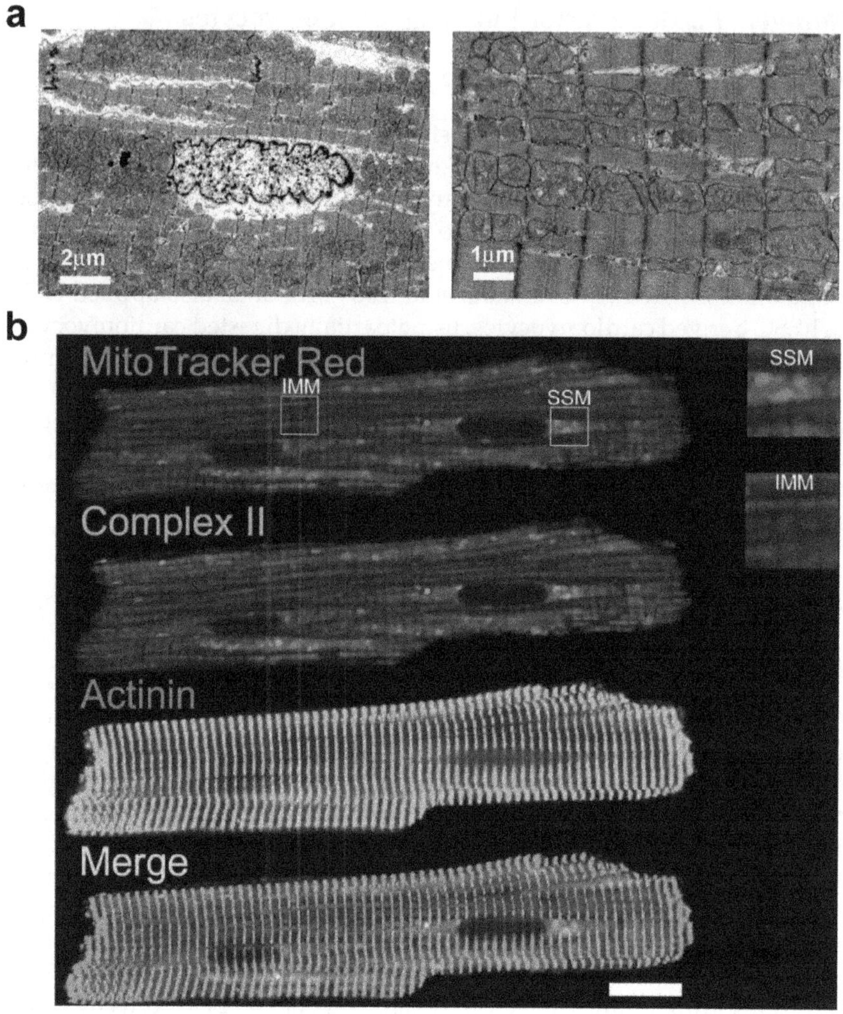

Figure 1. Mitochondrial abundance in cardiomyocytes. (a) Ultrastructure of adult human cardiomyocytes. Note the large numbers of mitochondria in the cardiomyocytes. (b) Fluorescent staining of cardiac mitochondria by MitoTracker Red, immunohistochemical staining for mitochondrial electron transport chain complex II and sarcomeric actinin, and a merged image, as indicated. Cardiomyocytes contains two distinguishable mitochondria: subsarcomeric (SSM) and intermyofibrillar mitochondria (IFM). Scale bar, 20 μm.

Color image of this figure appears in the color plate section at the end of the book.

potential. In contrast, non-nernstian dyes, such as the MitoTracker dyes JC-1 and nonyl acridine orange (NAO), cannot exit the mitochondrial matrix freely (Jakobs 2006). An investigation into the efficacy and applicability of various mitochondrial indicators, including MitoTracker Red, Green,

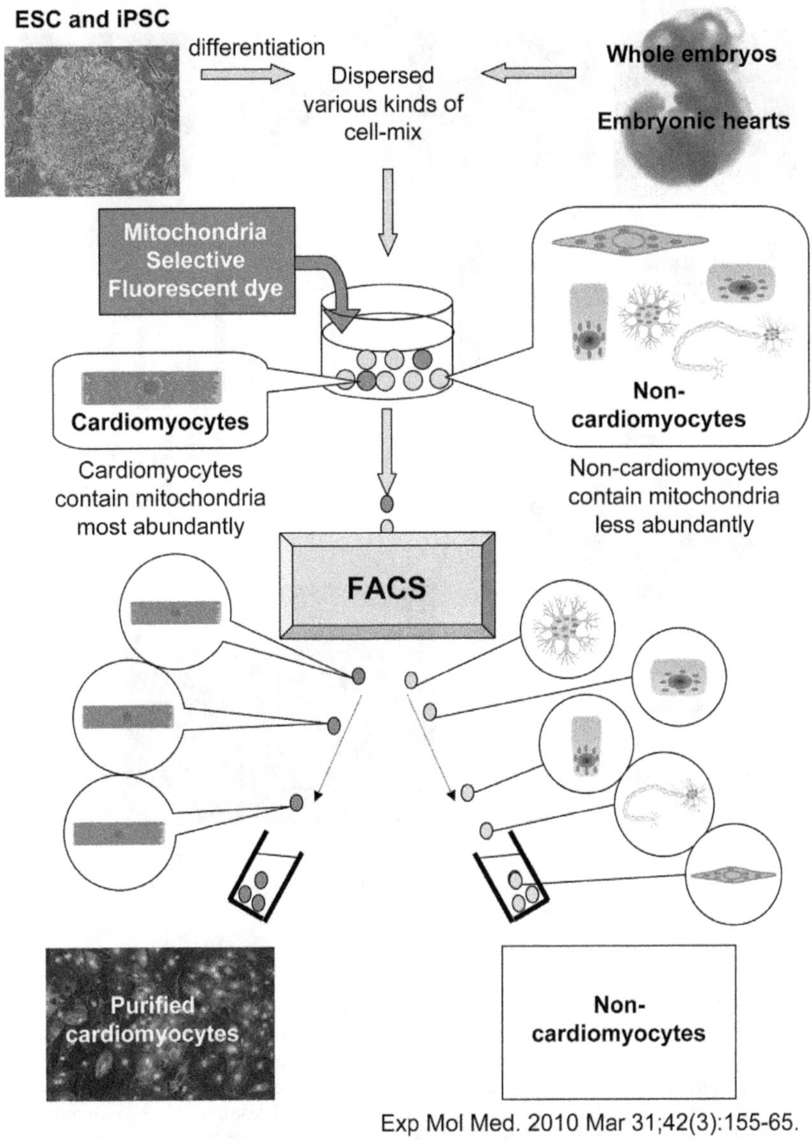

Figure 2. Schematic representation of the steps involved in the mitochondrial method of cardiomyocyte purification. Embryonic stem (ES) and induced pluripotent stem (iPS) cell derivatives or primary cells are dispersed and stained with tetramethylrhodamine methyl ester (TMRM) before being separated by fluorescence-activated cell sorting (FACS) (Fig. reproduced with permission from Hattori et al. 2010; Hattori and Fukuda 2010).

Color image of this figure appears in the color plate section at the end of the book.

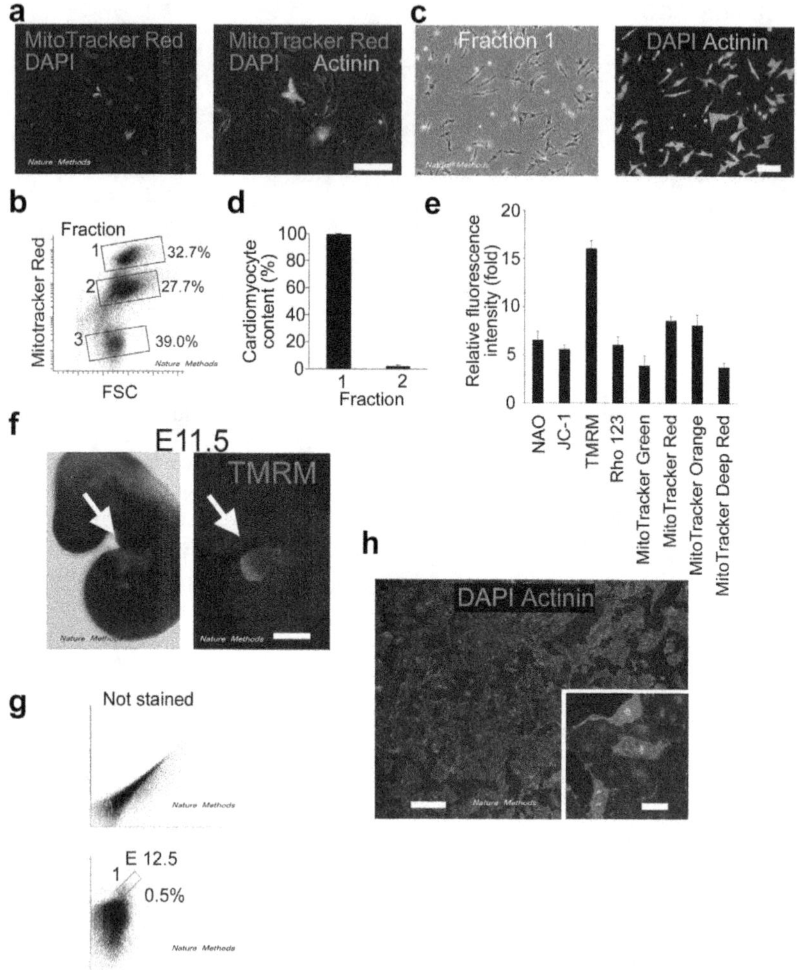

Figure 3. Characterization of mitochondria-selective dyes for primary cardiomyocyte purification. (a) Primary cultured neonatal rat heart cells prestained with MitoTracker Red (red) were immunostained for α-actinin (green). 4′,6′-Diamidino-2-phenylindole (DAPI; blue) was used to indicate nuclear staining. (b) Fluorescence-activated cell sorting (FACS) analysis of neonatal rat heart-derived cells stained with MitoTracker Red resulted in cells being divided into three fractions (1, 2, and 3). (c) Cells in Fraction 1 were sorted further (left) and stained immunofluorescently for α-actinin (green) and DAPI (blue), as shown on the right. (d) The cardiomyocyte content in Fractions 1 and 2 is shown. (e) Representative FACS charts of various mitochondrial indicators. NAO, nonyl acridine orange. (f) Tetramethylrhodamine methyl ester (TMRM)-stained living whole embryos [embryonic day (E) 11.5]. (g) FACS analysis of cells derived from whole embryos treated with (lower) or without (upper) TMRM. (h) Immunofluorescent staining for α-actinin in cells obtained from Fraction 1 in (g, lower panel). Inset shows proliferating cardiomyocytes. Scale bars, 100 μm (a,c,e,h); 20 μm (h, inset) (Figs. reproduced with permission from Hattori et al. 2010 (Hattori et al. 2010)).

Color image of this figure appears in the color plate section at the end of the book.

Orange, and Deep Red, TMRM, NAO, rhodamine 123, and JC-1, for cardiomyocyte enrichment revealed that TMRM was the most effective in separating the cardiomyocyte population from non-myocytes (Hattori et al. 2010) (Fig. 3e). Furthermore, 100 nM TMRM has no effect on cell viability, as determined by the 3-(4, 5-dimethyl-2 thiazoyl)-2, 5-diphenyl-2H-tetrazolium bromide (MTT) assay (Hattori et al. 2010). Based on these results, we believe that TMRM is the best of the currently available dyes.

Application of the mitochondrial method to embryonic hearts and whole-body embryos

Cardiomyocytes can be purified using the mitochondrial method from embryonic hearts from E11.5 rat embryos, through to hearts from postnatal day (P) 8 rats. For example, after cells from 30 embryonic hearts had been dispersed, treated with TMRM, and then subjected to FACS analysis, the mean TMRM fluorescence ratio of Fraction 1 to Fraction 2 was observed to increase gradually with increasing embryonic stage, with a rapid increased noted at the postnatal stage (Hattori et al. 2010). Preliminary, we successfully applied mitochondrial method to adult mouse cardiomyocytes isolated by Langendorff's perfusion protocol (unpublished).

The mitochondrial method can also be used to enrich cardiomyocytes from whole embryos. As shown in Fig. 3f, embryos stained with TMRM accumulate fluorescence in the heart. FACS analysis of dissociated cells obtained from stage E11.5, E12.5, and E13.5 whole rat embryos revealed that some of the embryo-derived cells were had high autofluorescence (Hattori et al. 2010). As reported previously (Monici 2005), the emission wavelength of cells exhibiting autofluorescence can be detected with both green and red band-pass filters, so pseudo two-dimensional (TMRM and empty green) separation can be used to eliminate autofluorescing cells from the assay, resulting in the isolation of cardiomyocytes (~0.2%), which exhibit the highest red fluorescence and the lowest green fluorescence (Hattori et al. 2010) (Fig. 3g). After two days culture, cells derived from E11.5 and E12.5 embryos were stained with anti-α-actinin antibody. Almost all the cells were found to be α-actinin-positive cardiomyocytes (Hattori et al. 2010) (99% purity; Fig. 3h).

Purification of mouse, marmoset, and human PSC-derived cardiomyocytes using the mitochondrial method

Cardiomyocytes that had differentiated from mESCs were first seen on Day 7 of differentiation; these cells exhibited marked accumulation of TMRM (Hattori et al. 2010). After TMRM staining, the mESC-derived cells were fixed and immunostained for Nkx2.5 and α-actinin (Fig. 4a). Analysis revealed complete colocalization of the Nkx2.5-, α-actinin-, and TMRM-positive areas in mESC-derived cardiomyocytes, although the intracellular localization of TMRM, Nkx2.5, and α-actinin clearly differed (Hattori et al. 2010). Of note, upon fixation, TMRM dissociates rapidly into the bulk solution compared with the other dyes, indicating that TMRM is unlikely to have any subsequent effects on the results of immunohistochemical analyses.

Some EB-derived differentiated cells also have high autofluorescence, as demonstrated by FACS analysis (Hattori et al. 2010). In this case, pseudo two-dimensional (TMRM and empty green) separation should be used to eliminate these cells (Fig. 4b). When using EB-derived cells, the cardiomyocyte population is first seen 7 days after EB formation (Hattori et al. 2010). The ratio of the mean TMRM fluorescence in Fraction 1 (cardiomyocytes) to Fraction 2 (non-cardiomyocytes) is elevated from Day 7 and increases gradually thereafter, reaching a plateau on Day 15 (Hattori et al. 2010) (Fig. 4c). The percentage of cells in Fraction 1 (cardiomyocytes) increases on Day 8, continues to increase gradually until peaking on Day 15 (Fig. 4d), and then declines thereafter, probably due to differences in the growth rate of cardiomyocytes and other cell types (Hattori et al. 2010). On the basis of these observations, it appears that the best time to obtain cardiomyocytes is on Day 15. Trypan blue-exclusion studies indicate that the viability of the sorted cells is 99.1 ± 1.5% (Hattori et al. 2010). This surprisingly high viability of the sorted cells may be due to the fact that living cells (i.e., active mitochondria-containing cells) were positively selected on the basis of their ability to accumulate TMRM.

Comparisons of various matrices used to coat the plastic dishes revealed that the best results, in terms of the cell viability of post-seeding purified PSC-derived cardiomyocytes, were obtained with fibronectin (Hattori et al. 2010). Approximately 5–9 × 10^5 cells from Day 15 EB were seeded directly into fibronectin-coated culture dishes equipped with flexiPERM conA. Immunofluorescent staining for α-actinin and Nkx2.5 confirmed that these cells were 99.5 ± 0.3% pure cardiomyocytes (Fig. 4e). We also succeeded in obtaining >99% pure cardiomyocytes from Day 12–25 EB (Hattori et al. 2010). Furthermore, when the mitochondrial method was used to enrich mouse induced PSC-derived cardiomyocytes, a pure cardiomyocyte cell sheet was obtained (Hattori et al. 2010) (Fig. 4e).

When common marmoset monkey and hESCs were differentiated into cardiomyocytes containing EB using conventional floating cell culture, EB-containing beating cells appeared between Days 12 and 15 for the marmoset cell culture and between Days 15 and 18 for the human cell culture (Hattori et al. 2010). Five to 15 days after the beating cells had appeared, the EB were loaded with 10 nM TMRM and transferred into cell-attachment dishes. As

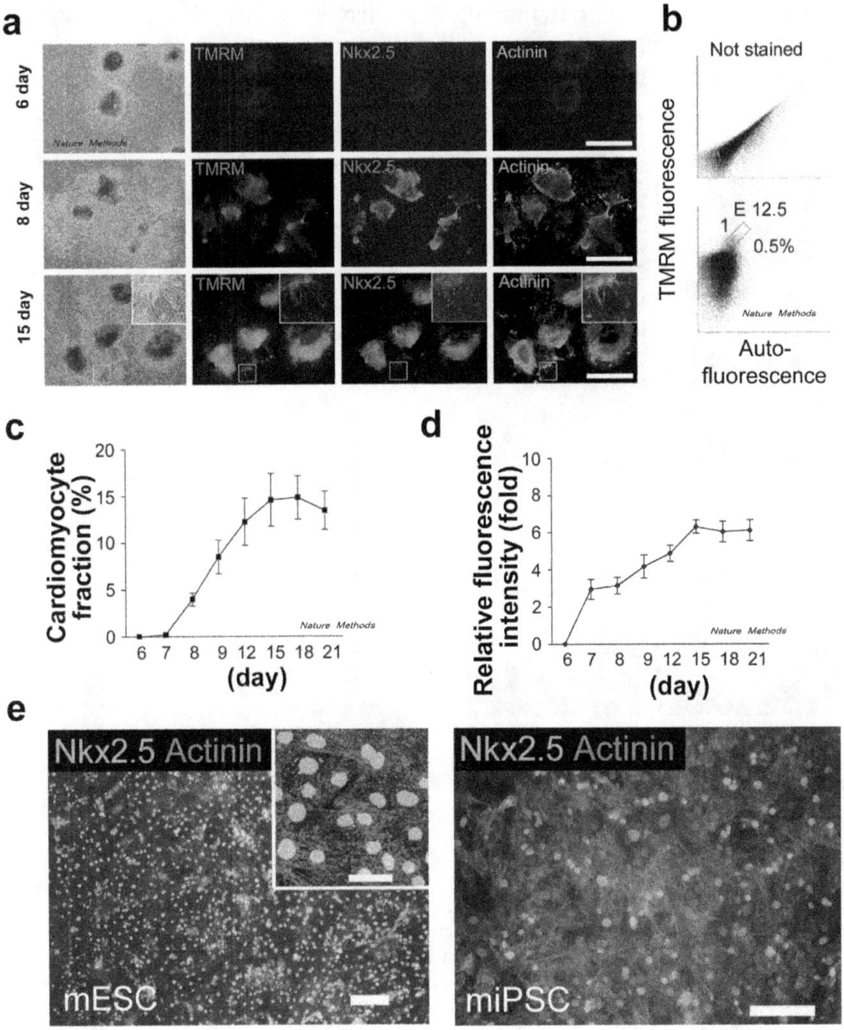

Figure 4. Purification of mouse pluripotent stem (PS) cell-derived cardiomyocytes. (a) Representative photographs of tetramethylrhodamine methyl ester (TMRM) fluorescence, as well as Nkx2.5 and α-actinin immunofluorescence, in differentiated mouse embryonic stem (ES) cells (on Days 6, 8, and 15). (b) Two examples of the fluorescence-activated cell sorting (FACS) analysis of embryoid body (EB)-derived cells 6–25 days after EB formation. (c) Time course of the relative fluorescence intensity of cells in Fraction 1 versus Fraction 2. (d) Time course of the percentage of cells in Fraction 1 (cardiomyocytes) 6–21 days after EB formation. (e) Immunofluorescent staining for Nkx2.5 (green) and α-actinin (red) in the sorted cells (Fraction 1) from EB derived from mESCs and induced PSCs (miPSC). Scale bars, 1 mm (a); 100 µm (e); 30 µm (e, inset) (Figs. reproduced with permission from Hattori et al. 2010 (Hattori et al. 2010)).

Color image of this figure appears in the color plate section at the end of the book.

shown in Fig. 5a, the EB in beating cells (arrows) exhibited extremely high TMRM fluorescence compared with EB in non-beating cells (arrowheads) for both marmoset and human cell cultures. The EB were then completely

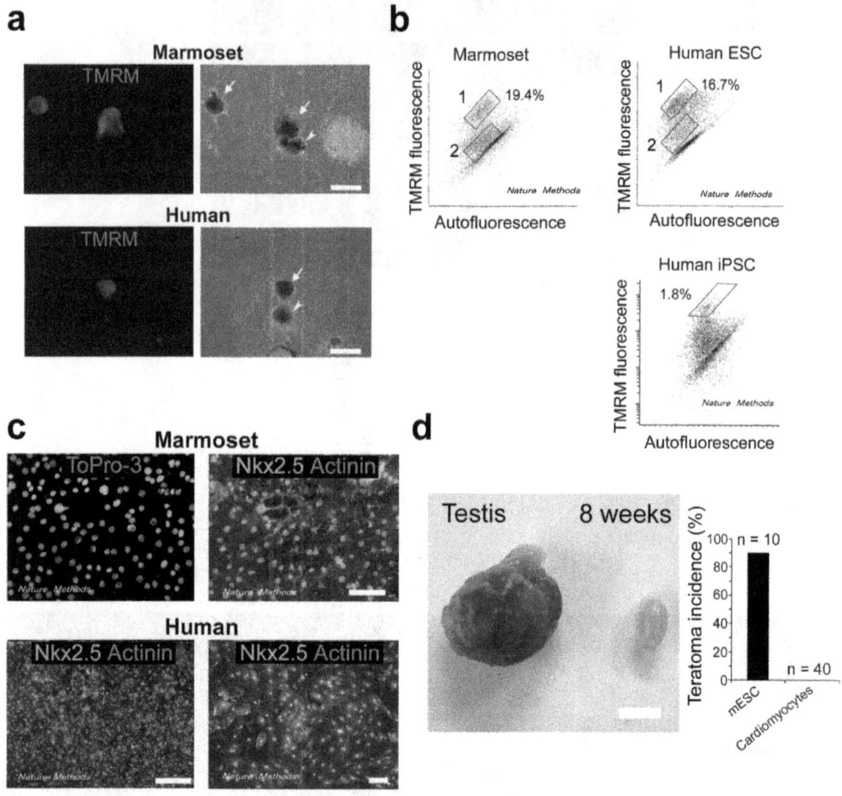

Figure 5. Purification of marmoset embryonic stem (ES) cell- and human pluripotent stem (PS) cell-derived cardiomyocytes. (a) Marmoset and human embryoid bodies (EB) treated with tetramethylrhodamine methyl ester (TMRM). The panels on the left show TMRM fluorescence, whereas the panels on the right show phase contrast photography of the same field. Arrows and arrowheads indicate spontaneously beating and non-beating areas, respectively. (b) Two-dimensional fluorescence-activated cell sorting (FACS) separation of TMRM-stained common marmoset ESC (cmESC) and hESC- and iPSC-derivatives. (c) The upper panels show immunofluorescent staining for α-actinin (red) and Nkx2.5 (green) in cells from cmESC derivatives. ToPro-3 was used to highlight nuclear staining. The lower panels show immunofluorescent staining for α-actinin (red) and Nkx2.5 (green) in cells from hESC and iPSC derivatives. (d) Transplantation of 250 undifferentiated ESC into testes resulted in teratoma formation, whereas the transplantation of 1.9×10^5 purified, sorted ESC-derived cardiomyocytes did not. The mean incidence of teratoma formation after transplantation of undifferentiated ESC and purified and sorted cardiomyocytes is summarized in the graph. Scale bars, 500 µm (a); 100 µm (c); 5 mm (d) (Figs. reproduced with permission from Hattori et al. 2010 (Hattori et al. 2010)).

Color image of this figure appears in the color plate section at the end of the book.

dispersed, stained with TMRM, and subjected to FACS analysis. Because both marmoset and hESC-derived cells autofluoresce, they were subjected to pseudo two-dimensional analysis to eliminate the autofluorescing cells (Fig. 5b). Immediately after sorting, the cells were fixed, stained immunologically for Nkx2.5, and subsequently subjected to a second round of FACS analysis. This clearly showed that over 99% of cells in Fraction 1 were cardiomyocytes. The sorted cells were then cultured and stained immunologically for Nkx2.5 and α-actinin (Fig. 5c). Quantitative analyses of three independent experiments revealed that the percentage of cardiomyocytes in Fraction 1 of common marmoset and human cell cultures was 99.0 ± 1.0% and 99.0 ± 0.9%, respectively. In contrast, the percentage of cardiomyocytes in Fraction 2 was 2.3 ± 0.6% and 2.5 ± 0.2%, respectively (Hattori et al. 2010).

In another study, we transplanted 1.9×10^5 aggregated mESC-derived cardiomyocytes, along with 250 undifferentiated mESCs as a control, into the testes of NOD-SCID mice (Hattori et al. 2010). Two months later, 90% of the control mice had developed teratomas (9/10 mice), but no teratomas were detected in any of the mice transplanted with purified mESC-derived cardiomyocytes (0/40 mice; Fig. 5d).

Mitochondrial method can be widely applicable, because it can obtain highly pure cardiomyocytes immediately and directly from various sources. This method has already been used in three reports (Narazaki et al. 2008; Shimoji et al. 2010; Tohyama et al. 2012). A detailed description of the protocol for the mitochondrial method of purifying cardiomyocytes is given in the following section.

Mitochondrial method of cardiomyocyte purification

Step 1 Dispersion of the embryo, heart and EB into single cells

As is often the case, once the EB are dispersed into single cells, the volume is markedly diminished, indicating that most of the content of the EB is comprised of extracellular matrix. However, as shown in the schematic diagram in Fig. 6, the cardiomyocytes within an EB seem to be in direct contact with one another. To decapsulate the cardiomyocyte aggregates, relatively long treatment with a collagenase is required (we routinely use collagenase type III; Worthington). To avoid grinding damage to the cells, we use microstirrers in 50-mL tubes. Irrespective of the cell source, the use of Ads buffer as a universal basic buffer during long (5–6 h total) periods of digestion is recommended. As the final step in cell dispersion, sometimes short (up to 10 min) treatment with trypsin–EDTA (TE) proves useful, because direct cardiomyocyte–cardiomyocyte connections via adherence and gap junctions depend on Ca^{2+}/Mg^{2+} ions (Fig. 7).

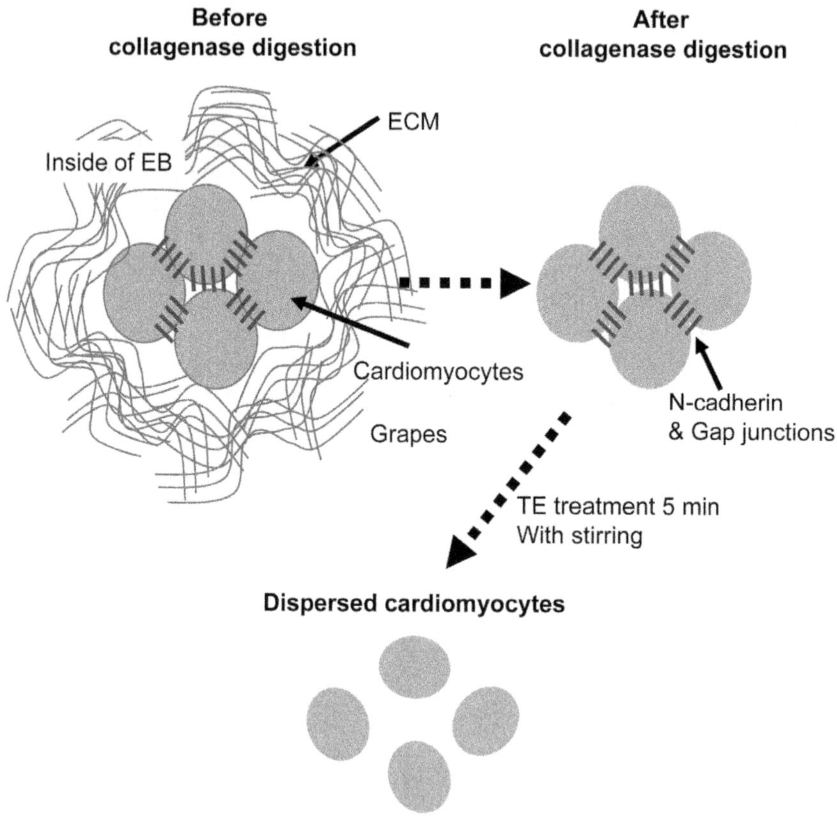

Figure 6. Schematic illustration of the nature of embryoid bodies (EB) and the procedures necessary for complete dispersion of the EB. Most of the volume of an EB is accounted for by the extracellular matrix (ECM). The ECM makes a hard shell that encapsulates the cells. Thus, if the EB are treated directly with trypsin–EDTA (TE), the cells will be damaged. This necessitates the treatment of EB with collagenase for a relatively long period of time (4–6 hrs) in our current setting. The small cell clumps that are resistant to collagenase treatment will form a residue. These cells are connected by adherence and gap junctions. For complete dispersion of these cell clumps, short-term (5–10 min) treatment with TE is effective.

Step 2 FACS machine and gate settings

In our laboratory, we use a FACSaria I (Becton Dickinson) equipped with three lasers. We routinely choose the high-pressure mode with 87 kHz and 70 psi using a 70-μm nodule and a 2.0 decay filter for cardiomyocytes. Drop shapes and stabilities are fundamental for good sorting; the drop delay setting is another important factor to consider. For ideal sorting of cardiomyocytes, every other setting needs to be optimized. For people new to the field, it is recommended that expert training is sought. The gate settings to

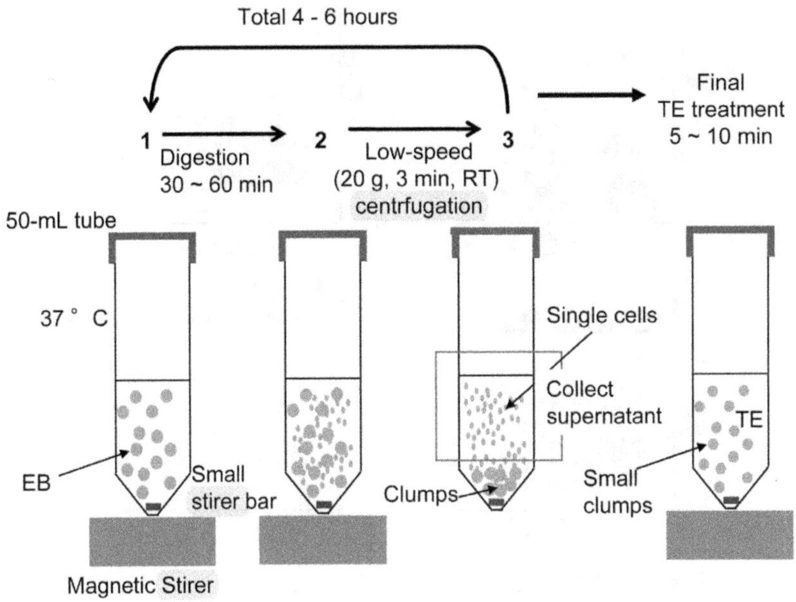

Figure 7. Step 1 of the mitochondrial method for the dispersion of hearts, embryos and embryoid bodies (EB). After dispersal, the EB are stained with 0.1% collagenase type 3 (Worthington), 0.25% Trypsin (DIFCO), and 10–20 nM tetramethylrhodamine methyl ester (TMRM) in Ads buffer [composition (in mM): NaCl 116; HEPES 20; NaH_2PO_4 12.5; glucose 5.6; KCl 5.4; $MgSO_4$ 0.8, pH 7.35] with stirring. Note that we use a 50-mL tube and small magnetic stirrer, with the tubes set up in a cell incubator. Digestion (30–60 min) is repeated several times, with the total digestion time ranging between 4 and 6 hours (this is determined after inspection of the tubes after each single bout of digestion). Finally, cell clumps consisting of a few cells are treated with 0.25% trypsin and 0.02% EDTA (TE) for 5–10 min, with stirring, to obtain single cells. All supernatants and TE-treated cells are combined and applied to fluorescence-activated cell sorting (FACS) sorting.

Color image of this figure appears in the color plate section at the end of the book.

eliminate doublets should be as per the manufacturer's instructions. The main population of eliminated doublets will be separated by red and green fluorescence. The tentative cardiomyocyte population will be gated as shown in the schematic in Fig. 8. Based on the results of the first-pass scan, the forward scatter (FSC), side scatter (SSC), and doublet elimination window parameters need to be reset to enable sorting of the selected cell population. Repeating these procedures will enable the gate for cardiomyocyte selection to be set. Depending on the intended use of the enriched cardiomyocytes, the gates can be set to wide for more general applications and strict for specific purposes (Fig. 8).

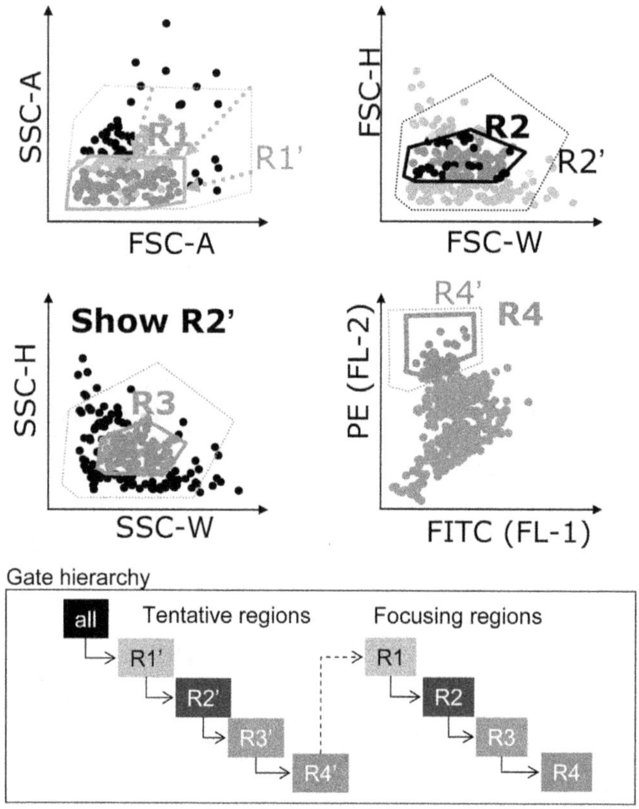

Figure 8. Step 2 of the mitochondrial method: gate settings. First, all data are displayed on a two-dimensional dot plot of forward scatter (FSC) area versus side scatter (SSC) area (FSC-A and SSC-A, respectively). Tentative gates are set for all cells (region 1: R1′), and then the events included in R1′ are dotted in a plot of FSC-height (H) and FSC-width (W). Another tentative gate is set (R2′) to eliminate doublets as per the instructions supplied by the manufacturer. Similarly, events included in R2′ are dotted in a plot of SSC-H and SSC-W and another tentative gate (R3′) is set to eliminate doublets. The events included in R3′ are dotted in a plot of phycoerythrin (PE) area and fluorescein isothiocyanate (FITC) area. The population of cardiomyocytes is displayed in the uppermost section for PE-A and furthest away from the diagonal line. Another tentative gate (R4′) is set for cardiomyocytes. The events included in R4′ are colored and displayed in the above four dot plots. Then, all regions are focused towards the population gated by R4′, meaning to make R1, 2 and 3. Close attention should be paid to the dot plot of PE-A and FITC-A; more distinguishable populations may be evident on this plot. Finally, the gate settings for R4 can be determined. Higher gating according to the PE-A axis results in a higher purity but lower yield; in turn, lower gating results in a lower purity but higher yield.

Color image of this figure appears in the color plate section at the end of the book.

Lactate method

Background

The mitochondrial method we developed (Hattori et al. 2010) and described above is able to purify cardiomyocytes to a high degree, but the yields obtained are insufficient for clinical-scale manufacturing. Thus, it remains necessary to develop another non-genetic method for the purification of cardiomyocytes. Encouraged by successful development of mitochondrial method, we further investigated for a new non-genetic purification method. We focused on characteristic lactate metabolism in cardiomyocytes, i.e., cardiomyocytes can effectively intake and utilize extrinsic lactate for ATP generation (Gertz et al. 1988). We hypothesized that PS cell-derived cardiomyocytes may be able to utilize extrinsic lactate, because we have confirmed PS cell-derived cardiomyocytes contain highly abundant functional mitochondria in which lactate-derivatives are metabolized as an energy fuel.

The heart produces energy from different sources, including glucose, fatty acids, and lactate. Although the adult myocardium uses fatty acids as a major source of energy, the fetal or newborn myocardium uses lactate, which is found in high levels in the fetal circulation (Binah et al. 2007; Fisher et al. 1981; Neely and Morgan 1974; Werner and Sicard 1987). PSC-derived cardiomyocytes exhibit characteristics typical of immature cells (Binah et al. 2007), suggesting that they may use lactate as their major source of energy. On the basis of these backgrounds, we hypothesized that undifferentiated PSCs and other proliferating cells that depend mainly on glycolysis may not be able to survive under glucose depleted- and lactate-abundant conditions, whereas PSC-derived cardiomyocytes would survive because they could use lactate as an alternative energy source. Our findings proved the hypothesis to be true, enabling us to obtain a highly purified cardiomyocyte population of cells from whole embryos and PSC-derived EB (Tohyama et al. 2011).

Purifying cardiomyocytes from the whole embryo using the lactate method

The lactate method simply uses glucose-free Dulbecco's minimal essential medium (DMEM) supplemented with physiological concentration of lactate. In the case of E11.5 rat embryos, glucose-free DMEM is supplemented with 1 mM lactate (Hattori et al. unpublished). After one week culture under these conditions, most cells making up the whole embryo were observed to have died; indeed, after several weeks of such selective culture, only the heart tubes were seen to be beating (Hattori et al. unpublished). The living cells were recovered and adhered to fibronectin-coated dishes; only living cardiomyocytes were found on the dishes (Hattori et al. unpublished; Fig. 9a).

Purifying cardiomyocytes from mESC-derived EB using the lactate method

Differentiated cardiomyocyte-containing EB derived from mESCs (Terami et al. 2007) were subjected to the culture conditions specified in the lactate method from the 9th day to the 16th day; then, the surviving cells, most of which were forming aggregates, were subcultured onto fibronectin-coated plates. After seven days, the cell-clumps which had spread to form 2D cell colonies, were stained immunofluorescently to determine the number of cardiomyocytes. Immunofluorescent staining revealed that all the surviving cells were positive for cardiomyocyte markers, indicating successful purification of cardiomyocytes (Tohyama et al. 2011; Fig. 9b). The purified cardiomyocytes maintained their ability to grow and underwent nuclear division, suggesting that the purification step did not cause significant damage to the cells.

Purifying cardiomyocytes from marmoset ESC and human PSC-derived EB using the lactate method

For the purification of marmoset ESC and human PSC-derived cardiomyocytes, the lactate concentration can be optimized from 1 to 4 mM, depending on the cell lines used and the EBs' conditions (Tohyama et al. 2011). After the initiation of cardiogenic differentiation by EB formation, marmoset EB were transferred to the specific culture conditions described above for periods ranging from one to three months. The cardiomyocytes were subsequently purified using subculture on fibronectin-coated plates for between seven and 14 days; the resulting cardiomyocytes are shown in Fig. 9c, d (Hattori et al. unpublished). The same procedure can be used for human PS-cell derived EB differentiated for between 25 days and three months (Tohyama et al. 2011). FACS analysis was used to determine the purity of the cardiomyocytes isolated and confirmed that the selected cardiomyocytes were 99.5% pure (Tohyama et al. 2011). To evaluate the tumorigenicity of the purified cells, we transplanted 2.0×10^5 hESC-derived cardiomyocytes, as well as 1000 undifferentiated hESCs as a control, into the testes of immunocompromised NOD-SCID mice (Tohyama et al. 2011). Two months later, nine of 10 control mice had developed teratomas (90%) that

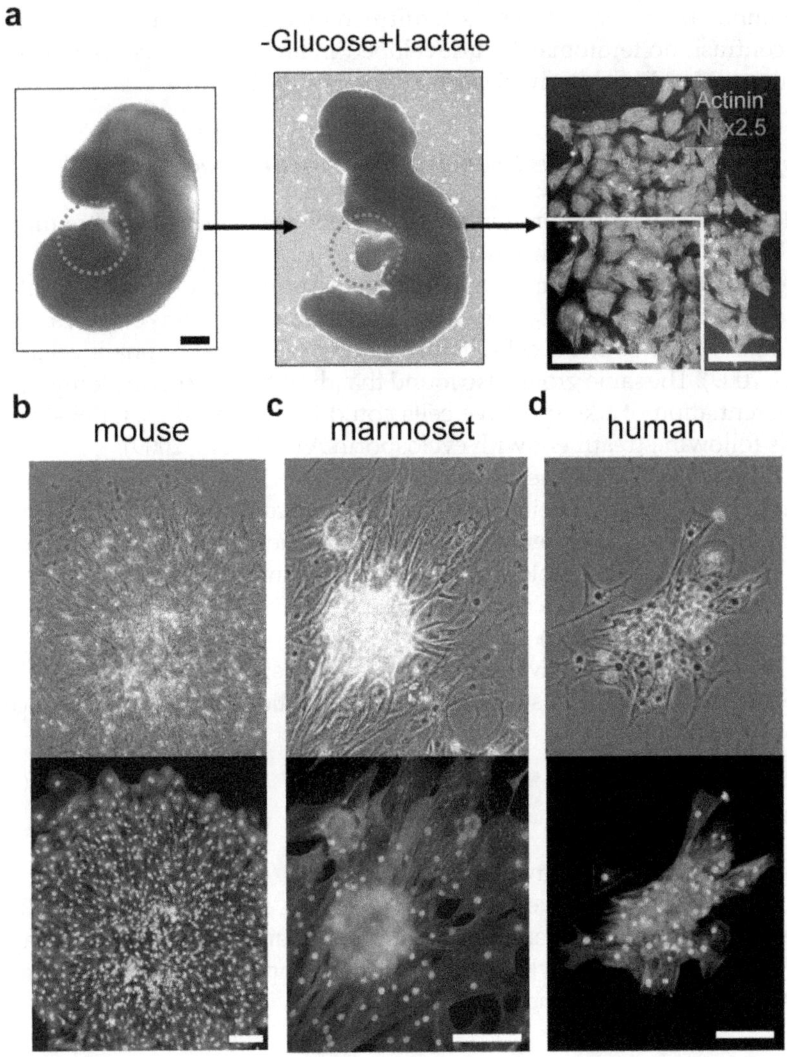

Figure 9. Purification of cardiomyocytes from various sources using the lactate method. Cardiomyocytes can be purified from rat embryonic day (E) 11.5 whole embryos. The E11.5 rat embryo is cultured with glucose-depleted, lactate-containing Dulbecco's minimal essential medium (DMEM) for 1 week. Following culture, all cells comprising the embryo appear to be dead, except for the heart tube, which is still beating. The surviving cells are recovered and seeded onto fibronectin-coated plates. The attached cells are fixed with 4% paraformaldehyde and stained immunohistochemically for actinin (green) and Nkx2.5 (red). (b) Mouse, marmoset and human embryonic stem cell-derived embryoid bodies were cultured in glucose-depleted, lactate-containing DMEM and the surviving cells transferred to fibronectin-coated plates before being stained for actinin (red) and Nkx2.5 (green). Scale bars, 100 μm.

Color image of this figure appears in the color plate section at the end of the book.

contained three germ layers, as confirmed by hematoxylin–eosin staining. In contrast, no teratomas had developed in any of the mice transplanted with purified hESC-derived cardiomyocytes (0/20 mice).

Non-genetic strategies to enrich cardiogenic progenitor cells

Selecting cardiogenic progenitor cells from PSC-derived differentiating cells may eventually enrich cardiomyocytes. It has been reported that flk-1-positive mesodermal cells obtained from mESC can differentiate into several types of mesodermal cells, including cardiomyocytes, smooth muscle cells, endothelial cells, hematopoietic cells, and lymphatic cells (Yan et al. 2009). The same group also found that the efficiency of cardiomyogenic differentiation of flk-1-positive cells could be enhanced by OP-9 stromal cells following treatment with cyclosporin A (Yan et al. 2009).

Hidaka et al. have reported that mESC-derived cells that are double positive for prion protein—a glycosylphosphatidylinositol (GPI)-linked protein known to be preferentially localized in membrane domains enriched in sphingolipids and cholesterol (lipid rafts)—and platelet-derived growth factor (PDGF) receptor α can differentiate into cardiac and smooth muscle cells (Hidaka et al. 2010).

Finally, Blin et al. have reported that stage-specific embryonic antigen (SSEA)-1 can be used to select multipotent cardiovascular progenitor cells (Blin et al 2010).

Discussion

The methods currently available to enrich and/or purify cardiomyocytes from various sources have been described in this chapter. As shown in Table 1, each of the different methods has various advantages and disadvantages. Thus, the decision as to which method is most appropriate needs to be made with the outcome in mind.

References

Anderson, D., Self, T., Mellor, I. R., Goh, G., Hill, S. J., and Denning, C. 2007. Transgenic enrichment of cardiomyocytes from human embryonic stem cells. Mol. Ther. 15, 2027–2036.

Binah, O., Dolnikov, K., Sadan, O., Shilkrut, M., Zeevi-Levin, N., Amit, M., Danon, A., and Itskovitz-Eldor, J. 2007. Functional and developmental properties of human embryonic stem cells-derived cardiomyocytes. J. Electrocardiol. 40, S192–196.

Blin, G., Nury, D., Stefanovic, S., Neri, T., Guillevic, O., Brinon, B., Bellamy, V., Rucker-Martin, C., Barbry, P., Bel, A. et al. 2010. A purified population of multipotent cardiovascular progenitors derived from primate pluripotent stem cells engrafts in postmyocardial infarcted nonhuman primates. J. Clin. Invest. 120, 1125–1139.

Doevendans, P. A., Kubalak, S. W., An, R. H., Becker, D. K., Chien, K. R., and Kass, R. S. 2000. Differentiation of cardiomyocytes in floating embryoid bodies is comparable to fetal cardiomyocytes. J. Mol. Cell. Cardiol. 32, 839–851.

Fijnvandraat, A. C., van Ginneken, A. C., de Boer, P. A., Ruijter, J. M., Christoffels, V. M., Moorman, A. F., and Lekanne Deprez, R. H. 2003. Cardiomyocytes derived from embryonic stem cells resemble cardiomyocytes of the embryonic heart tube. Cardiovasc. Res. 58, 399–409.

Fisher, D. J., Heymann, M. A., and Rudolph, A. M. 1981. Myocardial consumption of oxygen and carbohydrates in newborn sheep. Pediatr. Res. 15, 843–846.

Gassanov, N., Er, F., Zagidullin, N., and Hoppe, U. C. 2004. Endothelin induces differentiation of ANP-EGFP expressing embryonic stem cells towards a pacemaker phenotype. Faseb. J. 18, 1710–1712.

Gertz, E. W., Wisneski, J. A., Stanley, W. C., and Neese, R. A. 1988. Myocardial substrate utilization during exercise in humans. Dual carbon-labeled carbohydrate isotope experiments. J. Clin. Invest. 82, 2017–2025.

Hattori, F., Chen, H., Yamashita, H., Tohyama, S., Satoh, Y. S., Yuasa, S., Li, W., Yamakawa, H., Tanaka, T., Onitsuka, T. et al. 2010. Nongenetic method for purifying stem cell-derived cardiomyocytes. Nat. Methods 7, 61–66.

Hattori, F. and Fukuda, K. 2010. Strategies for ensuring that regenerative cardiomyocytes function properly and in cooperation with the host myocardium. Exp. Mol. Med. 42, 155–165.

Hidaka, K., Lee, J. K., Kim, H. S., Ihm, C. H., Iio, A., Ogawa, M., Nishikawa, S., Kodama, I., and Morisaki, T. 2003. Chamber-specific differentiation of Nkx2.5-positive cardiac precursor cells from murine embryonic stem cells. Faseb. J. 17, 740–742.

Hidaka, K., Shirai, M., Lee, J. K., Wakayama, T., Kodama, I., Schneider, M. D., and Morisaki, T. 2010. The cellular prion protein identifies bipotential cardiomyogenic progenitors. Circ. Res. 106, 111–119.

Jakobs, S. 2006. High resolution imaging of live mitochondria. Biochem. Biophys. Acta. 1763, 561–575.

Klug, M. G., Soonpaa, M. H., Koh, G. Y., and Field, L. J. 1996. Genetically selected cardiomyocytes from differentiating embryonic stem cells form stable intracardiac grafts. J. Clin. Invest. 98, 216–224.

Knaapen, M. W., Vrolijk, B. C., and Wenink, A. C. 1997. Ultrastructural changes of the myocardium in the embryonic rat heart. A. Nat. Rec. 248, 233–241.

Kolossov, E., Bostani, T., Roell, W., Breitbach, M., Pillekamp, F., Nygren, J. M., Sasse, P., Rubenchik, O., Fries, J. W., Wenzel, D. et al. 2006. Engraftment of engineered ES cell-derived cardiomyocytes but not BM cells restores contractile function to the infarcted myocardium. J. Exp. Med. 203, 2315–2327.

Laflamme, M. A., Chen, K. Y., Naumova, A. V., Muskheli, V., Fugate, J. A., Dupras, S. K., Reinecke, H., Xu, C., Hassanipour, M., Police, S. et al. 2007. Cardiomyocytes derived from human embryonic stem cells in pro-survival factors enhance function of infarcted rat hearts. Nat. Biotechnol. 25, 1015–1024.

Monici, M. 2005. Cell and tissue autofluorescence research and diagnostic applications. Biotechnol. Annu. Rev. 11, 227–256.

Muller, M., Fleischmann, B. K., Selbert, S., Ji, G. J., Endl, E., Middeler, G., Muller, O. J., Schlenke, P., Frese, S., Wobus, A. M. et al. 2000. Selection of ventricular-like cardiomyocytes from ES cells *in vitro*. Faseb. J. 14, 2540–2548.

Mummery, C., Ward-van Oostwaard, D., Doevendans, P., Spijker, R., van den Brink, S., Hassink, R., van der Heyden, M., Opthof, T., Pera, M., de la Riviere, A. B. et al. 2003. Differentiation of human embryonic stem cells to cardiomyocytes: role of coculture with visceral endoderm-like cells. Circulation 107, 2733–2740.

Narazaki, G., Uosaki, H., Teranishi, M., Okita, K., Kim, B., Matsuoka, S., Yamanaka, S., and Yamashita, J. K. 2008. Directed and systematic differentiation of cardiovascular cells from mouse induced pluripotent stem cells. Circulation 118, 498–506.

Neely, J. R. and Morgan, H. E. 1974. Relationship between carbohydrate and lipid metabolism and the energy balance of heart muscle. Annu. Rev. Physiol. 36, 413–459.

Nemir, M., Croquelois, A., Pedrazzini, T., and Radtke, F. 2006. Induction of cardiogenesis in embryonic stem cells via downregulation of Notch1 signaling. Circ. Res. 98, 1471–1478.

Pretlow, T. G., Glick, M. R., and Reddy, W. J. 1972. Separation of beating cardiac myocytes from suspensions of heart cells. Am. J. Pathol. 67, 215–226.

Shimoji, K., Yuasa, S., Onizuka, T., Hattori, F., Tanaka, T., Hara, M., Ohno, Y., Chen, H., Egasgira, T., Seki, T. et al. 2010. G-CSF promotes the proliferation of developing cardiomyocytes *in vivo* and in derivation from ESCs and iPSCs. Cell Stem Cell 6, 227–237.

Terami, H., Hidaka, K., Shirai, M., Narumiya, H., Kuroyanagi, T., Arai, Y., Aburatani, H., and Morisaki, T. 2007. Efficient capture of cardiogenesis-associated genes expressed in ES cells. Biochem. Biophys. Res. Commun. 355, 47–53.

Tohyama, S., Hattori, F., Sano, M., Hishiki, T., Nagahata, Y., Matsuura, T., Hashimoto, H., Suzuki, T., Yamashita, H., Satoh, Y. et al. 2013. Distinct metabolic flow enables large-scale purification of mouse and human pluripotent stem cell-derived cardiomyocytes. Cell Stem Cell 12: 127–137.

Werner, J. C. and Sicard, R. E. 1987. Lactate metabolism of isolated, perfused fetal, and newborn pig hearts. Pediatr. Res. 22, 552–556.

Xu, C., Police, S., Hassanipour, M., and Gold, J. D. 2006. Cardiac bodies: a novel culture method for enrichment of cardiomyocytes derived from human embryonic stem cells. Stem Cells Dev. 15, 631–639.

Yan, P., Nagasawa, A., Uosaki, H., Sugimoto, A., Yamamizu, K., Teranishi, M., Matsuda, H., Matsuoka, S., Ikeda, T., Komeda, M. et al. 2009. Cyclosporin-A potently induces highly cardiogenic progenitors from embryonic stem cells. Biochem. Biophys. Res. Commun. 379, 115–120.

Yuasa, S., Itabashi, Y., Koshimizu, U., Tanaka, T., Sugimura, K., Kinoshita, M., Hattori, F., Fukami, S., Shimazaki, T., Ogawa, S. et al. 2005. Transient inhibition of BMP signaling by Noggin induces cardiomyocyte differentiation of mouse embryonic stem cells. Nat. Biotechnol. 23, 607–611.

CHAPTER 9

Ion Channel Expression and Function of iPSC-derived Cardiomyocytes

Tetsushi Furukawa

Introduction

Induced pluripotent stem cells (iPSCs) can be differentiated into cardiomyocytes when they are cultured in floating condition. The human iPSC-derived cardiomyocytes provide new approaches to study the cellular properties of human cardiomyocytes, to screen drug efficacy and toxicity in human cardiomyocytes, to study pathogenesis and potential therapeutic strategy in human iPSC-derived cardiomyocytes from individuals with cardiac diseases, and for cell transplantation therapy. To efficiently achieve these future goals, it is indispensable to characterize the gene-expression profile and the functional properties of human iPSC-derived cardiomyocytes in details, which provide crucial information on whether human iPSC-derived cardiomyocytes are similar or different from adult human cardiomyocytes *in situ*. Cardiomyocytes *in situ* can be classified into several types, such as atrial muscle type, ventricular muscle type, nodal cell type, and His-Purkinje cell type. It is also essential to know which types of cardiomyocytes iPSC-derived cardiomyocytes belong to or whether iPSC-derived cardiomyocytes are a mixture of these various types of cardiomyocytes. Atrial muscle cells, ventricular muscle cells, nodal cells, and His-Purkinje cells each have a unique configuration of action potential, which is attributed to the expression of specific combination of ion channels.

Department of Bio-informational Pharmacology, Medical Research Institute, Tokyo Medical and Dental University, 1-5-45 Yushima, Bunkyo-ku, Tokyo 113-8510, Japan.
Email: t_furukawa.bip@mri.tmd.ac.jp

Thus, ion channels can be a specific marker to determine which types of cardiomyocytes iPSC-derived cardiomyocytes belong to. Therefore, it is worthwhile to t examine ion channel expression and function as the first step towards characterization of iPSC-derived cardiomyocytes.

Electrophysiological Characteristics in Human Cardiomyocytes

Before discussing ion channel expression and function of iPSC-derived cardiomyocytes, I will first briefly discuss electrophysiological features of human cardiomyocytes. This will facilitate understanding of the electrophysiological characteristics of iPSC-derived cardiomyocytes, and make it possible to compare characteristics between adult cardiomyocytes *in situ* and iPSC-derived cardiomyocytes. There are various types of cardiomyocytes in human adult hearts *in situ*, including conduction system cardiomyocytes and working cardiomyocytes. Conduction system cardiomyocytes are specilized in the generation of heartbeats, and the propagation of electrical signals throughout the heart. Conduction system cardiomyocytes are further classified into at least two types, nodal cell type including sino-atrial node cells and atrio-ventricular node cells, and the His-Purkinje system type, including the His bundle cells and the Purkinje fiber cells. Working cardiomyocytes are specilized in pumping out blood into the next chamber or artery, and can generate strong contraction force. Working cardiomyocytes are also further classified into at least two types—atrial myocytes and ventricular myocytes.

Figures 1–4 illustrate action potential configurations in the upper panel and ion currents flowing in each phase of action potential, in the lower panel—in a sino-atrial node cell, a Purkinje fiber cell, an atrial muscle cell, and ventricular muscle cells. An action potential of sino-atrial node cell has shallow maximum membrane potential during diastolic period, diastolic depolarization seen as an up-stroke of membrane potential between one action potential and the next (Phase 4), less sharp rising phase of action potential (Phase 0), and short duration of action potential. Low density of expression of inward rectifying potassium current (I_{K1}) results in shallow maximum diastolic membrane potential. Diastolic depolarization, which is assumed to be the main ionic mechanism for spontaneous firing of sino-atrial node cells, is caused by hyperpolarization-activated cation current, or funny current (I_f). Firing of action potential is induced by L-type voltage-dependent calcium current ($I_{Ca,L}$), resulting in less sharp Phase 0 depolarization compared to atrial or ventricular myocytes, in which voltage-dependent sodium current (I_{Na}) is responsible for the firing of action potential. Sino-atrial node cells barely participate in pumping out blood, and thus the contraction of sinus node cells is much less vigorous compared to that of atrial or ventricular myocytes. Therefore, Ca^{2+} entry

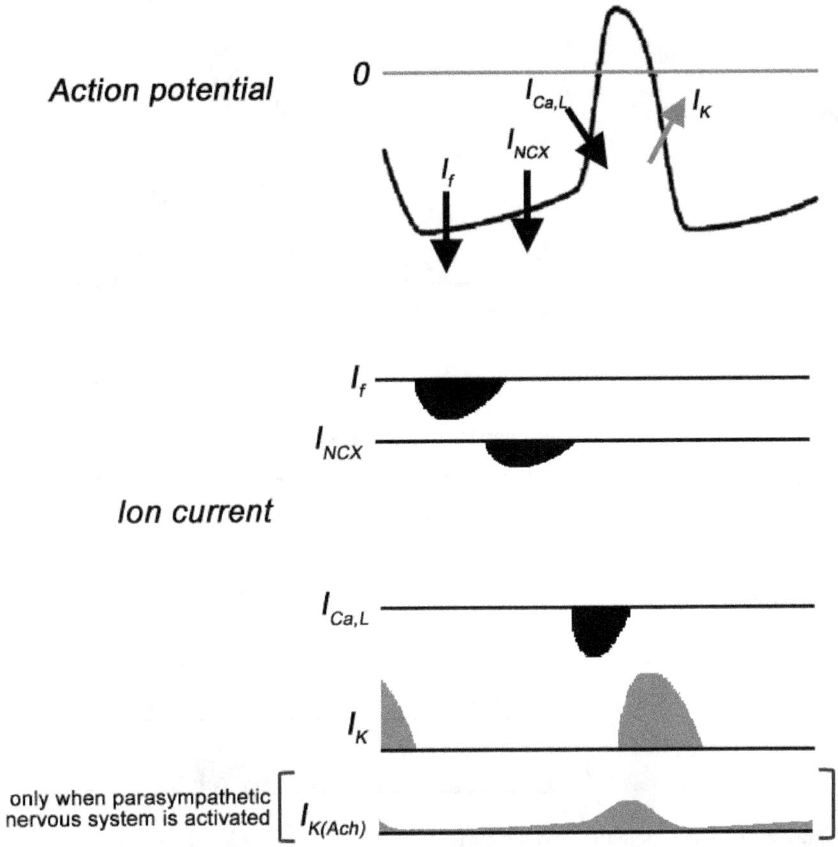

Figure 1. An action potential and ion currents in sino-atrial node cells. Upper panel: an action potential. Lower panel: membrane currents that flow at each phase of action potential. Inward currents are shown downward in black color, and outward currents are shown upward in gray color.

into cytoplasm is not so important and the density of $I_{Ca,L}$ is low, resulting in short duration of action potential plateau phase. Nodal cells, namely sino-atrial nodal cells and atrio-ventricular nodal cells, express a unique class of ion current, called acetylcholine-activated potassium current ($I_{K(Ach)}$). $I_{K(Ach)}$ is activated by acetylcholine that is the neurotransmitter of the secondary neuron of parasympathetic nerve. When parasympathetic nerve activity is increased, $I_{K(Ach)}$ is activated and the resting membrane potential becomes deep, rendering spontaneous firing of action potential in sino-atrial node slows and conduction in atrio-ventricular node also slows.

In a Purkinje fiber cell (Fig. 2), an action potential exhibits diastolic depolarization similar to the sino-atrial node cells, but the maximum diastolic membrane potential is quite deep, the rising phase of action

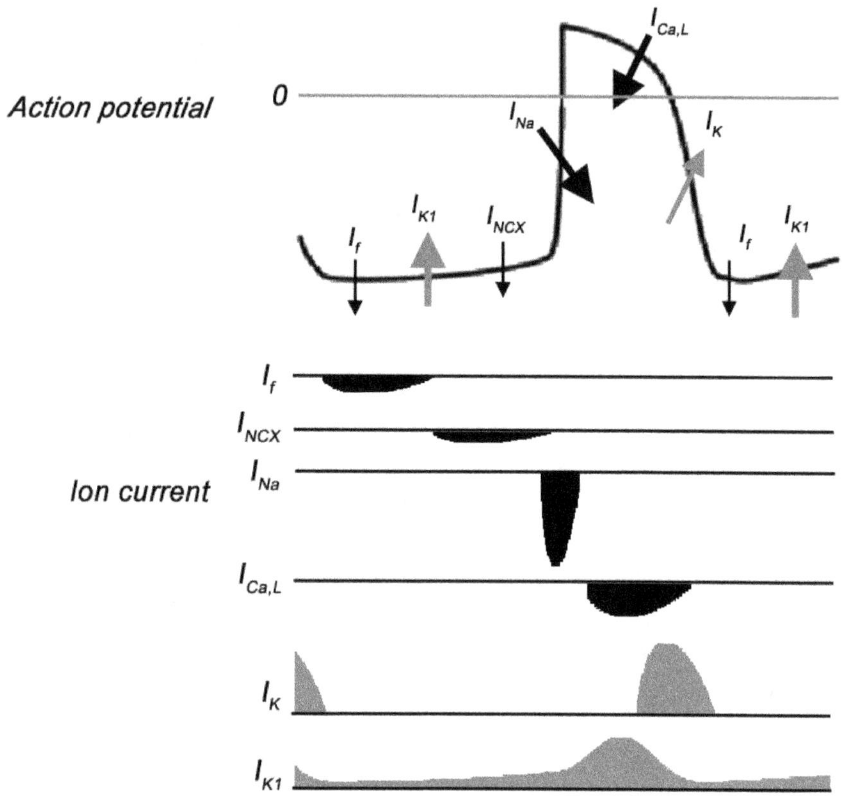

Figure 2. An action potential and ion currents in Purkinje fiber cells. Upper panel: an action potential. Lower panel: membrane currents that flow at each phase of action potential. Inward currents are shown downward in black color, and outward currents are shown upward in gray color.

potential (Phase 0) is sharp, and action potential duration is very long. The major differences in ionic currents between sino-atrial node cells and Purkinje fiber cells are the presence of decent density of I_{K1} in Purkinje fiber cells, which contributes to the deep maximum diastolic membrane potential, and the presence of I_{Na} in the rising phase of action potential (Phase 0), which reflects to the sharp Phase 0 depolarization.

In atrial myocytes (Fig. 3) and ventricular myocytes (Fig. 4), resting membrane potential is deep, which is attributed to high density of I_{K1}. Phase 0 depolarization is very sharp, because the firing of action potential depends on I_{Na}, and I_{Na} provide much larger current than $I_{Ca,L}$. Action potential duration is shorter in atrial myocytes than in ventricular myocytes, which is partly due to the expression of different sets of voltage-dependent potassium currents. Voltage-dependent potassium currents are classified

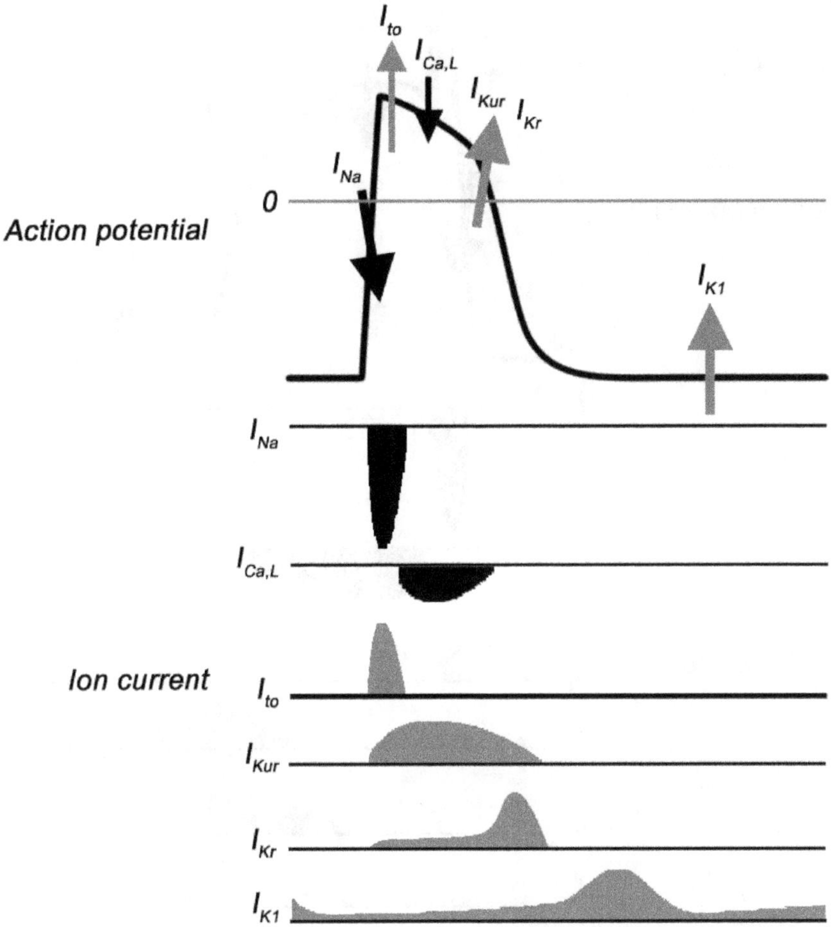

Figure 3. An action potential and ion currents in atrial myocytes. Upper panel: an action potential. Lower panel: membrane currents that flow at each phase of action potential. Inward currents are shown downward in black color, and outward currents are shown upward in gray color.

into two types; transient outward potassium current (I_{to}) that flow in the early repolarization phase (Phase 1) and delayed rectifier potassium current (I_K) that flow in the late repolarization phase (Phase 3). I_K can be further classified into three types depending on the speed of its activation: I_{Kur} ($I_{K\ ultra\text{-}rapid}$) that exhibits the fastest activation time course, I_{Kr} (I_K rapid) that exhibits fast activation time course, and I_{Ks} ($I_{K\ slow}$) that shows the slowest activation time course. Atrial myocytes have high density of I_{to} and I_{Kur}, both of which have relatively fast activation time courses, resulting in the short duration of action potential. In contrast, ventricular myocytes that

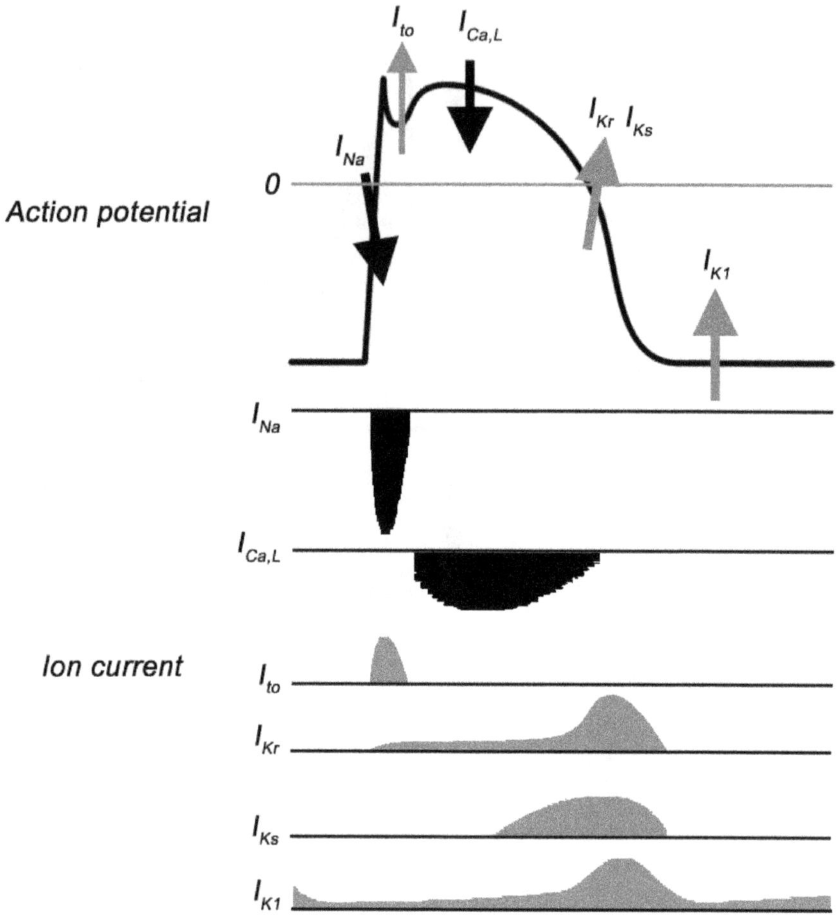

Figure 4. An action potential and ion currents in ventricular myocytes. Upper panel: an action potential. Lower panel: membrane currents that flow at each phase of action potential. Inward currents are shown downward in black color, and outward currents are shown upward in gray color.

have long duration of action potential have high density of I_{Kr} and I_{Ks}, both of which have relatively slow activation time course. The atrium pumps out blood to the ventricle, while the ventricle pumps out blood to the entire body from head to toe. Thus, much stronger contraction is needed in the ventricle than in the atrium. Therefore, the density of $I_{Ca,L}$ is higher in the ventricular myocytes than in the atrial myocytes, which also contributes, in part, to the longer action potential duration in the ventricular myocytes than in the atrial myocytes. Ventricular myocytes express a unique class of potassium current, called ATP-sensitive potassium current ($I_{K(ATP)}$).

$I_{K(ATP)}$ channels are inhibited by intracellular ATP in the normal condition. When intracellular concentration of ATP is decreased due to hypoxia of the ventricular myocytes, blockade of $I_{K(ATP)}$ channels is removed and the duration of action potential is remarkably shortened. Shortening of action potential duration in hypoxic condition causes decreased entry of Ca^{2+} from extracellular space into the cytoplasm, and elimination of cardiomyocyte contraction. This prevents consumption of ATP needed for maintenance of the basic function of cardiomyocytes, such as maintenence of proper intracellular ion concentrations and protects cells from hypoxic damage and cell death.

Expression of ion channels in human hearts

In the previous section, I have described that at least 11 ion currents contribute to the formation of action potentials in sino-atrial node cells, Purkinje fiber cells, atrial myocytes, and ventricular myocytes. In addition to $I_{Ca,L}$, voltage-dependent Ca^{2+} current (I_{Ca}) have T-type Ca^{2+} current ($I_{Ca,T}$). $I_{Ca,T}$ barely expresses in adult cardiomyocytes. However, in embryonic and neonatal cardiomyocytes, $I_{Ca,T}$ is the major I_{Ca} rather than $I_{Ca,L}$. Many ion channels consist of the α-subunit and the regulatory β-subunit. The α-subunit contains most of the key ion channel functional elements, including ion conducting pore, voltage sensor, and activation and inactivation gates. The β-subunit acts for trafficking of ion channel protein from its synthesizing place, endoplasmic reticulum, to the target place, plasma membrane. It also affects the voltage-dependency, and provides the target for protein modification, such as phosphorylation and *s*-nitrosylation, and drug modification. Table 1 summarizes the genes encoding the α-subunit and the β-subunit of channels generating each of these 11 ion currents. Please note that single ion channel gene sometimes has multiple nomenclatures. The nomenclature that I believe to be most frequently used is shown on top, and alternative nomenclatures are cited in parenthesis.

Functional Characteristics of iPSC-derived Cardiomyocytes

iPSC-derived cardiomyocytes exhibit diverse action potential configurations

Differentiation from iPSCs into cardiomyocytes is achieved by culture in the floating condition. When iPSC are cultured in the floating condition in ultra-low attachment dishes, they start to form spherical clusters of cells, called embryoid bodies. Two to three weeks after induction of cardiac differentiation in the floating condition, some embryoid bodies start contracting. Membrane potential was recorded either with the micro-electrode impalement technique

Table 1. Ion currents, ion channels, and encoding genes

Current	Ion channel	α-subunit	β-subunit
I_{Na}	voltage-dependent Na^+ channel	SCN5A(Nav1.5)	SCN1B(Navβ1)
			SCN3B(Navβ3) at neonatal stage
$I_{Ca,L}$	L-type voltage-dependent Ca^{2+} channel	Cav1.2(CACNA1C, α1C)	CACNB3(Cavβ2)
		minor contribution of Cav1.3(CACNA1D, α1D)	
$I_{Ca,T}$	T-type voltage-dependent Ca^{2+} channel	Cav3.1(CACNA1G, α1G)	
		Cav3.2(CACNA1H, α1H)	
I_f	hyperpolarization-activated cation channel	HCN4	
		minor contribution of HCN2	
I_{to}	transient outward channel	Kv4.2(KCND2)	KChIP2
		Kv4.3(KCND3)	
		Kv1.4(KCNA4)	
I_{Kur}	ultra-rapid activating delayed rectifier K^+ channel	Kv1.5(KCNH5)	KCNE3(MiRP3)
I_{Kr}	rapid activating delayed rectifier K^+ channel	KCNH2(hERG, Kv11.1)	KCNE2(MiRP1)
I_{Ks}	slow activating delayed rectifier K^+ channel	KCNQ1(KCNA8, KVLQT1, Kv7.1)	KCNE1(MinK)
I_{K1}	inward rectifier K^+ channel	KCNJ2(Kir2.1, IRK1)	
		minor contribution of KCNJ12(Kir2.2, IRK2)	
$I_{K(Ach)}$	Ach-activated K^+ channel	heteromer of	
		GIRK1(Kir3.1, KCNJ3)	
		and GIRK4(Kir3.4, KCNJ5)	
$I_{K(ATP)}$	ATP-sensitive K^+ channel	Kir6.2(KCNJ11)	SUR2A(ABCC9)

*Inside the parenthesis indicates the alternative nomenclature.

from beating embryoid bodies or with the patch-clamp technique in single iPSC-derived cardiomyocytes isolated from beating embryoid bodies. I will first discuss the data obtained with the microelectrode impalement technique. In the previous section, human adult cardiomyocytes are classified into four types, nodal type, His-Purkinje system type, atrial type, and ventricular type. In the developmental stage, ventricular myocytes often have the capacity to generate spontaneous beating with some expression of I_f and the presence of diastolic depolarization. Thus, in iPSC-derived cardiomyocytes that are assumed to be cardiomyocytes in the developmental stage, it is very difficult to differentiate between the ventricular myocytes and the His-Purkinje system cardiomyocytes. Thus, iPSC-derived cardiomyocytes are classified into three types, nodal type, atrial type, and ventricular type. From cells with spontaneous firing of action potentials, diverse configurations of action potentials were recorded (Kuzmenkin et al. 2009; Zhang et al. 2009). Action potentials with shallow maximum diastolic membrane potential, evident depolarization during diastolic period, and blunt rising phase (Phase 0) of action potential firing were classified into nodal type (Fig. 5 left). Among action potentials with deep maximum diastolic membrane potential, and sharp rising phase (Phase 0) of action potential firing, those with relatively short action potential duration were classified into atrial type (Fig. 5 middle), and those with relatively long action potential were classified into ventricular type (Fig. 5 right). Among action potentials recorded from human iPSC-derived cardiomyocytes with microelectrode impalement within 30 days after induction of cardiac differentiation, about a half of them fell into atrial type, and about a quarter of them fell into either of ventricular type or nodal type.

We also recorded membrane potentials from isolated human iPSC-derived cardiomyocytes with the patch-clamp technique. About 30 days after induction of cardiac differentiation, beating embryoid bodies were digested with collagenase and trypsin, and membrane potential was recorded with either the whole-cell patch configuration or the perforated patch-configuration of the patch-clamp technique. Some cells showed spontaneous firing and some did not. Even in those without spontaneous firing, electrical stimulation could produce action potentials with

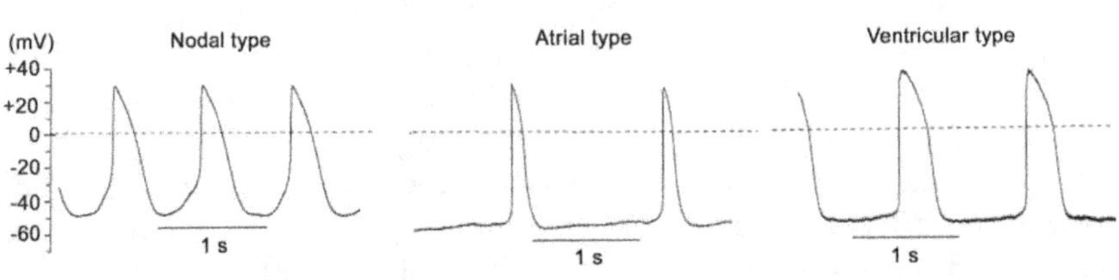

Figure 5. Three types of action potential configurations in human iPSC-derived cardiomyocytes.

configurations characteristic to cardiomyocytes. However, here we analyzed those with spontaneous firing in order to compare the data obtained in the microelectrode impalement experiments. About 60% of cells with spontaneous action potential firing showed action potential configuration classified into atrial type, about 25% showed ventricular type, and about 15% showed nodal type.

These data indicate that human iPSC-derived cardiomyocytes are a mixture of cells with atrial type, ventricular type, and nodal type action potential configurations. Fraction of cells with each type of action potential configurations was similar, although not identical between data obtained in microelectrode impalement experiments and those obtained in patch-clamp experiments. In both experiments, human iPSC-derived cardiomyocytes with atrial type action potential configuration were the most dominant ones; about one half or more than a half of cells fell into atrial type. In adult hearts, ventricular myocytes are dominant type of cardiomyocytes. At least two potential explanations could be possible to explain why cardiomyocytes with atrial type action potential were dominant in iPSC-derived cardiomyocytes. First, the timing of differentiation of human iPSC into cardiomyocytes is later in ventricular cardiomyocytes than in atrial cardiomyocytes. In development of the heart, the ventricle is originally the thin layer, referred to as the compact layer. In the later stage of heart development, cardiomyocytes of the compact layer proliferate, generating the thick ventricular wall, referred to as the trabecular layer. Thus, it is not surprising if in the early period after cardiac differentiation, the fraction of ventricular myocytes is not dominant as seen in adult hearts. An alternative possibility is that atrial myocytes and ventricular myocytes originate from the same lineage cells. They first differentiate into atrial myocytes, and thereafter transform into ventricular myocytes. Which of above two possibilities is the case in iPSC is the fundamental biological question, and is important to address, probably by analyzing temporal change of the fraction of three types of action potential configurations.

Ion channel genes expression

Expression of ion channel genes in human iPSC-derived cardiomyocytes was mostly assessed with the reverse transcription-polymerase chain reaction (PCR) technique. In table 1, at least 26 major ion channel genes are present in human adult cardiomyocytes. Unfortunately, comprehensive study to analyze expression of all those 26 genes in iPSC-derived cardiomyocytes has not been reported so far. Honda et al. examined expression of ion channel genes at 38 days after initiation of cardiac differentiation (Honda et al. 2011). They showed expression of mRNA for SCN5A, $Ca_V1.2$(CACNA1C or α1C), $Ca_V1.3$(CACNA1D or α1D), Kv4.3(KCND3), KCNH2(hERG or

Kv11.1), KCNQ1(KCNA8, KVLQT1 or Kv7.1), KCNJ2(Kir2.1 or IRK1), KCNJ12(Kir2.2 or IRK2), HCN4, and HCN2. Although they did not use quantitative analytic approach, expression level of Kv4.3(KCND3), and KCNJ2(Kir2.1 or IRK1) appeared to be low compared to the expression level in human adult left ventricle. In human left ventricles, Ca_V1.3(CACNA1D or α1D) and KCNJ12(Kir2.2) were almost undetectable with the reverse transcription-PCR technique, while they were detectable in human iPSC-derived cardiomyocytes.

Otsuji et al. examined expression of ion channel genes at one month after initiation of cardiac differentiation with the real-time reverse transcription-PCR technique (Otsuji et al. 2010). They showed expression of SCN1B(Na_Vβ1), Ca_V1.2(CACNA1C, α1C), CACNB3(Ca_Vβ2), KCNQ1 (KCNA8, KVLQT1, or Kv7.1) and KCNH2(hERG or Kv11.1) at the level comparable to the expression level in the adult hearts. The expression level of Ca_V3.2(CACNA1H,α1H) was much higher than that in human adult hearts. Although they found expression of SCN5A(Na_V1.5), KCNJ2(Kir2.1 or IRK1) and Kir6.2(KCNJ11) in human iPSC-derived cardiomyocytes, the expression level was much lower than the level in adult hearts. Expression of Kv1.4 (KCNA4) was almost undetectable.

Otsuji et al. also examined the time course of expression of ion channel genes by comparing expression level between one month and eight months after initiation of cardiac differentiation (Otsuji et al. 2010). Expression of KCNQ1(KCNA8, KVLQT1, or Kv7.1), KCNH2(hERG or Kv11.1), and Kv1.4 (KCNA4) were greater at eight months than at one month, while expression of KCNJ2(Kir2.1 or IRK1) was still very low at eight months after initiation of cardiac differentiation.

Thus, human iPSC-derived cardiomyocytes express most of the ion channel genes expressed in the adult hearts. Some potassium channel genes increase their expression over the course of eight months of differentiation, suggesting that human iPSC-derived cardiomyocytes maturate progressively during the course of differentiation in the culture medium. The most striking difference between human iPSC-derived cardiomyocytes and human adult hearts is the low level of expression of KCNJ2 (Kir2.1 or IRK1) in human iPSC-derived cardiomyocytes. Thus, it may be the key to induce expression of KCNJ2 (Kir2.1 or IRK1) to obtain mature, adult-like ventricular myocytes.

Ion currents

Recording of ionic currents from iPSC-derived cardiomyocytes was carried out using the patch-clamp experiments. In human iPSC-derived cardiomyocytes, Honda et al. reported the expression of I_{Na}, I_{Ca}, I_{Kr}, and I_{Ks} at 38 days after initiation of cardiac differentiation (Honda et al. 2011). Otsuji et al. reported the expression of I_f, I_{Na}, I_{Ca}, and I_{Ks} at one month after initiation of cardiac differentiation (Otsuji et al. 2010). Otsuji et al. could record very small I_{K1} as a barium-sensitive current. They could not record E4031-sensitive I_{Kr} at one month after differentiation. They showed that the amplitude of I_f, I_{K1}, I_{Na}, and I_{Ca} was increased at eight months after initiation of differentiation than at one month. Although I_{Kr} could not be recorded at one month after initiation of cardiac differentiation, it could be recorded at eight months with a current level comparable to that of I_{Ks}. Up to now, recording of I_{to} or separate analysis of T-type Ca^{2+} current ($I_{Ca,T}$) from $I_{Ca,L}$ has not been reported from human iPSC-derived cardiomyocytes.

Data on recording functional ion channel currents are mostly compatible with the data for ion channel gene expression data with some exception. At one month after initiation of cardiac development, KCNH2(hERG or Kv11.1) mRNA was expressed to a level comparable to the adult human heart, whereas KCNH2 current, I_{Kr}, was not detectable. No plausible explanation for the discrepancy between gene expression and current expression for KCNH2 (hERG or Kv11.1) have been made so far. hERG channel expression to the plasma membrane is complicated, and trafficking defect of hERG channel to the plasma membrane frequently occurs by KCNH2 mutations, drugs, or some intervention. Full glycosylation of the hERG channels is mandatory for its plasma membrane expression (Gong et al. 2002). Thus, it is interesting to examine if trafficking system of the hERG channel, such as its glycosylation, is already equipped or not in human iPSC-derived cardiomyocytes. LQT2 is caused by mutation in KCNH2(hERG or Kv11.1) genes, and many of KCNH2(hERG or Kv11.1) mutations cause trafficking defect of the hERG channel protein from the endoplasmic reticulum to the plasma membrane (Ficker et al. 2000). Thus, it would be important to determine if trafficking of the hERG channels in human iPSC-derived cardiomyocytes is similar to that in human adult cardiomyocytes in order to make a judgment if human iPSC-derived cardiomyocytes are suitable for analysis of disease model of human iPSC-derived cardiomyocytes for long QT syndrome type 2. I_{to} increases its density during the course of development (Wang and Duff 1997), and thus it is possible that expression of I_{to} is under the detectable level in human iPSC-derived cardiomyocytes during the early stage of cardiac differentiation. It is consistent with the low expression of Kv4.3(KCND3) and Kv1.4(KCNA4) mRNAs at one month after initiation of cardiac differentiation. Although the expression level of $Ca_V3.2$ (CACNA1H,α1H) was much higher than in human adult hearts, recording of $I_{Ca,T}$ separate from $I_{Ca,L}$ has not been reported so far. In the preliminary experiments, we found that among voltage-dependent Ca^{2+} currents, $I_{Ca,T}$ was dominant at an early stage after initiation of cardiac differentiation and $I_{Ca,L}$ was dominant at a later stage, which is consistent with the gene expression data in human iPSC-derived cardiomyocytes, and *in situ* change of voltage-sensitive Ca^{2+} currents in murine hearts.

Ma et al. tried to accelerate selection of human iPSC-derived cardiomyocytes by expressing blasticidin resistance gene under the control of cardiac *myosin heavy chain 6* (*MYH6*) (Ma et al. 2011). They thoroughly examined electrophysiological characteristics of human iPSC-derived cardiomyocytes obtained with this selection technique. Human iPSC-derived cardiomyocytes obtained with *MYH6*-dependent blasticidin selection showed robust expression of almost all currents expressed in human adult cardiomyocytes, including I_{Na}, $I_{Ca,T}$, $I_{Ca,L}$, I_f, I_{to}, I_{Kr}, I_{Ks}, and I_{K1}. They also examined current kinetics of I_{Na}, $I_{Ca,L}$, I_{to}, I_{Kr}, I_{Kr}, and I_f, which are comparable to those in guinea-pig ventricular myocytes and/or those of human ion currents expressed heterologously in cell lines. *MYH6*-dependently selected human iPSC-derived cardiomyocytes appear to be more mature than conventionally prepared human iPSC-derived cardiomyocytes, since *MYH6*-dependently selected human iPSC-derived cardiomyocytes express robust I_{to} and I_{K1}. This finding is not surprising, because cardiomyocytes dominantly express β-myosin heavy chain encoded by *MYH7* at an early stage of development, and α-myosin heavy chain encoded by *MYH6* at a later stage of development (Bouvagnet et al. 1987). Selection based on *MTH6* expression, thus, could select cardiomyocytes at a later stage of cardiac development.

Ca^{2+} handling and contraction

Ca^{2+} handling underlies contractility of cardiomyocytes, a primary task of cardiomyocytes. Thus, it is important to characterize contractile function and Ca^{2+} handling in iPSC-derived cardiomyocytes and compare with those in adult cardiomyocytes. In adult cardiomyocytes, Ca^{2+} entering through L-type Ca^{2+} channels on the plasma membrane bind to the ryanodine receptor type 2 (RYR2) on the sarcoplasmic reticulum, resulting in opening of RYR2 and release of Ca^{2+} from the sarcoplasmic reticulum lumen into the cytoplasm. Released Ca^{2+} bind to the troponin-C, causing interaction of actin and myosin and generating contraction force.

Itzhaki et al. examined Ca^{2+} handling of human iPSC-derived cardiomyocytes (Itzhaki et al. 2011). They found expression of mRNA for RYR2, inositol-(1,4,5)-triphosphate (IP_3) receptor 2 (IP3R2), sarco/endoplasmic reticulum Ca^{2+}-ATPase 2a (SERCA2a), $Ca_V1.2$, calsequestrin, and phospholamban. Ca^{2+} transient observed with Ca^{2+}indicator dye, fluo-4 fluorescent under a confocal imaging system was abolished by elimination of extracellular Ca^{2+} or application of a L-type Ca^{2+} channel blocker, nifedipine, indicating an important role of Ca^{2+} entry through L-type Ca^{2+} channels in generating Ca^{2+} transient. Ca^{2+} transient was markedly enhanced by caffeine, an activator of RYR2, and inhibited by ryanodine, an inhibitor of RYR2, confirming the role of RYR2. Ca^{2+} transient was also inhibited by application of thapsigargin, an inhibitor of SERCA. From these data, human iPSC-derived cardiomyocytes are equipped with Ca^{2+} handling machinery similar to the human adult cardiomyocytes. An interesting finding is that Ca^{2+} transients in human iPSC-derived cardiomyocytes were inhibited by 2-APB, a well-known cell-permeable IP3R antagonist. IP3 is generated by break-down of phosphatidylinositol-(4,5)-biphosphate (PIP2) with phospholipase C. Ca^{2+} transient was also diminished with U73122, a phospholipase C inhibitor. Thus, in addition to the L-type Ca^{2+} channel/RYR2 pathway, the phospholipase C/IP3R pathway plays a role in Ca^{2+} transient in human iPSC-derived cardiomyocytes, which is different from human adult cardiomyocytes.

Pfannkuche et al. measured development of contractility using avitalized murine ventricular tissue (Pfannkuche et al. 2009). Neonatal murine ventricles were cut with a vibratome generating 300 μm slices and were completely avitalized by simulated hypoxia and ischemia (oxygen and glucose deprivation) for 20 hours. In this way, the ventricular slices lost their contractile function and could be further used as a dead, natural and soft cardiac matrix tissue for transplantation experiments and recording force function. When avitalized tissue sections were transplanted with beating cell clusters from human iPSC embryoid bodies, spontaneous beatings were observed and mechanical force up to 1.52 μN was generated.

Xi et al. characterized contractile function of murine iPSC-derived cardiomyocytes in avitalized ventricular slices by oxygen and glucose deprivation (Xi et al. 2010). Contractile force up to 1.2 μN was generated. They compared the data obtained in murine iPSC-derived cardiomyocytes and native embryonic murine heart tissue. The contractile force was dependent on extracellular Ca^{2+} concentration, which is not different between iPSC-derived cardiomyocytes and native embryonic murine heart tissue. Three quarters of murine iPSC-derived cardiomyocytes ceased contraction with 1 μM nifedipine, and the remaining one quarter with 10 μM nifedipine, while 20% of native embryonic murine heart tissues stopped beating even at the highest dose of nifedipine (100 μM), though and in the remaining 80%, the fource was quite low but present. $NiCl_2$ at 5 and 10 mM, concentrations that block the Na^+-Ca^{2+} exchanger (NCX), stopped beating in both murine iPSC-derived cardiomyocytes and native embryonic murine heart tissue, indicating a significant role of NCX for contractile force, which is also demonstrated in human embryonic stem cell-derived cardiomyocytes (Fu et al. 2010).

Thus, iPSC-derived cardiomyocytes can develop contractile force, and the Ca^{2+} transient depends on L-type Ca^{2+} channel/RYR2 pathway, phospholipase C/IP3R pathway, and possibly NCX-involved pathway.

Sympathetic and Parasympathetic Regulation

In human adult cardiomyocytes, both electrophysiological and contractile functions are regulated by the autonomic nervous system. Sympathetic nervous system activation accelerates beating rates of cardiomyocytes and increases their contractile force. In contrast, parasympathetic nervous system activation slows down beating rates and weakens their contractile force.

Human iPSC-derived cardiomyocytes appear to possess autonomic regulation of both electrophysiological and contractile functions. Beating rates were accelerated by application of isoproterenol at 1 μM, a sympathetic nerve β-receptor agonist, and were slowed down by application of carbamylcholine at 1 μM, a parasympathetic nerve muscarinic receptor agonist (Zwi et al. 2009). Isoproterenol at 1 μM also increased the force developed in human iPSC-derived cardiomyocytes to 120% (Xi et al. 2010).

In murine iPSC-derived cardiomyocytes, it is interesting that the timing of initiation of sympathetic and parasympathetic influence was different. Kuzmenkin et al. compared the effects of isoproterenol 1 μM and carbacol 1 μM, a parasympathetic nerve muscarinic receptor agonist at an early stage of cardiac differentiation (10–12 days) and at a late stage of cardiac differentiation (16–19 days) (Kuzmenkin et al. 2009). Isoproterenol significantly accelerated spontaneous beating rates both at 10–12 days and at 16–19 days. In contrast, carbacol slowed down spontaneous beating rates at 16–19 days, but not at 10–12 days. Thus, at least in murine iPSC-derived cardiomyocytes, receptors and/or downstream signaling pathway was installed for sympathetic nervous system earlier than for parasympathetic nervous system.

Responses to Drugs

In this section, we briefly discuss responses to drugs; more detailed discussions follow in the next section entitled "Pharmacology and Future Application of iPSC-Derived Cardiomyocytes".

a. QT prolongation and Torsade de pointes type arrhythmias

Three major potential medical applications have been proposed for human iPSC: (1) drug efficacy and toxicity testing in human cells; (2) study of therapeutic strategy in diseased model iPS cells; and (3) cell implantation.

QT prolongation, Torsade de pointes type arrhythmias, and associated syncope and sudden death are one of the most common reasons for retraction of drugs from the market. Furthermore, many chemical compounds were withdrawn from the line of drug development due to this side action (Wilke et al. 2007). Currently used *in vitro* assay for this side action is the hERG test, since most drugs causing QT prolongation and Torsade de pointes type arrhythmias target the hERG channels. In the hERG test, hERG channel gene is heterologously expressed in some cell lines, such as human embryonic kidney (HEK) cells or Chinese hamster ovary (CHO) cells, and effects of chemical compounds on hERG currents are examined with the patch-clamp technique or with the automated patch-clamp system. The sensitivity and the specificity of the hERG test are not sufficiently high, so that some chemical compounds that have passed the hERG test exhibit QT prolongation and/or Torsade de pointes type arrhythmias later in *in vivo* assay or in humans thorough QT assay. Conversely, it is speculated that some chemical compounds with highly beneficial effects could be withdrawn from the line because the very wide safety margin is set in the hERG test. To improve the sensitivity and the specificity in *in vitro* assay, the use of human iPSC-derived cardiomyocytes in *in vitro* assay has strong promise.

b. Effects of drugs on human iPSC-derived cardiomyocytes

The effects of drugs on electrophysiological characteristics of human iPSC-derived cardiomyocytes were assessed on action potential configurations recorded with the microelectrode impalement technique and on action potential configurations and membrane currents recorded with the patch-clamp technique. However, they were most frequently assessed by the field potentials recorded from the embryoid body with the microelectrode assay (MEA) system. Field potentials recorded with the MEA system have an initial negative peak reflecting I_{Na}, a following sustained negative phase reflecting I_{Ca}, and a final positive phase reflecting I_K. The duration between the initial peak to the peak of the final positive phase is referred to as the field potential duration (FPD), which is believed to reflect the action potential duration and the QT interval on electrocardiogram. As expected, Tanaka et al. showed that quinidine, a Na^+ channel blocker, decreased the amplitude of the initial negative peak, verapamil, a Ca^{2+} channel blocker, shortened the FPD, and E4031, a hERG channel blocker, prolonged the FPD, and all of these effects were reversible and dose-dependent (Fig. 6) (Tanaka et al. 2009).

In addition to E4031, Otsuji et al. examined another two drugs that are known to induce QT prolongation and Torsade de pointes, dl-sotalol and nifekalant (Otsuji et al. 2010). Both dl-sotalol and nifekalant prolonged the FPD in a dose-dependent manner. It is worth noting that although dl-sotalol is well known to exhibit false negative results in *in vitro* hERG assays, the

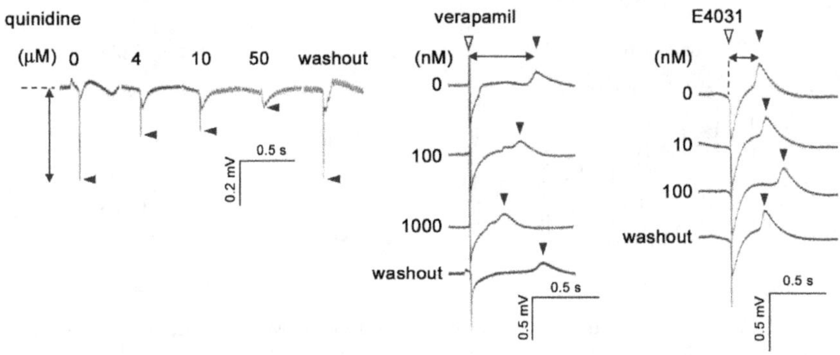

Figure 6. Effects of quinidine, verapamil, and E4031 on field potentials.
◀:initial negative peak
▼:final positive peak
∇:start of field potential
↔field potential duration
(cited from ref with permission)

results obtained in the MEA system corresponded well with clinical and *in vivo* information.

These data suggest a strong promise for use of human iPSC-derived cardiomyocytes in *in vitro* assay of drug efficacy and toxicity. However, there are some uncertainties to be clarified. The FPD is recorded in the embryoid bodies, consisting of various types of cardiomyocytes including nodal type, atrial type and ventricular type. One is not sureif mixture of these cell types affects the result of *in vitro* FPD assay. It is also not certain whether the FPD accurately reflects action potential duration and/or the QT interval. The FPD could be influenced by cell-to-cell coupling. No studies so far examined the expression and the function of the gap-junction channels in iPSC-derived cardiomyocytes. Thus, the influence of the gap-junction channels on *in vitro* FPD assay should be examined.

c. Effects on contractility

Cardiac toxicity—suppression of cardiac contraction—is in the secondary group as the reasons for drug withdrawal from the market. Up to now, *in vitro* and high-content screening system to assess contractility of cardiomyocytes has not been developed. Pfannkuche et al. developed a system to measure contraction force on avitalized soft tissue (Pfannkuche et al. 2009). However, this system is not suitable for high- or intermediate-throughput screening system. Recently, two potential new technologies have been introduced. One is the impedance sensor system: impedance sensing of cells could be used to assess various cellular events, including beating

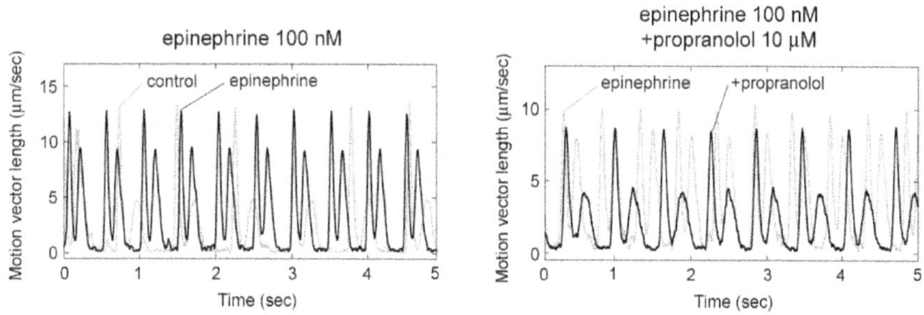

Figure 7. Contraction and relaxation speed obtained by the MVP technique. Left: effects of epinephrine at 100 nM. Right: effects of epinephrine 100 nM plus propranolol 10 μM.

frequency and contraction speed (Schwarzenberger et al. 2011). The other technique is the motion vector prediction (MVP) technique developed by Sony Co. (Hayakawa et al. 2012). The MVP system is based on the calculation of motion vector from video imaging at a high capturing rate (125 per second). The speed of cardiomyocyte contraction and relaxation can be separately assessed noninvasively. Hayakawa et al. examined the effects of epinephrine, an agonist of sympathetic nerve α- and β-receptors, and propranolol, an inhibitor of a sympathetic nerve β-receptor (Hayakawa et al. 2012). Epinephrine at 100 nM accelerated beating rates and the relaxation speed, and additional application of propranolol at 10 μM decreased beating rates and the relaxation speed to the level before epinephrine application.

The challenge to the assessment of contractile function in human iPSC-derived cardiomyocytes has just begun, and further technological improvement is desirable.

Acknowledgements

I wish to thank Dr. Junko Kurokawa, Dr. Yasunari Kanda, Dr. Fernando Lopez-Redondo, Miss. Li Min, and Mr. Shinichiro Ohkata in the Department of Bio-informational Pharmacology, Medical Research Institute, Tokyo Medical and Dental University for performing experiments, to Miss Mari Imaizumi, Mrs. Reiko Ohsumi, Mrs. Wakako Mukai, and Mrs. Miho Takekawa for technical assistance, and to Mrs. Kuniko Yamaguchi for secretarial support. I also thank to Dr. Kenji Yasuda, Dr. Tomoyuki Kaneko, Dr. Norimasa Nomura, and Dr. Tomoyo Hamada in Department of Biomedical Information, Institute of Biomaterials and Bioengineering, Tokyo Medical and Dental University for collaboration in iPSC study, and to Dr. Akio Yasuda, Dr. Eriko Matsui, Dr. Tomohiro Hayakawa, Dr. Takeshi Kunihiro, and Miss. Hatsume Uno in Advanced Materials Laboratories, Life Science Laboratory, Sony Co. for collaboration in MVP technique. This work was supported in part by Grants-in-Aid for Scientific Research from the Ministry of Education, Culture, Sports, Science and Technology (MEXT) of Japan, and project "Fundamental Technology Development for Promoting the Industrial Application of iPS Cells" (program leader Dr. Kenji Yasuda) from New Energy and Industrial Technology Development Organization (NEDO).

References

Bouvagnet, P., Neveu, S., Montoya, M., and Leger, J. 1987. Development changes in the human cardiac isomyosin distribution: an immunohistochemical study using monoclonal antibodies. Circ. Res. 61, 329–336.

Ficker, E., Dennis, A. T., Obejero-Paz, C. A., Castaldo, P., Taglialatela, M., and Brown, A. M. 2000. Retention in the endoplasmic reticulum as a mechanism of dominant-negative current suppression in human long QT syndrome. J. Mol. Cell Cardiol. 32, 2327–2337.

Fu, J. D., Jiang, P., Rushing, S., Liu, J., Chiamvimonvat, N., and Li, R. A. 2010. Na+/Ca2+ exchanger is a determinant of excitation-contraction coupling in human embryonic stem cell-derived ventricular cardiomyocytes. Stem Cells Dev. 19, 773–782.

Gong, Q., Anderson, C. L., January, C. T., and Zhou, Z. 2002. Role of glycosylation in cell surface expression and stability of HERG potassium channels. American Journal of Physiology—Heart and Circulatory Physiology 283, H77–H84.

Hayakawa, T., Kunihiro, T., Dowaki, S., Uno, H. , Matsui, E., Uchida, M., Kobayashi, S., Yasuda, A., Shimizu, T., and Okano, T. 2012. Noninvasive evaluation of contractile behavior of cardiomyocyte monolayers based on motion vector analysis. Tissue Eng. Part C Methods 18, 21–32.

Honda, M., Kiyokawa, J., Tabo, M., and Inoue, T. 2011. Electrophysiological characterization of cardiomyocytes derived from human induced pluripotent stem cells. J. Pharmacol. Sci. 117, 149–159.

Itzhaki, I., Maizels, L., Huber, I., Zwi-Dantsis, L., Caspi, O., Winterstern, A., Feldman, O., Gepstein, A., Arbel, G., Hammerman, H., Boulos, M., and Gepstein, L. 2011. Modelling the long QT syndrome with induced pluripotent stem cells. Nature 471, 225–229.

Kuzmenkin, A., Liang, H., Xu, G., Pfannkuche, K., Eichhorn, H., Fatima, A., Luo, H., Saric, T., Wernig, M., Jaenisch, R., and Hescheler, J. 2009. Functional characterization of cardiomyocytes derived from murine induced pluripotent stem cells *in vitro*. FASEB J. 23, 4168–4180.

Ma, J., Guo, L., Fiene, S. J., Anson, B. D., Thomson, J. A., Kamp, T. J., Kolaja, K. L., Swanson, B. J., and January, C. T. 2011. High purity human-induced pluripotent stem cell-derived cardiomyocytes: electrophysiological properties of action potentials and ionic currents. American Journal of Physiology—Heart and Circulatory Physiology 301, H2006–H2017.

Otsuji, T. G., Minami, I., Kurose, Y., Yamauchi, K., Tada, M., and Nakatsuji, N. 2010. Progressive maturation in contracting cardiomyocytes derived from human embryonic stem cells: Qualitative effects on electrophysiological responses to drugs. Stem Cell Research 4, 201–213.

Pfannkuche, K., Liang, H., Hannes, T., Xi, J., Fatima, A., Nguemo, F., Matzkies, M., Wernig, M., Jaenisch, R., Pillekamp, F. , Halbach, M., Schunkert, H., Saric, T., Hescheler, J., and Reppel, M. 2009. Cardiac myocytes derived from murine reprogrammed fibroblasts: intact hormonal regulation, cardiac ion channel expression and development of contractility. Cell Physiol. Biochem. 24, 73–86.

Schwarzenberger, T., Wolf, P., Brischwein, M., Kleinhans, R., Demmel, F., Lechner, A., Becker, B., and Wolf, B. 2011. Impedance sensor technology for cell-based assays in the framework of a high-content screening system. Physiol. Meas. 32, 977–993.

Tanaka, T., Tohyama, S., Murata, M., Nomura, F., Kaneko, T., Chen, H., Hattori, F., Egashira, T., Seki, T., Ohno, Y., Koshimizu, U., Yuasa, S., Ogawa, S., Yamanaka, S., Yasuda, K., and Fukuda, K. 2009. *In vitro* pharmacologic testing using human induced pluripotent stem cell-derived cardiomyocytes. Biochem. Bioph. Res. Co. 385, 497–502.

Wang, L. and Duff, H. J. 1997. Developmental Changes in Transient Outward Current in Mouse Ventricle. Circ. Res. 81, 120–127.

Wilke, R. A., Lin, D. W., Roden, D. M., Watkins, P. B., Flockhart, D., Zineh, I., Giacomini, K. M., and Krauss, R. M. 2007. Identifying genetic risk factors for serious adverse drug reactions: current progress and challenges. Nat. Rev. Drug Discov. 6, 904–916.

Xi, J., Khalil, M., Shishechian, N., Hannes, T., Pfannkuche, K., Liang, H., Fatima, A., Haustein, M., Suhr, F., Bloch, W., Reppel, M., Saric, T., Wernig, M., Janisch, R., Brockmeier, K., Hescheler, J., and Pillekamp, F. 2010. Comparison of contractile behavior of native murine ventricular tissue and cardiomyocytes derived from embryonic or induced pluripotent stem cells. FASEB J. 24, 2739–2751.

Zhang, J., Wilson, G. F. , Soerens, A. G., Koonce, C. H. , Yu, J., Palecek, S. P. , Thomson, J. A., and Kamp, T. J. 2009. Functional cardiomyocytes derived from human induced pluripotent stem cells. Circ. Res. 104, e30–e41.

Zwi, L., Caspi, O., Arbel, G., Huber, I., Gepstein, A., Park, I. H., and Gepstein, L. 2009. Cardiomyocyte differentiation of human induced pluripotent stem cells. Circulation 120, 1513–1523.

CHAPTER 10

Pharmacology and Future Application of iPS Cell-derived Cardiomyocytes

Mitsushige Murata

Introduction

There has been remarkable progress to date in the identification, derivation, and characterization of stem cells, including embryonic stem (ES) cells and induced-pluripotent stem (iPS) cells. The proliferative capacity and pluripotency of mouse ES cells, which were first established by Evans and colleagues in 1981, enabled the establishment of early developmental model systems and the generation of genetically modified mice, including knockout mice (Evans and Kaufman 1981). The progress of ES cell research slowed markedly once it moved into experimental animals such as *in vivo* models for cell-transplantation therapy (Fukuda and Yuasa 2006), although the establishment of human ES cells in 1998 reinvigorated attention toward ES cells as a source for regenerative therapy (Thomson et al. 1998). Since then, further progress has been made in directing the differentiation of stem cells into various cell types and tissues with enhanced survival properties following transplantation, although several problems have to be overcome to realize successful clinical application of these cells (Passier et al. 2008). Specifically, ethical and legal considerations have inhibited the development of human ES cells as a potential regenerative source because of the associated destruction of early human embryos; furthermore, ES cells do not display the autologous genotype of the donor patient (Evans and Kaufman 1981).

Department of Laboratory Medicine, Keio University, 35 Shinanomachi, Shinjuku-ku, Tokyo 160-8582, Japan.
Email: muratam@a7.keio.jp

To avoid these problems and to maintain stem cell characteristics, one possible solution is the reprogramming of differentiated cell nuclei into an ES cell-like, pluripotent state. This would involve generating murine and human iPS cells from somatic cells through overexpressing defined stem cell transcription factors (Nakagawa et al. 2008; Takahashi et al. 2007a; Takahashi et al. 2007b; Yu et al. 2007). ES cells and iPS cells are similar in morphology, proliferative ability, cell surface marker expression, and gene expression profiles, while iPS cells also have the capacity to differentiate into the three germ lines. Thus, iPS cells are an alternative to ES cells as a source for regenerative therapy that avoids ethical problems and immune rejection. In addition, iPS cells could be used to create patient and disease-specific stem cells. Among the many target diseases for regenerative medicine, heart disease is one of the most important due to the high morbidity and mortality rates associated with it. Human iPS cells will provide excellent models for investigating mechanisms of heart disease and for drug and toxicology testing.

Here, we discuss the currently accepted pharmacological characteristics of human iPS-derived cardiomyocytes (hiPS-CMs) and their potential applications for cardiovascular medicine.

1. Pharmacological Features of Human iPS-derived Cardiomyocytes

Studies have demonstrated that human ES cells can be differentiated into cardiomyocytes using the EB method (Kehat et al. 2001; Xu et al. 2002) or other induction methods (Laflamme et al. 2007; Mummery et al. 2003), such that the derived cells show an unambiguous cardiac phenotype. Thus, human ES-derived cardiomyocytes (hES-CMs) express the genes and proteins for sarcomeric organization, i.e., α-actinin, cardiac troponins, myosin heavy chain, myosin light chain, desmin, and tropomyosin (He et al. 2003; Kehat et al. 2001; Lev et al. 2005; Mummery et al. 2003; Xu et al. 2006; Xu et al. 2002), cardiac-specific transcription factors, e.g., Nkx2.5, GATA4, myocyte enhancer factor 2C (MEF2C), Tbx5, and Tbx20 (Kehat et al. 2001; Lev et al. 2005; Norstrom et al. 2006; Snir et al. 2003; Tamargo et al. 2004; Xu et al. 2006; Xu et al. 2002), and gap junctions (Cui et al. 2007; Kehat et al. 2001; Mummery et al. 2003; Xu et al. 2002).

Mauritz and colleagues concluded that mouse iPS-CMs are similar to mouse ES-CMs based on immunocytochemical and electrophysiological results, together with calcium imaging studies (Mauritz et al. 2008). In addition, Narazaki and colleagues successfully directed the differentiation of mouse iPSCs into cardiovascular cells by sorting Flk1 (vascular endothelial growth factor receptor-2)-expressing cardiovascular progenitor cells using a method previously established for mouse ESCs (Narazaki et al. 2008; Yamashita et al. 2000). Some of the miPS-CMs expressed HCN4 and the T-type calcium channel Cav3.2, both of which are localized at the sinoatrial node and play important roles in the automaticity of mouse ES-CMs (Yanagi et al. 2007), as well as sarcomeric formation factors and cardiac-specific transcription factors (Narazaki et al. 2008).

Following this, Zhang and colleagues were the first to describe hiPSC-derived cardiomyocytes (hiPS-CMs) that were generated from fetal and newborn fibroblasts transduced with *Oct4*, *Sox2*, *Nanog*, and *Lin28* and which expressed myofilament proteins and sarcomeric organization proteins (e.g., α-actinin, myosin light chain, and cardiac troponins) with the same efficiencies as cardiomyocytes derived from hESCs (hES-CMs) (Zhang et al. 2009). The same group also reported a marked reduction in proliferation of their hiPS-CMs, similar to hES-CMs (McDevitt et al. 2005; Snir et al. 2003), although generally lower owing to the continued expression of transgenes, as observed for miPS-CMs (Narazaki et al. 2008; Zhang et al. 2009). Our group reported that hiPS-CMs, which were generated from adult fibroblasts transduced with *Oct4*, *Sox2*, *Klf4* and *c-myc*, also expressed the genes and proteins for sarcomeric organization (e.g., α-actinin, myosin heavy chain, and tropomyosin), cardiac-specific transcription factors (e.g., Nkx2.5 and GATA4), chamber-specific proteins (e.g., ANP), and cardiac ion channel genes including: (1) the sodium channel α-subunit SCN5A, which determines cardiac excitability and conduction velocity, (2) the L-type calcium channel α-subunit $Ca_v1.2$, which contributes to cardiac contraction and (3) the rapidly activating delayed rectifier potassium channel KCNH2 (I_{kr}), which contributes to action potential (AP) repolarization (Tanaka et al. 2009). Furthermore, Zwi and colleagues reported that in addition to $Ca_v1.2$ and KCNH2, hiPS-CMs expressed genes for the slowly activating delayed rectifier potassium channel KCNQ1 (I_{ks}), which also contributes to AP repolarization, and for the hyperpolarization-activated cyclic nucleotide gated potassium channel HCN2, which is responsible for the If pacemaker current (Zwi et al. 2009). If these ion channels expressed in the hiPS-CMs are indeed functional, the responses of such stem cell-derived CMs to antiarrhythmic drugs might be identical to those of native cardiomyocytes. The pharmacological properties of hiPS-CMs could be tested *in vitro* using the On-Chip multi-electrode array (MEA) system to record extra-cellular field potentials (FP) (Suzuki and Yasuda 2007).

The effect of antiarrhythmic drugs on field potential configuration in hiPS-CMs

The sodium channel favors a sharp depolarizing upstroke of action potential (AP) (Phase 0), reflecting conduction velocity. Thus, sodium channel blockers can be used for the treatment of atrial as well as ventricular arrhythmias. Application of quinidine to the spontaneous beating embryoid bodies (EBs) of hiPS-CMs, dose dependently and reversibly attenuated the time span required for the decline of the voltage from the baseline to first negative peak (FP amplitude), which reflected the AP upstroke without any significant changes in the spontaneous beating rate (Tanaka et al. 2009; Yokoo et al. 2009). However, the effects of quinidine on FP configurations remain controversial probably due to its multichannel blocking properties (Mandel et al. 2012). Another depolarizing channel is the L-type calcium channel, whose inactivation activity is less rapid and less complete, resulting in residual plateau depolarization. Another calcium channel blocker, verapamil, dose dependently shortened the duration from first negative peak (open triangle) to first positive peak (closed triangle) (FPD), which is consistent with the calcium channel inhibitory effects on AP duration. Potassium channels disorders might therefore lead to prolongation of cardiac repolarization, and thus potentially to life-threatening arrhythmias, e.g., TdP and ventricular tachycardia. The fast component of the delayed rectifier potassium current (I_{Kr}) blocker, E-4031, as well as the slow component of I_{Ks} blocker, chromanol293B, dose dependently and reversibly shortened the FPD (Mandel et al. 2012). These results indicated that the main cardiac depolarizing and repolarizing ion channels are functionally expressed in hiPS-CMs, and that hiPS-derived cardiac tissue could also serve as a platform for drug testing, and specifically, for the effects of the drugs on repolarization (QT screening).

Chronotropic responses in hiPS-CMs

One of the most important properties of cardiac physiology is the ability to respond appropriately to neurohormonal regulation. Several studies revealed that hiPS-CMs could adequately respond to adrenergic and muscarinic stimuli, and that administration of the β-agonist isoproterenol resulted in a dose-dependent increase in the spontaneous beating frequency of hiPS EBs (Mandel et al. 2012; Tanaka et al. 2009; Zwi et al. 2009), while administration of the muscarinic agonist carbamylcholine induced a dose-related decrease. These effects of isoproterenol and carbamylcholine were blocked by metrprolol and atropine, respectively, indicating that hiPS-CMs show identical chronotropic responses to those of native cardiomyocytes. Recently, Mandel et al. reported that hiPS-CMs exhibit beat rate variability and power-law behavior as in humans, supporting the potential capability of these cells as biological pacemakers (Mandel et al. 2012).

2. Application of iPS Cells in Cardiac Regenerative Medicine

Cardiovascular diseases are important targets for regenerative medicine because they are associated with high morbidity and mortality. Most pathological processes that initiate irreversible heart dysfunction, such as myocardial infarction and cardiomyopathies, either result from or are exacerbated by a loss of heart cells. Because cardiomyocytes lack the capacity for self-repair, the recovery of heart dysfunction requires the replacement of damaged cells by transplantation with large quantities of healthy cardiomyocytes. To this end, human stem cells are a potential source for transplantation therapy because of their strong proliferative capacity and ability to differentiate reliably. Recent studies have shown that hiPS cells can differentiate into cardiomyocyte-like cells that are similar in terms of their gene expression profiles and physiological properties to native cardiomyocytes and ES-CMs (Tanaka et al. 2009; Yokoo et al. 2009; Zhang et al. 2009). These studies highlight the potential of hiPS cells in cardiovascular regenerative medicine (Murata et al. 2010).

The first application of iPS cell technology for cardiac regenerative medicine is likely to be the transplantation of iPS-CMs into diseased hearts to restore pump function. A considerable advantage of using hiPS cells is the possibility of creating isogenic cardiomyocytes that are genetically equivalent to the cells in the transplant recipient, thus avoiding the immune rejection that is likely to accompany allogenic transplants (i.e., ES-CMs may be rejected by the recipient's immune system). However, transplantation of stem cell-derived cardiomyocytes into animal models has raised many questions that need to be addressed before clinical transplantation into humans. Most studies have used a mixture of human ES cell derivatives containing a significant proportion of cardiomyocytes in addition to other cell types (Caspi et al. 2009; Laflamme et al. 2007). Because the transplantation of undifferentiated stem cells may lead to the formation of teratomas, it is necessary to establish stringent and reliable purification strategies for stem cell-derived cardiomyocytes. We recently reported an innovative non-genetic purification method for ES-CMs that uses a mitochondrial fluorescent dye (Hattori et al. 2009). A major obstacle to cardiomyocyte purification has been the lack of cardiac-specific cell marking systems; however, we found that a fluorescent mitochondrial-specific dye could be used to selectively mark rat cardiomyocytes, as well as murine, marmoset, and human stem cell-derived cardiomyocytes, so that the cells can be subsequently enriched (to > 99% purity) using FACS. This method does not require genetic modification of the cells and reduces concerns regarding tumor formation. Indeed, the ES-CMs purified using this method did not induce teratoma formation in either heart or testes.

This method is likely to be widely applicable beyond these species given that an abundance of cellular mitochondria is a common characteristic of cardiomyocytes, despite a reduction in mitochondrial content with aging. In addition, although the reported survival rate of purified and dispersed cardiomyocytes differentiated from stem cells is not particularly high (Hattan et al. 2005; Kolossov et al. 2006), we hypothesized that because ES-CMs exist as homophilic cell aggregates in mouse, marmoset, and human EBs, reaggregates of purified ES-CMs might be more stable and durable in the recipient heart. Based on this, we developed a more effective transplantation method by reaggregating purified ES- or iPS-CMs to produce increased long-term survival of the purified cardiomyocytes. Overall, these methods should increase the success rate for using stem cell-derived cardiomyocytes in basic and clinical applications.

Another likely and realistic application of cell transplantation therapy is to produce biological pacemakers from stem cell-derived cardiomyocytes. Currently, approximately three million people worldwide have an implantable pacemaker to control cardiac rhythm disturbances, including sick sinus syndrome and atrioventricular block (Wood and Ellenbogen 2002). Although these devices treat patients successfully, there are shortcomings associated with their use, including cost, patient discomfort, cosmetic problems, and increased susceptibility to infection. These issues have led to recent interest in the creation of a biological pacemaker, and two groups have already published proof-of-concept results using ES-CMs. Xue et al. demonstrated the pacemaking ability of transplanted human ES-CMs in a guinea pig preparation while Kehat et al. demonstrated that EBs injected into a swine model of complete heart block could function as an ectopic pacemaker (Xue et al. 2005; Kehat et al. 2004). These findings confirmed the capacity of transplanted, stem cell-derived cardiomyocytes to functionally couple with a host myocardium as an ectopic pacemaker, alluding to the potential of a biological pacemaker for clinical application. However, the follow-up period after transplantation in both studies was relatively short (< 3 weeks) despite the use of EBs containing non-cardiomyocytes or undifferentiated cells. A longer follow-up period is necessary to confirm that this technique is safe without any risk of tumor formation.

Application of iPS cells in the genetic analysis of cardiac disease

Human iPS-CMs could also be used in pharmacological testing. Many cardiac and non-cardiac-acting drugs prolong AP duration (APD), giving rise to acquired long QT syndrome (LQTS), which may itself result in the life-threatening arrhythmia, Torsades de pointes (TdP). The most frequent reason underlying the removal of drugs from the market is adverse cardiac side effects (Gussak et al. 2004; Roden 2004). Thus, it is essential that any proarrhythmic risk is identified at an early stage in the drug development process, so as to define an unacceptable safety profile and to mitigate costs.

Several potassium channels contribute to AP repolarization and duration (Snyders 1999; Tamargo et al. 2004). Early repolarization of the ventricular AP is attributable to the transient outward potassium channel (I_{to1}), whereas late repolarization is associated with the inward rectifier potassium channels (I_{K1}). Furthermore, both the rapid and slow delayed rectifier potassium channels (I_{Kr} and I_{Ks}) contribute to repolarization beyond the plateau phase, suggesting that dysfunction of either of these channels would lead to triggering activity in addition to QT prolongation. Excessive delayed repolarization is linked to arrhythmogenesis following the development of early after-depolarizations (EADs) (January and Moscucci 1992; January and Riddle 1989; January et al. 1988) and exacerbation of transmural dispersion of repolarization (TDR) (Antzelevitch 2005; Lankipalli et al. 2005; Shantsila et al. 2007). EADs may account for R-on-T ventricular extrasystoles, which trigger the initiation of TdP. In addition, increased TDR leads to increased heterogeneity of tissue refractoriness, which produces a substrate for re-entrant arrhythmias. Because the APD of the mid-myocardium is usually longer than that of the endocardium or epicardium due to decreased I_{Ks} channel density (Burashnikov and Antzelevitch 2002), the effect of suppressing I_{Kr} may be more prominent in the mid-myocardium than in other ventricular regions, implying that a blockade of I_{Kr} increases TDR. Thus, many drugs that block I_{Kr} may cause LQTS and TdP. In addition to the inhibitory effects of these drugs on I_{Kr}, several factors predispose to drug-induced TdP, including female gender, abnormal electrolytes (low plasma levels of K^+ and Mg^{2+}), bradycardia, and heart disease. Pharmacological interactions between I_{Kr}-blocking drugs and other drugs administered simultaneously may also interfere with the drug biotransformation and excretion, and thus exacerbate the arrhythmia. For example, the antifungal agent ketoconazole interferes with the biotransformation of the antihistamine terfenadine into a metabolite that does not prolong APD. Thus, the co-administration of these two drugs results in a high concentration of terfenadine, which leads to acquired LQTS (Priori et al. 2001a; Zechnich et al. 1994). This type of interaction applies to many drugs that inhibit cytochrome P450 enzymes. LQTS can also be acquired as an adverse effect of drug therapy or an electrolyte abnormality that alters the electrochemical conditions needed for normal cardiac excitability. Furthermore, genetic variations and mutations in I_{Kr}, I_{Ks}, or I_{K1} modify the susceptibility of these ion channels to drug interactions (Paulussen et al. 2004). As discussed, arrhythmogenesis could also result from a complex interaction between these related ion channels.

Thus, a predictive, high-throughput, cell-based, *in vitro* QT assay system is highly desirable for cardiotoxicity screening. Although patch-clamp experiments are the accepted method for investigating AP parameters and the precise electrophysiological properties of ion channels, the technique is time consuming and requires a skilled operator. Furthermore, even though currently available drug screening in heterologous expression systems is high throughput, these systems do not represent the native cardiac context and so may lack important accessory proteins or secondary targets that could mediate relevant adverse effects. Taking these issues into consideration, it appears that a combination of MEAs and stem cell-derived cardiomyocytes may represent the best system in which to measure the surface electrogenic activities of cell clusters. MEAs could be useful for recording the electrical activity of the various derivatives of human ES and iPS cells (Harding et al. 2007; Tanaka et al. 2009); importantly, they are easy to perform and can be adapted to automated high-throughput systems. However, such a combination also carries issues that need to be resolved. Stem cell-derived cardiomyocytes have a similar electrophysiological phenotype to embryonic cardiomyocytes. Thus, the implications of significant phenotypic differences between these cells and adult cardiomyocytes should be considered. Moreover, hiPS-CMs include distinct nodal and working cardiac subtypes, leading to electrophysiological heterogeneity that may not reflect the situation in native cardiac tissues. Most cardiac ion currents undergo developmental maturation in terms of current density and properties, despite different gene expression patterns for each ion channel (Sartiani et al. 2007). Maturation of the physiological phenotypes of hiPS-CMs could therefore be critical for determining the best cellular phase to be used for drug screening. Further investigations are needed to resolve these issues.

Finally, iPS-CMs could be used as models of cardiac diseases. Genetic cardiovascular diseases include channelopathies and cardiomyopathies, which are related to abnormal electrophysiology and impaired contractility. Cardiac channelopathies arise from genetic alterations that lead to dysfunctional cardiac ion channels. Common channelopathies include LQTS, Brugada syndrome, catecholaminergic polymorphic ventricular tachycardia (CPVT), and short QT syndrome, all of which are inherited arrhythmogenic diseases caused by mutations in the genes that encode ion channels or their related proteins (Marban 2002; Priori et al. 2001b). The electrical instability inherent in channelopathies (i.e., QT prolongation and triggering activity) increases the risk of fatal arrhythmias, which may lead to sudden cardiac death. One of the merits of iPS cells is that they can be used as patient or disease-specific stem cells, particularly in patients with genetic cardiac diseases. Because iPS cells derived from patients retain both their original genotype and phenotype, hiPS cells provide an excellent model for investigating the mechanisms underlying heart disease, as well as for drug testing and toxicology. Recently, Moretti et al. reported that patient-specific iPS cells maintained the disease genotype of the patient and recapitulated the functional features of the disorder (Moretti et al. 2009). In this study, fibroblasts derived from members of a family with autosomal-dominant LQTS Type 1 were reprogrammed to create LQT1 patient-specific iPS-CMs, which expressed cardiac-specific markers including cardiac troponin T, α-actinin, and atrial and ventricular isoforms of myosin light chain 2 (MLC2a and MLC2v). These iPS-CMs also exhibited particular electrophysiological characteristics including disease-specific abnormalities in APD, action potential rate adaptation, and I_{Ks}. Furthermore, precise characterization of a mutation in *KCNQ1* (R190Q) revealed a dominant-negative trafficking defect associated with a reduction in the I_{Ks} current, highlighting the benefit of using disease-specific hiPS-CMs to investigate the pathogenesis of a genetic disease. To date, insights into the pathogenesis of genetic arrhythmias have come primarily from heterologous expression systems or genetic animal models; however, such studies often produce conflicting results depending on the cell type and animal species used. In contrast, patient-derived hiPS-CMs precisely recapitulated the phenotypic and mechanistic features of native cardiomyocytes, precluding the discrepancies obtained using diverse experimental systems. Furthermore, it may be possible to investigate the therapeutic action of drugs to treat specific patients based on the *in vitro* analysis of their own hiPS-CMs, resulting in the establishment of patient-specific drug-screening systems, as well as customized therapies.

Although many causative mutations in ion channels and related genes have been identified in patients with channelopathies, the genetic mutations in approximately 40% of LQTS patients and 70% of Brugada syndrome patients have not been identified, suggesting that many of the genetic abnormalities that cause channelopathies remain unidentified. Furthermore, the severity of channelopathies may vary among family members, implying an association between phenotypic manifestation and epigenetic factors, in addition to genetic factors. In this regard, drug sensitivity testing or genetic screening of disease-specific hiPS-CMs could enable the identification of novel genetic targets for treatment.

References

Antzelevitch, C. 2005. Role of transmural dispersion of repolarization in the genesis of drug-induced Torsades de pointes. Heart Rhythm 2, S9–15.

Burashnikov, A. and Antzelevitch, C. 2002. Prominent I (Ks) in epicardium and endocardium contributes to development of transmural dispersion of repolarization but protects against development of early afterdepolarizations. J. Cardiovasc. Electrophysiol. 13, 172–177.

Caspi, O., Itzhaki, I., Kehat, I., Gepstein, A., Arbel, G., Huber, I., Satin, J., and Gepstein L. 2009. *In vitro* electrophysiological drug testing using human embryonic stem cell derived cardiomyocytes. Stem Cells Dev. 18, 161–172.

Cui, L., Johkura, K., Takei, S., Ogiwara, N., and Sasaki, K. 2007. Structural differentiation, proliferation, and association of human embryonic stem cell-derived cardiomyocytes *in vitro* and in their extracardiac tissues. J. Struct. Biol. 158, 307–317.

Evans, M. J. and Kaufman, M. H. 1981. Establishment in culture of pluripotential cells from mouse embryos. Nature 292, 154–156.

Fukuda, K. and Yuasa, S. 2006. Stem cells as a source of regenerative cardiomyocytes. Circ. Res. 98, 1002–1013.

Gussak, I., Litwin, J., Kleiman, R., Grisanti, S., and Morganroth, J. 2004. Drug-induced cardiac toxicity: emphasizing the role of electrocardiography in clinical research and drug development. J. Electrocardiol. 37, 19–24.

Harding, S.E., Ali, N. N., Brito-Martins, M., and Gorelik, J. 2007. The human embryonic stem cell-derived cardiomyocyte as a pharmacological model. Pharmacol. Ther. 113, 341–353.

Hattan, N., Kawaguchi, H., Ando, K., Kuwabara, E., Fujita, J., Murata, M., Suematsu, M., Mori, H., and Fukuda, K. 2005. Purified cardiomyocytes from bone marrow mesenchymal stem cells produce stable intracardiac grafts in mice. Cardiovasc. Res. 65, 334–344.

Hattori, F., Chen, H., Yamashita, H., Tohyama, S., Satoh, Y. S., Yuasa, S., Li, W., Yamakawa, H., Tanaka, T., Onitsuka, T. et al. 2009. Nongenetic method for purifying stem cell-derived cardiomyocytes. Nat. Methods.

He, J. Q., Ma, Y., Lee, Y., Thomson, J. A., and Kamp, T. J. 2003. Human embryonic stem cells develop into multiple types of cardiac myocytes: action potential characterization. Circ. Res. 93, 32–39.

January, C. T. and Moscucci, A. 1992. Cellular mechanisms of early afterdepolarizations. Ann. N. Y. Acad. Sci. 644, 23–32.

January, C. T. and Riddle, J. M. 1989. Early afterdepolarizations: mechanism of induction and block. A role for L-type Ca2+ current. Circ. Res. 64, 977–990.

January, C.T., Riddle, J. M., and Salata, J. J. 1988. A model for early afterdepolarizations: induction with the Ca2+ channel agonist Bay K 8644. Circ. Res. 62, 563–571.

Kehat, I., Kenyagin-Karsenti, D., Snir, M., Segev, H., Amit, M., Gepstein, A., Livne, E., Binah, O., Itskovitz-Eldor, J., and Gepstein, L. 2001. Human embryonic stem cells can differentiate into myocytes with structural and functional properties of cardiomyocytes. J. Clin. Invest. 108, 407–414.

Kehat, I., Khimovich, L., Caspi, O., Gepstein, A., Shofti, R., Arbel, G., Huber, I., Satin, J., Itskovitz-Eldor, J., and Gepstein, L. 2004. Electromechanical integration of cardiomyocytes derived from human embryonic stem cells. Nat. Biotechnol. 22, 1282–1289.

Kolossov, E., Bostani, T., Roell, W., Breitbach, M., Pillekamp, F., Nygren, J. M., Sasse, P., Rubenchik, O., Fries, J. W., Wenzel, D. et al. 2006. Engraftment of engineered ES cell-derived cardiomyocytes but not BM cells restores contractile function to the infarcted myocardium. J. Exp. Med. 203, 2315–2327.

Laflamme, M. A., Chen, K. Y., Naumova, A. V., Muskheli, V., Fugate, J. A., Dupras, S. K., Reinecke, H., Xu, C., Hassanipour, M., Police, S. et al. 2007. Cardiomyocytes derived from human embryonic stem cells in pro-survival factors enhance function of infarcted rat hearts. Nat. Biotechnol. 25, 1015–1024.

Lankipalli, R. S., Zhu, T., Guo, D., and Yan, G. X. 2005. Mechanisms underlying arrhythmogenesis in long QT syndrome. J. Electrocardiol. 38, 69–73.

Lev, S., Kehat, I., and Gepstein, L. 2005. Differentiation pathways in human embryonic stem cell-derived cardiomyocytes. Ann. N. Y. Acad. Sci. 1047, 50–65.

Mandel, Y., Weissman, A., Schick, R., Barad, L., Novak, A., Meiry, G., Goldberg, S., Lorber, A., Rosen, M. R., Itskovitz-Eldor, J. et al. 2012. Human embryonic and induced pluripotent stem cell-derived cardiomyocytes exhibit beat rate variability and power-law behavior. Circulation 125, 883–893.

Marban, E. 2002. Cardiac channelopathies. Nature 415, 213–218.

Mauritz, C., Schwanke, K., Reppel, M., Neef, S., Katsirntaki, K., Maier, L. S., Nguemo, F., Menke, S., Haustein, M., Hescheler, J. et al. 2008. Generation of functional murine cardiac myocytes from induced pluripotent stem cells. Circulation 118, 507–517.

McDevitt, T. C., Laflamme, M. A., and Murry, C. E. 2005. Proliferation of cardiomyocytes derived from human embryonic stem cells is mediated via the IGF/PI 3-kinase/Akt signaling pathway. J. Mol. Cell Cardiol. 39, 865–873.

Moretti, A., Bellin, M., Jung, C. B., Thies, T. M., Takashima, Y., Bernshausen, A., Schiemann, M., Fischer, S., Moosmang, S., Smith, A. G. et al. 2009. Mouse and human induced pluripotent stem cells as a source for multipotent Isl1+ cardiovascular progenitors. FASEB J.

Mummery, C., Ward-van Oostwaard, D., Doevendans, P., Spijker, R., van den Brink, S., Hassink, R., van der Heyden, M., Opthof, T., Pera, M., de la Riviere, A. B. et al. 2003. Differentiation of human embryonic stem cells to cardiomyocytes: role of coculture with visceral endoderm-like cells. Circulation 107, 2733–2740.

Murata, M., Tohyama, S., and Fukuda, K. 2010. Impacts of recent advances in cardiovascular regenerative medicine on clinical therapies and drug discovery. Pharmacol. Ther. 126, 109–118.

Nakagawa, M., Koyanagi, M., Tanabe, K., Takahashi, K., Ichisaka, T., Aoi, T., Okita, K., Mochiduki, Y., Takizawa, N., and Yamanaka, S. 2008. Generation of induced pluripotent stem cells without Myc from mouse and human fibroblasts. Nat. Biotechnol. 26, 101–106.

Narazaki, G., Uosaki, H., Teranishi, M., Okita, K., Kim, B., Matsuoka, S., Yamanaka, S., and Yamashita, J. K. 2008. Directed and systematic differentiation of cardiovascular cells from mouse induced pluripotent stem cells. Circulation 118, 498–506.

Norstrom, A., Akesson, K., Hardarson, T., Hamberger, L., Bjorquist, P., and Sartipy, P. 2006. Molecular and pharmacological properties of human embryonic stem cell-derived cardiomyocytes. Exp. Biol. Med. Maywood 231, 1753–1762.

Passier, R., van Laake, L. W., and Mummery, C. L. 2008. Stem-cell-based therapy and lessons from the heart. Nature 453, 322–329.

Paulussen, A. D., Gilissen, R. A., Armstrong, M., Doevendans, P. A., Verhasselt, P., Smeets, H. J., Schulze-Bahr, E., Haverkamp, W., Breithardt, G., Cohen, N. et al. 2004. Genetic variations of KCNQ1, KCNH2, SCN5A, KCNE1, and KCNE2 in drug-induced long QT syndrome patients. J. Mol. Med. 82, 182–188.

Priori, S. G., Aliot, E., Blomstrom-Lundqvist, C., Bossaert, L., Breithardt, G., Brugada, P., Camm, A. J., Cappato, R., Cobbe, S. M., Di Mario, C. et al. 2001a. Task Force on Sudden Cardiac Death of the European Society of Cardiology. Eur. Heart J. 22, 1374–1450.

Priori, S. G., Bloise, R., and Crotti, L. 2001b. The long QT syndrome. Europace 3, 16–27.

Roden, D. M. 2004. Drug-induced prolongation of the QT interval. N. Engl. J. Med. 350, 1013–1022.

Sartiani, L., Bettiol, E., Stillitano, F., Mugelli, A., Cerbai, E., and Jaconi, M. E. 2007. Developmental changes in cardiomyocytes differentiated from human embryonic stem cells: a molecular and electrophysiological approach. Stem Cells 25, 1136–1144.

Shantsila, E., Watson, T., and Lip, G.Y. 2007. Drug-induced QT-interval prolongation and proarrhythmic risk in the treatment of atrial arrhythmias. Europace 9 Suppl. 4, iv37–44.

Snir, M., Kehat, I., Gepstein, A., Coleman, R., Itskovitz-Eldor, J., Livne, E., and Gepstein, L. 2003. Assessment of the ultrastructural and proliferative properties of human embryonic stem cell-derived cardiomyocytes. Am. J. Physiol. Heart Circ. Physiol. 285, H2355–2363.

Snyders, D. J. 1999. Structure and function of cardiac potassium channels. Cardiovasc. Res. 42, 377–390.

Suzuki, I. and Yasuda, K. 2007. Detection of tetanus-induced effects in linearly lined-up micropatterned neuronal networks: application of a multi-electrode array chip combined with agarose microstructures. Biochem. Biophys. Res. Commun. 356, 470–475.

Takahashi, K., Okita, K., Nakagawa, M., and Yamanaka, S. 2007a. Induction of pluripotent stem cells from fibroblast cultures. Nat. Protoc. 2, 3081–3089.

Takahashi, K., Tanabe, K., Ohnuki, M., Narita, M., Ichisaka, T., Tomoda, K., and Yamanaka, S. 2007b. Induction of pluripotent stem cells from adult human fibroblasts by defined factors. Cell 131, 861–872.

Tamargo, J., Caballero, R., Gomez, R., Valenzuela, C., and Delpon, E. 2004. Pharmacology of cardiac potassium channels. Cardiovasc. Res. 62, 9–33.

Tanaka, T., Tohyama, S., Murata, M., Nomura, F., Kaneko, T., Chen, H., Hattori, F., Egashira, T., Seki, T., Ohno, Y. et al. 2009. *In vitro* pharmacologic testing using human induced pluripotent stem cell-derived cardiomyocytes. Biochem. Biophys. Res. Commun. 385, 497–502.

Thomson, J. A., Itskovitz-Eldor, J., Shapiro, S. S., Waknitz, M. A., Swiergiel, J. J., Marshall, V. S., and Jones, J. M. 1998. Embryonic stem cell lines derived from human blastocysts. Science 282, 1145–1147.

Wood, M. A. and Ellenbogen, K. A. 2002. Cardiology patient pages. Cardiac pacemakers from the patient's perspective. Circulation 105, 2136–2138.

Xu, C., He, J. Q., Kamp, T. J., Police, S., Hao, X., O'Sullivan, C., Carpenter, M. K., Lebkowski, J., and Gold, J. D. 2006. Human embryonic stem cell-derived cardiomyocytes can be maintained in defined medium without serum. Stem Cells Dev. 15, 931–941.

Xu, C., Police, S., Rao, N., and Carpenter, M. K. 2002. Characterization and enrichment of cardiomyocytes derived from human embryonic stem cells. Circ. Res. 91, 501–508.

Xue, T., Cho, H. C., Akar, F. G., Tsang, S. Y., Jones, S. P., Marban, E., Tomaselli, G. F., and Li, R. A. 2005. Functional integration of electrically active cardiac derivatives from genetically engineered human embryonic stem cells with quiescent recipient ventricular cardiomyocytes: insights into the development of cell-based pacemakers. Circulation 111, 11–20.

Yamashita, J., Itoh, H., Hirashima, M., Ogawa, M., Nishikawa, S., Yurugi, T., Naito, M., and Nakao, K. 2000. Flk1-positive cells derived from embryonic stem cells serve as vascular progenitors. Nature 408, 92–96.

Yanagi, K., Takano, M., Narazaki, G., Uosaki, H., Hoshino, T., Ishii, T., Misaki, T., and Yamashita, J. K. 2007. Hyperpolarization-activated cyclic nucleotide-gated channels and T-type calcium channels confer automaticity of embryonic stem cell-derived cardiomyocytes. Stem Cells 25, 2712–2719.

Yokoo, N., Baba, S., Kaichi, S., Niwa, A., Mima, T., Doi, H., Yamanaka, S., Nakahata, T., and Heike, T. 2009. The effects of cardioactive drugs on cardiomyocytes derived from human induced pluripotent stem cells. Biochem. Biophys. Res. Commun. 387, 482–488.

Yu, J., Vodyanik, M. A., Smuga-Otto, K., Antosiewicz-Bourget, J., Frane, J. L., Tian, S., Nie, J., Jonsdottir, G. A., Ruotti, V., Stewart, R. et al. 2007. Induced pluripotent stem cell lines derived from human somatic cells. Science 318, 1917–1920.

Zechnich, A. D., Hedges, J. R., Eiselt-Proteau, D., and Haxby, D. 1994. Possible interactions with terfenadine or astemizole. West. J. Med. 160, 321–325.

Zhang, J., Wilson, G. F., Soerens, A. G., Koonce, C. H., Yu, J., Palecek, S. P., Thomson, J. A., and Kamp, T. J. 2009. Functional cardiomyocytes derived from human induced pluripotent stem cells. Circ. Res. 104, e30–41.

Zwi, L., Caspi, O., Arbel, G., Huber, I., Gepstein, A., Park, I. H., and Gepstein, L. 2009. Cardiomyocyte differentiation of human induced pluripotent stem cells. Circulation 120, 1513–1523.

CHAPTER 11

Cardiac Regeneration Using Isl1-positive Cardiac Progenitor Cells

Yasuhiro Nakashima and Atsushi Nakano[a,*]

Introduction

The mammalian heart is a complex multi-chambered pump consisting of two ventricles, two atria, outflow and inflow tracts, conduction system, valves, endocardium, and coronary vessels. Proper functioning of the heart requires a highly coordinated process of assembly and integration of all these components. It has long been believed that the identity of each heart chamber is determined in each segment at the linear heart tube stage. However, this classical segmental prespecification view of cardiogenesis has been challenged by the identification of the second heart field (SHF). The cardiac progenitors in this SHF are delineated by the expression of Isl1, a LIM homeodomain transcription factor that plays a key role during cardiogenesis in vertebrates and non-vertebrates (Cai et al. 2003; Dodou et al. 2004; Lin et al. 2006; Lin et al. 2007; Mann et al. 2009; Moretti et al. 2006; Sun et al. 2007). In this chapter, we will review the current understanding of the biology of Isl1$^+$ cardiac progenitors and their potential application to cardiac regenerative medicine.

Department of Molecular Cell & Developmental Biology, Eli and Edythe Broad Center of Regenerative Medicine and Stem Cell Research, Jonsson Comprehensive Cancer Center, Molecular Biology Institute, University of California Los Angeles, 615 Charles E. Young Dr. South, Los Angeles, CA 90095.
[a]Email: anakano@ucla.edu
*Corresponding author

Isl1 in Heart Development

Isl1 expression in heart development

Cardiogenesis begins with the formation of the cardiac crescent, the first morphologically distinguishable structure of cardiac primordia during early embryogenesis. However, the existence of an extra-crescent population of cardiac progenitors was speculated more than 30 years ago (de la Cruz et al. 1977). This second source of cardiac progenitors was revisited and further characterized by several groups using different techniques; dye labeling, specific molecular markers, and retrospective single cell fate tracing experiments determined similar but slightly different populations of cardiac progenitors (Cai et al. 2003; Dodou et al. 2004; Kelly et al. 2001; Meilhac et al. 2004; Mjaatvedt et al. 2001; Waldo et al. 2001). In this review, "SHF" is used in a broader sense including the anterior and posterior subsets of the SHF, which corresponds to the *Isl1*$^+$ domains.

Isl1 is expressed in the SHF region at the cardiac crescent stage. At E7.0–E8.0 of mouse development, *Isl1* is expressed in the dorsomedial region of the first heart field (FHF), classically known as the cardiac crescent (Cai et al. 2003). As the bilateral cardiac crescent fuses and forms the single linear primitive heart tube, Isl1$^+$ progenitors migrate from the splanchnic mesenchyme toward the primitive heart tube. *In situ* hybridization revealed that *Isl1*-positive domains at the crescent and early linear heart tube stages are distinct from the FHF domain expressing Mlc2a, a marker of differentiated cardiomyocytes (Cai et al. 2003), suggesting that *Isl1* labels immature cardiac progenitors. Until the looping of the linear heart tube, the primitive heart tube is continuous to the body wall at any level through the dorsal mesocardium, the mesentery that suspends the linear heart tube. Thus, Isl1$^+$ progenitors migrate to the primitive heart tube at any level at this stage. However, upon breakdown of the dorsal mesocardium, the heart tube becomes connected to the body wall only at the arterial and the venous poles. From this point on, the migration path of Isl1$^+$ cardiac progenitors becomes restricted to these two poles (Fig. 1). The addition of cardiac progenitors at these two poles triggers rightward looping of the heart tube. *Isl1* expression is detected in the outflow tract (OFT), right atria (RA) and right ventricle (RV), but not in the remainder of the myocardium (Sun et al. 2007). At E9.5, *Isl1* expression becomes restricted to the several cardiac regions including the sinus venosus and OFT. *Isl1* continues to be expressed in ventral endoderm, splanchnic mesoderm, dorsal mesenchymal protrusion (DMP), and cardiac inflow tract (IFT) at the boundary of the myocardium and vascular smooth muscle layer, until mid-gestational stages (Galli et al. 2008; Snarr et al. 2007a; Sun et al. 2007). At fetal stages, *Isl1* expression persists in a restricted region within the OFT, aorta, pulmonary artery, RV,

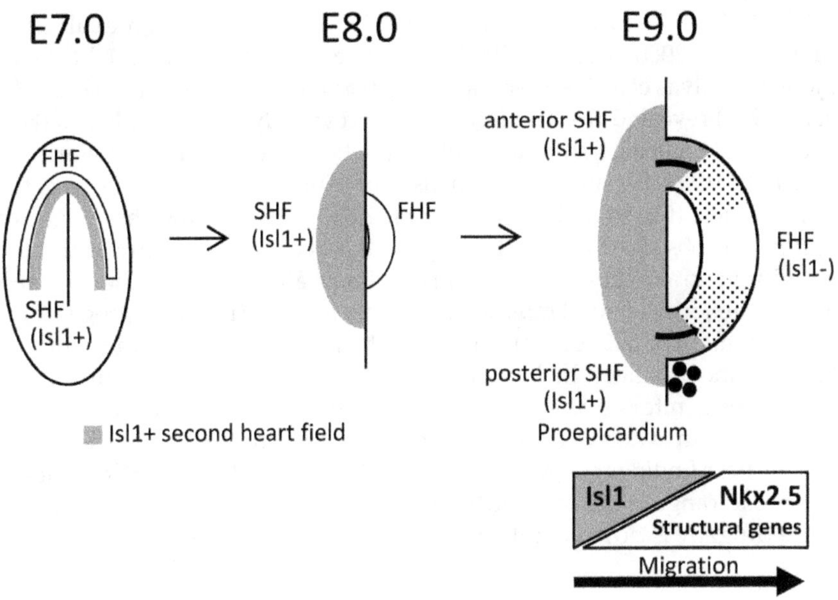

Figure 1. Migration of Isl1⁺ second heart field progenitors. Isl1⁺ progenitors appear in the anterior lateral plate mesoderm dorsomedial to the cardiac crescent first heart field. After the formation of primitive heart tube and breakdown of dorsal mesocardium, Isl1⁺ progenitors migrate through arterial and venous poles of the heart tube. Whereas arterial (anterior) population acquires ventricular phenotype and contributes to RV and OFT, venous (posterior) population gives rise to atrial myocytes in atria and inflow tracts.

venous valves, atrial septum, the sinoatrial (SA) and atrioventricular (AV) nodes. In the neonatal heart, *Isl1* is still expressed in the region of the SA node and at the base of the aorta/pulmonary artery, but less extensively, with the exception of cardiac ganglia, which exhibit strong expression. *Isl1* expression is also observed in scattered cells within the myocardium at the base of the aorta and pulmonary artery. Together, *Isl1* expression specifically marks immature cardiac progenitors in the SHF during early cardiac development, and remains only within a restricted region of the heart in later stages.

The progeny of Isl1⁺ cardiac progenitors in myocardial lineage

In vivo cell lineage tracing using the Cre-loxP strategy in the mouse has been an invaluable tool for precisely delineating the specific cell types that derive from *Isl1⁺* precursors, as well as for defining their specific locations in the components of the heart, vasculature and conduction system (Laugwitz et al. 2008). Several lines of data using Cre-loxP method revealed the contribution of Isl1⁺ cardiac progenitors to SHF-derived tissues (Cai et al.

2003; Dodou et al. 2004; Lin et al. 2006; Lin et al. 2007; Mann et al. 2009; Moretti et al. 2006; Sun et al. 2007). The mice crossed with an *Isl1-IRES-Cre* mouse (Srinivas et al. 2001) and a *CMV β-actin-lacZ* indicator mouse (Zinyk et al. 1998) revealed that the cells which previously expressed *Isl1* make a substantial contribution to the embryonic heart, comprising a majority of cells in the OFT, RV, and atria, and also contribute to specific regions of the left ventricle (Cai et al. 2003; Sun et al. 2007). A β-galactosidase negative region was observed within the ventral aspect of the left ventricle and the anterior ventral region of the left atria. Almost all cells within the OFT and RV, and majority of atrial muscle at E9.75 are derived from Isl1$^+$ progenitors. The migration of Isl1$^+$ progenitors to the heart tube continues until mid- to late-gestational stages. In the conduction system, a remarkable contribution of Isl1$^+$ progenitors was seen in the SA node, while only a few cells of the AV node seem to derive from Isl1$^+$ progenitors (Moretti et al. 2006; Sun et al. 2007). These findings suggest that Isl1$^+$ cardiac progenitors in SHF contribute to a broad range of the myocardium in the heart and that *Isl1* expression in these progeny is down regulated after they complete migration.

Contribution of Isl1$^+$ cardiac progenitors beyond the myocardial lineages

Isl1$^+$ cardiac progenitors in the SHF also contribute broadly to the other cardiovascular cell types. Cre-loxP lineage tracing experiments suggest that, besides the myocardial lineage, the progenies of Isl1$^+$ cardiac progenitors are detected in the endocardium and the endothelial cells lining the aortic arch arteries (Cai et al. 2003). Additionally, *Isl1*-derived cells were found throughout the smooth muscle wall of the proximal aorta, the trunk of the pulmonary artery, the stems of the main left and right coronary arteries, and connective tissue in the aortic and pulmonary valve leaflets. Although Cre-loxP lineage tracing by itself cannot tell whether these various types of *Isl1*-progenies arise from a common origin, a clonal analysis of ES-derived cardiac progenitors clearly demonstrates that a single Isl1$^+$ progenitor gives rise to three major cardiovascular cell lineages—cardiac, smooth muscle and endothelial cells (Moretti et al. 2006). These observations were the first demonstration that a subset of endocardial/endothelial cells may share common origin with the cardiac lineage during mammalian embryogenesis (Kattman et al. 2006; Moretti et al. 2006).

Smooth muscle cells are recruited locally from a wide range of tissues in response to specific local signals during embryogenesis. While the endocardial/endothelial lineage seems to separate from cardiac lineage in the early stages, cardiac and smooth muscle lineages are closely related until later stages. Lineage tracing of *Isl1-Cre* or *AHF-Cre* lines (a *Cre* line

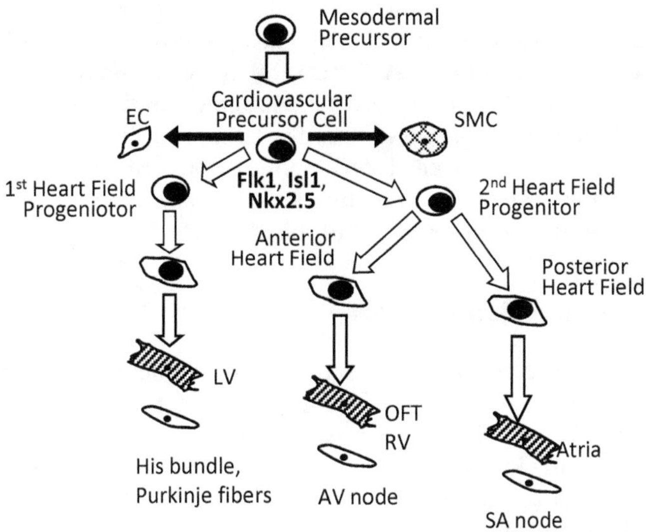

Figure 2. A model for lineage diversification of cardiovascular progenitors. Cardiovascular progenitors are specified from mesodermal precursors shortly after gastrulation. Putative common cardiovascular precursors give rise to FHF and SHF progenitors. The SHF progenitors are further subdivided into anterior and posterior SHF subpopulations.

driven by specific Mef2c promoter) and retrospective clonal assay suggest that a fraction of smooth muscle cells at the junction of OFT and ascending aorta arise from cardiac progenitors (Esner et al. 2006; Verzi et al. 2005; Waldo et al. 2005). Likewise, lineage analysis using atrial-specific *Sln-Cre* mice and *ex vivo* clonal analysis on feeder cells indicate that atrial subset of Isl1$^+$ progenitors clonally contribute to atrial working cardiomyocytes and smooth muscle cells at the heart-vessel junction in the IFT (Nakano et al. 2011). Interestingly, these Isl1$^+$ progenitors retain smooth muscle competency until late gestational stages. These observations lead to the speculation that the cardiac-smooth muscle bipotency of later Isl1$^+$ cardiac progenitors may serve to form a smooth boundary between the heart chamber and great vessels at outflow and inflow tracts.

Another population of the smooth muscle cells in the heart, the coronary wall, arises from the proepicardium, grape-like clusters that form between the sinus venosus and liver primordium. Using retroviral tagging methods, Mikawa and Gourdie demonstrated that the proepicardium is the source of the coronary vasculature (Mikawa and Gourdie 1996). Interestingly, these pro-epicardial progenitor cells are positive for several cardiac transcription factors including *Isl1* and *Nkx2-5*. Furthermore, a combinatorial approach using mouse genetic models (*Tbx18-Cre* and *Wt1-Cre*) and the ES system showed that pro-epicardial cells also contribute to a fraction of cardiomyocytes (Cai et al. 2008; Zhou et al. 2008). These reports provide the basis for the concept of epicardial-derived cells (EPDCs) that contribute to both coronary vasculature and myocardium (Gittenberger-de Groot et al. 2010; Lie-Venema et al. 2007; Munoz-Chapuli et al. 2002; Wessels and Perez-Pomares 2004). Given that the proepicardium is spatially and temporarily continuous to the posterior population of SHF, the proepicardial progenitors may share a common developmental origin with Isl1$^+$ cardiac progenitors in the posterior subset of the SHF.

Isl1 expression in human heart

Isl1$^+$ cardiac progenitors constitute the SHF in human hearts as well. *Isl1*$^+$ cells were found in the RA and OFT as well as in the left atrial wall and appendage in 11 and 18 weeks human fetal heart (Bu et al. 2009), which is consistent with the known contribution of Isl1$^+$ cardiac progenitors to SHF derived cells during murine cardiogenesis. Double-positive cells, expressing ISL1 and lineage-specific differentiation markers, cTNT, SM-MHC and PECAM1, were also identified in these regions. At later stage fetal hearts of 18 weeks of gestation, the relative and total numbers of Isl1$^+$ cardiac progenitors were significantly reduced, consistent with rodent models (Bu et al. 2009; Cai et al. 2003). Thus, the lineage diversification of Isl1$^+$ cardiac progenitors is evolutionally conserved in rodents and humans.

Isl1 expression in the FHF and the significance of the identification of SHF

Many of the analyses of the lineage contribution in mouse embryos are heavily based on the Cre-loxP recombination system. While this system is highly useful for tracing the lineage contribution of specific cell types in mice, it does have limitations: (1) it inevitably turns the gradient of the expression level of the Cre driver gene into a black-and-white image of the reporter gene, which potentially causes oversight of weak and short expression of the Cre drivers, (2) the readout depends on the ubiquity of the reporter gene and reporter, and (3) the recombination process causes a time gap between the expression of the endogenous gene and the reporter gene. These limitations can be critical when analyzing early embryonic events and sometimes complicate the interpretation of the results. For example, recent investigations have raised the possibility that *Isl1* is also expressed in the FHF: Xenopus embryos show the expression of *Isl1* in the cardiac crescent (Brade et al. 2007). Immunostaining on the mouse heart indicates that *Isl1* protein is expressed at E7.5 throughout the anterior intraembryonic coelomic walls (Prall et al. 2007), even though *Isl1-Cre* lineage tracing does not show significant contribution of $Isl1^+$ progenitors to the left ventricle (Cai et al. 2003; Sun et al. 2007). Furthermore, the analyses of the same *Isl1-Cre* line with another reporter line, *Gata4-flap* reporter mice, in which alkaline phosphatase reporter cassette is knocked into *Gata4* locus, demonstrated robust expression of the reporter gene in the entire heart including the left ventricle and proepicardium (Ma et al. 2008). These observations suggest that *Isl1* may be a pan-cardiac marker that labels both the FHF and SHF. The identification of specific markers for the FHF would be useful for resolving the apparently contradicting results regarding the lineage diversification during early cardiogenesis. Importantly, these results raise the possibility that Rosa26-lacZ reporter line, the most commonly used reporter line, may not be a perfect standard for tracing the earliest events during embryogenesis.

The possibility that *Isl1* is a pan-cardiac marker leads to the argument whether the FHF and SHF are "distinct" populations or the same population at slightly different stages. Obviously, a cell population should not be defined only by the expression of marker genes but also by its differentiation capability and behavior. *In vivo* retrospective clonal assay predicted that the FHF and SHF lineages likely originate from putative common precursors (Meilhac et al. 2004). Indeed, there is no evidence that these two populations are functionally different or mutually unexchangeable. Furthermore, there are no morphological differences between the progenitors in the FHF and SHF at the cellular level, nor are there clear morphological boundaries at the cardiac crescent stage. Even in the post-natal heart, there are no essential morphological differences among cardiomyocytes in OFT/RV and LV at the cellular level. Thus, the argument still remains as to whether it is relevant to define so-called SHF as another "organ field" of the heart (Abu-Issa et al. 2004; Moorman et al. 2007).

The significance of the identification of SHF is, rather, the discovery that cardiac progenitors continue to migrate to the heart tube until unexpectedly later stages. The cardiac crescent is a dynamic structure to which the progenitors are continuously supplied from surrounding extra-crescent tissues. The identification of $Isl1^+$ progenitors suggests that this supplementation continues throughout the process of heart tube formation, rightward looping, chamber ballooning and septation. Even at the late gestational stages, the addition of myocardia continues at specific domains of the venous pole. Thus, cardiogenesis is an even more dynamic process than was traditionally envisioned. In this regard, of great interest is the mechanism underlying the recruitment of cardiomyocytes from the yet-to-be committed progenitors at the later stages.

Molecular Network Regulating Isl1-positive Progenitors

Isl1 transcription factor

LIM homeodomain (HD) proteins are a group of homeodomain-containing transcription factors that are defined on the basis of a common LIM domain. The acronym of LIM is derived from the first identified members of this family, namely LIN-11 from *Caenorhabditis elegans* (Freyd et al. 1990), *Isl1* from rats (Karlsson et al. 1990), and MEC-3 from *C. elegans* (Way and Chalfie, 1988). LIM domain consists of a conserved cysteine- and histidine-rich structure of two tandemly repeated zinc fingers. LIM homeodomain proteins form a complex with nuclear LIM only proteins and nuclear Lim Interactor proteins (Gill 2003). Studies of LIM/HD proteins in embryonic motor neurons identified the combinations of several LIM/HD proteins which induce distinct motor neurons (Gill 2003; Thaler et al. 2004; Thor et al. 1999; Tsuchida et al. 1994). Aside from its expression in the embryonic cardiovascular system, *Isl1* is expressed in a variety of cell lineages including motor neurons and pancreatic endocrine islet cells.

As discussed above, *Isl1* is expressed widely in the cardiovascular lineages. Hence, Isl1^{+} progenitors represent a pool of cells in various cardiac sub-lineages, which likely arise from a common source. The common function of the *Isl1* transcription factor in these diverse populations is of interest. *Isl1* expression strongly correlates with the cell cycle in the postnatal myocardium, such that *Isl1* expression can be quickly turned on and off upon mitotic stimulation in the various sub-lineages of the cardiac progenitors (Nakano, unpublished). Therefore, *Isl1* may well be regarded as a marker for mitotic cardiac progenitors/myocytes. Many signaling pathways are known to influence the differentiation and expansion of Isl1^{+} cardiac progenitors during development and cell culture. The precise analysis of the function of *Isl1* transcription factor is a challenge as it is dependent on stage, sub-lineage, and expression dosage.

The role of Isl1 in cardiac development

Genetic analyses suggest a critical role of the *Isl1* transcription factor during cardiogenesis. Mice that are homozygous null for *Isl1* exhibit growth retardation at approximately E9.5 and die at approximately E10.5 (Ahlgren et al. 1997; Pfaff et al. 1996). *Isl1*-null hearts lack OFT and RV, and fail to undergo heart looping (Cai et al. 2003). Histological analysis showed that mutant hearts show down regulation of SHF markers including *Fgf10* and *Wnt11*, whereas the expression of FHF markers, *Tbx5* (Bruneau et al. 1999) and *eHAND*, were preserved, suggesting that *Isl1* is required for the development of OFT and RV. Atrial tissue in *Isl1* mutants also showed a significant reduction in size. Therefore, *Isl1* deficiency results in the loss of complete segments of SHF in the heart, suggesting that *Isl1* is required for their development. In this mutant, the cell fate of *Isl1*-null SHF progenitors can be traced by the expression of the remnant of the disrupted *Isl1* gene. This experiment revealed that the SHF progenitors were progressively lost in *Isl1*-null background. Proliferative indices with BrdU labeling were significantly less in *Isl1* mutants, with 76% reduction in splanchnic mesoderm and 39% reduction in pharyngeal endoderm relative to their controls where *Isl1* is normally expressed. *Isl1*–null cardiac progenitors underwent extensive apoptosis, whereas no apoptosis was seen in wild-type mice. *Isl1* lineage tracing analysis in *Isl1* null background showed the defect of migration of SHF-derived myocytes into the ventricle. These findings suggest that *Isl1* deletion results in failure of the expansion and migration of early SHF progenitors and that *Isl1* is essential for cell proliferation and survival in $Isl1^+$ cardiac progenitors.

In the ES *in vitro* differentiation model, *Isl1* knockdown in *Nkx2.5–GFP* ES cell line showed increase in the $Nkx2.5^+$ cardiac progenitor population (Kwon et al. 2009). In contrast, cells differentiated from the *Isl1* knockdown ES cells showed severely reduced beating frequencies and compromised expression of cardiac sarcomeric genes (*Myh6, Myh7, Mlc2a, and Mlc2v*). When differentiated by re-aggregation in suspension after purification of *Nkx2.5*$^+$ cardiac progenitors, expression of cardiomyocyte and smooth muscle cell genes were markedly down regulated while expression levels of the endothelial genes (CD31, Flk1) remained normal. Consistently, *Isl1*-overexpression resulted in the significant increase of cardiomyocytes in the ES system. Although this model does not allow for the analysis of *Nkx2.5*-negative cardiac progenitor population, these *in vitro* findings implicate that *Isl1* not only suppresses the expansion of *Nkx2.5*$^+$ cardiac progenitors, but is also necessary for proper differentiation of cardiac progenitors into myocardial and smooth muscle cell lineages.

Transcriptional network during $Isl1^+$ SHF development

The *Isl1* gene has an essential enhancer region at the evolutionarily conserved non-coding region located downstream of the translational stop codon, which is responsible for expression in the cardiac crescent until E8.0 and in the pharyngeal mesoderm and OFT at E8.5 in the mouse (Kang et al. 2009). This region includes several binding sites for forkhead (Fox) transcription factors, several members of which are expressed in the cardiogenic region during the early development in a highly conserved manner among humans, mice and chickens. *FoxA2, FoxC2* and *FoxF1* bind to this enhancer region; mutation in these Fox binding sites greatly diminishes its enhancer activity in the SHF *in vivo*, supporting the notion that *Isl1* is directly regulated by Fox transcription factors in the SHF. These Fox transcription factors are implicated in early cardiac development (Papanicolaou et al. 2008). *Foxc1* and *Foxc2* are expressed in the SHF at the cardiac crescent stage (Seo et al. 2006) and knockout of either *Foxc1* or *Foxc2* in the mouse results in perinatal lethality with a spectrum of cardiac defects in aortic arches, ventricular septal defects, and abnormal semi-lunar valves (Winnier et al. 1999). Moreover, mice lacking both *Foxc1* and *Foxc2* genes showed severe defects in RV and OFT formation, and *Isl1* expression in the double mutants was reduced in the pharyngeal and splanchnic mesoderm at E8.5 but not in the pharyngeal endoderm.

The enhancer region of the *Mef2c* gene that is responsible for its expression in the anterior domain of the SHF contains two consensus sites for *Isl1*, together with two GATA binding sites in an evolutionarily conserved manner. Mutation of either *Isl1* or GATA binding sites abrogates the expression of *Mef2c* transgene *in vivo*, suggesting an essential contribution of these factors to *Mef2c* expression in anterior SHF (Dodou et al. 2004). Mice lacking *Mef2c* are embryonic lethal with hypoplastic hearts and a single, linear ventricular chamber. The preserved expression of this enhancer in the common ventricular chamber of *Mef2c* null embryos implicates that the role of *Mef2c* might be the proper differentiation of the progenitor cells in anterior SHF rather than their migration to the heart tube (Verzi et al. 2005).

Wnt signaling and SHF progenitor

Wnt/β-catenin pathway has a highly context-dependent role during cardiogenesis. During early avian cardiogenesis, canonical Wnt signals from the neuroectoderm inhibit the specification of cardiac progenitors, resulting in a crescent-shaped heart-forming region (Marvin et al. 2001; Schneider and Mercola 2001; Tzahor and Lassar 2001). Mouse ES based data mostly suggest a positive effect of Wnt/β-catenin signaling on the appearance of beating cardiomyocytes (Paige et al. 2010; Qyang et al. 2007; Tzahor 2007; Ueno et al. 2007). *In vivo* loss- and gain-of-function experiments using various Cre mice mostly resulted in the conclusion that Wnt/β-catenin is involved in expansion of the cardiac progenitor population (Cohen et al. 2007; Klaus et al. 2007; Kwon et al. 2007; Lin et al. 2007; Naito et al. 2006; Qyang et al. 2007; Ueno et al. 2007).

Wnt/β-catenin signaling directly regulates *Isl1* expression in distal OFT, aortic sac, and pharyngeal mesoderm (Lin et al. 2007). The proximal promoter region of *Isl1* gene has an evolutionarily conserved LEF-1 binding site and is directly up regulated by LEF-1/β-catenin (Cohen et al. 2007; Lin et al. 2007). Wnt/β-catenin signaling reporter mice also show Wnt/β-catenin activation within the SHF. The gain of function of Wnt/β-catenin signaling in anterior subpopulation of SHF results in the enlargement of the OFT and pharyngeal mesoderm with over-proliferation of Isl1$^+$ cardiac progenitors (Cohen et al. 2007; Kwon et al. 2007; Qyang et al. 2007). Loss of function of Wnt/β-catenin signaling in the anterior SHF region results in defective RV development and decreased Isl1$^+$ cardiac progenitors in the OFT and pharyngeal mesoderm. These data indicate that canonical Wnt signaling promotes the expansion of Isl1$^+$ cardiac progenitors.

Wnt2 is expressed in a restricted region in the heart, namely, the posterior domain of the SHF. In *Wnt2*$^{-/-}$ mutants, nuclear translocation of β-catenin and the expression of downstream targets were both reduced in the IFT, the dorsal mesocardium and DMP. The number of *Isl1*$^+$ SHF progenitors is significantly reduced in the dorsal mesocardium and DMP of *Wnt2*$^{-/-}$ mutants at E9.5 and E10.5 (Tian et al. 2010). These data indicate that Wnt2 regulates the expansion of *Isl1*$^+$ SHF progenitors in the posterior pole of the heart. To sum up, Wnt/β-catenin signaling expands Isl1$^+$ cardiac progenitors and play pivotal roles in the SHF development in a spatiotemporally dependent manner *in vivo*. The Wnt/β-catenin pathway is clearly one of the most promising factors for cardiac regenerative purposes.

FGFs and BMPs

FGFs and BMPs are implicated in specification, growth and survival of cardiogenic precursors in both the vertebrate and invertebrate heart (Kirby 2002; Yutzey and Kirby 2002). The combination of BMP2 and FGF4 – but neither alone—induces cardiogenesis from the pre-cardiac mesoderm (Barron et al. 2000; Lough et al. 1996). These factors influence the expansion and differentiation of Isl1$^+$ cardiac progenitors. In *Isl1* null mice, the expression of *Fgf10*, *Bmp4*, and *Bmp7* is severely down regulated or undetectable in pharyngeal endoderm and splanchnic mesoderm, and the domains expressing *Fgf4*, *Fgf8*, *Bmp2*, and *Bmp6* are decreased in regions overlapping or contiguous with *Isl1* expression (Cai et al. 2003). Mouse knockouts of *Fgf4* or *Fgf8* are early embryonic lethal, and mice that are hypomorphic for *Fgf8* die neonatally due to cardiovascular defects, including malformations of the OFT (Abu-Issa et al. 2002; Feldman et al. 1995; Frank et al. 2002; Sun et al. 1999). Although *Fgf10* knockout mice die neonatally showing the lung phenotype, additional deletion of *Fgf10* over *Fgf8* knockout mice shows more severe anomalies in OFT development (Watanabe et al. 2010). *Isl1-Cre* specific deletion of *Bmpr1a* shows abnormal looping of the OFT at E9.5, thinning of the ventricular wall by E11.5, and septal defects in the OFT, ventricles and atria by E13.5, suggesting the critical role of BMP signaling in cardiogenesis (Yang et al. 2006). BMP and FGF signaling pathways constitute a complex regulatory mechanism during cardiogenesis (Tirosh-Finkel et al. 2010; van Wijk et al. 2009; Watanabe et al. 2010), and it remains to be seen how precisely *Isl1* is involved in this regulatory network.

Isl1 mutation and congenital heart diseases

A recent case-control study of 300 patients demonstrates that two different *Isl1* haplotypes contribute to the risk of congenital heart diseases in Caucasian and African American populations (Stevens et al. 2010). These data suggest that *Isl1* mutation is involved in human diseases and that congenital heart disease is consistent with the common variant–common disease hypothesis in two ethnically distinct populations.

Other studies suggest that impaired development of the DMP is involved in the pathogenesis of atrial and AV septal defects (Sharratt et al. 2003; Snarr et al. 2007b; Wirrig et al. 2007), particularly in patients with Down syndrome and in Trisomy 16 (Ts16) mouse model for this condition (Blom et al. 2003; Webb et al. 1999). Examination of E12 Ts16 mouse embryos with AV canal defects confirmed that the DMP was severely underdeveloped, and no *Isl1*$^+$ cells were found in the DMP in Ts16 embryos. These data suggest a possible role of Isl1$^+$ progenitors in the formation of the atrial septum.

Isolation of Isl1-positive Progenitors

Sources of Isl1$^+$ cardiac progenitor cells

In the early stage of cardiac development, *Isl1*$^+$ SHF progenitors emerge in the extra-crescent cardiac field and migrate to the heart at anterior and posterior poles before differentiating into cardiomyocytes, endothelial cells and smooth muscle cells. These ***embryonic Isl1$^+$ cardiac progenitors*** can be isolated and cultured *in vitro*. However, although most accurately reflecting the *in vivo* progenitors in SHF, this type of Isl1$^+$ cardiac progenitors usually display limited proliferation capability. On the other hand, ***postnatal*** and ***ES-derived Isl1$^+$ cardiac progenitors*** display robust proliferation, which have led to the idea of repairing damaged heart muscle by internal or external cardiac progenitor cells (Fig. 3).

Although the survival and growth of Isl1$^+$ cardiac progenitors on the plastic dish is very low, the use of a feeder layer (Laugwitz et al. 2005; Moretti et al. 2006) or re-aggregation method (Kwon et al. 2009) significantly improve the propagation of Isl1$^+$ cardiac progenitor cells. Cardiac mesenchymal feeder cells isolated from neonatal or embryonic mouse hearts promote the expansion of Isl1$^+$ cardiac progenitors without differentiation, and serve as a prespecification matrix toward an early cardiac precursor state.

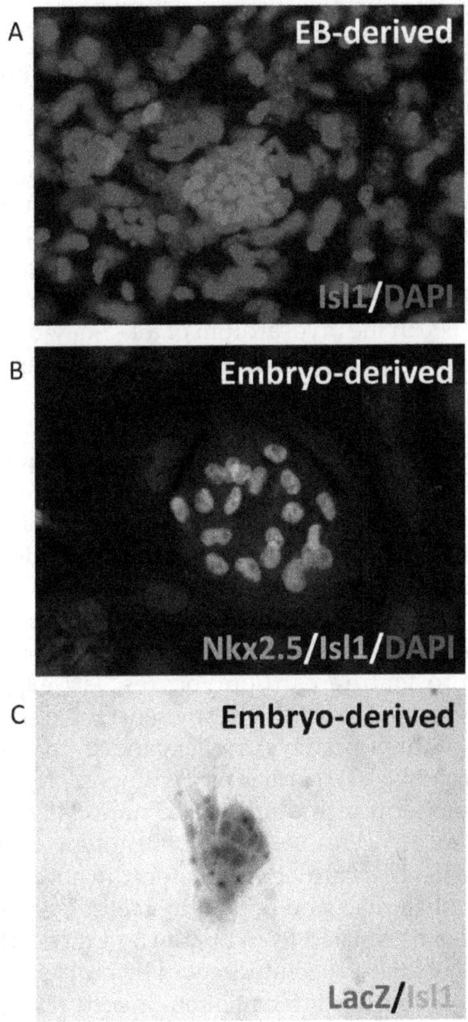

Figure 3. Isl1⁺ cardiac progenitors isolated on cardiac mesencymal feeder cells. (A) ES-derived Isl1⁺ cardiac progenitor colony on cardiac mesenchymal feeder layer. (B) Mouse embryo-derived Isl1⁺ cardiac progenitor colony on cardiac mesenchymal feeder layer. Cells in the colony express both Isl1 and Nkx2.5. (C) A colony of Isl1⁺ cardiac cells derived from posterior SHF at later stage. Progenitors were isolated from *Sln-Cre; R26R* mouse embryo at E10.5 and grown on feeder layer. LacZ staining indicates that these progenitors are derived from the posterior sublineage. Note that these Isl1⁺ cells display myofibrils.

Color image of this figure appears in the color plate section at the end of the book.

Postnatal Isl1⁺ cardiac progenitors

Postnatal $Isl1^+$ cardiac progenitors can be isolated and cultured *in vitro* from mouse hearts (Laugwitz et al. 2005). After *Isl1* expression is down regulated as the cardiac formation proceeds, a subset of *Isl1*$^+$ undifferentiated progenitors remains embedded in the sub-domain of SHF in the heart in an evolutionally conserved manner in mice, rats and humans (Genead et al. 2010; Laugwitz et al. 2005; Sun et al. 2007). The number of *Isl1* positive cells decreases progressively from embryonic day 12.5 to 18.5. After birth, relatively few $Isl1^+$ cardiac progenitors were still detectable, averaging from about 500 to 600 in the myocardium of a 1–5-day-old rat. This type of cardiac progenitor is presumably the developmental remnant of embryonic $Isl1^+$ cardiac progenitors (Laugwitz et al. 2005).

Postnatal $Isl1^+$ cardiac progenitors were isolated using inducible *Isl1-Cre* mice harboring tamoxifen-dependent Cre recombinase (*Isl1-mER-Cre-mER*) into the *Isl1* locus. Administration of Tamoxifen induces a rapid nuclear translocation of the mER-Cre-mER protein in this mouse line, and irreversibly marks *Isl1* expressing cells in a temporally-dependent manner (Laugwitz et al. 2005). After digestion of the neonatal heart tissues and culture, $Isl1^+$ progenitors were enriched in cardiac mesenchymal fraction. They expanded rapidly on the cardiac mesenchymal feeder layer without differentiating into mature cardiomyocytes. *Isl1*-expressing cardiac progenitors corresponded to 0.5% of the mesenchymal cell fraction during an early stage in culture (4–6 days) and expanded up to ~8% after 15–18 days in culture. Postnatal $Isl1^+$ cardiac progenitors expressed early markers of the cardiac mesoderm including Nkx2.5 and GATA-4, while lacking transcripts of mature myocytes, suggesting phenotypic characteristics of an undifferentiated state. $Isl1^+$ cardiac progenitors did not correspond to $Sca1^+$ cells, $cKit^+$ cells and cardiac side population cells. Postnatal $Isl1^+$ cardiac progenitors can also be isolated from human and rat hearts.

Postnatal $Isl1^+$ cardiac progenitors could be propagated for more than two weeks without losing the differentiation capacity (Laugwitz et al. 2005). Thirty percent of these $Isl1^+$ cardiac progenitor-derived cardiomyocytes expanded and differentiated into cardiomyocytes in culture, and displayed mature cardiac phenotype including the formation of the myocytic syncytium with surrounding cardiomyocytes, and the expression of cardiac specific proteins in completely organized pattern when cultured with neonatal cardiac myocytes. A subset of them showed contractile activity and intact electromechanical coupling to neighboring cells, as documented by the expression of the gap-junction protein, connexin 43. Physiologically, they showed a differentiated cardiac phenotype in real-time intracellular calcium $[Ca^{2+}]_i$ imaging, isoproterenol responsiveness, and action potential characteristics to neonatal cardiac myocytes. Repeated impalements of individual differentiated progenitors revealed reproducible action potentials with distinct morphologies, nodal-like and atrial-like, supporting the concept that these cells are able to differentiate into a distinct subset of cardiac lineages, including conduction system myocytes (He et al. 2003). Although their role in the *in vivo* heart is still unknown, postnatal Isl1⁺ cardiac progenitors can be a potential source for cell-based therapy for the damaged heart.

ES cell-derived Isl1⁺ cardiac progenitors

Isl1⁺ cardiac progenitors were also successfully isolated from mouse and human ES cell *in vitro* differentiation system and could be efficiently expanded on the cardiac mesenchymal feeder layer (Moretti et al. 2006). In the ES cell *in vitro* differentiation model, *Isl1* is expressed from day 4–6, shortly after Bry expression, followed by late cardiac maker expression such as MHC and Mlc2a/2v in a manner recapitulating cardiac development (Kwon et al. 2007; Moretti et al. 2006). Isl1⁺ cardiac progenitors can be clonally expanded by dissociating the EBs at day 5 into single cells and plating them at low density on cardiac mesenchymal feeder layer. Single or dividing *Isl1*-expressing cells are observed on the cardiac mesenchymal feeder after one day and expand to form an Isl1⁺ colony within five days. Around 40 ± 10% of colonies present *Isl1* expression in a characteristic focal pattern. These Isl1⁺ colonies express early cardiac markers including *GATA4, Tbx20, Isl1, Nkx2.5,* and *Flk1*. Interestingly, the clones with the expression signature of *Isl1, Nkx2.5* and *Flk1* are able to differentiate into cardiac, endothelial/endocardial and smooth muscle lineages. *In vitro* clonal analysis using cardiac mesenchymal feeder suggest that most of the single cell-derived colonies expressing *Isl1⁺/Nkx2.5⁺/Flk1⁺* differentiated into all three cardiovascular lineages expressing cardiac troponin T (cTnT), SM-MHC, or VE-cadherin. These findings indicate that ES cell-derived Isl1⁺ progenitor possesses the potential to serve as master cardiovascular progenitor cells *in vitro* that are able to give rise to three major cell types in the working myocardium and the heart vasculature.

The role of Wnt signaling in in vitro expansion and differentiation of Isl1⁺ cardiac progenitors

Attempts are being made to apply what we have learned from mouse development to *in vitro* manipulation of Isl1⁺ cardiac progenitors. BMP, FGF and the Nodal/Activin, and Wnt pathways are implicated in promoting cardiac induction in several mammalian models and ES cell differentiation. The renewal factors of Isl1⁺ cardiac progenitors were examined using a high-throughput chemical screening system with postnatal Isl1$^+$ cardiac progenitors (Qyang et al. 2007). Isl1$^+$ cardiac progenitors were expanded for four days in the presence of small molecules. From this screening, three candidates were identified as substantially increasing the number of Isl1$^+$ progenitors. One of the small molecules was 6-bromoindirubin-3′-oxime (BIO), previously shown to be an inhibitor of GSK-3 (Meijer et al. 2003). BIO increases the number of Isl1$^+$ progenitors in a dose-dependent manner up to seven-fold versus control. BIO also works on expanding human neonatal Isl1$^+$ progenitors isolated from the biopsies of patients with congenital heart defects. As GSK-3 is a negative regulator of the Wnt signaling pathway, treatment with Wnt3a-conditioned medium and co-culture with a feeder layer stably secreting Wnt3a resulted in a ~two-fold and a nearly six-fold increase of Isl1$^+$ progenitors compared with the control respectively. A potent extracellular inhibitor of Wnt signaling, Dickkopf-1 (Dkk1) (Logan and Nusse 2004) showed nearly 40% reduction in the number of Isl1$^+$ progenitor cells. Wnt3a also has a marked expanding effect on the Isl1$^+$ progenitors in the mouse embryo (E8.5). These findings suggest that Wnt/β-catenin signaling is one of the specific pathways to expand Isl1$^+$ cardiac progenitors *in vitro*. BIO and Wnt3A were reported to induce cardiogenesis in a stage specific manner in mouse ES cells when registered in early stages (Naito et al. 2006; Ueno et al. 2007), supporting that Isl1$^+$ cardiac progenitors emerging in the early stage of ES cell differentiation expand with the stimulation of Wnt/β-catenin signaling.

While Wnt3a induces the expansion of Isl1$^+$ cardiac progenitors isolated from mouse embryonic heart, it also has a negative effect on the maturation of Isl1$^+$ cardiac progenitors. EB-derived Mef2c-GFP$^+$ cells (rich in Isl1$^+$ cardiac progenitors) can be expanded by Wnt3a, whereas their cardiac differentiation is rather inhibited by Wnt3a (Qyang et al. 2007). They undergo spontaneous differentiation only when the Wnt3a feeder cells are removed. Very early registration of Wnt3a also inhibits augmentation of the initial number of ES cell-derived Isl1$^+$ clones. These findings suggest that Wnt/β-catenin signaling expands the clonal expansion of Isl1$^+$ cardiac progenitors, while it has an inhibitory role for their prespecification and differentiation. Thus, Wnt signaling induces cardiogenesis in a stage specific manner in mouse ES cells.

iPS cell-derived Isl1 cardiac progenitors

Isl1 cardiac progenitors have also been investigated in iPS cells. The expression pattern and time course of *Isl1* is similar in the embryoid bodies derived from iPS cells from day 3.5 after differentiation (Schenke-Layland et al. 2008; Narazaki et al. 2008). iPS cells from *Isl1-Cre* reporter mice showed that iPS cell-derived Isl1$^+$ cardiac progenitors are able to differentiate into cardiomyocytes, smooth muscle cells and endothelial cells *in vitro* comparably to ES cell system (Moretti et al. 2010). When injected into the left ventricular wall of the adult mice, these iPS-derived Isl1$^+$ cardiac progenitors showed the engraftment and differentiation into three cardiovascular lineages, including cardiomyocytes, with a more organized sarcomeric structure after 12 weeks. While the injection of undifferentiated iPS cells generated teratoma, iPS-derived Isl1$^+$ cardiac progenitors did not produce microscopic evidence for tumor formation at 12 weeks after injection. Human iPS cells have also represented a source of ISL$^+$ cardiac progenitors displaying cardiogenic potential. Upon cardiac induction by Bmp2 and Fgf receptor inhibitor, human iPS cells differentiated *ISL1*-positive cells with NKX2.5 and/or FLK1 co-expression. After successive cardiovascular induction with ascorbic acid and VEGF, these cells differentiated into three cardiovascular lineages.

Isl1-positive Cardiac Progenitors in Cardiac Regeneration

The use of Isl1$^+$ cardiac progenitors in cardiac regeneration

At present, there are various options for cell types and transplantation methods potentially applicable in cardiac regeneration. Their efficiency has been extensively tested in animal models. Some of the present methods have shown considerable but marginal results. The mechanism of cardiac regeneration is still unclear (Segers and Lee 2008; Sturzu and Wu 2011). Several characteristics of Isl1$^+$ cardiac progenitors that are advantageous as a cell source for cardiac regeneration are as follows:

Developmental biology-based: Isl1$^+$ cardiac progenitors were identified as a cardiac progenitor population during embryogenesis that appears in the early SHF region during a restricted period. It has been well characterized from developmental perspectives, which could further foster a deeper understanding of the link between the cardiac development program and regeneration mechanism.

Isl1 as a marker: *Isl1* is a discrete marker for immature cardiogenic cells during cardiac development, which is useful for identifying proliferative populations.

Expandability: Isl1$^+$ cardiac progenitors are highly expandable *ex vivo* and *in vitro* with defined factors including Wnt ligands and GSK inhibitors.

Multilineage differentiation: Isl1$^+$ cardiac progenitors can be a source of three major cardiovascular lineages (cardiac, endothelial, and smooth muscle cells). The endothelial and smooth muscle differentiation potential may be an advantage as vascularization of the implanted myocardium is beneficial for the survival and engraftment of the transplanted progenitors (Stevens et al. 2009).

Maturation: The cardiomyocytes differentiated from Isl1$^+$ cardiac progenitors are functional with electrophysiological characteristics of neonatal cardiomyocytes.

Contribution to sinoatrial node: The sinoatrial node is also a specific progeny of Isl1$^+$ cardiac progenitors that is a potential target for cell-based therapy.

Human biopsy: Isl1$^+$ cardiac progenitors can be propagated from human biopsies of congenital heart defect patients (Qyang et al. 2007).

Potential issues for cardiac regeneration using Isl1$^+$ cardiac progenitors

On the other hand, to realize the cardiac regenerative medicine using Isl1$^+$ cardiac progenitors, several hurdles need to be overcome.

Paucity of surface markers: The identification of a specific surface marker for Isl1$^+$ cardiac progenitors is critical for isolating Isl1$^+$ cardiac progenitors without genetic marking.

Low efficiency of engraftment: The low efficiency of engraftment is a common issue in cell-based therapy for the heart. After transplantation or infusion, only a few cardiogenic cells engraft for more than a few weeks, indicating that the primary effect of previous cell therapies would be due to a paracrine effect. Methods to reinforce the survival of cardiac progenitor cells are yet to be developed.

Maturation: It is also important to obtain mature and functional myocardial tissue that contributes to cardiac function in the long-term. While several factors that expand Isl1$^+$ cardiac progenitors are identified, the mechanism regulating the differentiation of Isl1$^+$ cardiac progenitors into myocardial tissue is still unclear. As multipotent Isl1$^+$ cardiac progenitors can give rise to SHF tissue consisting of cardiomyocytes and blood vessels, elucidating the developmental mechanism underlying the differentiation of Isl1$^+$ cardiac progenitors into the assembly of myocardial and vascular tissues would provide a clue towards attaining well-organized myocardial tissue from Isl1$^+$ cardiac progenitors *in vitro*.

Safety: Finally, evidence for safety in the long-term use of Isl1$^+$ cardiac progenitors needs to be accumulated. Application of Isl1$^+$ cardiac progenitors to cardiac regeneration therapy requires a deeper understanding of the cellular function of Isl1^{+} cardiac progenitors *in vitro*, the developmental process of organized cardiac tissues from Isl1^{+} cardiac progenitors, and the mechanism of the engraftment in animal models.

Approaches to regenerate cardiomyocytes using Isl1^{+} progenitors

Besides transplanting exogenous Isl1^{+} cardiac progenitors, encouraging the endogenous Isl1^{+} cardiac progenitors would be an alternative approach. Although putative developmental remnants of Isl1^{+} cardiac progenitors reside in a subset of SHF region including OFT and RV, it is still unclear whether postnatal residential Isl1^{+} cardiac progenitors expand *in situ* and contribute to cardiac repair in the damaged heart.

EPDCs (epicardial-derived progenitor cells) have been reported to contribute to cardiomyocytes during mouse cardiogenesis (Cai et al. 2008; Zhou et al. 2008). A recent report revealed that *Wt1*$^{+}$ epicardial-derived progenitor cells can be re-activated in a mouse model of myocardial infarction by administration of Thymosin β4, and contribute to the generation of *de novo* cardiomyocytes (Smart et al. 2011). In this model, *Isl1*$^{+}$/*Wt1*$^{+}$ cells are induced by priming with Thymosin β4 in *ex vivo* heart culture and in the epicardium *in vivo*, suggesting that endogenous epicardial-derived progenitor cells can be encouraged to contribute to cardiac repair. Remarkably, transplanted *Wt1*$^{+}$ progenitors are integrated into the surrounding host myocardium through gap junctions and synchronize *in vivo*, offering the promise of the usage of EPDC as a source for cell-based therapy. These findings raise the possibility that *Wt1*$^{+}$ cells and other progenitor populations potentially play a significant role in the repair of adult hearts. Investigating the Isl1^{+} cardiac progenitor function in adults may also provide a clue to the activation of residential progenitors that display cardiac regenerative potential.

Contractile myocardial patches have been generated from ES-derived Isl1^{+} cardiac progenitors (Domian et al. 2009). Differential labeling of FHF and SHF in ES *in vitro* differentiation system identified a subset of *Isl1*$^{+}$ cardiac progenitors that give rise to the ventricular cardiomyocytes. When cultured on micro patterned surfaces, these cells formed anisotropic tissue consisting of longitudinally aligned myocardial fibers. Ninety per cent of the differentiated cardiomyocytes display ventricular-type action potential, and muscular thin film using these progenitors generated fully functional ventricular tissue with synchronized contraction. These data represent the combination of bioengineering and stem cell biology as an approach to development of a human *in vitro* model for cardiac disease and potential therapeutic strategy using an ES-derived myocardial patch.

Conclusion

Clinical trials suggest that the cardiac regeneration potential of transplanted cardiac progenitor cells is still controversial and that the functional benefits are modest. Further understanding of the basic biology of cardiac differentiation is required before designing larger-scale trials. Effective cardiac regenerative medicine cannot be realized without organic integration of cardiac developmental biology, stem cell biology and bio-engineering approaches. While great progress has been made in the developmental biology of the heart in the past decade, we are still at a primitive stage in understanding the cellular basis of cardiac progenitor biology and cardiac regenerative medicine. Major hurdles that have hampered a better understanding of cardiac stem cell biology include: (1) lack of an efficient culture system, (2) paucity of markers (especially surface markers) and (3) lack of an efficient engraftment method. As discussed in this chapter, attempts are being made to overcome these hurdles. These efforts will hopefully take this research field to another level in the near future.

Acknowledgments

We thank Haruko Nakano, Andrew W. Harmon and Armin Arshi for the critical feedback on the manuscript.

References

Abu-Issa, R., Smyth, G., Smoak, I., Yamamura, K., and Meyers, E. N. 2002. Fgf8 is required for pharyngeal arch and cardiovascular development in the mouse. Development 129, 4613–4625.

Abu-Issa, R., Waldo, K., and Kirby, M. L. 2004. Heart fields: one, two or more? Dev. Biol. 272, 281–285.

Ahlgren, U., Pfaff, S. L., Jessell, T. M., Edlund, T., and Edlund, H. 1997. Independent requirement for ISL1 in formation of pancreatic mesenchyme and islet cells. Nature 385, 257–260.

Barron, M., Gao, M., and Lough, J. 2000. Requirement for BMP and FGF signaling during cardiogenic induction in non-precardiac mesoderm is specific, transient, and cooperative. Dev. Dyn. 218, 383–393.

Blom, N. A., Ottenkamp, J., Wenink, A. G., and Gittenberger-de Groot, A. C. 2003. Deficiency of the vestibular spine in atrioventricular septal defects in human fetuses with down syndrome. Am. J. Cardiol. 91, 180–184.

Brade, T., Gessert, S., Kuhl, M., and Pandur, P. 2007. The amphibian second heart field: Xenopus islet-1 is required for cardiovascular development. Dev. Biol. 311, 297–310.

Bruneau, B. G., Logan, M., Davis, N., Levi, T., Tabin, C. J., Seidman, J. G., and Seidman, C.E. 1999. Chamber-specific cardiac expression of Tbx5 and heart defects in Holt-Oram syndrome. Dev. Biol. 211, 100–108.

Bu, L., Jiang, X., Martin-Puig, S., Caron, L., Zhu, S., Shao, Y., Roberts, D. J., Huang, P. L., Domian, I. J. and Chien, K. R. 2009. Human ISL1 heart progenitors generate diverse multipotent cardiovascular cell lineages. Nature 460, 113–117.

Cai, C. L., Liang, X., Shi, Y., Chu, P. H., Pfaff, S. L., Chen, J., and Evans, S. 2003. Isl1 identifies a cardiac progenitor population that proliferates prior to differentiation and contributes a majority of cells to the heart. Dev. Cell 5, 877–889.

Cai, C. L., Martin, J. C., Sun, Y., Cui, L., Wang, L., Ouyang, K., Yang, L., Bu, L., Liang, X., Zhang, X. et al. 2008. A myocardial lineage derives from Tbx18 epicardial cells. Nature 454, 104–108.

Cohen, E. D., Wang, Z., Lepore, J. J., Lu, M. M., Taketo, M. M., Epstein, D. J., and Morrisey, E. E. 2007. Wnt/beta-catenin signaling promotes expansion of Isl-1-positive cardiac progenitor cells through regulation of FGF signaling. J. Clin. Invest 117, 1794–1804.

de la Cruz, M. V., Sanchez Gomez, C., Arteaga, M. M., and Arguello, C. 1977. Experimental study of the development of the truncus and the conus in the chick embryo. J. Anat. 123, 661–686.

Dodou, E., Verzi, M. P., Anderson, J. P., Xu, S. M., and Black, B. L. 2004. Mef2c is a direct transcriptional target of ISL1 and GATA factors in the anterior heart field during mouse embryonic development. Development 131, 3931–3942.

Domian, I. J., Chiravuri, M., van der Meer, P., Feinberg, A. W., Shi, X., Shao, Y., Wu, S. M., Parker, K. K., and Chien, K. R. 2009. Generation of functional ventricular heart muscle from mouse ventricular progenitor cells. Science 326, 426–429.

Esner, M., Meilhac, S. M., Relaix, F., Nicolas, J. F., Cossu, G., and Buckingham, M. E. 2006. Smooth muscle of the dorsal aorta shares a common clonal origin with skeletal muscle of the myotome. Development 133, 737–749.

Feldman, B., Poueymirou, W., Papaioannou, V. E., DeChiara, T. M., and Goldfarb, M. 1995. Requirement of FGF-4 for postimplantation mouse development. Science 267, 246–249.

Frank, D. U., Fotheringham, L. K., Brewer, J. A., Muglia, L. J., Tristani-Firouzi, M., Capecchi, M. R., and Moon, A. M. 2002. An Fgf8 mouse mutant phenocopies human 22q11 deletion syndrome. Development 129, 4591–4603.

Freyd, G., Kim, S. K., and Horvitz, H. R. 1990. Novel cysteine-rich motif and homeodomain in the product of the Caenorhabditis elegans cell lineage gene lin-11. Nature 344, 876–879.

Galli, D., Dominguez, J. N., Zaffran, S., Munk, A., Brown, N. A., and Buckingham, M. E. 2008. Atrial myocardium derives from the posterior region of the second heart field, which acquires left-right identity as Pitx2c is expressed. Development 135, 1157–1167.

Genead, R., Danielsson, C., Andersson, A. B., Corbascio, M., Franco-Cereceda, A., Sylven, C., and Grinnemo, K. H. 2010. Islet-1 cells are cardiac progenitors present during the entire lifespan: from the embryonic stage to adulthood. Stem Cells Dev. 19, 1601–1615.

Gill, G. N. 2003. Decoding the LIM development code. Trans Am. Clin. Climatol. Assoc. 114, 179–189.

Gittenberger-de Groot, A. C., Winter, E. M., and Poelmann, R. E. 2010. Epicardium-derived cells (EPDCs) in development, cardiac disease and repair of ischemia. J. Cell Mol. Med. 14, 1056–1060.

He, B., Li, G., and Zhang, X. 2003. Noninvasive imaging of cardiac transmembrane potentials within three-dimensional myocardium by means of a realistic geometry anisotropic heart model. IEEE Trans Biomed. Eng. 50, 1190–1202.

Kang, J., Nathan, E., Xu, S. M., Tzahor, E., and Black, B. L. 2009. Isl1 is a direct transcriptional target of Forkhead transcription factors in second heart field-derived mesoderm. Dev. Biol. 334, 513–522.

Karlsson, O., Thor, S., Norberg, T., Ohlsson, H., and Edlund, T. 1990. Insulin gene enhancer binding protein Isl-1 is a member of a novel class of proteins containing both a homeo- and a Cys-His domain. Nature 344, 879–882.

Kattman, S. J., Huber, T. L., and Keller, G. M. 2006. Multipotent flk-1+ cardiovascular progenitor cells give rise to the cardiomyocyte, endothelial, and vascular smooth muscle lineages. Dev. Cell 11, 723–732.

Kelly, R. G., Brown, N. A., and Buckingham, M. E. 2001. The arterial pole of the mouse heart forms from Fgf10-expressing cells in pharyngeal mesoderm. Dev. Cell 1, 435–440.

Kirby, M. L. 2002. Molecular embryogenesis of the heart. Pediatr Dev. Pathol. 5, 516–543.

Klaus, A., Saga, Y., Taketo, M. M., Tzahor, E., and Birchmeier, W. 2007. Distinct roles of Wnt/beta-catenin and Bmp signaling during early cardiogenesis. Proc. Natl. Acad. Sci. USA 104, 18531–18536.

Kwon, C., Arnold, J., Hsiao, E. C., Taketo, M. M., Conklin, B. R., and Srivastava, D. 2007. Canonical Wnt signaling is a positive regulator of mammalian cardiac progenitors. Proc. Natl. Acad. Sci. USA 104, 10894–10899.

Kwon, C., Qian, L., Cheng, P., Nigam, V., Arnold, J., and Srivastava, D. 2009. A regulatory pathway involving Notch1/beta-catenin/Isl1 determines cardiac progenitor cell fate. Nat. Cell Biol. 11, 951–957.

Laugwitz, K. L., Moretti, A., Caron, L., Nakano, A., and Chien, K. R. 2008. Islet1 cardiovascular progenitors: a single source for heart lineages? Development 135, 193–205.

Laugwitz, K. L., Moretti, A., Lam, J., Gruber, P., Chen, Y., Woodard, S., Lin, L. Z., Cai, C. L., Lu, M. M., Reth, M. et al. 2005. Postnatal isl1+ cardioblasts enter fully differentiated cardiomyocyte lineages. Nature 433, 647–653.

Lie-Venema, H., van den Akker, N. M., Bax, N. A., Winter, E. M., Maas, S., Kekarainen, T., Hoeben, R. C., deRuiter, M. C., Poelmann, R. E., and Gittenberger-de Groot, A. C. 2007. Origin, fate, and function of epicardium-derived cells (EPDCs) in normal and abnormal cardiac development. ScientificWorldJournal 7, 1777–1798.

Lin, L., Bu, L., Cai, C. L., Zhang, X., and Evans, S. 2006. Isl1 is upstream of sonic hedgehog in a pathway required for cardiac morphogenesis. Dev. Biol. 295, 756–763.

Lin, L., Cui, L., Zhou, W., Dufort, D., Zhang, X., Cai, C. L., Bu, L., Yang, L., Martin, J., Kemler, R. et al. 2007. Beta-catenin directly regulates Islet1 expression in cardiovascular progenitors and is required for multiple aspects of cardiogenesis. Proc. Natl. Acad. Sci. USA 104, 9313–9318.

Logan, C. Y. and Nusse, R. 2004. The Wnt signaling pathway in development and disease. Annu. Rev. Cell Dev. Biol. 20, 781–810.

Lough, J., Barron, M., Brogley, M., Sugi, Y., Bolender, D. L., and Zhu, X. 1996. Combined BMP-2 and FGF-4, but neither factor alone, induces cardiogenesis in non-precardiac embryonic mesoderm. Dev. Biol. 178, 198–202.

Ma, Q., Zhou, B., and Pu, W. T. 2008. Reassessment of Isl1 and Nkx2-5 cardiac fate maps using a Gata4-based reporter of Cre activity. Dev. Biol. 323, 98–104.

Mann, T., Bodmer, R., and Pandur, P. 2009. The Drosophila homolog of vertebrate Islet1 is a key component in early cardiogenesis. Development 136, 317–326.

Marvin, M. J., Di Rocco, G., Gardiner, A., Bush, S. M., and Lassar, A. B. 2001. Inhibition of Wnt activity induces heart formation from posterior mesoderm. Genes Dev. 15, 316–327.

Meijer, L., Skaltsounis, A. L., Magiatis, P., Polychronopoulos, P., Knockaert, M., Leost, M., Ryan, X. P., Vonica, C. A., Brivanlou, A., Dajani, R. et al. 2003. GSK-3-selective inhibitors derived from Tyrian purple indirubins. Chem. Biol. 10, 1255–1266.

Meilhac, S. M., Esner, M., Kelly, R. G., Nicolas, J. F., and Buckingham, M. E. 2004. The clonal origin of myocardial cells in different regions of the embryonic mouse heart. Dev. Cell 6, 685–698.

Mikawa, T. and Gourdie, R. G. 1996. Pericardial mesoderm generates a population of coronary smooth muscle cells migrating into the heart along with ingrowth of the epicardial organ. Dev. Biol. 174, 221–232.

Mjaatvedt, C. H., Nakaoka, T., Moreno-Rodriguez, R., Norris, R. A., Kern, M. J., Eisenberg, C. A., Turner, D., and Markwald, R. R. 2001. The outflow tract of the heart is recruited from a novel heart-forming field. Dev. Biol. 238, 97–109.

Moorman, A. F., Christoffels, V. M., Anderson, R. H., and van den Hoff, M. J. 2007. The heart-forming fields: one or multiple? Philos. Trans. R. Soc. Lond. B. Biol. Sci. 362, 1257–1265.

Moretti, A., Bellin, M., Jung, C. B., Thies, T. M., Takashima, Y., Bernshausen, A., Schiemann, M., Fischer, S., Moosmang, S., Smith, A. G. et al. 2010. Mouse and human induced pluripotent stem cells as a source for multipotent Isl1+ cardiovascular progenitors. Faseb J. 24, 700–711.

Moretti, A., Caron, L., Nakano, A., Lam, J. T., Bernshausen, A., Chen, Y., Qyang, Y., Bu, L., Sasaki, M., Martin-Puig, S. et al. 2006. Multipotent embryonic isl1+ progenitor cells lead to cardiac, smooth muscle, and endothelial cell diversification. Cell 127, 1151–1165.

Munoz-Chapuli, R., Macias, D., Gonzalez-Iriarte, M., Carmona, R., Atencia, G., and Perez-Pomares, J. M. 2002. [The epicardium and epicardial-derived cells: multiple functions in cardiac development]. Rev. Esp. Cardiol. 55, 1070–1082.

Naito, A. T., Shiojima, I., Akazawa, H., Hidaka, K., Morisaki, T., Kikuchi, A., and Komuro, I. 2006. Developmental stage-specific biphasic roles of Wnt/beta-catenin signaling in cardiomyogenesis and hematopoiesis. Proc. Natl. Acad. Sci. USA 103, 19812–19817.

Nakano, H., Williams, E., Hoshijima, M., Sasaki, M., Minamisawa, S., Chien, K. R., and Nakano, A. 2011. Cardiac origin of smooth muscle cells in the inflow tract. J. Mol. Cell Cardiol. 50, 337–345.

Narazaki, G., Uosaki, H., Teranishi, M., Okita, K., Kim, B., Matsuoko, S., Yamanaka, S., and Yamashita, J. K. 2008. Directed and systematic differentiation of cardiovascular cells from mouse induced pluripotent stem cells. Circulation 29: 498–506

Paige, S. L., Osugi, T., Afanasiev, O. K., Pabon, L., Reinecke, H., and Murry, C. E. 2010. Endogenous Wnt/beta-catenin signaling is required for cardiac differentiation in human embryonic stem cells. PLoS One 5, e11134.

Papanicolaou, K. N., Izumiya, Y., and Walsh, K. 2008. Forkhead Transcription Factors and Cardiovascular Biology. Circ. Res. 102: 16–31.

Pfaff, S. L., Mendelsohn, M., Stewart, C. L., Edlund, T., and Jessell, T. M. 1996. Requirement for LIM homeobox gene Isl1 in motor neuron generation reveals a motor neuron-dependent step in interneuron differentiation. Cell 84, 309–320.

Prall, O. W., Menon, M. K., Solloway, M. J., Watanabe, Y., Zaffran, S., Bajolle, F., Biben, C., McBride, J. J., Robertson, B. R., Chaulet, H. et al. 2007. An Nkx2-5/Bmp2/Smad1 negative feedback loop controls heart progenitor specification and proliferation. Cell 128, 947–959.

Qyang, Y., Martin-Puig, S., Chiravuri, M., Chen, S., Xu, H., Bu, L., Jiang, X., Lin, L., Granger, A., Moretti, A. et al. 2007. The renewal and differentiation of Isl1+ cardiovascular progenitors are controlled by a Wnt/beta-catenin pathway. Cell Stem Cell 1, 165–179.

Schenke-Layland, K., Rhodes, K. E., Angelis, E., Butylkova, Y., Heydarkhan-Hagvall, S., Gekas, C., Zhang, R., Goldhaber, J. I., Mikkola, H. K., Plath, K. et al. 2008. Reprogrammed mouse fibroblasts differentiation into cells of the cardiovascular and hematopoietic lineages. Stem Cells 26: 1537–1546.

Schneider, V. A. and Mercola, M. 2001. Wnt antagonism initiates cardiogenesis in Xenopus laevis. Genes Dev. 15, 304–315.

Segers, V. F. and Lee, R. T. 2008. Stem-cell therapy for cardiac disease. Nature 451, 937–942.

Seo, S., Fujita, H., Nakano, A., Kang, M., Duarte, A., and Kume, T. 2006. The forkhead transcription factors, Foxc1 and Foxc2, are required for arterial specification and lymphatic sprouting during vascular development. Dev. Biol. 294, 458–470.

Sharratt, G. P., Webb, S., and Anderson, R. H. 2003. The vestibular defect: an interatrial communication due to a deficiency in the atrial septal component derived from the vestibular spine. Cardiol. Young 13, 184–190.

Smart, N., Bollini, S., Dube, K. N., Vieira, J. M., Zhou, B., Davidson, S., Yellon, D., Riegler, J., Price, A. N., Lythgoe, M. F. et al. 2011. De novo cardiomyocytes from within the activated adult heart after injury. Nature.

Snarr, B. S., O'Neal, J. L., Chintalapudi, M. R., Wirrig, E. E., Phelps, A. L., Kubalak, S. W., and Wessels, A. 2007a. Isl1 expression at the venous pole identifies a novel role for the second heart field in cardiac development. Circ. Res. 101, 971–974.

Snarr, B. S., Wirrig, E. E., Phelps, A. L., Trusk, T. C., and Wessels, A. 2007b. A spatiotemporal evaluation of the contribution of the dorsal mesenchymal protrusion to cardiac development. Dev. Dyn. 236, 1287–1294.

Srinivas, S., Watanabe, T., Lin, C. S., William, C. M., Tanabe, Y., Jessell, T. M., and Costantini, F. 2001. Cre reporter strains produced by targeted insertion of EYFP and ECFP into the ROSA26 locus. BMC Dev. Biol. 1, 4.

Stevens, K. N., Hakonarson, H., Kim, C. E., Doevendans, P. A., Koeleman, B. P., Mital, S., Raue, J., Glessner, J. T., Coles, J. G., Moreno, V. et al. 2010. Common variation in ISL1 confers genetic susceptibility for human congenital heart disease. PLoS One 5, e10855.

Stevens, K. R., Kreutziger, K. L., Dupras, S. K., Korte, F. S., Regnier, M., Muskheli, V., Nourse, M. B., Bendixen, K., Reinecke, H., and Murry, C. E. 2009. Physiological function and transplantation of scaffold-free and vascularized human cardiac muscle tissue. Proc. Natl. Acad. Sci. USA 106, 16568–16573.

Sturzu, A. C. and Wu, S. M. 2011. Developmental and regenerative biology of multipotent cardiovascular progenitor cells. Circ. Res. 108, 353–364.

Sun, X., Meyers, E. N., Lewandoski, M., and Martin, G. R. 1999. Targeted disruption of Fgf8 causes failure of cell migration in the gastrulating mouse embryo. Genes Dev. 13, 1834–1846.

Sun, Y., Liang, X., Najafi, N., Cass, M., Lin, L., Cai, C. L., Chen, J., and Evans, S. M. 2007. Islet 1 is expressed in distinct cardiovascular lineages, including pacemaker and coronary vascular cells. Dev. Biol. 304, 286–296.

Thaler, J. P., Koo, S. J., Kania, A., Lettieri, K., Andrews, S., Cox, C., Jessell, T. M., and Pfaff, S. L. 2004. A postmitotic role for Isl-class LIM homeodomain proteins in the assignment of visceral spinal motor neuron identity. Neuron 41, 337–350.

Thor, S., Andersson, S. G., Tomlinson, A., and Thomas, J. B. 1999. A LIM-homeodomain combinatorial code for motor-neuron pathway selection. Nature 397, 76–80.

Tian, Y., Yuan, L., Goss, A. M., Wang, T., Yang, J., Lepore, J. J., Zhou, D., Schwartz, R. J., Patel, V., Cohen, E. D. et al. 2010. Characterization and *in vivo* pharmacological rescue of a Wnt2-Gata6 pathway required for cardiac inflow tract development. Dev. Cell 18, 275–287.

Tirosh-Finkel, L., Zeisel, A., Brodt-Ivenshitz, M., Shamai, A., Yao, Z., Seger, R., Domany, E., and Tzahor, E. 2010. BMP-mediated inhibition of FGF signaling promotes cardiomyocyte differentiation of anterior heart field progenitors. Development 137, 2989–3000.

Tsuchida, T., Ensini, M., Morton, S. B., Baldassare, M., Edlund, T., Jessell, T. M., and Pfaff, S. L. 1994. Topographic organization of embryonic motor neurons defined by expression of LIM homeobox genes. Cell 79, 957–970.

Tzahor, E. 2007. Wnt/beta-catenin signaling and cardiogenesis: timing does matter. Dev. Cell 13, 10–13.

Tzahor, E. and Lassar, A. B. 2001. Wnt signals from the neural tube block ectopic cardiogenesis. Genes Dev. 15, 255–260.

Ueno, S., Weidinger, G., Osugi, T., Kohn, A. D., Golob, J. L., Pabon, L., Reinecke, H., Moon, R. T., and Murry, C. E. 2007. Biphasic role for Wnt/beta-catenin signaling in cardiac specification in zebrafish and embryonic stem cells. Proc. Natl. Acad. Sci. USA 104, 9685–9690.

van Wijk, B., van den Berg, G., Abu-Issa, R., Barnett, P., van der Velden, S., Schmidt, M., Ruijter, J. M., Kirby, M. L., Moorman, A. F., and van den Hoff, M. J. 2009. Epicardium and myocardium separate from a common precursor pool by crosstalk between bone morphogenetic protein- and fibroblast growth factor-signaling pathways. Circ. Res. 105, 431–441.

Verzi, M. P., McCulley, D. J., De Val, S., Dodou, E., and Black, B. L. 2005. The right ventricle, outflow tract, and ventricular septum comprise a restricted expression domain within the secondary/anterior heart field. Dev. Biol. 287, 134–145.

Waldo, K. L., Hutson, M. R., Ward, C. C., Zdanowicz, M., Stadt, H. A., Kumiski, D., Abu-Issa, R., and Kirby, M. L. 2005. Secondary heart field contributes myocardium and smooth muscle to the arterial pole of the developing heart. Dev. Biol. 281, 78–90.

Waldo, K. L., Kumiski, D. H., Wallis, K. T., Stadt, H. A., Hutson, M. R., Platt, D. H., and Kirby, M. L. 2001. Conotruncal myocardium arises from a secondary heart field. Development 128, 3179–3188.

Watanabe, Y., Miyagawa-Tomita, S., Vincent, S. D., Kelly, R. G., Moon, A. M., and Buckingham, M. E. 2010. Role of mesodermal FGF8 and FGF10 overlaps in the development of the arterial pole of the heart and pharyngeal arch arteries. Circ. Res. 106, 495–503.

Way, J. C., and Chalfie, M. 1988. mec-3, a homeobox-containing gene that specifies differentiation of the touch receptor neurons in C. elegans. Cell 54, 5–16.

Webb, S., Anderson, R. H., Lamers, W. H., and Brown, N. A. 1999. Mechanisms of deficient cardiac septation in the mouse with trisomy 16. Circ. Res. 84, 897–905.

Wessels, A., and Perez-Pomares, J. M. 2004. The epicardium and epicardially derived cells (EPDCs) as cardiac stem cells. Anat. Rec. A. Discov. Mol. Cell Evol. Biol. 276, 43–57.

Winnier, G. E., Kume, T., Deng, K., Rogers, R., Bundy, J., Raines, C., Walter, M. A., Hogan, B. L., and Conway, S. J. 1999. Roles for the winged helix transcription factors MF1 and MFH1 in cardiovascular development revealed by nonallelic noncomplementation of null alleles. Dev. Biol. 213, 418–431.

Wirrig, E. E., Snarr, B. S., Chintalapudi, M. R., O'Neal J., L., Phelps, A. L., Barth, J. L., Fresco, V. M., Kern, C. B., Mjaatvedt, C. H., Toole, B. P. et al. 2007. Cartilage link protein 1 (Crtl1), an extracellular matrix component playing an important role in heart development. Dev. Biol. 310, 291–303.

Yang, L., Cai, C. L., Lin, L., Qyang, Y., Chung, C., Monteiro, R. M., Mummery, C. L., Fishman, G. I., Cogen, A., and Evans, S. 2006. Isl1Cre reveals a common Bmp pathway in heart and limb development. Development 133, 1575–1585.

Yutzey, K. E. and Kirby, M. L. 2002. Wherefore heart thou? Embryonic origins of cardiogenic mesoderm. Dev. Dyn. 223, 307–320.

Zhou, B., Ma, Q., Rajagopal, S., Wu, S. M., Domian, I., Rivera-Feliciano, J., Jiang, D., von Gise, A., Ikeda, S., Chien, K. R. et al. 2008. Epicardial progenitors contribute to the cardiomyocyte lineage in the developing heart. Nature 454, 109–113.

Zinyk, D. L., Mercer, E. H., Harris, E., Anderson, D. J., and Joyner, A. L. 1998. Fate mapping of the mouse midbrain-hindbrain constriction using a site-specific recombination system. Curr. Biol. 8, 665–668.

CHAPTER 12

Cell Sheet Engineering for Heart Tissue Repair

Katsuhisa Matsuura[a] and Tatsuya Shimizu[b,*]

Introduction

Within the rapidly advancing field of regenerative medicine, cell-based therapies have emerged as a promising alternative to cardiac transplant for the management of damaged heart tissue. Various transplant methods have been examined in subjects with acute or prior myocardial infarction, including direct myocardium injection or intracoronary infusion of cell suspension (Segers and Lee 2008). Cell suspension transplantation methods are particularly easy to apply as they do not require particular manipulation of tissues, but this strategy is limited by the inability to precisely control the shape, size and location of transplanted cells, thereby possibly leading to marginal engraftment and suboptimal outcomes. Furthermore, isolated cell transplantation is not well suited for the treatment of myocardial tissue defects.

These obstacles have led to ongoing efforts to create bioengineered heart tissue grafts through the use of biodegradable scaffolds (Freed et al. 1994). While some studies describe the use of these tissues for defects of cartilage or the urinary bladder (Chung 2006), the body of literature describing the clinical use of tissue engineering based on biodegradable scaffolds is extremely limited. This is mainly due to problems with the re-creation of tissue structures that incorporate diffuse vascular networks that are needed

Institute of Advanced Biomedical Engineering and Science, Tokyo Women's Medical University, 8-1 Kawada-cho, Shinjuku, Tokyo, 162-8666, Japan.
[a]Email: kmatsuura@abmes.twmu.ac.jp
[b]Email:tshimizu@abmes.twmu.ac.jp
*Corresponding author

to provide critical metabolic elements, including oxygen and glucose. Other studies suggest that biodegradable scaffolds may be unsuitable for the regeneration of cell-dense tissues, including the myocardium (Shimizu et al. 2006). Although the degradation of biocompatible and biodegradable polymers, such as poly(lactic acid), poly(glycolic acid), and their copolymer, theoretically allows proliferating cells to migrate into the space previously occupied by the degraded biomaterials, large amounts of extracellular matrix (ECM) components are often present within the scaffolding structures and can actually interfere with cell migration. Additionally, the significant inflammatory responses commonly observed in association with degradation of the polymers can result in damage to the implanted scaffolds and associated cells (Yang et al. 2007). Finally, techniques that can re-create adequate vascular networks are essential to reconstitute tissues that possess the structure and function of the native organs (Fillinger et al. 1990; Shimizu et al. 2006).

In order to overcome these problems, an innovative approach to tissue engineering was developed using a temperature-responsive culture surface (Takezawa et al. 1990; Yamada et al. 1990). Poly (*N*-isopropylacrylamide, PIPAAm) and its copolymer exhibit a critical solution temperature (LCST) in aqueous media in the vicinity of 32°C (Heskins 1968). At a temperature below the LCST, this substance hydrates and forms an expanded structure in aqueous media. By contrast, at a temperature above the LCST, this substance dehydrates and forms a compact structure. Such a conformational change in response to temperature has been used to modulate the physicochemical properties of polymeric materials (Kobayashi et al. 2001; Nakayama et al. 2007; Nakayama et al. 2005; Yoshida 1995). For example, at 37°C, PIPAAm-grafted surfaces are slightly hydrophobic, allowing cells to proliferate under normal conditions. A decrease in temperature to below 32°C, however, results in the rapid hydration of the polymer surface, leading to the spontaneous detachment of the cells as a single cell and/or as uniform tissue sheets. Because PIPAAm is covalently immobilized onto the culture surfaces, PIPAAm remains bound to the surfaces even after cell detachment. This strategy can therefore be utilized to yield a non-invasive harvest of cultured cells, as an intact layer cell sheet containing deposited ECM (Kushida et al. 1999) that can be collected in a non-enzymatic process by simply reducing the culture temperature to below 32°C for less than 1 hr (Fig. 1). This process allows for the transplantation of cell sheets that do not contain biodegradable scaffolds, thereby avoiding any limitations associated with degradation of these scaffolds *in vivo*. Thus far, the sheets of cells generated in this manner and transplanted into subjects include those composed of including keratinocytes (Yamato et al. 2001), retinal pigmental epithelial cells (von Recum et al. 1999), corneal epithelial cells (Nishida et al. 2004), oral mucosal epithelial cells (Nishida et al. 2004; Hayashida et

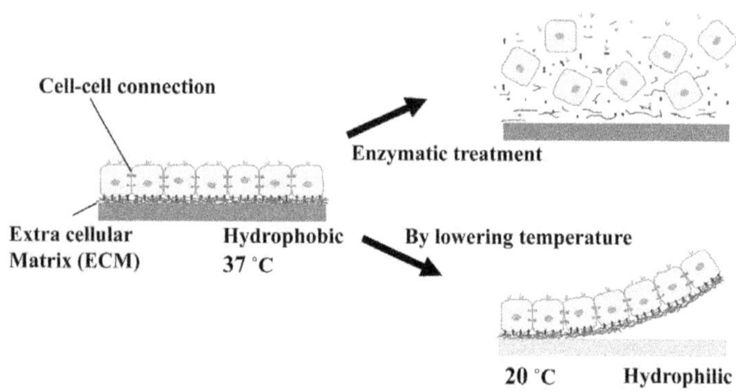

Figure 1. Cell harvest from temperature-responsive culture dishes. At 37°C, the surface of the temperature-responsive culture dishes is hydrophobic, so cells can attach and proliferate onto them. By lowering the temperature below 32°C, the surface reversibly changes to the hydrophilic state, which leads to the detachment of cells from the dishes. Typical cell harvest using enzymatic digestion results in the disruption of both adhesive proteins and membrane receptors. When using temperature-responsive culture dishes, cell sheets with preserved cell to cell connection and extracellular matrixes and intact membrane receptors are harvested.

Color image of this figure appears in the color plate section at the end of the book.

al. 2005), urothelial cells (Shiroyanagi et al. 2003; Shiroyanagi et al. 2004), periodontal ligament cells (Hasegawa et al. 2005; Akizuki et al. 2005), aortic endothelial cells (Kushida et al. 1999; Hirose et al. 2000), corneal endothelial cells (Ide et al. 2006; Sumide et al. 2006), cardiomyocytes (Shimizu et al. 2006; Shimizu et al. 2002), cardiac progenitor cells (Matsuura et al. 2009), vascular smooth muscle cells (Hobo et al. 2008), skeletal myoblasts (Memon et al. 2005) and kidney epithelial cells (Kushida et al. 2005).

Myocardial Tissue Engineering by Layering Cell Sheets

As reviewed above, cell sheet engineering can avoid the limitations associated with the use of biodegradable polymers. When cells are cultured to confluence on the temperature-responsive surfaces, they can be harvested as intact cell sheets without disruption of cell-to-cell connections. Layering of these cell sheets also allows for the establishment of rapid cell-to-cell connections between layers due to the ECM deposited on the basal sheet surface. These properties, in combination with the lack of inflammatory response otherwise associated with degradation of biopolymers, means that this method is particularly well-suited for generating dense, thicker tissues, such as those required for myocardial tissue engineering.

We have already generated sheets of neonatal rat cardiomyocytes from temperature-responsive culture dishes (Shimizu et al. 2002). When two cell sheets were layered, the cardiomyocytes began to pulsate spontaneously and simultaneously. To determine the time course of electrical communication, the electrical potential of layered cardiomyocyte sheets were monitored using a multiple-electrode extracellular recording system (Haraguchi et al. 2006). While the layered cardiomyocytes sheets initially showed independent pulsations, electrical potentials of these two cardiomyocyte sheets had synchronized within 1 hr. Immunohistological analyses and dye transfer assays also showed that the formation of the gap junction occurred rapidly between the layered cardiomyocyte sheets. The presence of gap junction precursor proteins, such as connexin 43, on the cell surface of cardiomyocyte sheets may allow for rapid electrical coupling between the layered sheets. Furthermore, ECM on the basal surface of the cell sheets can be used as adhesive components to promote initiate attachment between the layered cell sheets and to accelerate the docking of the gap junction precursor proteins.

These results demonstrate that electrically communicative 3D cardiac constructs can be achieved by layering monolayer cardiomyocyte sheets. Indeed, when four cardiac cell sheets were layered, macroscopic pulsations were observed. *In vivo*, layered cardiac cell sheets were transplanted onto the dorsal subcutaneous tissues of nude rats, and after three weeks, surface electrograms originating from the grafts were detected independently from the host's electrocardiogram. Additionally, this spontaneous beating could be clearly observed within the transplanted grafts. Histological studies showed that this graft had structures that were characteristic of cardiac tissues, including elongated sarcomeres, desmosomes, and gap junctions. In one study, after layered cardiac sheets were transplanted into the subcutaneous tissues of athymic rats, rapid organization of the microvasculature within the myocardium tissues was apparent within a few days (Shimizu et al. 2006). Three days after transplantation, the beating of transplanted grafts could be observed macroscopically and continued for up to one year. The size, conduction velocity, and contractile force of transplanted grafts also increased in proportion to host rat growth.

Optimal cardiac function will ultimately require synchronized beating between myocardial tissue grafts and the host heart. Fortunately, in one study, cardiomyocytes were noted to have bridged between the grafts and the host heart following transplantation of layered cardiac sheets onto infarcted rat hearts (Sekine et al. 2006). Morphological analyses demonstrated the existence of gap junctions and intercalated disks between the bridging cardiomyocytes. Additionally, a dye transfer assay revealed small molecules passing through functional gap junctions. Finally, transplantation of layered cardiac sheets was also able to repair damaged hearts, with improvement in host ejection fraction, and inhibition of left ventricular dilatation (Miyagawa et al. 2005).

Cell Sources for Cellular Transplant Treatments

Although tissue-engineering methods have enabled reconstruction of myocardial tissue, problems regarding the identification/choice of cell sources persist. Thus far, cells derived from fetal or neonatal hearts have been applied to regenerate myocardial tissues in animal studies, but a similar approach in humans presents a complicated ethical dilemma and may be insufficient to provide the large amount of cardiomyocytes needed for clinical application. We have had recent success in the creation of ES-derived cardiac cell sheets (Matsuura et al. 2011). When mouse ES cells that express the neomycin-resistance gene under the control of alpha myosin heavy chain promoter were treated with neomycin after expansion and cardiac induction in a stirred suspension culture, almost all the cells showed spontaneous beating and expression of sarcomeric proteins. Further, when ES-derived cardiomyocytes were cultured on a temperature-responsive surface, they showed spontaneous but non-synchronized beating, and were able to be detached from the culture surface in intact sheets by modification of culture temperature. However, when cardiac fibroblasts isolated from neonatal hearts were co-cultured with ES-derived cardiomyocytes, cardiomyocytes beat synchronously with one another, intact cell sheets were generated through low temperature-induced detachment from the culture surface, and connexin43 expression was observed at the edge of adjacent cardiomyocytes in cell sheets. Furthermore, when two cell sheets were partially overlaid, electro-propagation was observed between the cell sheets. These observations suggest that stem cell-derived cardiomyocytes are a viable source for the creation of layered cardiac grafts.

Vascularization of 3D-layered Cardiac Graft

Poor vascularization can limit the viability of bioengineered constructs, due to hypoxia, nutrient insufficiency, and waste accumulation, and this problem is particularly pronounced when engineering cell-dense thick tissues, Therefore, new methods for fabricating functional tissue with a well organized vasculature are needed.

In cell sheet-based engineering, the thickness limit of layered cardiac cell sheet in subcutaneous tissue is approximately 80 μm (three layers). However, we recently demonstrated that ~1-mm thick cardiac tissue grafts with well-organized microvascular networks could be fabricated by using a multi-step transplantation process (Shimizu et al. 2006) in which repeated transplantation of triple-layer grafts was performed. This process resulted in synchronization of tissue layers as well as complete graft perfusion when the functional multilayer graft was fabricated over a surgically connectable artery and vein with subsequent ectopic transplantation of the artery, vein, and transplanted layers en bloc.

Control of the vascularization process *in vitro* may also facilitate the creation of thicker functional tissues. For example, when endothelial cells were co-cultured within cardiac cell sheets, angiogenesis-related genes expression and the formation of the endothelial networks were observed *in vitro* (Sekiya et al. 2006). The endothelial cell networks were maintained within the cell sheets after harvest from temperature-responsive culture dishes, and these endothelial cell networks matured to form tubularized vascular networks after *in vivo* transplantation. Moreover, all the graft vessels originated from the grafts themselves, and these vessels migrated to connect with the host's blood vessels. Finally, by changing the ratio of endothelial cells, it was possible to control the formation of blood vessels within the grafts. These observations suggest that regulated neovascularization can be achieved in myocardial tissue grafts engineered by cell sheet engineering.

Cell Sheet Transplantation for Heart Failure

As mentioned above, single cell suspension, direct intra-myocardium injection, and intra-coronary infusion have been used as cell therapies for heart failure. However transplanted cells tend to disappear soon after transplantation when using those cell delivery methods. By contrast, more transplanted cells were noted to persist when using cell sheet transplantation in comparison with direct cell injection (Matsuura et al. 2009; Hamdi et al. 2011).

Various types of cells have been transplanted onto hearts damaged by myocardial infarction and cardiomyopathy. For example, skeletal myoblast cell sheets have been transplanted into a rat model of myocardial infarction (Sekiya et al. 2009), a hamster model of dilated cardiomyopathy (Kondoh et al. 2006; Hata et al. 2006), and a canine model of dilated cardiomyopathy. Implantation of myoblast sheets was shown to be more effective than direct injection of myoblasts into the myocardium in suppressing fibrosis and in improving left ventricular ejection fraction in a rat model of ischemia (Memon et al. 2005). The relative benefit of these treatments on cardiac function in these models has led to clinical trials of myoblast sheet transplantation in humans (Miyagawa et al. 2010). Adipose tissue-derived mesenchymal cell sheet transplantation also produced beneficial effects on cardiac function in animal models of remote myocardial infarction (Miyahara et al. 2006).

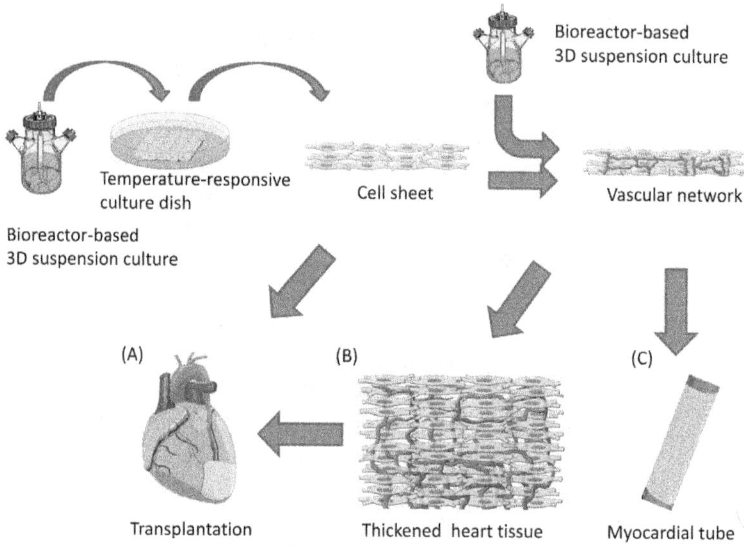

Figure 2. Future views of myocardial tissue reconstruction using cell sheet-based tissue engineering. Cell sheets are created using stem cell-derived cardiomyocytes and fibroblasts after the expansion and purification in the bioreactor system. (A) Transplantation of the layered constructs to repair the injured heart is one of the directions. (B) Thickened heart tissue as disease heart models and for reconstructing the broad heart injury. (C) Myocardial tubes as independent cardiac assist device.

Color image of this figure appears in the color plate section at the end of the book.

Studies have also demonstrated that Sca-1(+) cardiac progenitor cell (CPCs) sheet transplantation improved systolic function and inhibited left ventricular dilatation in a model of acute myocardial infarction (Matsuura et al. 2009). Cell sheet-based delivery enabled CPCs to survive in the infarcted heart and allowed some of engrafted cells to differentiate into cardiomyocytes.

Paracrine effects are also important for the beneficial effect of cell sheet transplantation on cardiac function. Indeed, transplantation of CPCs results in an increase in capillary density and suppression of fibrosis in the damaged portion of cardiac tissues near the transplanted tissues, suggesting the involvement of paracrine factors. Comprehensive analysis by cytokine antibody array revealed that the culture supernatants of CPCs contained significantly higher levels of soluble vascular cell adhesion molecule (VCAM)-1 than those of control cells. Also, VCAM-1 expression remained high in mice that had received CPCs, but did not remain high in control mice. Furthermore, the culture supernatants of CPCs were shown to have angiogenic effects, suppress cardiomyocyte death, and promote migration of the CPCs themselves *in vivo*. When expression of the VCAM-1 receptor α4β1 integrin was blocked after transplantation of the CPC sheets, the effects of transplantation, including improvement in cardiac function, enhanced angiogenesis, and suppression of fibrosis, were diminished, and the engraftment of CPCs was suppressed. Taken together, these data suggest that transplantation of CPCs improves cardiac function by pluripotent mechanisms involving promotion of angiogenesis, suppression of fibrosis, protection of cardiomyocytes, and promotion of cell migration and survival.

Future Challenges in Cell Sheet-based Myocardial Tissue Engineering

One particularly advanced therapeutic goal for the treatment of severe heart failure is the creation of functional organ-like structures that can act as independent cardiac assist devices. As proof of concept, we previously described the *in vitro* fabrication of pulsatile myocardial tubes using as novel cell sheet-wrapping device (Kubo et al. 2007). The tubular constructs, which consisted of three neonatal rat cardiomyocytes sheets wrapped around fibrin tubes, demonstrated spontaneous, synchronized pulsation. Moreover, inner pressure changes constructs. The mean inner pressure gradient of each beat was 0.11 ± 0.01 mmHg. Histological studies also demonstrated the presence of well-differentiated sarcomeres and diffuse gap junctions within the tubular constructs that resembled native cardiac muscle.

These constructs have also been studied *in vivo* (Sekine et al. 2006) by replacing the abdominal aorta of athymic rats with a resected adult rat thoracic aorta sequentially wrapped with six neonatal cardiomyocyte sheets. Spontaneous and synchronous pulsation and myocardial tube pulsation was observed that was independent of the host's native tissues. In this case, the inner pressure gradient originating from the aortic replacement myocardial tubes was 5.9±1.7 mmHg, which was much greater than the values generated by the *in vitro* myocardial tubes. Histological analysis and transmission electron microscopy studies of the grafts implanted in the abdominal cavity revealed that the thickness of the aorta and expression of brain natriuretic peptide and myosin heavy chain had significantly increased. These data suggest that functional myocardial tubes created by cell sheet engineering have the potential to provide circulatory support and act as an independent cardiac assist device.

Conclusion

Myocardial tissue engineering based on cell sheet engineering can provide new *in vitro* heart models and might be useful for cardiovascular tissue repair.

References

Akizuki, T., Oda, S., Komaki, M., Tsuchioka, H., Kawakatsu, N., Kikuchi, A. et al. 2005. Application of periodontal ligament cell sheet for periodontal regeneration: a pilot study in beagle dogs. J. Periodontal Res. 40, 245–51.

Chung, S. Y. 2006. Bladder tissue-engineering: a new practical solution? Lancet 367, 1215–6.

Freed, L. E., Vunjak-Novakovic, G., Biron, R. J., Eagles, D. B., Lesnoy, D. C., Barlow, S. K. et al. 1994. Biodegradable polymer scaffolds for tissue engineering. Biotechnology. 12, 689–93.

Fillinger, M. F., Reinitz, E. R., Schwartz, R. A., Resetarits, D. E., Paskanik, A. M., Bruch, D. et al. 1990. Graft geometry and venous intimal-medial hyperplasia in arteriovenous loop grafts. J. Vasc. Surg. 11, 556–66.

Hamdi, H., Planat-Benard, V., Bel, A., Puymirat, E., Geha, R., Pidial, L. et al. 2011. Epicardial adipose stem cell sheets results in greater post-infarction survival than intramyocardial injections. Cardiovasc. Res. 12 First Published Online.

Haraguchi, Y., Shimizu, T., Yamato, M., Kikuchi, A., and Okano, T. 2006. Electrical coupling of cardiomyocyte sheets occurs rapidly via functional gap junction formation. Biomaterials 27, 4765–74.

Hasegawa, M., Yamato, M., Kikuchi, A., Okano, T., and Ishikawa, I. 2005. Human periodontal ligament cell sheets can regenerate periodontal ligament tissue in an athymic rat model. Tissue Eng. 11, 469–78.

Hata, H., Matsumiya, G., Miyagawa, S., Kondoh, H., Kawaguchi, N., Matsuura, N. et al. 2006. Grafted skeletal myoblast sheets attenuate myocardial remodeling in pacing-induced canine heart failure model. J. Thorac. Cardiovasc. Surg. 132, 918–24.

Hayashida, Y., Nishida, K., Yamato, M., Watanabe, K., Maeda, N., Watanabe, H. et al. 2005. Ocular surface reconstruction using autologous rabbit oral mucosal epithelial sheets fabricated *ex vivo* on a temperature-responsive culture surface. Invest. Ophthalmol. Vis. Sci. 46, 1632–9.

Heskins, M. 1968. Solution properties of poly (N-isopropylacrylamide). J. Macromol. Sci. Pure. 2, 1441.

Hobo, K., Shimizu, T., Sekine, H., Shin'oka, T., Okano, T., and Kurosawa, H. 2008. Therapeutic angiogenesis using tissue engineered human smooth muscle cell sheets. Arterioscler. Thromb. Vasc. Biol. 28, 637–43.

Hirose, M., Kwon, O. H., Yamato, M., Kikuchi, A., and Okano, T. 2000. Creation of designed shape cell sheets that are noninvasively harvested and moved onto another surface. Biomacromolecules 1, 377–81.

Ide, T., Nishida, K., Yamato, M., Sumide, T., Utsumi, M., Nozaki, T. et al. 2006. Structural characterization of bioengineered human corneal endothelial cell sheets fabricated on temperature-responsive culture dishes. Biomaterials 27, 607–14.

Kobayashi, J., Kikuchi, A., Sakai, K., and Okano, T. 2001. Aqueous chromatography utilizing pH-/temperature-responsive polymer stationary phases to separate ionic bioactive compounds. Anal. Chem. 73, 2027–33.

Kondoh, H., Sawa, Y., Miyagawa, S., Sakakida-Kitagawa, S., Memon, I. A., Kawaguchi, N. et al. 2006. Longer preservation of cardiac performance by sheet-shaped myoblast implantation in dilated cardiomyopathic hamsters. Cardiovasc. Res. 69, 466–75.

Kubo, H., Shimizu, T., Yamato, M., Fujimoto, T., and Okano, T. 2007. Creation of myocardial tubes using cardiomyocyte sheets and an *in vitro* cell sheet-wrapping device. Biomaterials 28, 3508–16.

Kushida, A., Yamato, M., Konno, C., Kikuchi, A., Sakurai, Y., and Okano, T. 1999. Decrease in culture temperature releases monolayer endothelial cell sheets together with deposited fibronectin matrix from temperature-responsive culture surfaces. J. Biomed. Mater. Res. 45, 355–62.

Kushida, A., Yamato, M., Isoi, Y., Kikuchi, A., and Okano, T. 2005. A non-invasive transfer system for polarized renal tubule epithelial cell sheets using temperature-responsive culture dishes. Eur. Cell Mater.10, 23–30.

Langer, R. S. and Vacanti, J. P. 1993. Tissue engineering. Science 260, 920–6.

Langer, R. S. and Vacanti, J. P. 1999. Tissue engineering: the challenges ahead. Sci. Am. 280, 86–9.

Matsuura, K., Honda, A., Nagai, T., Fukushima, N., Iwanaga, K., Tokunaga, M. et al. 2009. Transplantation of cardiac progenitor cells ameliorates cardiac dysfunction after myocardial infarction in mice. J. Clin. Invest. 119, 2204–17.

Matsuura, K., Masuda, S., Haraguchi, Y., Yasuda, N., Shimizu, T., Zandstra, P., Hagiwara, N., and Okano, T. 2011. Creation of mouse embryonic stem cell-derived cell sheets. Biomaterials (in press).

Memon, I. A., Sawa, Y., Fukushima, N., Matsumiya, G., Miyagawa, S., Taketani, S. et al. 2005. Repair of impaired myocardium by means of implantation of engineered autologous myoblast sheets. J. Thorac. Cardiovasc. Surg. 130, 1333–41.

Miyagawa, S., Saito, A., Sakaguchi, T., Yoshikawa, Y., Yamauchi, T., Imanishi, Y. et al. 2010. Impaired myocardium regeneration with skeletal cell sheets—a preclinical trial for tissue-engineered regeneration therapy. Transplantation 90, 364–72.

Miyagawa, S., Sawa, Y., Sakakida, S., Taketani, S., Kondoh, H., Memon, I. A. et al. 2005. Tissue cardiomyoplasty using bioengineered contractile cardiomyocyte sheets to repair damaged myocardium, their integration with recipient myocardium. Transplantation 80, 1586–95.

Miyahara, Y., Nagaya, N., Kataoka, M., Yanagawa, B., Tanaka, K., Hao, H. et al. 2006. Monolayered mesenchymal stem cells repair scarred myocardium after myocardial infarction. Nat. Med. 12, 459–65.

Nakayama, M., Chung, J. E., Miyazaki, T., Yokoyama, M., Sakai, K., and Okano, T. 2007. Thermal modulation of intracellular drug distribution using thermoresponsive polymeric micelles. Reactive and Functional Polymers 67, 1398–407.

Nakayama, M. and Okano, T. 2005. Polymer terminal group effects on properties of thermoresponsive polymeric micelles with controlled outer-shell chain lengths. Biomacromolecules 6, 2320–7.

Nishida, K., Yamato, M., Hayashida, Y., Watanabe, K., Maeda, N., Watanabe, H. et al. 2004. Functional bioengineered corneal epithelial sheet grafts from corneal stem cells expanded *ex vivo* on a temperature-responsive cell culture surface. Transplantation 77, 379–85.

Nishida, K., Yamato, M., Hayashida, Y., Watanabe, K., Yamamoto, K., Adachi, E. et al. 2004. Corneal reconstruction with tissue-engineered cell sheets composed of autologous oral mucosal epithelium. N. Engl. J. Med. 351, 1187–96.

Segers, V. F. and Lee, R. T. 2008. Stem-cell therapy for cardiac disease. Nature 451, 937–42.

Sekine, H., Shimizu, T., Kosaka, S., Kobayashi, E. and Okano, T. 2006. Cardiomyocyte bridging between hearts and bioengineered myocardial tissues with mesenchymal transition of mesothelial cells. J. Heart Lung Transplant. 25, 324–32.

Sekine, H., Shimizu, T., Yang, J., Kobayashi, E., and Okano, T. 2006. Pulsatile myocardial tubes fabricated with cell sheet engineering. Circulation 114(1 Suppl.), I87–93.

Sekiya, N., Matsumiya, G., Miyagawa, S., Saito, A., Shimizu, T., Okano, T. et al. 2009. Layered implantation of myoblast sheets attenuates adverse cardiac remodeling of the infarcted heart. J. Thorac. Cardiovasc. Surg. 138, 985–93.

Sekiya, S., Shimizu, T., Yamato, M., Kikuchi, A., and Okano, T. 2006. Bioengineered cardiac cell sheet grafts have intrinsic angiogenic potential. Biochem. Biophys. Res. Commun. 341, 573–82.

Shimizu, T., Sekine, H., Isoi, Y., Yamato, M., Kikuchi, A., and Okano T. 2006. Long-term survival and growth of pulsatile myocardial tissue grafts engineered by the layering of cardiomyocyte sheets. Tissue Eng. 12, 499–507.

Shimizu, T., Sekine, H., Yang, J., Isoi, Y., Yamato, M., Kikuchi, A. et al. 2006. Polysurgery of cell sheet grafts overcomes diffusion limits to produce thick, vascularized myocardial tissues. FASEB J. 20, 708–10.

Shimizu, T., Yamato, M., Isoi, Y., Akutsu, T., Setomaru, T., Abe, K. et al. 2002. Fabrication of pulsatile cardiac tissue grafts using a novel 3-dimensional cell sheet manipulation technique and temperature-responsive cell culture surfaces. Circ. Res. 90, e40.

Shiroyanagi, Y., Yamato, M., Yamazaki, Y., Toma, H., and Okano, T. 2003. Transplantable urothelial cell sheets harvested noninvasively from temperature-responsive culture surfaces by reducing temperature. Tissue Eng. 9, 1005–12.

Shiroyanagi, Y., Yamato, M., Yamazaki, Y., Toma, H., and Okano, T. 2004. Urothelium regeneration using viable cultured urothelial cell sheets grafted on demucosalized gastric flaps. BJU Int. 93, 1069–75.

Sumide, T., Nishida, K., Yamato, M., Ide, T., Hayashida, Y., Watanabe, K. et al. 2006. Functional human corneal endothelial cell sheets harvested from temperature-responsive culture surfaces. FASEB J. 20, 392–4.

Takezawa, T., Mori, Y., and Yoshizato, K. 1990. Cell culture on a thermo-responsive polymer surface. Biotechnology 8, 854–6.

von Recum, H., Kikuchi, A., Yamato, M., Sakurai, Y., Okano, T., and Kim, S. W. 1999. Growth factor and matrix molecules preserve cell function on thermally responsive culture surfaces. Tissue Eng. 5, 251–65.

Yamada, N., Okano, T., Sakai, H., Karikusa, F., Sawasaki, Y., and Sakurai, Y. 1990. Thermo-responsive polymeric surfaces; control of attachment and detachment of cultured cells. Die Makromolekulare Chemie, Rapid Communications 11, 571–6.

Yamato, M., Utsumi, M., Kushida, A., Konno, C., Kikuchi, A., and Okano T. 2001. Thermo-responsive culture dishes allow the intact harvest of multilayered keratinocyte sheets without dispase by reducing temperature. Tissue Eng. 7, 473–80.

Yang, J., Yamato, M., Shimizu, T., Sekine, H., Ohashi, K., Kanzaki, M. et al. 2007. Reconstruction of functional tissues with cell sheet engineering. Biomaterials 28, 5033–43.

Yoshida, R., Uchida, K., Kaneko, Y., Sakai, K., Kikuchi, A., Sakurai, Y. et al. 1995. Comb-type grafted hydrogels with rapid deswelling response to temperature changes. Nature 374, 240–2.

CHAPTER 13

Construction of Functional Cardiovascular Tissues Using Magnetic Nanoparticles

Jong-Kook Lee,[1,*] Akira Ito[2] and Hiroyuki Honda[3]

Introduction

Recent advances in stem cell biology, such as generation of induced pluripotent stem cells (iPS cells) (Takahashi et al. 2007; Yu et al. 2007) and embryonic stem cells (ES cells) (Thomson et al. 2007), has progressed to the feasibility of induction of cardiac differentiation from these cell sources. Since these important breakthroughs, a substantial numbers of studies have been conducted to explore the possibility of cardiac regeneration therapy to treat incurable heart diseases using stem cell derived cardiac cells. Most of the studies have been attempted by local injection of stem cell derived cardiomyocytes into diseased hearts; however, the functional outcomes were not necessarily as good as expected. These discouraging results have been attributed, at least in part, to low survival rate or hampered function of the transplanted cells *in vivo*. These findings suggest that simple injection of isolated cells can only yield aggregation of cells with the loss of extracellular matrix and cell-to-cell connections that play an important role in maintaining survival and function of cells.

[1]Department of Cardiovascular Regenerative Medicine, Graduate School of Medicine, Osaka University.
Email: jlee@cardiology.med.osaka-u.ac.jp
[2]Department of Chemical Engineering, Faculty of Engineering, Kyushu University.
Email: akira@chem-eng.kyushu-u.ac.jp
[3]Department of Biotechnology, Graduate School of Engineering, Nagoya University.
Email: honda@nubio.nagoya-u.ac.j
*Corresponding author

Based upon these observations, engineering of three dimensional (3D) functional tissues/organs *in vitro* before transplantation has been considered as a key technology for regenerative medicine. To construct 3D functional cardiovascular tissues, various techniques have been explored extensively.

First, 3D tissues were constructed using artificial scaffolds made from biomaterials. The technique resulted in much better function compared to simple injection of isolated cells. However, immunogenicity or inflammation, potentially caused by the biomaterials of scaffolds, could be major obstacles in clinical application.

Instead of using immunogenic artificial scaffolds, cardiac sheets have been fabricated using temperature-responsive culture dishes (Shimizu et al. 2002). With this technique, cardiac cell sheets can be peeled from culture dishes by only lowering temperature. Because enzymes are not used in this technique, cells are protected from damage and extracellular matrices are preserved. A detailed explanation of the technique will be available in the chapter by Shimizu et al. in this book.

Recently, we have developed a novel tissue engineering technique by magnetic nanoparticles (magnetite: Fe_3O_4) and magnetic force (Mag-TE). The technique has been applied in various types of cells such as cardiomyocytes, keratinocytes, retinal epithelial cell, dermal fibroblast, and mesenchymal stem cell (MSC) sheets. In this chapter, we have discussed application of the Mag-TE to construct 3D cardiovascular tissues.

Tissue Engineering Using Magnetic Nanoparticles and Magnetic Force

In Mag-TE, two lines of functional magnetic nanoparticles have been developed: 1) magnetic cationic liposomes (MCLs); 2) antibody-conjugated magnetoliposome (AMLs) (Ito et al. 2005). In both particles, magnetites (diameter of single magnetite particles: 10 nm) are wrapped with liposomes (Fig. 1).

MCLs consist of magnetites and surrounding cationic liposomes. Diameter of MCLs is approximately 150 nm. Because of electrostatic interaction between positively-charged liposomes and negatively-charged cell membranes, magnetites are easily incorporated across the membrane and accumulated in cytoplasms of target cells such as cardiomyocytes (CM), endothelial cells, smooth muscle cells and fibroblasts (magnetite labeling). Once labeled with magnetic nanoparticles, labeled cells can be manipulated without direct contact with the application of magnetic force (magnetic manipulation). The MCLs can be used to construct multilayered tissues with homogenous patterns (Shimizu et al. 2007) (Fig. 1 and Fig. 2).

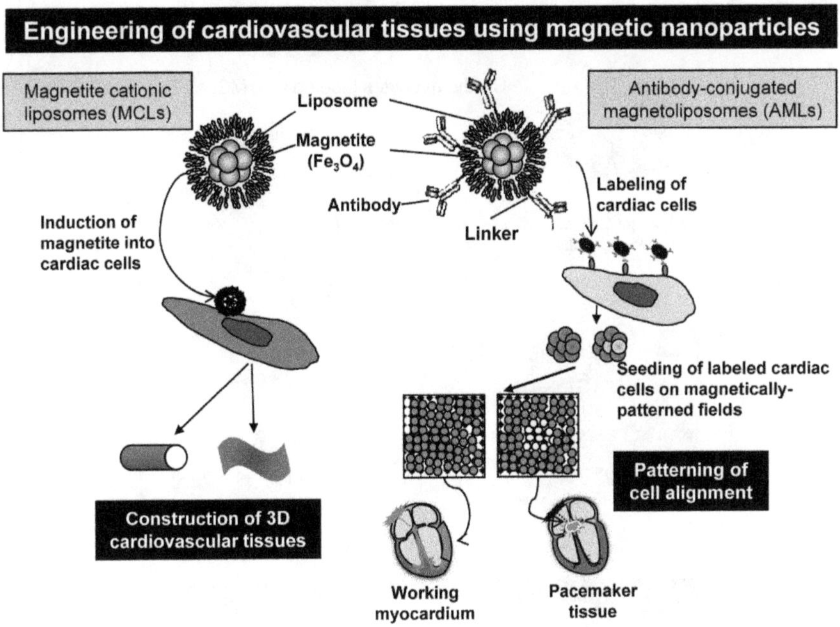

Figure 1. Schematic explanation of tissue engineering of cardiovascular tissues using magnetic nanoparticles and magnetic force. Magnetite cationic liposomes (MCLs) and antibody-conjugated magnetoliposomes (AMLs). MCLs are mainly used in multi-layering of cardiac tissues such as myocardial and blood vascular tissues. On the other hand, AMLs are suitable to label target cells specifically.

Color image of this figure appears in the color plate section at the end of the book.

On the other hand, in AMLs, antibodies are conjugated with magnetoliposomes (Ito et al. 2005). With specific antibodies on the surface, AMLs will selectively bind with the surface antigen of target cells. Thus, the techniques with AMLs can be used to construct tissues with heterogenous types of cells such as patterning of cell allocation and alignments (Fig. 1).

Multilayering of Cardiomyocytes to Construct 3D Cardiac Tissues

Engineering of cardiac tissues has been extensively studied. 3D tissues were first constructed using biodegradable porous scaffolds (Zammaretti et al. 2004; Zimmermann et al. 2006). However, these materials often caused inflammatory reaction, which in turn seriously damaged both host and grafted myocardium. In addition, loss of cell-to-cell contact prevents the formation of gap junctions which are important to maintain electrophysiological properties. To avoid these problems in grafts, 3D cardiac tissues must be constructed without scaffolds. Recently, Shimizu

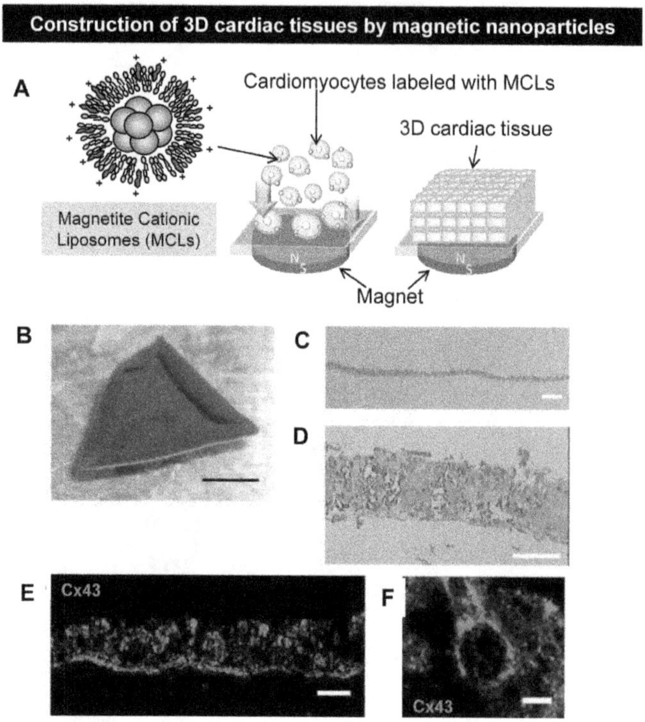

Figure 2. (A) Schemes to explain multilayering technique using magnetite cationic liposomes (MCLs) and magnetic force. To label cardiomyocytes (CMs) magnetically, MCLs were added to the culture medium of CMs at the concentration of 100 pg/cell and incubated for 4 hr. Then, the magnetically labeled CMs were recovered and used to construct the CM sheets. The CM sheets were fixed in 10% formalin solution and embedded in paraffin. (B) Bright-field photograph of CM sheet constructed by Mag-TE; scale bar, 5 mm. (C) Cross-sections of CM sheet; scale bar, 500 mm (Hematoxylin and eosin stain). (D) Magnified view of C (scale bar, 100 mm). (E) Connexin43expression in engineered myocardial tissues (Immunofluorescence stained with anti- Cx43 antibody; scale bar, 100 mm. (F) Magnified view of (E). (Cited Shimizu et al. Biotechnol. Bioeng. 2007) {Shimizu, 2006 2351/id}

Color image of this figure appears in the color plate section at the end of the book.

et al. reported alternative methods to engineer cardiac tissues using temperature-responsive culture dishes (Shimizu et al. 2002). Because either scaffolds or enzymes are not used, the technique has been clinically applied using myoblast sheets (Sawa et al. 2012).

Mag-TE is another "scaffold-less" technique applied in the construction of cardiac tissues without scaffolds and enzymatic dispersion. Fig. 2A explains the procedure of Mag-TE-induced construction of 3D cardiac sheets. First, MCLs were applied to cardiomyocytes. The positively charged surface of the nanoparticles enhances transfer of magnetites into cytoplasms. Then, magnetically-labeled cardiomyocytes were seeded on culture dishes with low attachment surface profile. While magnetic forces (4,000 Gauss) were applied, the cardiac sheet remained attached at the bottom of culture dishes; on the other hand, when the magnet was removed, the cell sheets were detached. The obtained cardiac cell sheets showed thickness of approximately 120 μm (8 layers), regular, spontaneous beating, and electrical propagation with conduction velocity of 9.1±1.8 cm/s. Histological examination revealed expression of connexin43 between ventricular cardiomyocytes.

Toxicity of MCLs against CMs is an important issue. Our previous studies showed that MCLs did not inhibit the proliferation of several cell types within the magnetite concentration tested, including human keratinocytes, human MSC, human aortic endothelial cells, human aortic smooth muscle cells, normal human dermal fibroblast and human umbilical vascular endothelial cells. It was also reported that human mesenchymal stem cells maintained an *in vitro* ability to differentiate into osteoblasts, adipocytes, or chondrocytes, with magnetite particles inside cells. When cardiac sheets were constructed at a magnetite concentration of 100 pg/cell, no decrease was observed in the percentage of viable CMs within 24 hr of culture. Moreover, the magnetite nanoparticles inside the cells did not affect the electrical conduction properties *in vitro*.

This data suggests that Mg-TE based construction of cardiac 3D sheets could be feasible, at least *in vitro*. Further investigation regarding safety of magnetites will be required.

Construction of Blood Vessels

Although larger blood vessels (diameter >about 5 mm) have been successfully engineered using synthetic materials such as polyethylene terephthalate (PET) or expanded polyetrafluoroethylene (ePTEE), these materials have not been applicable to smaller diameter blood vessels (< about 5 mm). Instead of artificial biomaterial, scaffolds obtained from decellularized native vascular tissues have been used to construct tissue-engineered blood vessels (TEBV). These scaffolds are not only non-antigenic but also facilitate cell attachment. In this technique, however, reseeding of cells (cell alignment or allocation) is also an important step. Recently, we have reported that blood vessels were successfully engineered using Mag-TE (Shimizu et al. 2007), in which a magnet force-induced technique was applied to a tubular scaffold obtained from porcine decellularized common carotid arteries (dCCA). Human smooth muscle cells (SMCs) and fibroblasts labeled with MCLs were seeded onto dCCA. The summary of the procedures has been shown in Fig. 3A. As a result, vascular grafts were aligned in an orderly fashion with human SMCs and fibroblasts which were

ig.3

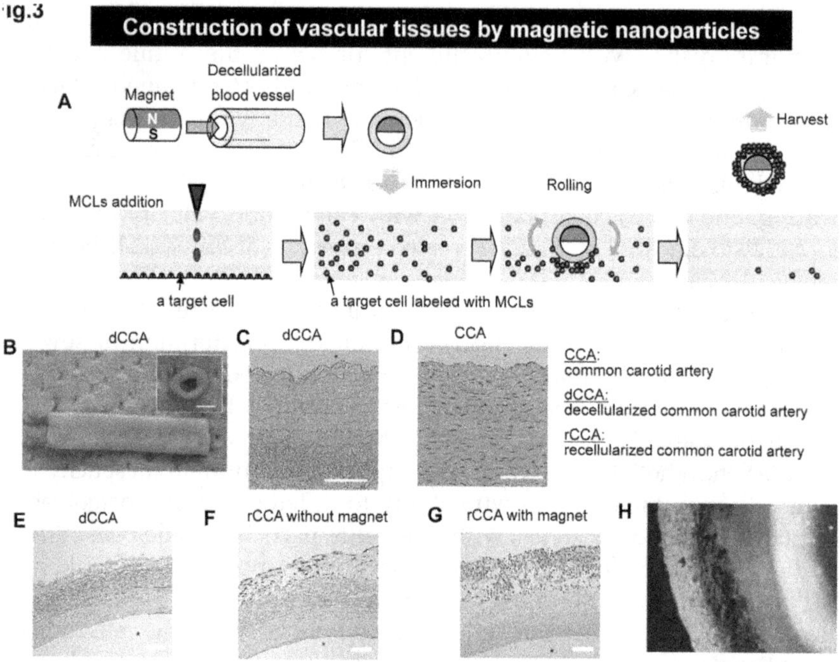

Figure 3. (A) A schematic illustration of Mag-seeding onto tubular scaffolds. MCLs were added to the medium to label target cells magnetically. Decellularized blood vessel into which a magnet was inserted was immersed in the cell suspension and rolled in the suspension to seed cell evenly. (B) A gross view of decellularized common carotid arteries (dCCA). The dCCAs were cut into 20 mm for subsequent experiments. Scale bar in the right upper box, 2 mm. (C, D) Cross-sectional view of dCCA (C) and CCA (D) with HE stain (E, F, G). Cross section of dCCA seeded with 3T3s with (F, G) or without (E) magnet. * Lumen. (H) Fluorescence microscopy of the cross section of dCCA seeded with Smooth muscle cells (SMCs) and Normal Human Dermal Fibroblasts (NHDFs). SMCs and NHDFs were probed with blue and green dye, respectively. * Lumen (Cited from Shimizu et al. J. Biosci. Bioeng. 2007) {Shimizu, 2007 2755 /id}.

Color image of this figure appears in the color plate section at the end of the book.

seeded onto scaffolds by magnetic force sequentially. In combination with further seeding of magnetically-labeled endothelial cells, it is conceivable that graftable blood vascular tissues can be constructed.

Patterning of Cell Alignment and Allocation

Certain types of cardiac tissues such as sinus node (pacemaker tissue in native heart) and conduction system show unique structures, consisting of heterogenous types of cells, unlike homogenous working myocardium such as ventricular muscle. For instance, native sinus node tissue has the efficient alignment of pacemaker cells and intermediate cells to generate sufficient electrical potential to pace surrounding static working myocardium. Therefore, to engineer an ideal biological pacemaker, cell patterning techniques might be required, as well as the purification of pacemaker cells.

To construct specifically designed structures, cell patterning technique by Mag-TE can be used in the construction of 3D tissues with heterogenous cell alignment with heterotypic cells.

So far, several cell patterning methods have been developed including "microcontact printing" (Dike et al. 1999) and "lithography" (Bhatia et al. 1997). Although these techniques are highly precise, the methods require large scale devices and chemical and biological modification on cells, which could be a substantial limitation in conducting the procedure.

In contrast, in the cell patterning method using magnetic force, the manipulation applied on cells is usual antibody-to-antigen reaction, which is almost compatible with those used in the MACS (magnetic-activated cell sorting) technique. Thus, cell damage would be minimized in Mag-TE, compared to the above mentioned techniques such as "microcontact printing" or "lithography".

Conclusion

3D tissues of myocardium and blood vessels have been reported to show better function compared to simple aggregation of isolated cells. To construct better and larger 3D tissues, technologies have been continuously upgraded.

In this chapter, as one of the tissue engineering technologies, we illustrated how 3D tissues of myocardium and blood vessels can be constructed by magnetic nanoparticles and magnetic force. Although the technique has an advantage in being able to induce tissue construction under non-contact manner without expensive devices, the long-term effects of accumulated nanoparticles of magnetites remain to be cleared.

References

Bhatia, S. N., Yarmush, M. L., and Toner, M. 1997. Controlling cell interactions by micropatterning in co-cultures: Hepatocytes and 3T3 fibroblasts. Journal of Biomedical Materials Research 34,189 –199.

Dike, L. E., Chen, C. S., Mrksich, M., Tien, J., Whitesides, G. M., and Ingber, D. E. 1999. Geometric control of switching between growth, apoptosis, and differentiation during angiogenesis using micropatterned substrates. *In vitro* cellular & developmental biology. Animal 35, 441–448.

Ito, A., Hibino, E., Kobayashi, C., Terasaki, H., Kagami, H., Ueda, M., Kobayashi, T., and Honda, H. 2005. Construction and delivery of tissue-engineered human retinal pigment epithelial cell sheets, using magnetite nanoparticles and magnetic force. Tissue Engineering 11, 489–496.

Ito, A., Hibino, E., Shimizu, K., Kobayashi, T., Yamada, Y., Hibi, H., Ueda, M., and Honda, H. 2005. Magnetic force-based mesenchymal stem cell expansion using antibody-conjugated magnetoliposomes. Journal of Biomedical Materials Research Part B—Applied Biomaterials 75B, 320–327.

Sawa, Y., Miyagawa, S., Sakaguchi, T., Fujita, T., Matsuyama, A., Saito, A., Shimizu, T., and Okano, T. 2012. Tissue engineered myoblast sheets improved cardiac function sufficiently to discontinue LVAS in a patient with DCM: report of a case. Surgery Today 42, 181–184.

Shimizu, K., Ito, A., Arinobe, M., Murase, Y., Iwata, Y., Narita, Y., Kagami, H., Ueda, M., and Honda, H. 2007. Effective cell-seeding technique using magnetite nanoparticles and magnetic force onto decellularized blood vessels for vascular tissue engineering. Journal of Bioscience and Bioengineering 103, 472–478.

Shimizu, K., Ito, A., Lee, J. K., Yoshida, T., Miwa, K., Ishiguro, H., Numaguchi, Y., Murohara, T., Kodama, I., and Honda, H. 2006. Construction of multilayered cardiomyocyte sheets using magnetite nanoparticles and magnetic force. Biotechnol. Bioeng.

Shimizu, T., Yamato, M., Isoi, Y., Akutsu, T., Setomaru, T., Abe, K., Kikuchi, A., Umezu, M., and Okano, T. 2002. Fabrication of pulsatile cardiac tissue grafts using a novel 3-dimensional cell sheet manipulation technique and temperature-responsive cell culture surfaces. Circulation Research 90, e40–e48.

Takahashi, K., Tanabe, K., Ohnuki, M., Narita, M., Ichisaka, T., Tomoda, K., and Yamanaka, S. 2007. Induction of pluripotent stem cells from adult human fibroblasts by defined factors. Cell 131, 861–872.

Thomson, J. A., Itskovitz-Eldor, J., Shapiro, S. S., Waknitz, M. A., Swiergiel, J. J., Marshall, V. S., and Jones, J. M. 1998. Embryonic stem cell lines derived from human blastocysts. Science 282, 1145–1147.

Yu, J., Vodyanik, M. A , Smuga-Otto, K., Antosiewicz-Bourget, J., Frane, J. L., Tian, S., Nie, J., Jonsdottir, G. A., Ruotti, V., Stewart, R., Slukvin, I. I., and Thomson, J. A. 2007. Induced pluripotent stem cell lines derived from human somatic cells. Science 318, 1917–1920.

Zammaretti, P. and Jaconi, M. 2004. Cardiac tissue engineering, regeneration of the wounded heart. Current Opinion in Biotechnology 15, 430–434.

Zimmermann, W. H., Melnychenko, I., Wasmeier, G., Didie, M., Naito, H., Nixdorff, U., Hess, A., Budinsky, L., Brune, K., Michaelis, B., Dhein, S., Schwoerer, A., Ehmke, H., and Eschenhagen, T. 2006. Engineered heart tissue grafts improve systolic and diastolic function in infarcted rat hearts. Nat. Med. 12, 452–458.

CHAPTER 14

Cardiac Cell Therapy for Ischemic Heart Disease

Hideaki Kanazawa,[1,a] Konstantinos Malliaras,[1,b] Michelle Kreke[2] and Eduardo Marbán[1,c,*]

Introduction

Cardiovascular disease remains the most common cause of death and disability among Americans, claiming more lives each year than cancer, diabetes mellitus, HIV and accidents combined (Heron et al. 2009). Ischemic heart disease is the predominant contributor to cardiovascular morbidity and mortality; ~1 million myocardial infarctions (MIs) occur per year in the USA, while ~5 million patients suffer from chronic heart failure (Roger et al. 2011). Although implementation of cardiovascular prevention strategies has led to a significant decrease in the incidence of acute MI, and advances in medical and catheter-based therapies have reduced MI-associated mortality, new approaches would be most valuable for patients who go on to develop ventricular dysfunction and/or heart failure (White et al. 2005).

Over the past decade, stem cell transplantation has been explored as a therapeutic strategy for acute or chronic ischemic cardiomyopathy. Multiple candidate cell types used to regenerate the injured heart have been tested in preclinical and clinical trials, including: embryonic stem cells (ESCs)

[1]Cedars-Sinai Heart Institute, Los Angeles, California, USA.
[a]Email: hideaki_@mta.biglobe.ne.jp
[b]Email: Konstantinos.Malliaras@cshs.org
[c]Email: Eduardo.Marban@cshs.org
[2]Capricor Inc., Los Angeles, California, USA.
Email: Michelle.Kreke@cshs.org
*Corresponding author

(Caspi et al. 2007; Dai et al. 2007; Laflamme et al. 2007), induced pluripotent stem cells (iPSCs) (van Laake et al. 2010), neonatal cardiomyocytes (Müller-Ehmsen et al. 2002; Reffelmann et al. 2003), skeletal myoblasts (SKMs) (Menasché et al. 2008), endothelial progenitor cells (EPCs) (Katritsis et al. 2005), bone marrow mononuclear cells (BMMNCs) (Wollert et al. 2004; Schächinger et al. 2006a; Lunde et al. 2006), mesenchymal stem cells (MSCs) (Chen et al. 2004) and most recently cardiac stem cells (CSCs) (Beltrami et al. 2003; Smith et al. 2007). Although no consensus has yet emerged, the ideal cell type for the treatment of heart disease should:

1. Be safe, i.e., not create tumors (a complication of undifferentiated ESCs or iPSCs delivered to the heart (Murry and Keller 2008)) or arrhythmias (a risk of SKM transplantation (Menasché et al. 2003)).
2. Improve heart function.
3. Create healthy and functional cardiac muscle and vasculature, integrated into host tissue.
4. Be easily harvested using routine clinical methods, readily grown in large numbers, and amenable to delivery by minimally invasive approaches.
5. Be available as a standardized "off the shelf" reagent.
6. Not elicit an immune reaction.
7. Circumvent societal ethical concerns.

In this chapter, we provide a critical assessment of the various cell types used for heart regeneration and the insights arising from the first generation of clinical trials. We also touch upon the future directions of cell therapy for heart disease with the focus on cells that are already in the clinic, or soon will be (Malliaras et al. 2011; Malliaras and Marbán 2011).

Cell Types Used for Cardiac Regeneration Therapy

1. Skeletal myoblasts

Skeletal myoblasts (SKMs) are conceptually attractive for cellular cardiomyoplasty: they have a contractile phenotype, can be harvested for autologous transplantation, and are resistant to ischemia. After a decade of experimental studies (Jain et al. 2001; Leobon et al. 2003; He et al. 2005), SKMs were the first cell type to enter the clinical arena for heart regeneration. In June 2000, autologous SKMs, isolated and expanded from a thigh muscle biopsy, were intramyocardially injected in a patient with severe ischemic heart failure as an adjunct to coronary bypass grafting (CABG) surgery (Menasché et al. 2001). Several small, non-randomized phase I trials ensued, demonstrating a functional benefit, albeit with a high incidence of ventricular arrhythmias (Menasché et al. 2001; Siminiak et al. 2004; Hagège et al. 2006). SKMs differentiate into multinucleated myotubes (not cardiomyocytes) after injection into the heart. These myotubes lack gap junctions and form islands of conduction block in the heart, resulting in electrical inhomogeneities that slow conduction velocity and predispose to reentrant ventricular arrhythmias (Abraham et al. 2005). In sharp contrast to the functional benefit observed in the early uncontrolled studies (Menasché et al. 2003), the first prospective randomized placebo-controlled phase II SKM trial (MAGIC trial), exhibited lack of efficacy and was discontinued prematurely (Menasché et al. 2008). In addition, despite the use of prophylactic amiodarone, a trend towards excess arrhythmias was observed in myoblast-treated patients, thus confirming the safety concern that had already been raised by earlier phase I trials (Menasché et al. 2003). On the other hand, the recently-published SEISMIC trial (Duckers et al. 2011) argued that injection of autologous SKMs in heart failure (HF) patients is safe and may provide symptomatic relief (a trend towards increased exercise tolerance was observed in the cell-treated group); nevertheless, no significant effect on global left ventricular ejection fraction (LVEF) was detected. Taken together, the trajectory of SKMs is instructive and argues against premature enthusiasm solely on the basis of preclinical studies.

2. Bone marrow-derived cells

Unlike SKMs, bone marrow-derived cells moved into patients without the benefit of a convincing preclinical development program; in fact, the first report of clinical application of bone marrow-derived cells for heart regeneration (Strauer et al. 2001) surfaced within four months of the publication of a rodent study showing extensive engraftment and cardiogenic transdifferentiation of bone marrow-derived cells in mice (Orlic et al. 2001). Clinical application was catalyzed by the relative accessibility of bone marrow and the extensive clinical experience with bone marrow transplantation for the treatment of various hematologic disorders. In addition, in contrast to SKMs, the large numbers of unfractionated autologous cells that can be obtained without *ex vivo* expansion make bone marrow–derived cells suitable for treatment of recent AMI.

The bone marrow is a highly heterogeneous tissue, containing several different cell populations including rare hematopoietic, endothelial and mesenchymal stem cells. Human hematopoietic stem cells (HSCs) can traditionally be defined as rare CD34+ cells capable of reconstituting all blood lineages (Civin and Gore 1993) and, possibly, the ability to transdifferentiate into cardiomyocytes, endothelial cells and smooth muscle cells *in vivo* (Sussman and Murry 2008). EPCs are a subset of hematopoietic cells that promote neovascularization either directly (differentiation into endothelial cells) (Crosby et al. 2000) or indirectly (secretion of pro-angiogenic cytokines) (Rehman et al. 2003). MSCs can be roughly defined as CD105 + CD90+ cells, isolated by preferential adherence to plastic in tissue culture, which are capable of osteogenic, chondrogenic and adipogenic differentiation (Dominici et al. 2006). MSCs purportedly exhibit low immunogenicity (Aggarwal et al. 2005), rendering allogeneic applications plausible. It should be noted that BMMNCs, isolated by density centrifugation following bone marrow aspiration, actually contain very few stem cells (~2–4% HSCs/EPCs and ~0.01% MSCs); the vast majority of BMMNCs comprise committed hematopoietic cells at various stages of maturation (Dimmeler and Zeiher 2009).

In the clinical setting, autologous BMMNCs are by far the most frequently used cell type for treatment of acute MI. To date, >1000 patients have been treated with bone marrow-derived cells (either unfractionated or enriched in progenitor subpopulations) in numerous clinical trials worldwide.

In general, two major conclusions can be drawn:

1. So far, no safety concerns regarding delivery of BMMNCs through a recently reopened infarct-related coronary artery have emerged.
2. Efficacy has been inconsistent, and primary end points in the majority of large randomized controlled studies have not been met. However, overall clinical outcomes have been generally modestly positive.

The legacy of these studies has left the field with cautious optimism, while motivating a search for better cell types. It is plausible (but still conjectural) that a stem cell source with a higher propensity to regenerate myocardium, directly and indirectly, might increase the benefits to patients.

Finally, bone marrow-derived cells have also been used for the treatment of refractory angina (Losordo et al. 2007; Tse et al. 2007; Van Ramshorst et al. 2009) and chronic heart failure (Mocini et al. 2006; Hendrikx et al. 2006; Stamm et al. 2007), albeit on a much smaller scale compared with acute MI. Early, small clinical studies have shown some hints of efficacy; however, primary efficacy endpoints have not been met in these underpowered studies and results have been inconsistent.

3. Adipose-derived cells

Adipose tissue has been investigated as a source of adult progenitor/ stem cells for the purposes of cardiac repair as this tissue contains a heterogeneous mixture of endothelial, haematopoietic and mesenchymal progenitor cells which can be easily harvested by liposuction. Preclinical studies have shown that adipose-derived stem cells (ASCs) are associated with improvement in the ejection fraction in animal models of myocardial infarction and neoangiogenesis via paracrine factors has been postulated as a potential mechanism of action of ASCs (Valina et al. 2007; Wang et al. 2009). Clinical trials are currently ongoing and include the APOLLO Study (ClinicalTrials.gov Identifier: NCT00442806) and PRECISE Study (ClinicalTrials.gov Identifier: NCT00426868) which are assessing the safety and efficacy of ASCs in AMI and chronic myocardial ischemia, respectively (Mozid et al. 2011).

4. Heart-derived cells

Although the mammalian heart has traditionally been viewed as a terminally differentiated postmitotic organ, the concept of endogenous cardiac regeneration has now been firmly established. Through use of radiocarbon dating of human postmortem cardiac tissue, it has been argued that cardiomyocyte turnover in the adult human heart occurs at an age-dependent rate of 0.4–1% per year, with ~40% of the mature heart composed of post-natally generated myocytes (Bergmann et al. 2009) (but a recent study using immunohistochemistry reported strikingly higher turnover rates) (Kajstura et al. 2010).

In addition, multiple populations of putative endogenous CSCs have now been identified. CSCs presumably function physiologically to offset a low, but finite basal rate of cardiomyocyte loss (It should be noted, however, that an obligatory role for CSCs in cardiomyocyte renewal has yet to be demonstrated; an alternative path to regeneration is via re-activation of the cell cycle in adult cardiomyocytes, with or without partial dedifferentiation) (Zhang et al. 2010; Mercola et al. 2011).

The intrinsically low number of CSCs (one estimate posits 1 CSC per ~10,000 cardiomyocytes) (Beltrami et al. 2003), helps to rationalize why endogenous repair does not suffice to reverse major injury. However, because CSCs are resident in the heart and pre-programmed to regrow heart tissue and reconstitute all cardiac lineages (but not extracardiac tissues), their isolation, *ex vivo* expansion, and subsequent cardiac delivery represent a logical strategy to regenerate the heart iatrogenically.

Historically, the term 'cardiac stem cells' was first used by Deisher to describe multipotent small, round, slowly replicating, non-adherent cells isolated from the hearts of adult P53-deficient mice, with apparent multipotent potential (Deisher 2000). This seminal work received little attention at the time and remains largely neglected. In 2003–2004, several studies advanced the notion that the adult heart contains its own reservoir of antigenically distinctive stem cells. CSCs, defined by an ability to differentiate into multiple cardiac lineages *in vitro* and *in vivo*, were identified in rodents by stem cell-related markers and other phenotypic properties, including c-Kit (CD117, the receptor for stem cell factor) (Beltrami et al. 2003), Sca-1 (stem cell antigen-1) (Oh et al. 2003) and sphere-forming ability (the ability to self-organize into three-dimensional microtissues of CSCs and supporting cells). The investigators showed that such cell products, when injected into the heart in post-MI models, produced multilineage differentiation (cardiomyocytes, endothelial cells, vascular smooth muscle cells) and functional benefits in some studies (Beltrami et al. 2003; Messina et al. 2004).

Our own attention to the regenerative potential of the human heart and its possible therapeutic application was focused by the work of Messina et al. (Messina et al. 2004), who first reported the isolation of CSCs from human myocardium. Working with large human cardiac surgical specimens as the source tissue, these investigators described a technically straightforward approach to generate CSCs and supporting cells. Biopsies minced (Fig. 1A) and placed in primary culture were found, spontaneously, to shed cells (Fig. 1B and C), which could be harvested by gentle enzymatic digestion. When placed in suspension culture, these cells self-organized into spherical clusters termed 'cardiospheres' (CSps) (Fig. 1D), by analogy to neurospheres formed by neural stem cells (Reynolds and Weiss 1992). This study also showed that CSps provide an environment favoring upregulation of stemness in the proliferative core of the sphere as well as increased angiogenesis and cardiogenesis. Human CSps, when injected into post-MI SCID mouse hearts, engrafted, exhibited cardiogenic and vasculogenic differentiation and improved heart function. This exciting work was limited by the requirement for open surgical biopsies, and by the fact that CSps would appear, from first principles (Feigl 1983), to be too large (50–200 μm) to be safely delivered via the clinically routine intracoronary route.

Our laboratory adapted and miniaturized the CSp culture method to enable the use of minimally invasive percutaneous endomyocardial biopsies as the source tissue (Smith et al. 2007). CSps were re-plated and further expanded in monolayer culture (Fig. 1E) to yield therapeutically relevant numbers (tens of millions) of cardiosphere derived cells (CDCs) in a timely manner (4–6 weeks), despite the small amount of starting tissue material. Figure 1 depicts the key steps in the method. CDCs (in contrast to antigenically-purified cells) are a naturally heterogeneous cardiac-derived cell population rich in CSCs and cardiac mesenchymal cells (Fig. 1F). When grown according to established methods, CDCs are clonogenic and exhibit multilineage potential (Davis et al. 2009), thus fulfilling key criteria for stem cells. Moreover, CDCs can be safely delivered via the intracoronary route within a defined dosage range (Johnston et al. 2009).

Over the past six years, we have demonstrated that CDCs can engraft, differentiate and improve cardiac function post-MI in mice (Smith et al.

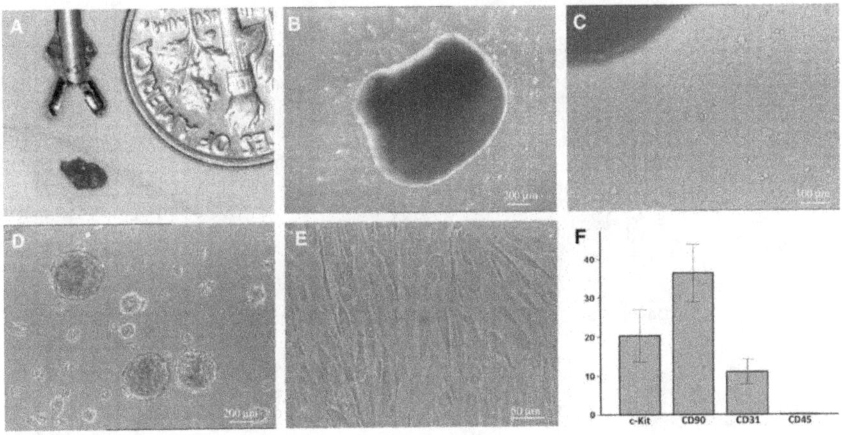

Figure 1. Specimen processing for human cardiosphere growth and CDC expansion. Cardiac biopsies (A) are minced into fragments termed explants. Explants are placed in primary culture and spontaneously shed outgrowth cells (B) which upon confluency (C) can be harvested by gentle enzymatic digestion. When placed in suspension culture, these cells self-organized into multicellular spherical clusters termed cardiospheres (D). CSps are collected and plated onto fibronectin-coated dishes, generating CDCs (E). Flow cytometry experiments demonstrate that CDCs are a naturally heterogeneous population of non-hematologic origin (CD45−), comprising endogenous cardiac stem cells (c-Kit+) and cardiac mesenchymal stem cells (CD90+). Reproduced from reference (Malliaras and Marbán 2011).

2007; Li et al. 2010; Chimenti et al. 2010), rats (Terrovitis et al. 2009; Davis et al. 2010; Cheng et al. 2010) and pigs (Johnston et al. 2009; Lee et al. 2011). Figure 2 provides a synopsis of the functional benefit produced by CDC therapy in various animal models. With regard to safety, it should be noted that no tumors have been detected in >1000 experiments, no increases in toxicology signals or in arrhythmias have been observed, nor has there been excess mortality or morbidity in cell-treated groups relative to placebo controls.

The question of whether CDCs outperform other cell types is under active investigation. In small comparative studies, CDCs outperformed MSCs *in vitro* (Koninckx et al. 2011) and *in vivo* (Takehara et al. 2008). In a head-to-head comparison of four different cell types (CDCs, BMMNCs, bone marrow-derived MSCs and adipose-derived MSCs) in the same animal model in the same laboratory, CDCs emerged as superior in terms of paracrine factor secretion, angiogenesis, cardiomyogenic differentiation, ischemic tissue preservation, anti-remodeling effects and functional benefit post-MI (Cheng et al. 2010). Interestingly, the natural mixture of CDCs also outperformed purified c-Kit+ CSCs (Smith et al. 2008). This finding suggests that the non-CSC sub-population supports the survival and engraftment of the CSC sub-population, translating into greater efficacy of the mixture relative to the purified fractions. However, it is equally possible that the c-Kit+ subpopulation is irrelevant to the mechanism of benefit.

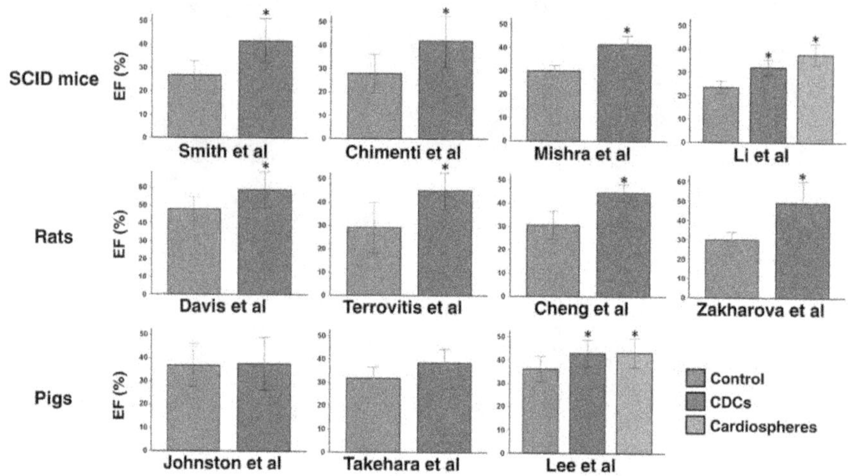

Figure 2. CDCs and CSps improve cardiac function post-MI in mice, rats and pigs, compared with sham-injected controls (*$P < 0.05$ compared with controls; in the study by Zakharova et al. (Zakharova et al. 2010) CDCs were transplanted as a sheet, while in the study by Takehara et al. (Takehara et al. 2008) CDCs were injected with a basic fibroblast growth factor incorporating hydrogel). Reproduced from reference (Malliaras and Marbán 2011).

Color image of this figure appears in the color plate section at the end of the book.

With regard to clinical translation, a phase I/II clinical study of CDCs—the CADUCEUS (CArdiosphere-Derived aUtologous stem CElls to reverse ventricUlar dySfunction) study, is under way (ClinicalTrial.gov Identifier: NCT00893360); human subjects with recent acute MI (2–3 months post-MI) and left ventricular dysfunction (EF, 25–45%) receive intracoronary infusion of 12.5–25 million autologous CDCs harvested from endomyocardial biopsies, or conventional treatment, by prospective randomized assignment. Enrollment is complete, and full 6-month follow-up data have been presented recently (Makkar et al. 2011). The MRI efficacy results are highly positive, showing reductions in scar mass mirrored by increases in viable myocardial mass in CDC-treated subjects (but not controls). These data represent the first clinical demonstration of therapeutic regeneration. Another trial—the SCIPIO (ClinicalTrial.gov Identifier: NCT00474461) of heart-derived cells is ongoing in post-CABG patients with heart failure; in this single-center study, a total of 0.5–1 million c-Kit-purified heart-derived cells are infused intracoronarily four months after tissue harvesting during surgery. The natural tendency of function to improve slowly and progressively after CABG (Ragosta et al. 1993; Ugander et al. 2010) may complicate the interpretation of this trial.

Emergent Insights

Cell therapy targeted at heart disease has now been in clinical use for more than 10 years. Although the overall outcomes have been modestly positive, cell therapy has yet to live up to its potential; the goal of true myocardial regeneration remains elusive, with the exception of the recent CADUCEUS data (Makkar et al. 2011). In brief, a number of lessons arise from the AMI studies with BMMNCs, which by far are the most numerous today:

1. The safety profile has been quite favorable.
2. Efficacy (as gauged by an increase in LVEF) has been inconsistent and, overall, modest. However, EF might not be the most suitable index for assessing cell therapy, especially in patient populations that are not very ill at baseline, thus having little room for improvement. To that end, it is intriguing that significant benefits on clinical end points have been reported, despite the small effect of cell therapy on EF. Other surrogate markers may need to be employed.
3. Acute retention and long-term engraftment of cells is disappointingly low. Delivery methods need to be optimized and combined with strategies to boost cardiac cell retention.
4. The cardiomyogenic potential of cells used to date is low. However, functional benefit can be achieved through indirect non-myogenic pathways, via the release of paracrine factors.
5. Autologous cell therapy poses significant logistic, economic, and timing constraints and is limited by variable and unpredictable cell potency. Highly standardized, "off the shelf" readily available allogeneic cell products, if proved safe and effective, would be particularly attractive.

These points are considered individually below.

1. Safety of intracoronary delivery

Intracoronary delivery of BMMNCs through a recently reopened artery (days to weeks after revascularization) has an excellent safety profile. Cell infusion does not inflict additional ischemic damage to the myocardium (Wollert et al. 2004). In addition, no increased incidence of arrhythmias, increased arrhythmia risk variables (as assessed by heart-rate variability, microvolt T-wave alternans, and signal-averaged electrocardiography measures) (Huikuri et al. 2008) or increased arrhythmia inducibility (Wollert et al. 2004) has been reported. Importantly, the fear of increased in-stent/culprit artery restenosis after cell infusion (arising from one small, non-randomized study in which CD133+ cells were administered (Bartunek et al. 2005)) was not confirmed in studies employing intravascular ultrasonography (Huikuri et al. 2008) and in recent meta-analyses (Martin-Rendon et al. 2008; Zhang et al. 2009) showing no excess proclivity to target-vessel restenosis or repeat revascularization in recipients of BMMNC therapy. Finally, there is no evidence of cardiac calcifications or increased tumorigenesis after BMMNC therapy. Figure 3 shows the excellent profile of intracoronary infusion of BMMNCs, with respect to several clinical safety endpoints.

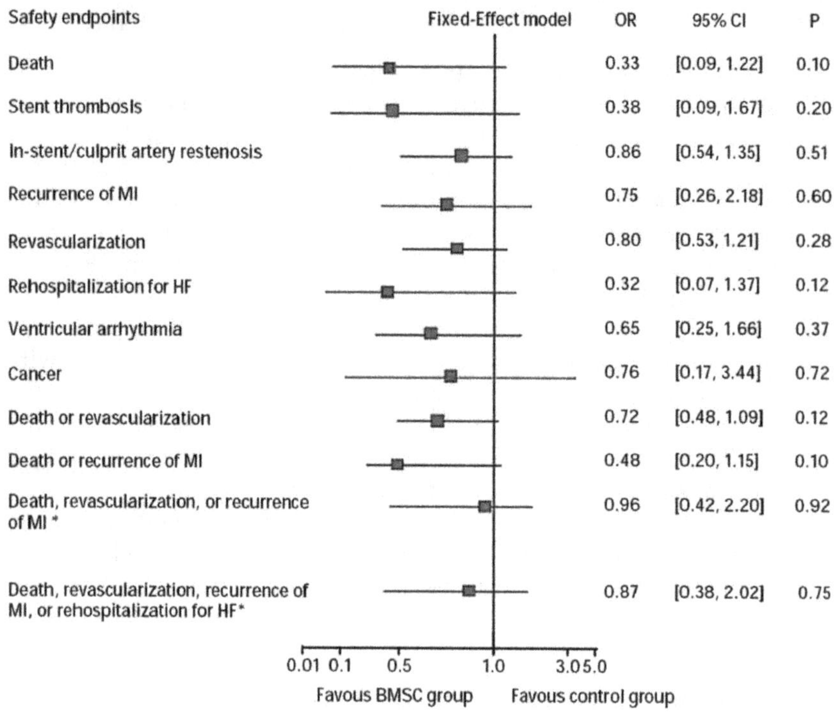

Figure 3. Excellent safety profile of intracoronary delivery of BMMNCs. Odds ratio of BMMNC infusion therapy with respect to several safety endpoints. Five trials (REPAIR-AMI (Schächinger et al. 2006a,b), BOOST (Wollert et al. 2004), ASTAMI (Lunde et al. 2006), Janssens et al. (Ugander wt al. 2010) and Yao et al. (Yao et al. 2006)) were included in the meta-analysis. Reproduced from reference (Zhang et al. 2009).

2. Inconsistent efficacy in clinical endpoints for acute myocardial infarction

As with SKMs, early small and uncontrolled pilot studies of BMMNC delivery after acute MI demonstrated consistently positive results (Fernandez-Aviles et al. 2004; Bartunek et al. 2005). However, the results of later large randomized studies (at least with regard to effects on EF) have been conflicting (Table 1). The first large-scale randomized, but not placebo-controlled study (BOOST) demonstrated a transient functional

Table1. Representative clinical trials of stem cell therapy for myocardial ischemia.

Trial name	Publish year	Number of patients	Cell type	Delivery route	Days after MI	Outcomes	Imaging modalities
Acute myocardial infarction							
BOOST	2004	60	BMMNCs	Intracoronary	6 ±1 days	LVEF ↑ LVEDV →	MRI
REPAIR-AMI	2006	187	BMMNCs	Intracoronary	3 ~6 days	LVEF ↑ LVEDV →	LVG
ASTAMI	2006	97	BMMNCs	Intracoronary	6 ±1 days	LVEF → LVEDV →	MRI echo, SPECT
LEUVEN-AMI *Janssens et al.*	2006	67	BMMNCs	Intracoronary	1 day	LVEF →, LVEDV → regional contractility ↑	MRI echo, PET
FINCELL	2008	77	BMMNCs	Intracoronary	3 days	LVEF ↑	LVG echo
REGENT	2009	200	BMMNCs	Intracoronary	3 ~12 days	LVEF ↑	MRI LVG
HEBE	2010	200	BMMNCs vs PBMMNCs	Intracoronary	3 ~8 days	LVEF → regional contractility →	MRI
Ischemic heart failure							
TOPCARE-CHD	2006	58	BMMNCs	Intracoronary	>1 month	LVEF ↑	LVG
MAGIC	2008	97	SKMs	Transepicardial during CABG	>4 months	LVEF → LVEDV ↓, LVESD ↓	echo
CAuSMIC	2009	23	SKMs	Transendocardial (NOGA)	CIHF	LVEDD ↓ QOL ↑	echo SPECT
STAR-heart	2010	391	BMSCs	Intracoronary	8.5±3.2 years	LVEF ↑ Survival ↑	LVG
SEISMIC	2011	40	SKMs	Transendocardial (MyoCath)	CIHF with ICD	LVEF → 6MWT ↑	MUGA scan

Abbreviations: SKMs, skeletal myoblasts; BMMNCs, bone marrow mononuclear cells; BMSCs, bone marrow stem/progenitor cells; PBMMNCs, peripheral bone marrow mononuclear cells; CIHF, chronic ischemic heart failure; CABG, coronary artery bypass grafting; LVEF, left ventricular ejection fraction; LVESV, left ventricular end-systolic volume; LVEDV, left ventricular end-diastolic volume; LVEDD, left ventricular end-diastolic diameter; MRI, magnetic resonance imaging; echo, echocardiography; SPECT, single photon emission computed tomography; LVG, left ventricular angiography; PET, positron emission tomography; ICD, implanted cardioverter-defibrillator; 6MWT, six-minute walk test; MUGA, multiple gated acquisition

benefit (increase in LVEF) at six months after BMMNC infusion (Wollert et al. 2004), but it was not sustained at 18 months (Meyer et al. 2006) or five years after therapy (Meyer et al. 2009). Two subsequent double-blind placebo-controlled studies yielded varied results: REPAIR-AMI showed a significant improvement in EF at four months (Schächinger et al. 2006b). However, a study by Jannsens et al. (Janssens et al. 2006) (in which BMMNCs were infused within 24 hours after reperfusion, earlier than in the other studies) showed no improvement in EF at four months but a more profound decrease in scar size in the cell-treated group (which, however, was not accompanied by an increase of viable myocardium). In contrast to the previous studies, results of the randomized but not placebo-controlled ASTAMI trial were resoundingly negative (Lunde et al. 2006), a difference that has since been attributed to lower numbers and impaired quality of the administered cells (Seeger et al. 2007).

In subsequent large randomized studies, functional effects of BMMNC therapy were:

a. Positive: FINCELL, the first study of BMMNC infusion in patients treated with thrombolysis, showed improvement of LVEF at six months (Huikuri et al. 2008).
b. Marginally positive: In the REGENT study LVEF improved in the cell-treated groups but not in controls at six months; however, no difference was observed when the treatment effects were compared among groups (Tendera et al. 2009).
c. Mixed: In BONAMI, there was no difference in EF but an increase in myocardial viability (Roncalli et al. 2011).
d. Completely negative: HEBE (Hirsch et al. 2011) and SCAMI (Wöhrle et al. 2010).

A synopsis of the cell therapy treatment effects on LVEF, LV volumes, and infarct size, as measured with cardiac MRI (the gold standard for evaluation of LV structure, function, and infarct size) in the aforementioned trials, where available is provided in Fig. 4.

The reason for the inconsistent outcomes from these clinical trials is unclear but may be explained by differences in cell isolation protocols, timing of cell infusion, patient selection, and the imaging modalities used to measure the treatment effect. To address these issues, several clinical trials are underway for acute MI (Table 2). Specifically, these studies aim to determine the optimal timing, dosage, and route of cell delivery. The REGENERATE-AMI study (ClinicalTrial.gov Identifier: NCT00765453) aims to deliver bone marrow stem cells within six hours of successful primary percutaneous coronary intervention (PCI) and limited to anterior wall AMI only. The TIME trial aims to elucidate the optimal timing of cell delivery and compare intracoronary infusion of BMMNCs administered at three days versus seven days after reperfusion (ClinicalTrial.gov Identifier: NCT00684021) (Traverse et al. 2009). In addition, the Bone Marrow Cells in Acute Myocardial Infarction (BAMI) study (Phase III), aiming to recruit 3,000 patients with an ejection fraction below 45% after AMI and successful reperfusion, was recently launched in Europe. The aim of this study is to evaluate the effect of intracoronary infusion of BMMNCs four days after reperfusion on all-cause mortality in acute MI.

A recent meta-analysis (including 811 patients participating in 13 randomized trials) showed modest but significant benefit following BMMNC therapy: LVEF increased by ~3%, end-systolic volume decreased by ~5 ml and scar size decreased by ~3.5% in the cell-treated groups compared with controls (Martin-Rendon et al. 2008). Sub-group analysis revealed that the benefit of cell therapy was greater when cells were infused within seven days following infarction (the optimal time window seems to be 5–7 post-MI (Martin-Rendon et al. 2008; Dimmeler et al. 2009)) and when the dose administered was >100 million BMMNCs (Martin-Rendon et al. 2008).

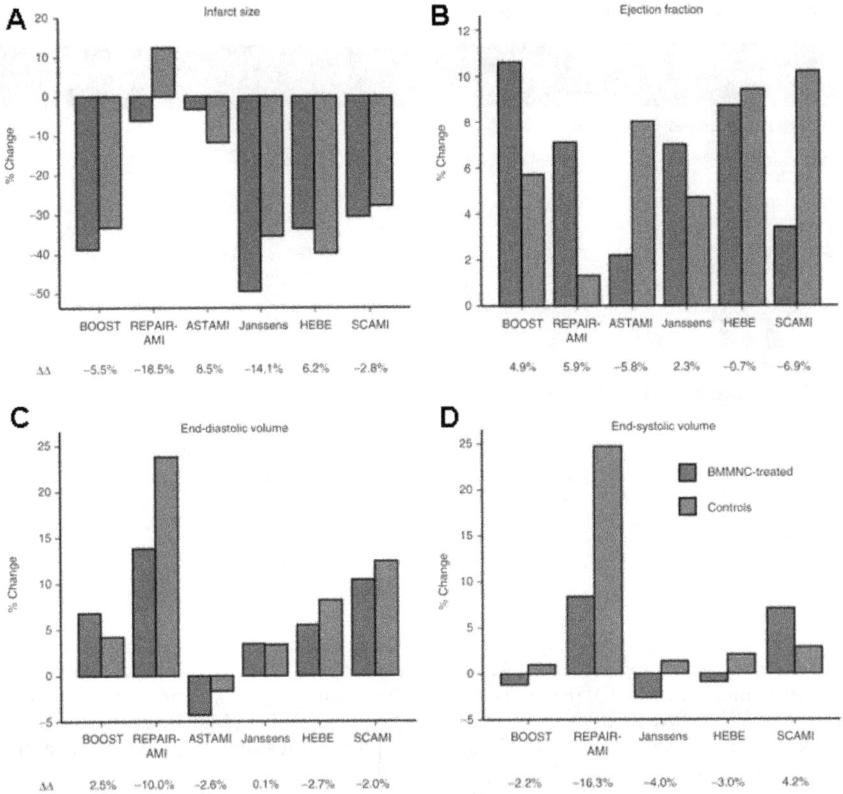

Figure 4. BMMNC-therapy treatment effects on cardiac structure and function as measured with magnetic resonance imaging (MRI). A synopsis of the cell therapy treatment effects on infarct size (A), left-ventricular ejection fraction (B), end-diastolic volume (C), and end-systolic volume (D) as measured with cardiac MRI in the large randomized controlled intracoronary BMMNC trials is provided. BOOST (Meyer et al. 2006), REPAIR-AMI (Dill et al. 2009), ASTAMI (Lunde et al. 2006), HEBE (Hirsch et al. 2011), SCAMI (Wöhrle et al. 2010), and the study by Janssens et al. (Janssens et al. 2006) are included in the analysis. The relative treatment effect normalized to baseline (% change) is presented for both treated patients and controls; the difference in relative treatment effects between treated patients and controls (ΔΔ) is also provided. BMMNC, bone marrow mononuclear cell. Reproduced from reference (Malliaras et al. 2011).

Color image of this figure appears in the color plate section at the end of the book.

3. *Targeting a sicker patient population*

The main patient population used in the first generation of clinical trials has been a first-infarct population, with little ventricular dysfunction and a projected low mortality and morbidity even without adjunctive biological therapy. However, data from various different trials suggest that patients

Table 2. Ongoing cell therapy trials for myocardial ischemia.

Trial name (Identifier)	Number of patients	Cell type	Delivery route	Days after MI	Primary outcomes	Imaging modalities
Acute myocardial infarction						
REGENERATE-AMI (NCT00765453)	102	BMSCs	Intracoronary	6 hours	LVEF	MRI
TIME (NCT00684021)	120	BMMNCs	Intracoronary	3 or 7 days	LVEF	MRI
SWISS-AMI (NCT00355186)	150	BMSCs	Intracoronary	5~7 vs 21~28 days	LVEF	MRI
APOLLO (NCT00442806)	48	ADRCs	NS	NS	Safety	MRI SPECT echo
BAMI	3,000	BMMNCs	Intracoronary	4 days	Mortality	MRI, echo
Ischemic heart failure						
MARVEL (NCT00526253)	330	SKMs	Transendocardial	>3 months	6MWT QOL, LVEF	MRI
TAC-HFT (NCT00768066)	60	MSCs or BMMNCs	Transendocardial	CIHF	Safety	MRI
PRECISE (NCT00426868)	36	ADRCs	Transepicardial	CIHF	Safety	MRI SPECT echo
PROMETHEUS (NCT00587990)	45	MSCs	Transepicardial during CABG	CIHF	Safety	MRI

Abbreviations: SKMs, skeletal myoblasts; BMMNCs, bone marrow mononuclear cells; BMSCs, bone marrow stem/progenitor cells; ADRCs, adipose-derived stem/regenerative cells; MSCs, mesenchymal stem cells; CIHF, chronic ischemic heart failure; CABG, coronary artery bypass grafting; LVEF, left ventricular ejection fraction; QOL, quality of life; MRI, magnetic resonance imaging; echo, echocardiography; SPECT, single photon emission computed tomography; 6MWT, six-minute walk test; NS, not specified

with more severe MI (the ones with the worse prognosis) benefit most from cell therapy. In the REPAIR-AMI study, patients with a lower baseline EF (<48.9%) showed a significant, threefold higher recovery in global EF than seen in the converse group. In addition, the beneficial effect on clinical endpoints was also preferentially observed in those patients with a lower baseline EF after myocardial infarction (Schächinger et al. 2006a). In the REGENT study, significant functional benefit was observed only in cell-treated patients who had baseline LVEF <37% (Tendera et al. 2009). In BOOST, only patients with larger infarcts and greater infarct transmurality demonstrated sustained functional improvement at later time points (Meyer et al. 2006), while in the study by Janssens et al. (Janssens et al. 2006), cell therapy led to enhanced recovery of regional function only in the most severely infarcted myocardial segments (characterized by the greatest infarct transmurality). Finally, in the TOPCARE-CHD registry, NT-proBNP serum levels >735 pg/ml at baseline were an independent predictor of a favorable response three months after intracoronary administration of BMMNCs (Assmus et al. 2007). Taken together, the results indicate that the greatest benefits of stem cell therapy occur in patients with the greatest infarct-induced myocardial damage. This finding has major implications for the design of future clinical studies: cell therapy can maximize its potential for successful myocardial repair and regeneration by targeting a sicker patient population.

The paradigm that sicker patient populations receive greater benefit by cell-therapy has been challenged by the findings of two recently published randomized clinical trials. The HEBE (Hirsch et al. 2011) and the BONAMI (Roncalli et al. 2011) trials enrolled patients with large infarcts based on the conclusions of the REPAIR-AMI or REGENT trials; however, no difference in global LV function recovery could be detected between the treatment and control groups. In addition, in the HEBE trial, no functional benefit was observed despite the fact that cells where infused between three and eight days after primary PCI—a time period that is very similar to the "optimal time window" of 5–7 days—as identified by the results of the REPAIR-AMI trial (Schächinger et al. 2006) and data from meta-analyses (Martin-Rendon et al. 2008; Dimmeler et al. 2009). Thus, many questions with regard to selection of patients, timing of infusion, and potency of various cell populations, surrogate endpoints and clinical endpoints still remain unanswered (Janssens et al. 2011).

4. Clinical trials for chronic ischemic cardiomyopathy

Table 1 shows landmark clinical trials for chronic ischemic cardiomyopathy. The MAGIC trial (Menasche et al. 2008) was the first randomized, placebo-controlled study for patients with ischemic heart failure. Autologous SKMs were injected in and around akinetic segments during CABG surgery. Although no difference was observed between SKM and placebo groups in the improvement of LV function after six months, significant reductions in LV end-diastolic and -systolic volumes were seen after therapy.

BMMNCs transplantation has also been assessed in patients with chronic ischemic heart failure by intracoronary infusion. In the TOPCARE-CHD trial (randomized trial), an absolute change in LVEF of 2.9% in the treatment group was observed with no adverse events at three months after cell transplantation (Assmus et al. 2007). The STAR-heart study (prospective study) is the largest study to date investigating the effect of bone marrow stem cell therapy in chronic ischemic heart failure (Strauer et al. 2010). Over three months to five years after intracoronary bone marrow stem cell therapy, there was a significant improvement in hemodynamics, exercise capacity and LV contractility. Importantly, there was a significant decrease in long-term mortality in the cell-treated patients compared with the control group. This is the first time that a stem cell study with long-term follow-up has shown a survival benefit in patients receiving cell therapy (Mozid et al. 2011).

The results of these previous trials have prompted the randomized controlled trials including the MARVEL Trial, randomized, Phase II/III study (ClinicalTrials.gov Identifier: NCT00526253) which will assess the safety and efficacy of intramyocardial injection SKMs in patients with chronic ischemic heart failure or the REGENERATE-IHD study, randomised controlled, Phase II/III study (ClinicalTrials.gov Identifier: NCT00747708) (Yeo and Mathur 2009) which have been designed to compare the efficacy and safety of delivering BMMNCs using three different routes: peripheral infusion versus intracoronary or intramyocardial delivery. These trials are ongoing studies (Table 2).

5. Low cell retention and engraftment

Low cell retention and engraftment after cell delivery to the heart are persistent obstacles to successful myocardial regeneration. Numerous studies have investigated short-term cell retention and long-term engraftment both in experimental animals (Aicher et al. 2003, Hou et al. 2005) and in humans (Hofmann et al. 2005; Blocklet et al. 2006) and the results have been overwhelmingly disappointing; acute cell retention (i.e., within 24 hours of delivery) in the heart is generally <10%, regardless of the cell type or delivery route. Cells acutely lost from the heart are washed out via the coronary venous system or mechanically ejected via the injection site, while retention rates in beating hearts are markedly lower than in non-beating hearts (Terrovitis et al. 2009). Importantly, >90% of the successfully retained cells die within the first week (Müller-Ehmsen et al. 2002) most probably due to ischemia, inflammation or anoikis (apoptosis due to detachment from the extra-cellular matrix), and <1% of transplanted cells can be identified four weeks after transplantation (Zeng et al. 2007). Taking into account that studies of MSCs (Quevedo et al. 2009) and CDCs (Oh et al. 2003; Terrovitis et al. 2009) reveal a strong correlation between engraftment rate, even if low, and long-term functional benefit, there is good reason to believe that the development of more effective delivery methods, combined with successful means of boosting transplanted cell retention and engraftment, would significantly enhance the utility of cell therapy.

6. Methods of cell delivery

So far, stem cells for heart repair have been delivered clinically via three routes: systemic intravenous infusion, intracoronary infusion, and intramyocardial injection (either by direct open-chest injections as an adjunct to CABG or by transendocardial catheter-based injections) (Beeres et al. 2008), while retrograde coronary venous (Hou et al. 2005), transvenous intramyocardial (Thompson et al. 2003; Siminiak et al. 2005) and intrapericardial approaches (Azene et al. 2011) have been used mostly under experimental settings.

Intravenous infusion, albeit simple in execution, is hampered by trapping of cells in the lungs, resulting in acute retention rates of <1% (Freyman et al. 2006). Intracoronary delivery of cells into a recanalized infarct-related artery is safe and convenient (can be performed with standard balloon catheters) and has the inherent advantage that cells are homogeneously infused into myocardial regions with preserved oxygen and nutrient supply. However, retention of cells is sub-optimal (most cells are washed away before they can migrate into the surrounding tissue), unperfused regions of the myocardium are inaccessible, and the high dependence on endogenous homing signals makes this approach more attractive for use relatively early after AMI, when such signals are highly expressed. In addition, safety concerns are raised during infusion of larger cells because of potential capillary plugging and microinfarction; this problem, if recognized, can be overcome by appropriate cell dosing and optimization of the infusate (Johnston et al. 2009). It should be noted, however, that in the vast majority of pre-clinical and clinical studies to date, dosing has by and large been arbitrary, guided more by feasibility and accessibility than by systematic dosage optimization; a remarkable 6,700-fold range in cell dose is observed in human bone marrow-derived cell trials (Murry et al. 2005). All things considered, the intracoronary delivery route seems to be best suited for the treatment of AMI.

On the other hand, direct intramyocardial cell injections result in better cardiac retention of cells (and less washout to off-target organs) compared to intracoronary (Hou et al. 2005) or systemic approaches (Freyman et al. 2006), which may account for the superior functional benefit associated with this route of administration (Lee et al. 2011). Freshly infarcted myocardium might engender a higher risk of perforation (an unverified conjecture, as intramyocardial catheter-based delivery has been safely implemented as early as 10 days after MI (Krause et al. 2009)); therefore, intramyocardial cell injections are probably more suitable for patients with chronic ischemic cardiomyopathy and for patients with refractory ischemia lacking options for revascularization. Importantly, because open-chest injections would be unlikely to gain clinical acceptance unless adjunctive to clinically indicated surgery, less invasive catheter-mediated methods of intramyocardial delivery need to be explored. A particularly attractive approach is catheter-guided intramyocardial cell delivery, employing endoventricular electromechanical mapping technology; the three-dimensional NOGA® (Biologics Delivery Systems, a Johnson & Johnson company, Irwindale, CA, USA) (Gyöngyösi and Dib 2011). The acquisition of spatial, electrophysiologic, and mechanical information in real time enables (i) precise identification/localization of the target zone for cell delivery (viable myocardium in the border zone in the case of MI, as injections directly into the scar yield minimal engraftment (Lee et al. 2011); hibernating myocardium in refractory angina) and (ii) accurate catheter navigation and orientation (as the three-dimensional map offers much higher spatial resolution than that with two-plane fluoroscopy), resulting in optimal distribution of injection sites.

Traditionally, the NOGA navigation system has been used to identify viable myocardium for the treatment of refractory angina (Tse et al. 2007; Van Ramshorst et al. 2009). In the PROTECT-CAD trial, a randomized, blinded, placebo-controlled trial, direct endomyocardial implantation of autologous bone marrow cells significantly improved exercise time, modest LVEF and NYHA functional class in patients with severe chronic myocardial ischemia who failed conventional therapy (Tse et al. 2007). Van Ramshorst et al. reported on a randomized, double-blind, placebo-controlled trial of 50 patients who received twofold higher number of bone marrow cells than in the PROTECT-CAD trial. In this study, the cell-injected group showed significantly improvements in LV function, myocardial perfusion and angina symptoms (Van Ramshorst et al. 2009). In addition, a recently published double-blind, randomized, phase II study (ClinicalTrials.gov identifier: NCT00300053) (Losordo et al. 2011) explored the therapeutic potential of circulating CD34+ cells in patients with refractory angina who have exhausted all other treatment options. Direct intramyocardial injection of cells with NOGA mapping catheter was associated with an improvement of exercise tolerance and with reduction in angina frequency in the treated group, compared with control. However, one study subject died as a result of complications from the delivery procedure.

Furthermore, in the FOCUS-HF (First Bone Marrow Mononuclear Cell United State Study in Heart Failure), single-blind, placebo-controlled trial, transendocardial injection of BMMNCs was associated with improvements in symptoms, quality of life, and possibly perfusion in patients with chronic heart failure (Perin et al. 2011). This result provides further evidence that the NOGA system could also be used for chronic ischemic cardiomyopathy, enabling the targeting of viable myocardium in the scar border zone.

Finally, when it comes to cardiac-derived cell products, another advantage of intramyocardial injections is the potential for safe administration of CSps for heart repair. CSps, due to their larger size of 50–200 μm, are expected to embolize at the arteriolar level and thus may not be safe to administer intracoronarily. We have found that intramyocardial delivery of CSps disproportionally boosts cardiac function in small (Li et al. 2010) and large animal models of ischemic cardiomyopathy compared with monolayer-cultured CDCs (Lee et al. 2011). This effect can be attributed to the three-dimensional multilayer composition of CSps, yielding a three-dimensional microtissue product with an upregulation in 'stemness' and increased expression of adhesion molecules, enabling improved cardiomyogenesis and more robust engraftment (Li et al. 2010). These findings provide good reason to favor CSps in future direct injection studies.

7. Strategies to boost retention and increase engraftment

A major challenge to the effectiveness of cell therapy is the low percentage of cell retention after delivery into the heart, whether by direct intramyocardial injection or by intracoronary infusion. Multiple lines of evidence indicate that strategies that effectively boost acute retention translate into greater functional benefit downstream (Terrovitis et al. 2009; Cheng et al. 2010). Several methods have been used, including priming of host tissue to increase homing, pre-conditioning of transplanted cells with cytokines, pro-survival factors and physical stimuli, genetic engineering of cells, use of biomaterial scaffolds, and delivery of three-dimensional cellular aggregates (Christman and Lee 2006; Terrovitis et al. 2010). While much work has focused on improving transplanted cell quality and creating a more hospitable host environment, most cells are washed away too quickly to be effective. Physical approaches (capping the injection site with fibrin glue to prevent back-flux of injected cells (Terrovitis et al. 2009) and using magnetic targeting to retain iron-loaded CDCs within the heart (Cheng et al. 2010)) to retaining cells in the diseased tissue long enough to enable biological bonding may boost treatment efficacy and would be synergistic with efforts to improve cell quality or environmental receptiveness.

8. Cardiogenic potential of bone marrow-derived cells

The question of whether bone marrow derived cells in general, and hematopoietic cells in particular, can transdifferentiate into cardiomyocytes has been hotly debated for a decade. The controversy began with a report showing extensive engraftment and cardiogenic differentiation of bone marrow derived hematopoietic cells after injection in infarcted hearts of mice (Orlic et al. 2001). In addition, studies examining postmortem human cardiac tissue from male patients, who had received hearts from female donors, reported cardiac cellular chimerism; Y-chromosome positive cardiomyocytes (indicating extracardiac origin) were detected in the transplanted female hearts, but the extent of the phenomenon varied greatly (from 0.04 to 10%, a difference of ~250 fold) among studies (Quaini et al. 2002; Laflamme et al. 2002). However, several groups have since contested the claims that bone marrow cells can readily acquire a cardiac phenotype (Murry et al. 2004; Balsam et al. 2004), while subsequent studies have argued that what originally was interpreted as transdifferentiation may have arisen from cell fusion (Alvarez-Dolado et al. 2003). Currently, no consensus exists on whether bone-marrow derived progenitor cells can meaningfully differentiate into cardiomyocytes *in vivo* (Sussman et al. 2008), although skepticism prevails.

A major challenge in regenerative medicine is the sheer number of cells that need to be replaced; in a typical MI, ~1 billion cardiomyocytes are lost (Robey et al. 2008). As a result, when reviewing data from clinical studies, we are faced with a 'numbers paradox': cell transplantation translates into a range of beneficial effects, but these effects occur in a setting characterized by poor cell retention and minimal long-term survival as well as negligible cardiomyogenic transdifferentiation. For the sake of argument, let us consider the study by Janssens et al. (Janssens et al. 2006); there, intracoronary infusion of ~170 million BMMNCs was associated with a reduction in infarct size at four months (measured by cardiac MRI) compared with placebo infusion; cell therapy induced an additional decrease in scar mass of ~3 g (equivalent to ~60 million cardiomyocytes, since 1 g of human myocardium contains ~20 million myocytes). Taking into account that acute retention of bone marrow derived cells after intracoronary infusion in humans is 1–2% (Hofmann et al. 2005) (measured by administration of ^{18}F-FDG-labeled cells and PET), the 'numbers paradox' becomes clear: how can 2–4 million acutely retained non-myogenic cells (~90% of which will most probably die within the first week (Müller-Ehmsen et al. 2002) without compensatory proliferation, resulting in miniscule long-term engraftment (Zeng et al. 2007)) lead to the production of 60 million new cardiomyocytes four months after cell administration? In any case, it is not clear that any new cardiomyocytes have been created by BMMNC therapy; while several studies have shown reductions in scar size by MRI (Janssens et al. 2006), none has shown an increase in the mass of viable myocardium.

Potential mechanisms of benefit remain unclear but include cardioprotection of resident myocytes, recruitment of endogenous progenitors, upregulation of cardiomyocyte cell-cycle re-entry, stimulation of angiogenesis, modulation of inflammatory and fibrogenic processes resulting in better infarct healing, improved cardiac metabolism and contractility, and induction of secondary humoral effects in the host tissue, which may or may not be sustained after transplanted cells have been eliminated (Cho et al. 2007; Gnecchi et al. 2008). Nevertheless, clinical efficacy should not be dismissed just because mechanisms are not understood.

9. Paracrine hypothesis

In the vast majority of experimental studies, the number of differentiated myocytes derived from transplanted stem cells is too small to account for the observed improvements in cardiac function (Tang et al. 2010). This can be attributed to the combination of low long-term cardiac engraftment and the low cardiomyogenic potential of the majority of adult stem cell types used for heart regeneration. Thus, the prevailing concept of adult stem cell efficacy has shifted towards the 'paracrine hypothesis', according to which the transplanted cells are proposed to produce soluble factors that are beneficial to the infarcted heart (Gnecchi et al. 2008). Indeed, SKMs (Korf-Klingebiel et al. 2008), bone-marrow derived cells (Gnecchi et al. 2006) and cardiac-derived cells (Chimenti et al. 2010) produce and secrete a broad variety of cytokines, chemokines and growth factors that are known to be involved in cardiac repair; also, hypoxic stress increases the production of several of these factors. In addition, strong support of a paracrine mechanism for cardiac repair comes from experimental studies in which the administration of conditioned medium is able to recapitulate, at least partially, the beneficial effects observed after stem cell therapy (Dimmeler and Leri 2008). Potential effects of paracrine factors include cytoprotection of resident myocytes, upregulation of angiogenesis, modulation of inflammatory processes resulting in better infarct healing, improved cardiac metabolism and contractility, promotion of cardiomyocyte cell cycle re-entry, recruitment of endogenous stem cells, and induction of secondary humoral effects in the host tissue (Cho et al. 2007; Perez-Ilzarbe et al. 2008). It is encouraging that functional benefit can be achieved through indirect pathways; this mechanism of action rationalizes the persistence of benefit despite the evanescence of transplanted cell survival, in that the new tissue originates from the recipient heart rather than from the transplant.

10. Use of allogeneic cells as "universal donor cells"

In the first decade of cell therapy for human heart regeneration, most clinical trials have been conducted using autologous cells. Autologous sources are attractive because immunologic rejection is avoided by default. Nevertheless, autologous therapy is associated with serious limitations that complicate widespread clinical application. Specifically, autologous therapy necessitates patient-specific tissue harvesting, cell processing, and quality control, which pose significant logistic, economic, and timing constraints, thus limiting the scalability of heart stem cell therapy. In addition, stem cell growth properties and plasticity may be hampered by age and co-morbidities (Dimmeler and Leri 2008), resulting in interpatient variability in cell potency. The use of allogeneic cells, if safe and effective, would obviate such limitations, enabling the generation of highly standardized, "off the shelf" cellular products and thus opening up a new treatment paradigm. Stem cells could be grown in large numbers from allogeneic tissue sources in a central facility under strict quality control and banked for future use, enabling safe and effective myocardial repair in a timely, cost-efficient manner. The obvious disadvantage of allogeneic therapy is the risk of immune rejection, which may limit effectiveness (whether or not it poses safety hazards) and potential development of immune memory.

To that end, MSCs, which are purportedly immunoprivileged, have attracted interest; clinical trials involving the administration of proprietary allogeneic human MSCs to patients with heart disease are already under way, and the preliminary results have been encouraging (Dib et al. 2009; Hare et al. 2009). It should be noted that, without immunosuppression or HLA matching, most allogeneic cells (even MSCs) will eventually be rejected after *in vivo* transplantation (Poncelet et al. 2007; Zangi et al. 2009). Nevertheless, since the vast majority of the observed functional benefit is attributable to indirect pathways, rejection of allogeneic cells may not be an issue if it is delayed long enough to allow them to exert their protective and regenerative paracrine effects, resulting in sustained benefit without the requirement for stable engraftment of transplanted cells.

Conclusion and Future Directions

Stem cell therapy has emerged as a potential therapeutic strategy for myocardial infarction. A variety of cell types have been used in pre-clinical animal models and in clinical trials to repair or regenerate the injured heart either directly (through formation of new transplanted tissue) or indirectly (through paracrine effects activating endogenous regeneration). The results of the first generation of clinical trials have been mixed. This is the result of the non-systematic approach that has so often been employed so far regarding clinical translation of cellular cardiomyoplasty. This history motivates a search, for key missing steps include more effective delivery route and timing, more potent cell types, safe and effective "off the shelf" cell products and better strategies to boost cell engraftment, as well as pivotal studies in clinically relevant large-animal models before moving on to human subjects. Given the spotty trajectory of cell therapy to date, we believe that a more systematic approach will facilitate more effective clinical translation.

Acknowledgments

This work was supported by grants from California Institute for Regenerative Medicine and the Board of Governors of the Cedars-Sinai Medical Center. Eduardo Marbán occupies the Mark S. Siegel Family Foundation Chair of the Cedars-Sinai Medical Center. Figures 2 and 4 were adapted with permission from references (Malliaras and Marbán 2011; Malliaras et al. 2011).

References

Abraham, M. R., Henrikson, C. A., Tung, L. et al. 2005. Antiarrhythmic engineering of skeletal myoblasts for cardiac transplantation. Circ. Res. 97, 159–67.

Aggarwal, S. and Pittenger, M. F. 2005. Human mesenchymal stem cells modulate allogeneic immune cell responses. Blood 105, 1815–1822.

Aicher, A., Brenner, W., Zuhayra, M. et al. 2003. Assessment of the tissue distribution of transplanted human endothelial progenitor cells by radioactive labeling. Circulation 107, 2134–2139.

Alvarez-Dolado, M., Pardal, R., Garcia-Verdugo, J. M. et al. 2003. Fusion of bone-marrow-derived cells with Purkinje neurons, cardiomyocytes and hepatocytes. Nature 425, 968–973.

Assmus, B., Fischer-Rasokat, U., Honold, J. et al. 2007. TOPCARE-CHD Registry: transcoronary transplantation of functionally competent BMCs is associated with a decrease in natriuretic peptide serum levels and improved survival of patients with chronic postinfarction heart failure: results of the TOPCARE-CHD Registry. Circ. Res. 100, 1234–1241.

Azene, N., Ehtiati, T., Fu, Y. et al. 2011. Intrapericardial delivery of visible microcapsules containing stem cells using xfm (x-ray fused with magnetic resonance imaging). J. Cardiovasc. Magn. Reson. 13(Suppl. 1), P26.

Badillo, A. T., Beggs, K. J., Javazon, E. H. et al. 2007. Murine bone marrow stromal progenitor cells elicit an *in vivo* cellular and humoral alloimmune response. Biol. Blood Marrow Transplant. 13, 412–422.

Balsam, L. B., Wagers, A. J., Christensen, J. L. et al. 2004. Haematopoietic stem cells adopt mature haematopoietic fates in ischaemic myocardium. Nature 428, 668–673.

Bartunek, J. et al. 2005. Intracoronary injection of CD133-positive enriched bone marrow progenitor cells promotes cardiac recovery after recent myocardial infarction: feasibility and safety. Circulation 112, I178–I183.

Beeres, S. L., Atsma, D. E., van Ramshorst, J. et al. 2008. Cell therapy for ischaemic heart disease. Heart 94, 1214–1226.

Beltrami, A. P., Barlucchi, L., Torella, D. et al. 2003. Adult cardiac stem cells are multipotent and support myocardial regeneration. Cell 114, 763–776.

Bergmann, O. et al. 2009. Evidence for cardiomyocyte renewal in humans. Science 324, 98–102.

Blocklet, D., Toungouz, M., Berkenboom, G. et al. 2006. Myocardial homing of nonmobilized peripheral-blood CD34+ Cells after intracoronary injection. Stem Cells 24, 333–336.

CADUCEUS. CArdiosphere-Derived aUtologous stem CElls to reverse ventricUlar dySfunction (NCT00893360). www.clinicaltrials.gov.

Caspi, O., Huber, I., Kehat, I. et al. 2007. Transplantation of human embryonic stem cell-derived cardiomyocytes improves myocardial performance in infarcted rat hearts. J. Am. Coll. Cardiol. 50, 1884–1893.

Chen, S. L., Fang, W. W., Ye, F. et al. 2004. Effect on left ventricular function of intracoronary transplantation of autologous bone marrow mesenchymal stem cell in patients with acute myocardial infarction. Am. J. Cardiol. 94, 92–95.

Cheng, K., Li, T-S., Malliaras, K. et al. 2010. Paracrine and functional superiority of human cardiosphere-derived cells as compared to mesenchymal stem cells, adipose-derived stem cells or bone marrow mononuclear cells. Circulation 122, A20957.

Cheng, K., Li, T. S., Malliaras, K. et al. 2010. Magnetic targeting enhances engraftment and functional benefit of iron-labeled cardiosphere-derived cells in myocardial infarction. Circ. Res. 106, 1570–1581.

Chimenti, I., Smithm R. R., Li, T. S. et al. 2010. Relative roles of direct regeneration versus paracrine effects of human cardiosphere-derived cells transplanted into infarcted mice. Circ. Res. 106, 971–980.

Cho, H. J., Lee, N., Lee, J. Y. et al. 2007. Role of host tissues for sustained humoral effects after endothelial progenitor cell transplantation into the ischemic heart. J. Exp. Med. 204, 3257–3269.

Christman, K. L. and Lee, R. J. 2006. Biomaterials for the treatment of myocardial infarction. J. Am. Coll. Cardiol. 48, 907–913.

Civin, CI. and Gore, S. D. 1993. Antigenic analysis of hematopoiesis: a review. J. Hematother. 2, 137–144.

Crosby, J. R., Kaminski, W. E., Schatteman, G. et al. 2000. Endothelial cells of hematopoietic origin make a significant contribution to adult blood vessel formation. Circ. Res. 87, 728–730.

Dai, W., Field, L. J., Rubart, M. et al. 2007. Survival and maturation of human embryonic stem cell-derived cardiomyocytes in rat hearts. J. Mol. Cell Cardiol. 43, 504–516.

Davis, D. R, Kizana, E., Terrovitis, J. et al. 2010. Isolation and expansion of functionally competent cardiac progenitor cells directly from heart biopsies. J. Mol. Cell Cardiol. 49, 312–321.

Davis, D. R., Zhang, Y., Smith, R. R. et al. 2009. Validation of the cardiosphere method to culture cardiac progenitor cells from myocardial tissue. PLoS One 4, e7195.

Deisher, T. A. 2000. Cardiac-derived stem cells. IDrugs 3, 649–653.

Dib, N., Henry, T., DeMaria, A. et al. 2009. The first US study to assess the feasibility and safety of endocardial delivery of allogeneic mesenchymal stem precursor cells in patient with heart failure, three month interim analysis (abstr). Circulation 120 (Suppl.), 810.

Dill, T. et al. 2009. Intracoronary administration of bone marrow-derived progenitor cells improves left ventricular function in patients at risk for adverse remodeling after acute ST-segment elevation myocardial infarction: results of the Reinfusion of Enriched Progenitor cells and Infarct Remodeling in Acute Myocardial Infarction study (REPAIR-AMI) cardiac magnetic resonance imaging substudy. Am. Heart J. 157, 541–547.

Dimmeler, S. and Leri, A. 2008. Aging and disease as modifiers of efficacy of cell therapy. Circ. Res. 102, 1319–1330.

Dimmeler, S. and Zeiher, A. M. 2009. Cell therapy of acute myocardial infarction: open questions. Cardiology 113, 155–160.

Dominici, M., Le Blanc, K., Mueller, I. et al. 2006. Minimal criteria for defining multipotent mesenchymal stromal cells. The International Society for Cellular Therapy position statement. Cytotherapy 8, 315–317.

Duckers, H. J., Houtgraaf, J., Hehrlein, C. et al. 2011. Final results of a phase IIa, randomised, open-label trial to evaluate the percutaneous intramyocardial transplantation of autologous skeletal myoblasts in congestive heart failure patients: the SEISMIC trial. EuroIntervention 6, 805–812.

Eliopoulos, N., Stagg, J., Lejeune, L. et al. 2005. Allogeneic marrow stromal cells are immune rejected by MHC class I- and class II-mismatched recipient mice. Blood 106, 4057–4065.

Feigl, E. O. 1983. Coronary physiology. Physiol. Rev. 63, 1–205.

Fernandez-Aviles, F., San Roman, J. A., Garcia-Frade, J. et al. 2004. Experimental and clinical regenerative capability of human bone marrow cells after myocardial infarction. Circ. Res. 95, 742–748.

Freyman, T., Polin, G., Osman, H. et al. 2006. A quantitative, randomized study evaluating three methods of mesenchymal stem cell delivery following myocardial infarction. Eur. Heart J. 27, 1114–1122.

Gnecchi, M., He, H., Noiseux, N. et al. 2006. Evidence supporting paracrine hypothesis for Akt-modified mesenchymal stem cell-mediated cardiac protection and functional improvement. FASEB J. 20, 661–669.

Gnecchi, M., Zhang, Z., Ni, A. et al. 2008. Paracrine mechanisms in adult stem cell signaling and therapy. Circ. Res. 103, 1204–1219.

Gyöngyösi, M. and Dib, N. 2011. Diagnostic and prognostic value of 3D NOGA mapping in ischemic heart disease. Nat. Rev. Cardiol. 8, 393–404.

Hagège, A. A., Marolleau, J. P., Vilquin, J. T. et al. 2006. Circulation 114(1 Suppl.), 108–113.

Hare, J. M., Traverse, J. H., Henry, T. D. et al. 2009. A randomized, double-blind, placebo-controlled, dose-escalation study of intravenous adult human mesenchymal stem cells (prochymal) after acute myocardial infarction. J. Am. Coll. Cardiol. 54, 2277–2286.

He, K. L., Yi, G. H., Sherman, W. et al. 2005. Autologous skeletal myoblast transplantation improved hemodynamics and left ventricular function in chronic heart failure dogs. J. Heart Lung Transplant 24, 1940–1949.

Hendrikx, M., Hensen, K., Clijsters, C. et al. 2006. Recovery of regional but not global contractile function by the direct intramyocardial autologous bone marrow transplantation: results from a randomized controlled clinical trial. Circulation 114(Suppl. 1), I101–I107.

Heron, M., Hoyert, D. L., Murphy, S. L. et al. 2009. Deaths: final data for 2006. Natl. Vital Stat. Rep. 57, 1–134.

Hirsch, A. et al. 2011. Intracoronary infusion of mononuclear cells from bone marrow or peripheral blood compared with standard therapy in patients after acute myocardial infarction treated by primary percutaneous coronary intervention: results of the randomized controlled HEBE trial. Eur. Heart J. 32, 1736–1747.

Hofmann, M., Wollert, K. C., Meyer, G. P. et al. 2005. Monitoring of Bone Marrow Cell Homing Into the Infarcted Human Myocardium. Circulation 111, 2198–2202.

Hou, D., Youssef, E. A-S., Brinton, T. J. et al. 2005. Radio-labeled cell distribution after intramyocardial, intracoronary, and interstitial retrograde coronary venous delivery: implications for current clinical trials. Circulation 112(9 Suppl.), I150–6.

Huikuri, H. V. et al. 2008. Effects of intracoronary injection of mononuclear bone marrow cells on left ventricular function, arrhythmia risk profile, and restenosis after thrombolytic therapy of acute myocardial infarction. Eur. Heart J. 29, 2723–2732.

Jain, M., DerSimonian, H., Brenner, D. A. et al. 2001. Cell therapy attenuates deleterious ventricular remodeling and improves cardiac performance after myocardial infarction. Circulation 103, 1920–1927.

Janssens, S. et al. 2006. Autologous bone marrow-derived stem-cell transfer in patients with ST-segment elevation myocardial infarction: double-blind, randomised controlled trial. Lancet 367, 113–121.

Janssens, S.P. 2011. Cardiac bone marrow cell therapy: the proof of the pudding remains in the eating. Eur. Heart J. 32, 1697–1700.

Johnston, P. V., Sasano, T., Mills, K. et al. 2009. Engraftment, differentiation, and functional benefits of autologous cardiosphere-derived cells in porcine ischemic cardiomyopathy. Circulation 120, 1075–1083.

Kajstura, J. et al. 2010. Cardiomyogenesis in the adult human heart. Circ. Res. 107, 305–315.

Katritsis, D. G., Sotiropoulou, P. A., Karvouni, E. et al. 2005. Transcoronary transplantation of autologous mesenchymal stem cells and endothelial progenitors into infarcted human myocardium. Catheter Cardiovasc. Interv. 65, 321–329.

Koninckx, R., Daniëls, A., Windmolders, S. et al. 2011. Mesenchymal stem cells or cardiac progenitors for cardiac repair? A comparative study. Cell. Mol. Life Sci. 68, 2141–2156.

Korf-Klingebiel, M., Kempf, T., Sauer, T. et al. 2008. Bone marrow cells are a rich source of growth factors and cytokines: implications for cell therapy trials after myocardial infarction. Eur. Heart J. 29, 2851–2858.

Krause, K. et al. 2009. Percutaneous intramyocardial stem cell injection in patients with acute myocardial infarction: first-in-man study. Heart 95, 1145–1152.

Laflamme, M. A., Chen, K. Y., Naumova, A. V. et al. 2007. Cardiomyocytes derived from human embryonic stem cells in pro-survival factors enhance function of infarcted rat hearts. Nat. Biotechnol. 25, 1015–1024.

Laflamme, M. A., Myerson, D., Saffitz, J. E. et al. 2002. Evidence for cardiomyocyte repopulation by extracardiac progenitors in transplanted human hearts. Circ. Res. 90, 634–640.

Lee, S. T., White, A. J., Matsushita, S. et al. 2011. Intramyocardial injection of autologous cardiospheres or cardiosphere-derived cells preserves function and minimizes adverse ventricular remodeling in pigs with heart failure post-myocardial infarction. J. Am. Coll. Cardiol. 57, 455–465.

Leobon, B., Garcin, I., Menasche, P. et al. 2003. Myoblasts transplanted into rat infarcted myocardium are functionally isolated from their host. Proc. Natl. Acad. Sci. USA 100, 7808–7811.

Li, T. S., Cheng, K., Lee, S. T. et al. 2010. Cardiospheres recapitulate a niche-like microenvironment rich in stemness and cell-matrix interactions, rationalizing their enhanced functional potency for myocardial repair. Stem Cells 28, 2088–2098.

Losordo, D. W., Henry, T. D., Davidson, C., Sup Lee, J. et al. 2011. Intramyocardial, Autologous CD34+ Cell Therapy for Refractory Angina. Circ. Res. 109, 428–436.

Losordo, D. W., Schatz, R. A., White, C. J. et al. 2007. Intramyocardial transplantation of autologous CD34+ stem cells for intractable angina: a phase I/IIa double-blind, randomized controlled trial. Circulation 115, 3165–3172.

Lunde, K., Solheim, S., Aakhus, S. et al. 2006. Intracoronary injection of mononuclear bone marrow cells in acute myocardial infarction. N. Engl. J. Med. 355, 1199–1209.

Makkar, R., Smith, R. R., Cheng, K. et al. 2011. The CADUCEUS (CArdiosphere-Derived aUtologous Stem CElls to Reverse ventricUlar dySfunction) Trial. American Heart Association, 83th Scientific Meeting. 2011.11.12-16, Orlando, FL, USA.

Malliaras, K., Kreke M., and Marbán E. 2011. The stuttering progress of cell therapy for heart disease. Clin. Pharmacol. Ther. 90, 532–541.

Malliaras, K. and Marbán, E. 2011. Cardiac cell therapy: where we've been, where we are, and where we should be headed. Br. Med. Bull. 98, 161–185.

Martin-Rendon, E., Brunskill, S. J., Hyde, C. J., Stanworth, S. J., Mathur, A. and Watt, S. M. 2008. Autologous bone marrow stem cells to treat acute myocardial infarction: a systematic review. Eur. Heart J. 29, 1807–1818.

Menasché, P., Hagège, A. A., Scorsin, M. et al. 2001. Myoblast transplantation for heart failure. Lancet 357, 279–280.

Menasché, P. et al. 2003. Autologous skeletal myoblast transplantation for severe postinfarction left ventricular dysfunction. J. Am. Coll. Cardiol. 41, 1078–1083.

Menasché, P., Alfieri, O., Janssens, S. et al. 2008. The Myoblast Autologous Grafting in Ischemic Cardiomyopathy (MAGIC) trial: first randomized placebo-controlled study of myoblast transplantation. Circulation 117, 1189–1200.

Mercola, M., Ruiz-Lozano P., and Schneider, M. D. 2011. Cardiac muscle regeneration: lessons from development. Genes Dev. 25, 299–309.

Messina, E., De Angelis, L., Frati, G. et al. 2004. Isolation and expansion of adult cardiac stem cells from human and murine heart. Circ. Res. 95, 911–921.

Meyer, G. P., Wollert, K. C., Lotz, J. et al. 2006. Intracoronary bone marrow cell transfer after myocardial infarction: eighteen months' follow-up data from the randomized, controlled BOOST trial. Circulation 113, 1287–1294.

Meyer, G. P. et al. 2009. Intracoronary bone marrow cell transfer after myocardial infarction: 5-year follow-up from the randomized-controlled BOOST trial. Eur. Heart J. 30, 2978–2984.

Mishra, R., Vijayan K., Colletti, E.J. et al. 2011. Characterization and functionality of cardiac progenitor cells in congenital heart patients. Circulation 123, 364–373.

Mocini, D., Staibano M., Mele, L. et al. 2006. Autologous bone marrow mononuclear cell transplantation in patients undergoing coronary artery bypass grafting. Am. Heart. J. 151, 192–207.

Mozid, A. M., Arnous S., Sammut E. C., and Mathur A. 2011. Stem cell therapy for heart diseases. Br. Med. Bull. 98, 143–159.

Müller-Ehmsen, J., Peterson, K. L., Kedes, L. et al. 2002. Rebuilding a damaged heart: long-term survival of transplanted neonatal rat cardiomyocytes after myocardial infarction and effect on cardiac function. Circulation 105, 1720–1726.

Müller-Ehmsen, J., Whittaker, P., Kloner, R. A. et al. 2002. Survival and development of neonatal rat cardiomyocytes transplanted into adult myocardium. J. Mol. Cell Cardiol. 34, 107–116.

Murry, C. E., Field, L. J., and Menasché, P. 2005. Cell-based cardiac repair: reflections at the 10-year point. Circulation 112, 3174–3183.

Murry, C. E. and Keller, G. 2008. Differentiation of embryonic stem cells to clinically relevant populations: lessons from embryonic development. Cell 132, 661–680.

Murry, C., E, Soonpaa, M. H., Reinecke, H. et al. 2004. Haematopoietic stem cells do not transdifferentiate into cardiac myocytes in myocardial infarcts. Nature 428, 664–668.

Oh, H., Bradfute, S. B., Gallardo, T. D. et al. 2003. Cardiac progenitor cells from adult myocardium: homing, differentiation, and fusion after infarction. Proc. Natl. Acad. Sci. USA 100, 12313–12318.

Orlic, D., Kajstura, J., Chimenti, S. et al. 2001. Bone marrow cells regenerate infarcted myocardium. Nature 410, 701–705.

Perez-Ilzarbe, M., Agbulut, O., Pelacho, B. et al. 2008. Characterization of the paracrine effects of human skeletal myoblasts transplanted in infarcted myocardium. Eur. J. Heart Fail. 10, 1065–1072.

Perin, E. C., Silva, G. V., Henry, T. D., Cabreira-Hansen, M. G. et al. 2011. A randomized study of transendocardial injection of autologous bone marrow mononuclear cells and cell function analysis in ischemic heart failure (FOCUS-HF). Am. Heart J. 161, 1078–1087.

Poncelet, A. J., Vercruysse, J., Saliez, A. et al. 2007. Although pig allogeneic mesenchymal stem cells are not immunogenic *in vitro*, intracardiac injection elicits an immune response *in vivo*. Transplantation 83, 783–790.

Quaini, F., Urbanek, K., Beltrami, A. P. et al. 2002. Chimerism of the transplanted heart. N. Engl. J. Med. 346, 5–15.

Quevedo, H. C., Hatzistergos, K. E., Oskouei, B. N. et al. 2009. Allogeneic mesenchymal stem cells restore cardiac function in chronic ischemic cardiomyopathy via trilineage differentiating capacity. Proc. Natl. Acad. Sci. USA 106, 14022–14027.

Ragosta, M., Beller, G. A., Watson, D. D. et al. 1993. Quantitative planar rest-redistribution 201Tl imaging in detection of myocardial viability and prediction of improvement in left ventricular function after coronary bypass surgery in patients with severely depressed left ventricular function. Circulation 87, 1630.

Reffelmann, T., Dow J. S., Dai, W. et al. 2003. Transplantation of neonatal cardiomyocytes after permanent coronary artery occlusion increases regional blood flow of infarcted myocardium. J. Mol. Cell Cardiol. 35, 607–613.

Rehman, J., Li J., Orschell, C. M. et al. 2003. Peripheral blood 'endothelial progenitor cells' are derived from monocyte/macrophages and secrete angiogenic growth factors. Circulation 107, 1164–1169.

Robey, T. E., Saiget, M. K., Reinecke, H. et al. 2008. Systems approaches to preventing transplanted cell death in cardiac repair. J. Mol. Cell Cardiol. 45, 567–581.

Roger, V. L. et al. 2011. American Heart Association Statistics Committee and Stroke Statistics Subcommittee. Heart disease and stroke statistics–2011 update, a report from the American Heart Association. Circulation 123, e18–e209.

Roncalli, J. et al. 2011. Intracoronary autologous mononucleated bone marrow cell infusion for acute myocardial infarction, results of the randomized multicenter BONAMI trial. Eur. Heart J. 32, 1748–1757.

Reynolds, B. A. and Weiss, S. 1992. Generation of neurons and astrocytes from isolated cells of the adult mammalian central nervous system. Science 255, 1707–1710.

Schächinger, V., Erbs, S., Elsässer, A. et al. 2006a . Improved clinical outcome after intracoronary administration of bone-marrow-derived progenitor cells in acute myocardial infarction, final 1-year results of the REPAIR-AMI trial. Eur. Heart J. 27, 2775–2783.

Schächinger, V., Erbs, S., Elsässer, A. et al. 2006b. Intracoronary bone marrow-derived progenitor cells in acute myocardial infarction. N. Engl. J. Med. 355, 1210–1221.

SCIPIO. Cardiac stem cell infusion in patients with ischemic cardiOmyopathy (NCT00474461). www.clinicaltrials.gov.

Seeger, F. H., Tonn T., Krzossok N. et al. 2007. Cell isolation procedures matter: a comparison of different isolation protocols of bone marrow mononuclear cells used for cell therapy in patients with acute myocardial infarction. Eur. Heart J. 28, 766–772.

Siminiak, T., Fiszer, D., Jerzykowska, O. et al. 2005. Percutaneous trans-coronary-venous transplantation of autologous skeletal myoblasts in the treatment of post-infarction myocardial contractility impairment: The POZNAN trial. Eur. Heart J. 26, 1188–1195.

Siminiak, T., Kalawski, R., Fiszer, D. et al. 2004. Autologous skeletal myoblast transplantation for the treatment of postinfarction myocardial injury: phase I clinical study with 12 months of follow-up. Am. Heart J. 148, 531–537.

Smith, R. R., Chimenti, I., and Marbán, E. 2008. Unselected human cardiosphere-derived cells are functionally superior to c-Kit- or CD90-purified cardiosphere-derived cells. Circulation 118, S_420.

Smith, R. R., Barile, L., Cho, H. C. et al. 2007. Regenerative potential of cardiosphere-derived cells expanded from percutaneous endomyocardial biopsy specimens. Circulation 115, 896–908.

Stamm, C., Kleine, H. D., Choi, Y. H. et al. 2007. Intramyocardial delivery of CD133+ bone marrow cells and coronary artery bypass grafting for chronic ischemic heart disease: safety and efficacy studies. J. Thorac. Cardiovasc. Surg. 133, 717–725.

Strauer, B. E., Brehm, M., Zeus, T. et al. 2001. Intracoronary, human autologous stem cell transplantation for myocardial regeneration following myocardial infarction. Dtsch. Med. Wochenschr. 126, 932–938.

Strauer, B. E., Yousef, M., and Schannwell, C. M. 2010. The acute and long-term effects of intracoronary Stem cell Transplantation in 191 patients with chronic heARt failure: the STAR-heart study. Eur. J. Heart Fail. 12, 721–729.

Sussman, M. A. and Murry, C. E. 2008. Bones of contention: marrow-derived cells in myocardial regeneration. J. Mol. Cell Cardiol. 44, 950–953.

Takehara, N., Tsutsumi, Y., Tateishi, K. et al. 2008. Controlled delivery of basic fibroblast growth factor promotes human cardiosphere-derived cell engraftment to enhance cardiac repair for chronic myocardial infarction. J. Am. Coll. Cardiol. 52, 1858–1865.

Tang, X. L., Rokosh, G., Sanganalmath, S. K. et al. 2010. Intracoronary administration of cardiac progenitor cells alleviates left ventricular dysfunction in rats with a 30-day-old infarction. Circulation 121, 293–305.

Tendera, M., Wojakowski, W., Ruzyłło, W. et al. 2009. Intracoronary infusion of bone marrow-derived selected CD34+CXCR4+ cells and non-selected mononuclear cells in patients with acute STEMI and reduced left ventricular ejection fraction: results of randomized, multicentre Myocardial Regeneration by Intracoronary Infusion of Selected Population of Stem Cells in Acute Myocardial Infarction (REGENT) Trial. Eur. Heart J. 30, 1313–1321.

Terrovitis, J., Lautamäki, R., Bonios, M. et al. 2009. Noninvasive quantification and optimization of acute cell retention by *in vivo* positron emission tomography after intramyocardial cardiac-derived stem cell delivery. J. Am. Coll. Cardiol. 54, 1619–1926.

Terrovitis, J. V., Smith, R. R., and Marbán, E. 2010. Assessment and optimization of cell engraftment after transplantation into the heart. Circ. Res. 106, 479–794.

Thompson, C. A., Nasseri, B. A., Makower, J. et al. 2003. Percutaneous transvenous cellular cardiomyoplasty: a novel nonsurgical approach for myocardial cell transplantation. J. Am. Col. Cardiol. 41, 1964–1971.

Traverse, J. H., Henry, T. D., Vaughan, D. E., and Ellis S. G. 2009. Cardiovascular Cell Therapy Research Network (CCTRN). Rationale and design for TIME: A phase II, randomized, double-blind, placebo-controlled pilot trial evaluating the safety and effect of timing of administration of bone marrow mononuclear cells after acute myocardial infarction. Am. Heart J. 158, 356–363.

Tse, H. F. et al. 2007. Prospective randomized trial of direct endomyocardial implantation of bone marrow cells for treatment of severe coronary artery diseases (PROTECT-CAD trial). Eur. Heart J. 28, 2998–3005.

Ugander, M., Cain, P. A., Johnsson, P. et al. 2010. Chronic non-transmural infarction has a delayed recovery of function following revascularization. BMC Cardiovasc. Disord. 10, 4.

Valina, C., Pinkernell, K., Song, Y. H. et al. 2007. Intracoronary administration of autologous adipose tissue-derived stem cells improves left ventricular function, perfusion, and remodelling after acute myocardial infarction. Eur. Heart J. 28, 2667–2677.

Van Laake, L. W., Qian, L., Cheng, P. et al. 2010. Reporter-based isolation of induced pluripotent stem cell-and embryonic stem cell-derived cardiac progenitors reveals limited gene expression variance. Circ. Res. 107, 340–347.

Van Ramshorst, J., Bax, J. J., Jeroen, J. et al. 2009. Intramyocardial bone marrow cell injection for chronic myocardial ischemia: a randomized controlled trial. JAMA 301, 1997–2004.

Wang, L., Deng, J., Tian, W. et al. 2009. Adipose-derived stem cells are an effective cell candidate for treatment of heart failure: an MR imaging study of rat hearts. Am. J. Physiol. Heart Circ. Physiol. 297, H1020–1031.

White, H. D. et al. 2005. Mortality and morbidity remain high despite captopril and/or Valsartan therapy in elderly patients with left ventricular systolic dysfunction, heart failure, or both after acute myocardial infarction: results from the Valsartan in Acute Myocardial Infarction Trial (VALIANT). Circulation 112, 3391–3399.

Wöhrle, J., Merkle, N., Mailänder, V. et al. 2010. Results of intracoronary stem cell therapy after acute myocardial infarction. Am. J. Cardiol. 105, 804–812.

Wollert, K. C., Meyer, G. P., Lotz, J. et al. 2004. Intracoronary autologous bone-marrow cell transfer after myocardial infarction: the BOOST randomised controlled clinical trial. Lancet 364, 141–148.

Yao, K., Huang, R. C., Ge, L. et al. 2006. Observation on the safety: clinical trial on intracoronary autologous bone marrow mononuclear cells transplantation for acute myocardial infarction. Chin. J. Cardiol. 34, 577–581.

Yeo, C. and Mathur, A. 2009. Autologous bone marrow-derived stem cells for ischemic heart failure: REGENERATE-IHD trial. Regen. Med. 4, 119–127.

Zakharova, L., Mastroeni, D., Mutlu, N. et al. 2010. Transplantation of cardiac progenitor cell sheet onto infarcted heart promotes cardiogenesis and improves function. Cardiovasc. Res. 87, 40–49.

Zhang, S., Sun, A., Xu, D. et al. 2009. Impact of timing on efficacy and safety of intracoronary autologous bone marrow stem cells transplantation in acute myocardial infarction: a pooled subgroup analysis of randomized controlled trials. Clin. Cardiol. 32, 458–466.

Zhang, Y., Li, T. S., Lee, S. T. et al. 2010. Dedifferentiation and proliferation of mammalian cardiomyocytes. PLoS One 5, e12559.

Zangi, L., Margalit, R., Reich-Zeliger, S. et al. 2009. Direct imaging of immune rejection and memory induction by allogeneic mesenchymalstromal cells. Stem Cells 27, 2865–2874.

Zeng, L., Hu, Q., Wang, X. et al. 2007. Bioenergetic and functional consequences of bone marrow-derived multipotent progenitor cell transplantation in hearts with postinfarction left ventricular remodeling. Circulation 115, 1866–1875.

CHAPTER 15

Induced Cardiomyocytes: Direct Reprogramming for Cardiac Regeneration

Masaki Ieda

Introduction

Cardiovascular disease remains the leading cause of morbidity and mortality in developed countries. Because terminally differentiated cardiomyocytes lack the ability to regenerate, their malfunction or significant loss due to disease may lead to lethal consequences such as heart failure. Heart transplantation is an established therapy for heart failure, but is limited due to the number of donor organs available (Taylor et al. 2008). Cell replacement therapy is an attractive option for myocardial repair. In particular, advances in stem cell research during the past decades have brought a tremendous amount of excitement about the possibility of using adult or embryonic sources of stem cells to repair or regenerate damaged cardiac tissues (Mummery et al. 2002).

Embryonic stem (ES) cells possess the ability to remain undifferentiated and propagate infinitely in culture while maintaining their pluripotency to differentiate into all three embryonic germ layers and their derivatives, including cardiomyocytes. Therefore, it is theoretically possible to generate an infinite number of cardiomyocytes from differentiated cultures of ES

Department of Clinical and Molecular Cardiovascular Research; Department of Cardiology, Keio University School of Medicine; and JST, CREST—35 Shinanomachi, Shinjuku-ku, Tokyo 160-8582, Japan.
Email: mieda@z8.keio.jp

cells (Laflamme et al. 2007; Nussbaum et al. 2007; van Laake et al. 2008). Moreover, with the recent discovery that induced pluripotent stem (iPS) cells, which resemble ES cells, can be generated from human dermal fibroblasts, investigators have begun to examine the possibility of making cardiomyocytes from somatic cells through iPS cell generation (Takahashi and Yamanaka 2006; Zhang et al. 2009). However, the efficiency of cardiac differentiation, the risk of tumor formation, and the multiple steps needed to generate cardiomyocytes through iPS cell generation must be improved for clinical use (Hattori et al. 2010; Seki et al. 2010).

Pluripotent reprogramming of cells, by the four factors specified by Yamanaka et al. has revolutionized the field of regenerative medicine, with projected applications in cell therapy, disease modeling, and personalized medicine (Amabile and Meissner 2009; Carvajal-Vergara et al. 2010; Itzhaki et al. 2011; Kamp and Lyons 2009; Moretti et al. 2010; Nelson and Terzic 2011; Park et al. 2008a; Tiscornia et al. 2011; Wu and Hochedlinger 2011; Yazawa et al. 2011; Yoshida and Yamanaka 2010). The discovery of iPSCs further delivered the direct generation of specific cell types in an approach called lineage reprogramming. Such a strategy involves reprogramming mature cell types directly into another cell type without passing through a stem cell by the combination of several lineage specific factors. Recent studies demonstrated that direct lineage reprogramming could yield a diverse range of medically relevant cell types including pancreatic β cells, neurons, and hepatocytes (Fig. 1) (Huang et al. 2011a,b; Ieda et al. 2010; Sekiya and Suzuki 2011; Vierbuchen et al. 2010; Zhou et al. 2008). Our team, as well as others recently reported that a somatic cell could be directly reprogrammed into cardiomyocyte-like cells without first becoming a stem/progenitor cell (Ieda et al. 2010; Vierbuchen et al. 2010; Zhou et al. 2008). We found that a combination of three transcription factors, Gata4, Mef2c, and Tbx5, rapidly and efficiently induced cardiomyocyte-like cells from mouse cardiac and tail-tip fibroblasts. The induced cardiomyocytes (iCMs) were similar to neonatal cardiomyocytes and could be useful as a source of cardiomyocytes for regenerative purpose (Ieda et al. 2010). Here we review the background and recent studies of cellular reprogramming and fate conversion with their potential future applications in regenerative medicine.

The History of Cell Fate Conversion by Defined Factors

1. MyoD is a master regulator for skeletal muscle differentiation

MyoD is a classic example of a master gene for cell differentiation, in the sense that transduction of this gene is sufficient to activate the genetic program for muscle differentiation in non-muscle cells (Berkes and Tapscott 2005; Tapscott 2005). In 1979, Taylor and Jones (Constantinides et

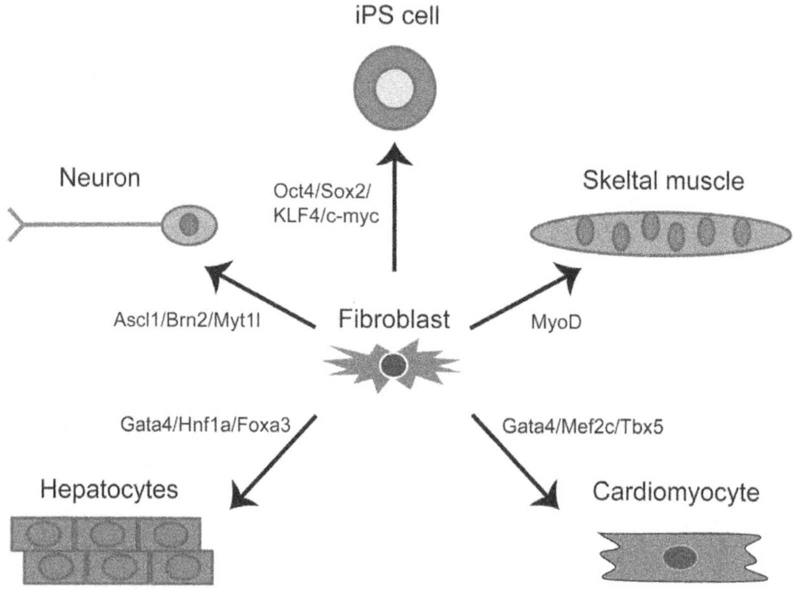

Figure 1. Transcription factor overexpression causes cell fate changes. Fibroblasts can be converted into iPS cells, skeletal muscle cells, cardiomyocytes, hepatocytes, and neurons by transduction of reprogramming factors.

al. 1977; Taylor and Jones 1979) found that treatment of the 10T1/2 mouse embryonic fibroblast cell line with 5-azacytidine (5-azaC), an inhibitor of DNA methylation, induced differentiation of these cells into skeletal muscle cells. This observation led to the hypothesis that DNA demethylation and the resulting activation of unidentified genes were responsible for the conversion to muscle cells (Lassar et al. 1986). A subtractive hybridization experiment comparing untreated and 5-azaC-treated fibroblasts led to the identification of a cDNA encoding MyoD (Davis et al. 1987). When overexpressed in primary fibroblasts or in a wide variety of other cell types, including pigment, nerve, fat, and liver cells, MyoD induced reprogramming into skeletal muscle cells (Choi et al. 1990; Weintraub et al. 1989). These findings provided the first evidence that a single gene can initiate a complex program of differentiation, thereby acting as a master switch.

The ability of MyoD to convert fibroblasts and other cell types into skeletal muscle strongly implicated a central role in myogenesis, and subsequent studies sought to determine the biological roles and molecular mechanisms of MyoD function in muscle development (Pownall et al. 2002). The MyoD protein contains a basic helix-loop-helix (bHLH) motif that is common to a large family of transcription factors (Ledent et al. 2002; Ledent and Vervoort, 2001; Pownall et al. 2002). Indeed, the strongly related bHLH proteins, Myf5, Mrf4, and myogenin, are also expressed in skeletal muscle, with each factor playing a sequential and synergistically crucial role in muscle cell specification and differentiation *in vivo* (Buckingham et al. 2003; Molkentin and Olson 1996; Perry et al. 2001). It was therefore proposed that lineage associated transcription factors, which help to establish and maintain cellular identity during development, could also change cell fate when ectopically expressed in certain heterologous cells.

2. Lineage conversion in hematopoietic cells by defined factors

Following the discovery of MyoD, conversion of one cell type into another was predominantly demonstrated among hematopoietic cells, which are an ideal cell type for lineage conversion experiments because the cellular lineages are well defined. Ectopic expression of GATA1, the erythroid-megakaryocyte-affiliated transcription factor, produced erythroid-megakaryocytic-eosinophil-basophils from granulocyte-macrophage progenitors freshly isolated from normal bone marrow (Heyworth et al. 2002; Kulessa et al. 1995; Visvader et al. 1992). C/EBPα, a transcription factor required for the formation of granulocyte-macrophage precursors, can convert committed B- and T-cell progenitors into functional macrophages at frequencies approaching 100% (Laiosa et al. 2006; Xie et al. 2004; Zhang et al. 2004). Mature immunoglobulin-producing B cells could also be switched into macrophages by C/EBPα, although at lower frequencies.

Inactivation of key regulators may also lead to the activation of earlier genetic programs in committed cells, resulting in their dedifferentiation and activation of multi-lineage potential. For instance, deletion of the Pax5 transcription factor induced dedifferentiation of B cells into common hematopoietic progenitor cells (Cobaleda et al. 2007; Nutt et al. 1999; Rolink et al. 1999). Under appropriate culture conditions, these cells may then differentiate into granulocyte/macrophage, T-cell, dendritic, natural killer, and osteoclast cell lineages.

Direct Reprogramming into the Desired Cell Types by Combinations of Defined Factors

1. Stem cell induction by the yamanaka factors

In the 1960s, John Gurdon and colleagues (Gurdon 1962, 2006) demonstrated that transferring the nucleus of a differentiated frog cell into an enucleated egg could reprogram the cell back to the totipotency of a zygote, and thus give rise to a whole new frog. Following this striking demonstration using frog cells, reprogramming into pluripotent states by somatic nuclear transfer in mammals was first reported in 1997 with the birth of Dolly the sheep, and subsequently accomplished in many other mammalian species (Eggan et al. 2001; Eggan et al. 2004; Gurdon and Byrne 2003; Gurdon et al. 2003; Hochedlinger and Jaenisch 2002; Inoue et al. 2005; Li et al. 2004; Tamashiro et al. 2002; Wilmut et al. 1997). These impressive demonstrations of the inherent plasticity and potency of cells led to a search for reprogramming factors able to convert differentiated cells back to a pluripotent stage.

Takahashi and Yamanaka (Takahashi and Yamanaka 2006) achieved a breakthrough in this field by overexpressing four embryonic stem cell (ESC)-specific transcription factors in somatic fibroblasts to induce a pluripotent state, resulting in the so-called induced pluripotent stem cell (iPSC). Using retroviral vectors, they found that iPSCs can be induced by the transduction of four genes, comprising Oct4, Sox2, Klf4, and c-Myc which are referred to as the Yamanaka factors. Many laboratories have since improved iPSC generation techniques to show that murine iPSCs share all defining features with naive mouse ESCs, including expression of pluripotency markers, reactivation of both X chromosomes, and the ability to generate chimeric mice (Boland et al. 2009; Kang et al. 2009; Maherali and Hochedlinger 2008; Maherali et al. 2007; Okita et al. 2007; Stadtfeld et al. 2008; Wernig et al. 2007; Zhao et al. 2009).

iPSCs can be derived from a number of different species, including humans, rats, and rhesus monkeys, by expression of the four Yamanaka factors (Li et al. 2009; Liu et al. 2008; Park et al. 2008b; Takahashi et al. 2007; Yu et al. 2007). Similarly, iPSCs have been derived from other somatic cell populations, such as keratinocytes, neural cells, stomach and liver cells, and melanocytes, as well as from terminally differentiated lymphocytes, demonstrating the universality of the induced pluripotency (Aasen et al. 2008; Aoi et al. 2008; Hanna et al. 2008; Kim et al. 2008; Seki et al. 2010). Importantly, reprogramming can be induced not only by Oct4, Sox2, Klf4, and c-Myc, but also by alternative combinations that employ Nanog, Lin28, ESRRB, NR5A2, and other genes that help to establish the core transcriptional circuitry of stem cells (Ichida et al. 2009; Mitsui et al. 2003; Yu et al. 2007).

2. In vivo reprogramming into pancreatic β cells

The generation of iPSCs sparked the new and related idea of converting mature cell types directly into another cell type without becoming a stem cell, using a combination of several lineage specific factors (Huang et al. 2011a; Kanazawa et al. 2010; Szabo et al. 2010; Vierbuchen et al. 2010; Zhou et al. 2008). If target cells can be obtained directly without passing through stem cells, the possibility of tumor formation after the cell graft is very low. Zhou et al. [32] showed that adenoviral gene transfer of transcription factors could efficiently reprogram pancreatic exocrine cells into functional β-cells in mouse (Zhou et al. 2008), using Neurogenin 3, Pdx1, and Mafa, which are known to be important in the embryonic development of pancreas and β-cells (Zhou et al. 2007). The induced β-cells were indistinguishable from endogenous islet β-cells in structure and expressed genes essential for β-cell function. Importantly, the induced β-cells ameliorated hyperglycemia in type 1 diabetic mice by remodeling local vasculature and secreting insulin. Thus, although the new β-cells were not reorganized into complete islet structures and this may limit their effectiveness, this study provided the first evidence of cellular reprogramming *in vivo* by defined factors.

3. Conversion of fibroblasts into neurons

It is conceivable then that while direct cell-type conversion within the same lineage may be easy, converting to a completely different lineage could be challenging. To this end, Vierbuchen et al. (Vierbuchen et al. 2010) converted mouse dermal fibroblasts into functional neurons *in vitro* using neuronal lineage-specific transcription factors, Ascl1, Brn2, and Myt1l. These induced neuronal (iN) cells expressed multiple neuron-specific proteins, generated action potentials, and formed functional synapses. The same group recently reported that when combined with the bHLH transcription factor NeuroD1, these three factors could also convert human fibroblasts into iN cells (Pang et al. 2011) that also showed typical neuronal morphologies and expressed multiple neuronal markers, even after downregulation of the exogenous transcription factors. However, the iN cells in this study represented a heterogeneous population of excitatory neurons and it was unclear whether direct lineage reprogramming could produce a specific neuronal subtype. Caiazzo et al. (Caiazzo et al. 2011) reported that a combination of three transcription factors, Ascl1, Nurr1, and Lmx1a, could generate functional dopaminergic neurons from mouse and human fibroblasts without reverting to a progenitor cell stage, as well as from Parkinson's disease patients. These studies thus revealed that lineage conversions are not restricted to within the same lineage or germ layer, since fibroblasts are mesodermal in origin and neurons are derived from ectoderm.

4. Induced hepatocyte-like cells repair damaged liver

Overexpression of lineage-specific transcription factors directly converts terminally differentiated cells into some other lineages; however, it remained unclear whether these lineage-converted cells could reconstitute damaged tissues *in vivo*. In 2011, two groups clearly demonstrated that induced hepatocyte-like (iHep) cells directly generated from fibroblasts reconstituted damaged hepatic tissues after cell transplantation (Huang et al. 2011a;

Sekiya and Suzuki 2011). Huang et al. (Huang et al. 2011) demonstrated the direct induction of functional iHep cells from mouse tail-tip fibroblasts by transduction of Gata4, Hnf1a, and Foxa3, and inactivation of p19Arf. These iHep cells showed typical epithelial morphology, expressed hepatic genes, and acquired hepatocyte functions. Notably, transplanting the iHep cells into fumarylacetoacetate-hydrolase-deficient mice repopulated the damaged liver and rescued almost half the recipients from death by restoring liver function. The authors proposed that Gata4 and Foxa3 probably act as 'pioneer factors' to trigger a global chromatin modification during the hepatic conversion (Cirillo et al. 2002; Cirillo and Zaret 1999; Zaret 2008). Similarly, Sekiya et al. (Sekiya and Suzuki 2011) identified three specific combinations of two transcription factors, comprising Hnf4a plus Foxa1, Foxa2, or Foxa3 that could convert mouse embryonic and adult fibroblasts into iHep cells, which were also functional and rescued damaged hepatic tissues after transplantation. These studies provide novel strategies to generate functional hepatocyte-like cells for the purpose of liver engineering and regenerative medicine. However, because the iHep cell transplantation experiments produced only partial rescue and cells that are still not identical to the bonafide hepatocytes, further modifications for generating iHep cells are needed to apply this approach in clinical settings. Importantly, both studies showed no tumor formation using the iHep cells.

5. Direct conversion of fibroblast cells into cardiomyocyte-like cells

Although embryonic mesoderm can be induced to differentiate into cardiomyocytes, no "master regulator" of cardiac differentiation, like MyoD for skeletal muscle, was identified (Bondue et al. 2008; Davis et al. 1987; Takeuchi and Bruneau 2009). As a cell source for cardiac reprogramming, we used cardiac fibroblasts. Cardiac fibroblasts are found throughout cardiac tissues along with cardiomyocytes, and account for more than half of the cells in the heart (Baudino et al. 2006; Camelliti et al. 2005; Ieda et al. 2009). In diseased hearts, fibroblasts proliferate and secrete extracellular matrix and growth factors which induce fibrosis and myocardial remodeling, leading to heart failure (Weber and Brilla 1991). To determine the candidate factors of cardiac reprogramming, we identified the genes which are specifically expressed in embryonic cardiomyocytes. We developed a novel cell purification system in which embryonic cardiomyocytes and cardiac fibroblasts can be purified using fluorescence activated cell sorting (FACS) (Ieda et al. 2009). Using this system, we selected 14 factors as candidates for cardiac reprogramming, which are specifically expressed in cardiomyocytes and are critical for heart development.

We next developed a screening system in which the induction of mature cardiomyocytes from fibroblasts could be analyzed quantitatively by FACS. We generated α myosin heavy chain (αMHC) promoter-driven EGFP-IRES-Puromycin transgenic mice (αMHC-GFP), in which only mature cardiomyocytes expressed the green fluorescent protein (GFP) (Gulick et al. 1991; Ieda et al. 2007). Transduction of all 14 factors into fibroblasts induced 1.7% of GFP^+ cells, and serial removal of individual factors demonstrated that a combination of three factors (Gata4, Mef2c and Tbx5) were sufficient for efficient GFP^+ cell induction (around 15%) from cardiac fibroblasts. The three cardiac reprogramming factors, Gata4, Mef2c, and Tbx5, are core cardiac transcription factors during early heart development (Olson, 2006; Srivastava 2006; Zhao et al. 2008) and are known to interact with each other, co-activate cardiac gene expression and promote cardiomyocyte differentiation (Garg et al. 2003; Ghosh et al. 2009; Morin et al. 2000). We designated these GFP^+ cardiomyocyte-like cells as induced cardiomyocytes (iCMs).

The iCMs expressed several cardiac marker proteins, such as sarcomeric α-actinin, cardiac troponin T and atrial natriuretic factor (Kattman et al. 2006). They also had well-defined sarcomeric structures similar to neonatal cardiomyocytes. The global gene expression profile of GFP^+ iCMs is not identical but similar to neonatal cardiomyocytes, and different from original cardiac fibroblasts. We also found that functionally important cardiac genes, such as ion channel genes and sarcomere genes were upregulated more in 4-week iCMs than in 2-week iCMs, suggesting full maturation occurred over several weeks. The chromatin state of iCMs was also similar to cardiomyocytes but different from fibroblasts in histone modifications and DNA methylation patterns (Li et al. 2007). These results suggest that iCMs are cardiomyocyte-like cells in gene expression and epigenetic states. We also demonstrated that iCMs possess functional properties which are characteristic of cardiomyocytes. The iCMs exhibited intracellular Ca^{2+} transient and action potentials after 2–4 weeks of culture. In addition, cardiac fibroblast-derived iCMs contracted spontaneously.

To determine if the iCMs could arise from cells in other organs, we transduced the three factors into mouse tail-tip fibroblasts. We found that tail-tip fibroblast-derived GFP^+ cells expressed cardiac markers and spontaneous intracellular Ca^{2+} transient, while the percentage of cardiac troponin T^+ cells was 50% less than that of cardiac fibroblast-derived iCMs. These results indicate that fibroblasts from other organs can be reprogrammed into cardiomyocytes, but the full reprogramming rate is lower than that from cardiac fibroblasts. These findings also exclude the possibility that iCMs arise from contamination of cardiomyocytes or cardiac progenitors in the cardiac fibroblast population (Andersen et al. 2009; Davis et al. 2009).

While direct conversion of somatic cells to β-cells and neurons were reported, the "route" of cell fate alteration was not clear. There are two possibilities in the reprogramming of fibroblasts into differentiated cardiomyocytes. One is a direct conversion of fibroblasts into differentiated cardiomyocytes and the other is that fibroblasts first revert to a cardiac progenitor/stem cell fate before further cardiac differentiation. We were able to genetically test these two possibilities by using the mice expressing Isl1-Cre-yellow fluorescent protein (YFP) and Mesp1-Cre-YFP obtained by crossing Isl1-Cre or Mesp1-Cre mice with R26R-EYFP mice (Srinivas et al. 2001). Isl1 and Mesp1 are transiently expressed in early cardiac progenitor cells before further cardiac differentiation (Bu et al. 2009; Laugwitz et al. 2005). We found that iCMs induced from Isl1-Cre-YFP and Mesp1-Cre-YFP fibroblasts did not express YFP, suggesting the fibroblasts directly reprogrammed into differentiated cardiomyocytes without passing through a cardiac progenitor cell state, with Gata4/Mef2c/Tbx5 transduction. Importantly, this is the first study representing the global gene expression profiles and epigenetic status of directly induced cells, and demonstrating direct evidence of cell fate in lineage conversion using several reporter mice.

We next investigated whether Gata4/Mef2c/Tbx5-transduced cardiac fibroblasts can be transplanted *in vivo* and be reprogrammed into cardiomyocytes within the heart. Cardiac fibroblasts infected with Gata4/Mef2c/Tbx5 expressed α-MHC GFP and sarcomeric markers, and converted into cardiomyocyte-like cells in the mouse heart after 2 weeks of cell transplantation. These results suggest that large amounts of an individual's own fibroblasts grown from a cardiac biopsy or skin biopsy can be transduced with the cardiac reprogramming factors *ex vivo* and be delivered to damaged hearts to regenerate cardiomyocytes.

Direct reprogramming of somatic cells to a pluripotent state using the Yamanaka factors entails extensive genetic and epigenetic resetting. Thus, it is possible that the reprogramming factors function to erase cell identity by epigenetic mechanisms and that subsequent exposure to external lineage-specific signals directs lineage specification and induction of terminal differentiation. Efe et al. showed that transient overexpression of the four Yamanaka factors, Oct4, Sox2, Klf4 and c-Myc, can convert mouse fibroblasts directly into spontaneously contracting cardiomyocytes mediating through cardiac progenitor cells but not through a pluripotent intermediate (Efe et al. 2011). Compared to our transdifferentiation protocol, they showed that the spontaneous contractions begin after 11 days and the efficiency of generating cTnT+ cardiomyocyte yield was six-fold higher. This latter result might be due to generating mitotically active cardiac precursor cells first rather than terminally differentiated cardiac cells in their protocol. For future clinical applications, they now need to confirm that their fibroblast-derived cardiomyocytes are functionally integrated into the host tissue and do not form teratoma.

The Impact and Challenges of Induced Cardiomyocyte Technology on Future Cardiac Regenerative Therapy

Damaged myocardium could be treated by iPS cell-derived cardiomyocytes in the future (Caspi et al. 2007; Yoshida and Yamanaka). For the process of cardiac regeneration using iPS cell-derived cardiomyocytes, first, the patient's own fibroblasts would be harvested from a skin biopsy, expanded *ex vivo*, and transduced with the four Yamanaka factors to generate iPS cells. Then, the cardiomyocytes would be differentiated from the iPS cells and purified by FACS to avoid contamination by undifferentiated cells and other cell types. iPS cell-derived cardiomyocytes could then be transplanted into the patient's heart by a surgical operation. In this case, multiple steps and a long process are necessary to accomplish cardiac regeneration. In contrast, cardiomyocytes could be directly generated from endogenous cardiac fibroblasts with the defined factors using iCM technology (Fig. 2). Introduction of the defined factors directly into the damaged heart may reprogram the endogenous fibroblast population, which represents more than 50% of the cells, into new functional cardiomyocytes. If this is possible as shown in pancreatic β cell generation, there are several advantages; first,

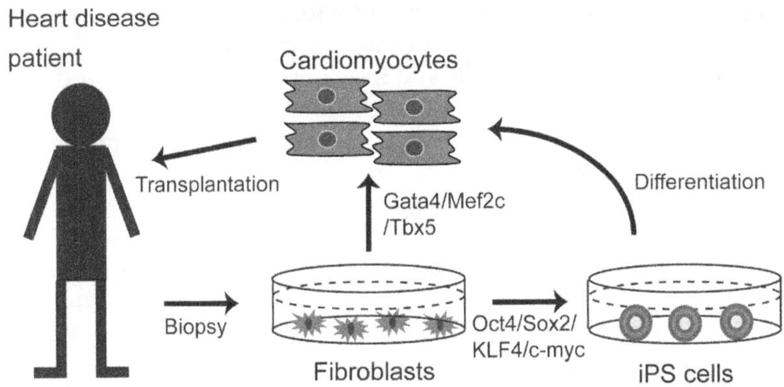

Figure 2. Potential application of cellular reprogramming technology for future regenerative medicine. Fibroblasts are harvested from heart disease patients and converted into iPS cells or induced cardiomyocytes by overexpression of Oct4/Sox2/KLF4/c-myc or Gata4/Mef2c/Tbx5, respectively. Patient's iPS cells are differentiated into cardiomyocytes *in vitro* and then transplanted into the patient's heart. Induced cardiomyocytes may be transplanted into the injured heart directly.

the process is simple and short; second, the avoidance of reprogramming to pluripotent cells before cardiac differentiation would greatly lower the risk of tumor formation; third, direct injection of defined factors can avoid cell transplantation in which long term cardiac cell survival is still challenging (Blum and Benvenisty 2008; Huber et al. 2007; Laflamme et al. 2007; Zhang et al. 2001). However, the functional properties of induced cardiomyocytes should be characterized more carefully and it is still unclear whether direct cardiac reprogramming is possible in human cells. Moreover, the efficiency of cardiac reprogramming, particularly generation of bonafide cardiac cells, is still low and should be improved in future research. Studies in human cells and understanding of the molecular mechanism of direct cardiac reprogramming are necessary to advance this technology for future clinical applications.

Conclusion

Cellular reprogramming and fate conversion have long been recognized as a possibility, although the impact of cell-type conversion by defined genes was most prominently exemplified only recently by the discovery of iPSCs. This landmark finding fundamentally altered approaches and ideas about regenerative medicine and also provided a broad strategy to induce desired cell types by introducing lineage-specific factors. The numerous reprogramming studies have taught us that cells have more plasticity and cell fate can be moulded more easily than previously thought once the right combination of transcription factors is identified. Notably, the issues of appropriate cellular differentiation, risk of tumor formation due to contaminating immature cells and insertional mutagenesis, and proper integration into the recipient organ require further and careful investigation (Dai et al. 2007; Miura et al. 2009; Yamashita et al. 2010) for future clinical applications to be realized.

Heart disease is still one of the most life-threatening diseases all over the world. Given the shortage of heart transplantation donors, cell replacement therapy might be one of the most promising and exciting research areas to be pursued. It is clear that the discovery of iPS cells has fundamentally altered the approach to regenerative medicine, but the field is still in its infancy. The work of numerous laboratories has led to significant therapeutic and scientific advances in cell therapy to cardiac regenerative medicine. Despite the exponentially increasing knowledge regarding stem cell-based cardiac regeneration, many questions and challenges remain. The issues of appropriate cardiac differentiation, risk of tumor formation due to contamination of immature cells, and proper integration into the recipient myocardium should be improved before clinical applications.

The new direct cardiac reprogramming technology has just emerged and much refinement and characterization of the reprogramming process will be necessary. We hope cardiac regeneration therapy will become a next generation therapy and help many heart disease patients in the future.

Acknowledgements

M. I. is supported in part by research grants from the Japan Science and Technology Agency's Core Research of Evolutional Science & Technology program. I apologize to authors whose work has not been cited due to space limitations.

References

Aasen, T., Raya, A., Barrero, M. J., Garreta, E., Consiglio, A., Gonzalez, F., Vassena, R., Bilic, J., Pekarik, V., Tiscornia, G. et al. 2008. Efficient and rapid generation of induced pluripotent stem cells from human keratinocytes. Nat. Biotechnol. 26, 1276–1284.

Amabile, G. and Meissner, A. 2009. Induced pluripotent stem cells: current progress and potential for regenerative medicine. Trends Mol. Med. 15, 59–68.

Andersen, D. C., Andersen, P., Schneider, M., Jensen, H. B., and Sheikh, S. P. 2009. Murine "cardiospheres" are not a source of stem cells with cardiomyogenic potential. Stem Cells 27, 1571–1581.

Aoi, T., Yae, K., Nakagawa, M., Ichisaka, T., Okita, K., Takahashi, K., Chiba, T., and Yamanaka, S. 2008. Generation of pluripotent stem cells from adult mouse liver and stomach cells. Science 321, 699–702.

Baudino, T. A., Carver, W., Giles, W., and Borg, T. K. 2006. Cardiac fibroblasts: friend or foe? Am. J. Physiol. Heart Circ. Physiol. 291, H1015–1026.

Berkes, C. A. and Tapscott, S. J. 2005. MyoD and the transcriptional control of myogenesis. Semin Cell Dev. Biol. 16, 585–595.

Blum, B. and Benvenisty, N. 2008. The tumorigenicity of human embryonic stem cells. Adv. Cancer Res. 100, 133–158.

Boland, M. J., Hazen, J. L., Nazor, K. L., Rodriguez, A. R., Gifford, W., Martin, G., Kupriyanov, S. and Baldwin, K. K. 2009. Adult mice generated from induced pluripotent stem cells. Nature 461, 91–94.

Bondue, A., Lapouge, G., Paulissen, C., Semeraro, C., Iacovino, M., Kyba, M., and Blanpain, C. 2008. Mesp1 acts as a master regulator of multipotent cardiovascular progenitor specification. Cell Stem Cell 3, 69–84.

Bu, L., Jiang, X., Martin-Puig, S., Caron, L., Zhu, S., Shao, Y., Roberts, D. J., Huang, P. L., Domian, I. J., and Chien, K. R. 2009. Human ISL1 heart progenitors generate diverse multipotent cardiovascular cell lineages. Nature 460, 113–117.

Buckingham, M., Bajard, L., Chang, T., Daubas, P., Hadchouel, J., Meilhac, S., Montarras, D., Rocancourt, D., and Relaix, F. 2003. The formation of skeletal muscle: from somite to limb. J. Anat. 202, 59–68.

Caiazzo, M., Dell'Anno, M. T., Dvoretskova, E., Lazarevic, D., Taverna, S., Leo, D., Sotnikova, T. D., Menegon, A., Roncaglia, P., Colciago, G., Russo, G., Carninci, P., Pezzoli, G., Gainetdinov, R. R., Gustincich, S., Dityatev, A., and Broccoli, V. 2011. Direct generation of functional dopaminergic neurons from mouse and human fibroblasts. Nature 476, 224–227.

Camelliti, P., Borg, T. K., and Kohl, P. 2005. Structural and functional characterisation of cardiac fibroblasts. Cardiovasc Res. 65, 40–51.

Carvajal-Vergara, X., Sevilla, A., D'Souza, S. L., Ang, Y. S., Schaniel, C., Lee, D. F., Yang, L., Kaplan, A. D., Adler, E. D., Rozov, R. et al. 2010. Patient-specific induced pluripotent stem-cell-derived models of LEOPARD syndrome. Nature 465, 808–812.

Caspi, O., Huber, I., Kehat, I., Habib, M., Arbel, G., Gepstein, A., Yankelson, L., Aronson, D., Beyar, R. and Gepstein, L. 2007. Transplantation of human embryonic stem cell-derived cardiomyocytes improves myocardial performance in infarcted rat hearts. J. Am. Coll Cardiol. 50, 1884–1893.

Choi, J., Costa, M. L., Mermelstein, C. S., Chagas, C., Holtzer, S., and Holtzer, H. 1990. MyoD converts primary dermal fibroblasts, chondroblasts, smooth muscle, and retinal pigmented epithelial cells into striated mononucleated myoblasts and multinucleated myotubes. Proc. Natl. Acad. Sci. USA 87, 7988–7992.

Cirillo, L. A., Lin, F. R., Cuesta, I., Friedman, D., Jarnik, M., and Zaret, K. S. 2002. Opening of compacted chromatin by early developmental transcription factors HNF3 (FoxA) and GATA-4. Mol. Cell 9, 279–289.

Cirillo, L. A. and Zaret, K. S. 1999. An early developmental transcription factor complex that is more stable on nucleosome core particles than on free DNA. Mol. Cell 4, 961–969.

Cobaleda, C., Jochum, W., and Busslinger, M. 2007. Conversion of mature B cells into T cells by dedifferentiation to uncommitted progenitors. Nature 449, 473–477.

Constantinides, P. G., Jones, P. A., and Gevers, W. 1977. Functional striated muscle cells from non-myoblast precursors following 5-azacytidine treatment. Nature 267, 364–366.

Dai, W., Field, L. J., Rubart, M., Reuter, S., Hale, S. L., Zweigerdt, R., Graichen, R. E., Kay, G. L., Jyrala, A. J., Colman, A. et al. 2007. Survival and maturation of human embryonic stem cell-derived cardiomyocytes in rat hearts. J. Mol. Cell Cardiol. 43, 504–516.

Davis, D. R., Zhang, Y., Smith, R. R., Cheng, K., Terrovitis, J., Malliaras, K., Li, T. S., White, A., Makkar, R., and Marban, E. 2009. Validation of the cardiosphere method to culture cardiac progenitor cells from myocardial tissue. PLoS One 4, e7195.

Davis, R. L., Weintraub, H., and Lassar, A. B. 1987. Expression of a single transfected cDNA converts fibroblasts to myoblasts. Cell 51, 987–1000.

Efe, J. A., Hilcove, S., Kim, J., Zhou, H., Ouyang, K., Wang, G., Chen, J., and Ding, S. 2011. Conversion of mouse fibroblasts into cardiomyocytes using a direct reprogramming strategy. Nat. Cell Biol. 13, 215–222.

Eggan, K., Akutsu, H., Loring, J., Jackson-Grusby, L., Klemm, M., Rideout, W. M., 3rd, Yanagimachi, R. and Jaenisch, R. 2001. Hybrid vigor, fetal overgrowth, and viability of mice derived by nuclear cloning and tetraploid embryo complementation. Proc. Natl. Acad. Sci. USA 98, 6209–6214.

Eggan, K., Baldwin, K., Tackett, M., Osborne, J., Gogos, J., Chess, A., Axel, R., and Jaenisch, R. 2004. Mice cloned from olfactory sensory neurons. Nature 428, 44–49.

Garg, V., Kathiriya, I. S., Barnes, R., Schluterman, M. K., King, I. N., Butler, C. A., Rothrock, C. R., Eapen, R. S., Hirayama-Yamada, K., Joo, K. et al. 2003. GATA4 mutations cause human congenital heart defects and reveal an interaction with TBX5. Nature 424, 443–447.

Ghosh, T. K., Song, F. F., Packham, E. A., Buxton, S., Robinson, T. E., Ronksley, J., Self, T., Bonser, A. J., and Brook, J. D. 2009. Physical interaction between TBX5 and MEF2C is required for early heart development. Mol. Cell Biol. 29, 2205–2218.

Gulick, J., Subramaniam, A., Neumann, J., and Robbins, J. 1991. Isolation and characterization of the mouse cardiac myosin heavy chain genes. J. Biol. Chem. 266, 9180–9185.

Gurdon, J. B. 1962. The developmental capacity of nuclei taken from intestinal epithelium cells of feeding tadpoles. J. Embryol. Exp. Morphol. 10, 622–640.

Gurdon, J. B. 2006. From nuclear transfer to nuclear reprogramming: the reversal of cell differentiation. Annu. Rev. Cell Dev. Biol. 22, 1–22.

Gurdon, J. B. and Byrne, J. A. 2003. The first half-century of nuclear transplantation. Proc. Natl. Acad. Sci. USA 100, 8048–8052.

Gurdon, J. B., Byrne, J. A., and Simonsson, S. 2003. Nuclear reprogramming and stem cell creation. Proc. Natl. Acad. Sci. USA 100 Suppl. 1, 11819–11822.

Hanna, J., Markoulaki, S., Schorderet, P., Carey, B. W., Beard, C., Wernig, M., Creyghton, M. P., Steine, E. J., Cassady, J. P., Foreman, R. et al. 2008. Direct reprogramming of terminally differentiated mature B lymphocytes to pluripotency. Cell 133, 250–264.

Hattori, F., Chen, H., Yamashita, H., Tohyama, S., Satoh, Y. S., Yuasa, S., Li, W., Yamakawa, H., Tanaka, T., Onitsuka, T. et al. 2010. Nongenetic method for purifying stem cell-derived cardiomyocytes. Nat. Methods 7, 61–66.

Heyworth, C., Pearson, S., May, G., and Enver, T. 2002. Transcription factor-mediated lineage switching reveals plasticity in primary committed progenitor cells. Embo. J. 21, 3770–3781.

Hochedlinger, K. and Jaenisch, R. 2002. Monoclonal mice generated by nuclear transfer from mature B and T donor cells. Nature 415, 1035–1038.

Huang, P., He, Z., Ji, S., Sun, H., Xiang, D., Liu, C., Hu, Y., Wang, X., and Hui, L. 2011. Induction of functional hepatocyte-like cells from mouse fibroblasts by defined factors. Nature 475, 386–389.

Huber, I., Itzhaki, I., Caspi, O., Arbel, G., Tzukerman, M., Gepstein, A., Habib, M., Yankelson, L., Kehat, I., and Gepstein, L. 2007. Identification and selection of cardiomyocytes during human embryonic stem cell differentiation. Faseb J. 21, 2551–2563.

Ichida, J. K., Blanchard, J., Lam, K., Son, E. Y., Chung, J. E., Egli, D., Loh, K. M., Carter, A. C., Di Giorgio, F. P., Koszka, K. et al. 2009. A small-molecule inhibitor of tgf-Beta signaling replaces sox2 in reprogramming by inducing nanog. Cell Stem Cell 5, 491–503.

Ieda, M., Fu, J. D., Delgado-Olguin, P., Vedantham, V., Hayashi, Y., Bruneau, B. G., and Srivastava, D. 2010. Direct reprogramming of fibroblasts into functional cardiomyocytes by defined factors. Cell 142, 375–386.

Ieda, M., Kanazawa, H., Kimura, K., Hattori, F., Ieda, Y., Taniguchi, M., Lee, J. K., Matsumura, K., Tomita, Y., Miyoshi, S. et al. 2007. Sema3a maintains normal heart rhythm through sympathetic innervation patterning. Nat. Med. 13, 604–612.

Ieda, M., Tsuchihashi, T., Ivey, K. N., Ross, R. S., Hong, T. T., Shaw, R. M., and Srivastava, D. 2009. Cardiac fibroblasts regulate myocardial proliferation through beta1 integrin signaling. Dev. Cell 16, 233–244.

Inoue, K., Wakao, H., Ogonuki, N., Miki, H., Seino, K., Nambu-Wakao, R., Noda, S., Miyoshi, H., Koseki, H., Taniguchi, M. et al. 2005. Generation of cloned mice by direct nuclear transfer from natural killer T cells. Curr. Biol. 15, 1114–1118.

Itzhaki, I., Maizels, L., Huber, I., Zwi-Dantsis, L., Caspi, O., Winterstern, A., Feldman, O., Gepstein, A., Arbel, G., Hammerman, H. et al. 2011. Modelling the long QT syndrome with induced pluripotent stem cells. Nature 471, 225–229.

Kamp, T. J. and Lyons, G. E. 2009. On the road to iPS cell cardiovascular applications. Circ. Res. 105, 617–619.

Kanazawa, H., Ieda, M., Kimura, K., Arai, T., Kawaguchi-Manabe, H., Matsuhashi, T., Endo, J., Sano, M., Kawakami, T., Kimura, T. et al. 2010. Heart failure causes cholinergic transdifferentiation of cardiac sympathetic nerves via gp130-signaling cytokines in rodents. J. Clin. Invest 120, 408–421.

Kang, L., Wang, J., Zhang, Y., Kou, Z., and Gao, S. 2009. iPS cells can support full-term development of tetraploid blastocyst-complemented embryos. Cell Stem Cell 5, 135–138.

Kattman, S. J., Huber, T. L., and Keller, G. M. 2006. Multipotent flk-1+ cardiovascular progenitor cells give rise to the cardiomyocyte, endothelial, and vascular smooth muscle lineages. Dev. Cell 11, 723–732.

Kim, J. B., Zaehres, H., Wu, G., Gentile, L., Ko, K., Sebastiano, V., Arauzo-Bravo, M. J., Ruau, D., Han, D. W., Zenke, M. et al. 2008. Pluripotent stem cells induced from adult neural stem cells by reprogramming with two factors. Nature 454, 646–650.

Kulessa, H., Frampton, J., and Graf, T. 1995. GATA-1 reprograms avian myelomonocytic cell lines into eosinophils, thromboblasts, and erythroblasts. Genes Dev. 9, 1250–1262.

Laflamme, M. A., Chen, K. Y., Naumova, A. V., Muskheli, V., Fugate, J. A., Dupras, S. K., Reinecke, H., Xu, C., Hassanipour, M., Police, S. et al. 2007. Cardiomyocytes derived from human embryonic stem cells in pro-survival factors enhance function of infarcted rat hearts. Nat. Biotechnol. 25, 1015–1024.

Laiosa, C. V., Stadtfeld, M., Xie, H., de Andres-Aguayo, L., and Graf, T. 2006. Reprogramming of committed T cell progenitors to macrophages and dendritic cells by C/EBP alpha and PU.1 transcription factors. Immunity 25, 731–744.

Lassar, A. B., Paterson, B. M., and Weintraub, H. 1986. Transfection of a DNA locus that mediates the conversion of 10T1/2 fibroblasts to myoblasts. Cell 47, 649–656.

Laugwitz, K. L., Moretti, A., Lam, J., Gruber, P., Chen, Y., Woodard, S., Lin, L. Z., Cai, C. L., Lu, M. M., Reth, M. et al. 2005. Postnatal isl1+ cardioblasts enter fully differentiated cardiomyocyte lineages. Nature 433, 647–653.

Ledent, V., Paquet, O., and Vervoort, M. 2002. Phylogenetic analysis of the human basic helix-loop-helix proteins. Genome Biol 3, RESEARCH0030.

Ledent, V. and Vervoort, M. 2001. The basic helix-loop-helix protein family: comparative genomics and phylogenetic analysis. Genome Res. 11, 754–770.

Li, B., Carey, M., and Workman, J. L. 2007. The role of chromatin during transcription. Cell 128, 707–719.

Li, J., Ishii, T., Feinstein, P., and Mombaerts, P. 2004. Odorant receptor gene choice is reset by nuclear transfer from mouse olfactory sensory neurons. Nature 428, 393–399.

Li, W., Wei, W., Zhu, S., Zhu, J., Shi, Y., Lin, T., Hao, E., Hayek, A., Deng, H., and Ding, S. 2009. Generation of rat and human induced pluripotent stem cells by combining genetic reprogramming and chemical inhibitors. Cell Stem Cell 4, 16–19.

Liu, H., Zhu, F., Yong, J., Zhang, P., Hou, P., Li, H., Jiang, W., Cai, J., Liu, M., Cui, K. et al. 2008. Generation of induced pluripotent stem cells from adult rhesus monkey fibroblasts. Cell Stem Cell 3, 587–590.

Maherali, N. and Hochedlinger, K. 2008. Guidelines and techniques for the generation of induced pluripotent stem cells. Cell Stem Cell 3, 595–605.

Maherali, N., Sridharan, R., Xie, W., Utikal, J., Eminli, S., Arnold, K., Stadtfeld, M., Yachechko, R., Tchieu, J., Jaenisch, R. et al. 2007. Directly reprogrammed fibroblasts show global epigenetic remodeling and widespread tissue contribution. Cell Stem Cell 1, 55–70.

Mitsui, K., Tokuzawa, Y., Itoh, H., Segawa, K., Murakami, M., Takahashi, K., Maruyama, M., Maeda, M. and Yamanaka, S. 2003. The homeoprotein Nanog is required for maintenance of pluripotency in mouse epiblast and ES cells. Cell 113, 631–642.

Miura, K., Okada, Y., Aoi, T., Okada, A., Takahashi, K., Okita, K., Nakagawa, M., Koyanagi, M., Tanabe, K., Ohnuki, M. et al. 2009. Variation in the safety of induced pluripotent stem cell lines. Nat. Biotechnol. 27, 743–745.

Molkentin, J. D. and Olson, E. N. 1996. Defining the regulatory networks for muscle development. Curr. Opin. Genet. Dev. 6, 445–453.

Moretti, A., Bellin, M., Welling, A., Jung, C. B., Lam, J. T., Bott-Flugel, L., Dorn, T., Goedel, A., Hohnke, C., Hofmann, F. et al. 2010. Patient-specific induced pluripotent stem-cell models for long-QT syndrome. N. Engl. J. Med. 363, 1397–1409.

Morin, S., Charron, F., Robitaille, L., and Nemer, M. 2000. GATA-dependent recruitment of MEF2 proteins to target promoters. Embo. J. 19, 2046–2055.

Mummery, C., Ward, D., van den Brink, C. E., Bird, S. D., Doevendans, P. A., Opthof, T., Brutel de la Riviere, A., Tertoolen, L., van der Heyden, M., and Pera, M. 2002. Cardiomyocyte differentiation of mouse and human embryonic stem cells. J. Anat. 200, 233–242.

Nelson, T. J. and Terzic, A. 2011. Induced pluripotent stem cells: an emerging theranostics platform. Clin. Pharmacol. Ther. 89, 648–650.

Nussbaum, J., Minami, E., Laflamme, M. A., Virag, J. A., Ware, C. B., Masino, A., Muskheli, V., Pabon, L., Reinecke, H., and Murry, C. E. 2007. Transplantation of undifferentiated

murine embryonic stem cells in the heart: teratoma formation and immune response. Faseb J. 21, 1345–1357.
Nutt, S. L., Heavey, B., Rolink, A. G., and Busslinger, M. 1999. Commitment to the B-lymphoid lineage depends on the transcription factor Pax5. Nature 401, 556–562.
Okita, K., Ichisaka, T., and Yamanaka, S. 2007. Generation of germline-competent induced pluripotent stem cells. Nature 448, 313–317.
Olson, E. N. 2006. Gene regulatory networks in the evolution and development of the heart. Science 313, 1922–1927.
Pang, Z. P., Yang, N., Vierbuchen, T., Ostermeier, A., Fuentes, D. R., Yang, T. Q., Citri, A., Sebastiano, V., Marro, S., Sudhof, T. C. et al. 2011. Induction of human neuronal cells by defined transcription factors. Nature 476, 220–223.
Park, I. H., Arora, N., Huo, H., Maherali, N., Ahfeldt, T., Shimamura, A., Lensch, M. W., Cowan, C., Hochedlinger, K., and Daley, G. Q. 2008a. Disease-specific induced pluripotent stem cells. Cell 134, 877–886.
Park, I. H., Lerou, P. H., Zhao, R., Huo, H., and Daley, G. Q. 2008b. Generation of human-induced pluripotent stem cells. Nat. Protoc. 3, 1180–1186.
Perry, R. L., Parker, M. H., and Rudnicki, M. A. 2001. Activated MEK1 binds the nuclear MyoD transcriptional complex to repress transactivation. Mol. Cell 8, 291–301.
Pownall, M. E., Gustafsson, M. K., and Emerson, C. P., Jr. 2002. Myogenic regulatory factors and the specification of muscle progenitors in vertebrate embryos. Annu. Rev. Cell Dev. Biol. 18, 747–783.
Rolink, A. G., Nutt, S. L., Melchers, F., and Busslinger, M. 1999. Long-term in vivo reconstitution of T-cell development by Pax5-deficient B-cell progenitors. Nature 401, 603–606.
Seki, T., Yuasa, S., Oda, M., Egashira, T., Yae, K., Kusumoto, D., Nakata, H., Tohyama, S., Hashimoto, H., Kodaira, M. et al. 2010. Generation of induced pluripotent stem cells from human terminally differentiated circulating T cells. Cell Stem Cell 7, 11–14.
Sekiya, S., and Suzuki, A. 2011. Direct conversion of mouse fibroblasts to hepatocyte-like cells by defined factors. Nature 475, 390–393.
Srinivas, S., Watanabe, T., Lin, C. S., William, C. M., Tanabe, Y., Jessell, T. M., and Costantini, F. 2001. Cre reporter strains produced by targeted insertion of EYFP and ECFP into the ROSA26 locus. BMC Dev. Biol. 1, 4.
Srivastava, D. 2006. Making or breaking the heart: from lineage determination to morphogenesis. Cell 126, 1037–1048.
Stadtfeld, M., Maherali, N., Breault, D. T., and Hochedlinger, K. 2008. Defining molecular cornerstones during fibroblast to iPS cell reprogramming in mouse. Cell Stem Cell 2, 230–240.
Szabo, E., Rampalli, S., Risueno, R. M., Schnerch, A., Mitchell, R., Fiebig-Comyn, A., Levadoux-Martin, M., and Bhatia, M. 2010. Direct conversion of human fibroblasts to multilineage blood progenitors. Nature 468, 521–526.
Takahashi, K., Tanabe, K., Ohnuki, M., Narita, M., Ichisaka, T., Tomoda, K., and Yamanaka, S. 2007. Induction of pluripotent stem cells from adult human fibroblasts by defined factors. Cell 131, 861–872.
Takahashi, K. and Yamanaka, S. 2006. Induction of pluripotent stem cells from mouse embryonic and adult fibroblast cultures by defined factors. Cell 126, 663–676.
Takeuchi, J. K. and Bruneau, B. G. 2009. Directed transdifferentiation of mouse mesoderm to heart tissue by defined factors. Nature 459, 708–711.
Tamashiro, K. L., Wakayama, T., Akutsu, H., Yamazaki, Y., Lachey, J. L., Wortman, M. D., Seeley, R. J., D'Alessio, D. A., Woods, S. C., Yanagimachi, R. et al. 2002. Cloned mice have an obese phenotype not transmitted to their offspring. Nat. Med. 8, 262–267.
Tapscott, S. J. 2005. The circuitry of a master switch: Myod and the regulation of skeletal muscle gene transcription. Development 132, 2685–2695.

Taylor, D. O., Edwards, L. B., Aurora, P., Christie, J. D., Dobbels, F., Kirk, R., Rahmel, A. O., Kucheryavaya, A. Y., and Hertz, M. I. 2008. Registry of the International Society for Heart and Lung Transplantation: twenty-fifth official adult heart transplant report 2008. J. Heart Lung Transplant *27*, 943–956.

Taylor, S. M. and Jones, P. A. 1979. Multiple new phenotypes induced in 10T1/2 and 3T3 cells treated with 5-azacytidine. Cell 17, 771–779.

Tiscornia, G., Monserrat, N., and Belmonte, J. C. 2011. Modelling Long QT Syndrome With iPS Cells: Be Still, My Beating Heart. Circ. Res. 108, 648–649.

van Laake, L. W., Passier, R., Doevendans, P. A., and Mummery, C. L. 2008. Human embryonic stem cell-derived cardiomyocytes and cardiac repair in rodents. Circ. Res. 102, 1008–1010.

Vierbuchen, T., Ostermeier, A., Pang, Z. P., Kokubu, Y., Sudhof, T. C., and Wernig, M. 2010. Direct conversion of fibroblasts to functional neurons by defined factors. Nature 463, 1035–1041.

Visvader, J. E., Elefanty, A. G., Strasser, A., and Adams, J. M. 1992. GATA-1 but not SCL induces megakaryocytic differentiation in an early myeloid line. Embo. J. 11, 4557–4564.

Weber, K. T. and Brilla, C. G. 1991. Pathological hypertrophy and cardiac interstitium. Fibrosis and renin-angiotensin-aldosterone system. Circulation 83, 1849–1865.

Weintraub, H., Tapscott, S. J., Davis, R. L., Thayer, M. J., Adam, M. A., Lassar, A. B., and Miller, A. D. 1989. Activation of muscle-specific genes in pigment, nerve, fat, liver, and fibroblast cell lines by forced expression of MyoD. Proc. Natl. Acad. Sci. USA 86, 5434–5438.

Wernig, M., Meissner, A., Foreman, R., Brambrink, T., Ku, M., Hochedlinger, K., Bernstein, B. E. and Jaenisch, R. 2007. *In vitro* reprogramming of fibroblasts into a pluripotent ES-cell-like state. Nature 448, 318–324.

Wilmut, I., Schnieke, A. E., McWhir, J., Kind, A. J., and Campbell, K. H. 1997. Viable offspring derived from fetal and adult mammalian cells. Nature 385, 810–813.

Wu, S. M. and Hochedlinger, K. 2011. Harnessing the potential of induced pluripotent stem cells for regenerative medicine. Nat. Cell Biol. 13, 734.

Xie, H., Ye, M., Feng, R., and Graf, T. 2004. Stepwise reprogramming of B cells into macrophages. Cell 117, 663–676.

Yamashita, T., Kawai, H., Tian, F., Ohta, Y., and Abe, K. 2010. Tumorigenic Development of Induced Pluripotent Stem Cells in Ischemic Mouse Brain. Cell Transplant.

Yazawa, M., Hsueh, B., Jia, X., Pasca, A. M., Bernstein, J. A., Hallmayer, J., and Dolmetsch, R. E. 2011. Using induced pluripotent stem cells to investigate cardiac phenotypes in Timothy syndrome. Nature 471, 230–234.

Yoshida, Y. and Yamanaka, S. 2010. Recent stem cell advances: induced pluripotent stem cells for disease modeling and stem cell-based regeneration. Circulation 122, 80–87.

Yu, J., Vodyanik, M. A., Smuga-Otto, K., Antosiewicz-Bourget, J., Frane, J. L., Tian, S., Nie, J., Jonsdottir, G. A., Ruotti, V., Stewart, R. et al. 2007. Induced pluripotent stem cell lines derived from human somatic cells. Science 318, 1917–1920.

Zaret, K. S. 2008. Genetic programming of liver and pancreas progenitors: lessons for stem-cell differentiation. Nat. Rev. Genet. 9, 329–340.

Zhang, J., Wilson, G. F., Soerens, A. G., Koonce, C. H., Yu, J., Palecek, S. P., Thomson, J. A., and Kamp, T. J. 2009. Functional cardiomyocytes derived from human induced pluripotent stem cells. Circ. Res. 104, e30–41.

Zhang, M., Methot, D., Poppa, V., Fujio, Y., Walsh, K., and Murry, C. E. 2001. Cardiomyocyte grafting for cardiac repair: graft cell death and anti-death strategies. J. Mol. Cell Cardiol. 33, 907–921.

Zhang, P., Iwasaki-Arai, J., Iwasaki, H., Fenyus, M. L., Dayaram, T., Owens, B. M., Shigematsu, H., Levantini, E., Huettner, C. S., Lekstrom-Himes, J. A. et al. 2004. Enhancement of hematopoietic stem cell repopulating capacity and self-renewal in the absence of the transcription factor C/EBP alpha. Immunity 21, 853–863.

Zhao, R., Watt, A. J., Battle, M. A., Li, J., Bondow, B. J., and Duncan, S. A. 2008. Loss of both GATA4 and GATA6 blocks cardiac myocyte differentiation and results in acardia in mice. Dev. Biol. 317, 614–619.

Zhao, X. Y., Li, W., Lv, Z., Liu, L., Tong, M., Hai, T., Hao, J., Guo, C. L., Ma, Q. W., Wang, L. et al. 2009. iPS cells produce viable mice through tetraploid complementation. Nature 461, 86–90.

Zhou, Q., Brown, J., Kanarek, A., Rajagopal, J., and Melton, D. A. 2008. *In vivo* reprogramming of adult pancreatic exocrine cells to beta-cells. Nature 455, 627–632.

Zhou, Q., Law, A. C., Rajagopal, J., Anderson, W. J., Gray, P. A., and Melton, D. A. 2007. A multipotent progenitor domain guides pancreatic organogenesis. Dev Cell 13, 103–114.

CHAPTER 16

Modeling Arrhythmogenic Heart Disease with Patient-Specific Induced Pluripotent Stem Cells

Daniel Sinnecker,* Alexander Goedel,[a] Ralf J. Dirschinger,[b] Alessandra Moretti[c] and Karl-Ludwig Laugwitz[d]

Modeling Cardiovascular Disease with Patient-specific Induced Pluripotent Stem Cells: An Emerging Field?

The reprogramming of somatic cells to induced pluripotent stem cells (iPSCs) by expression of a specific cocktail of transcription factors (Takahashi and Yamanaka 2006; Takahashi et al. 2007) offers the opportunity to generate patient-specific iPSC lines from somatic cells of patients affected by genetically-caused diseases. Since the patient-derived iPSC lines are genetically identical to the patient, they can be used to model heritable diseases by differentiating them to the cell types that are normally affected by the disease. These cells can then be used to study the pathomechanism of the disease *in vitro*. This concept appears especially beneficial for disciplines like cardiovascular biology, where samples of human tissue for research purposes are not easily obtained, often requiring invasive procedures like myocardial biopsy.

Klinikum rechts der Isar—Technische Universität München, I. Medizinische Klinik—Kardiologie, Ismaninger Strasse 22, 81675 München, Germany.
[a]Email: alexander.goedel@gmx.de
[b]Email: ralf.dirschinger@gmx.de
[c]Email: amoretti@med1.med.tum.de
[d]Email: klaugwitz@med1.med.tu-muenchen.de
*Corresponding author: sinnecker@tum.de

In principle, all types of chromosomally inherited diseases, ranging from monogenetic disorders to complex diseases which arise from the interplay between environmental, lifestyle and genetic factors can be studied with this approach. However, presumably because of the inherent simplicity, the early applications of this emerging technology have been models of monogenetic disorders. The first diseases successfully modeled with human iPSC technology were neurodegenerative (Ebert et al. 2009; Lee et al. 2009) and hematological (Ye et al. 2009) disorders.

The potential for cardiovascular biology was soon recognized. Due to the nature of the technology, cell-autonomous diseases, where the disease phenotype can be recapitulated by analyzing single cells affected by the disease-causing mutation, will give the clearest results. Consistently, several groups have first concentrated on modeling channelopathies, diseases of cardiac ion channels that display a paradigmatic cell-autonomous pathophysiology. The first reports of successful modeling of cardiovascular diseases using iPSC technology concerned different forms of the long-QT syndrome (Moretti et al. 2010b; Itzhaki et al. 2011; Yazawa et al. 2011; Matsa et al. 2011), a group of channelopathies with the common clinical presentation of a prolonged QT interval in the ECG and an increased susceptibility to ventricular arrhythmias.

Long-QT Syndrome as a Paradigm for Disease Modeling with Induced Pluripotent Stem Cells

The decision of several groups to use the new methodology of patient-specific stem cells to model the long-QT syndrome was likely based on the following aspects of the disease that make it a promising candidate for iPSC-based disease modeling: First, the long-QT syndrome is quite a common disorder. The incidence of the genetically caused congenital forms was classically estimated to range from 1:20 000 to 1:5000, while a recent report based on ECG screening of neonates suggests an even higher prevalence of 1:2000 (Schwartz et al. 2009). The acquired forms, typically unwanted side effects of pharmacotherapy, are even more frequent and represent a common problem in daily clinical practice. Second, based on what is known to date, the disease phenotype of congenital long-QT syndromes develops in a paradigmatically cell-autonomous manner caused by aberrations in the action potentials of single cardiomyocytes. Third, the molecules affected by the disease-causing mutations are plasma membrane ion channels that are easily accessible to physiological investigations, even (by means of patch clamp recordings) at the single-molecule level. Finally, despite invaluable insights gained by animal models, electrophysiological differences between human and non-human cardiomyocytes make disease models necessary that better represent the electrophysiology of a human heart. Thus, a new class of disease models based on human iPSC was expected to add new insights into the pathophysiology of these disorders.

Long-QT Syndrome: A Disorder of Cardiac Repolarization

The cardiac action potential and the importance of repolarization

The contraction of the heart is governed by electrical currents across the plasma membrane that lead to cellular de- and repolarizations which regulate the action of the contractile apparatus. The temporal pattern of membrane potential changes in a single myocyte during a heartbeat is called the action potential. The cardiac action potential is generated by the concerted action of a great number of plasma membrane ion channels, leading to depolarizing or repolarizing currents that shape the different phases of the action potential (Figs. 1A, B). To provide an ordered contraction of the whole organ, it is necessary that several characteristics of the action potential differ between myocytes from different anatomical regions of the heart. Furthermore, a strict regulation is required to ensure that, for example, the action potential becomes shorter under conditions of an increased heart rate. This is achieved by an adrenergically mediated upregulation of repolarizing potassium currents that leads to a shortening of the plateau phase of the action potential.

The common feature of the different types of the long-QT syndrome is that, at least under specific conditions, the plateau phase of the cardiac action potential is prolonged, leading to a prolonged QT interval in the surface ECG. This clinically apparent feature is linked to an increased susceptibility to life-threatening ventricular arrhythmias like ventricular tachycardia or ventricular fibrillation. The typical arrhythmia observed in patients with long-QT syndrome was described first by François Dessertenne in 1966 and termed "Torsade de pointes" (Dessertenne 1966). This terminus, which may be translated as "twisting spikes", was chosen to describe the nature of this polymorphic ventricular tachycardia, which is characterized by a progressively changing amplitude and shape of the QRS complex that gives the impression of the electrical axis rotating around the isoelectric line. The precise mechanism by which a prolonged cardiac action potential results in this type of polymorphic tachycardia is not completely understood. However, a large body of evidence indicates that two basic mechanisms contribute to arrhythmogenesis in individuals affected by the long-QT syndrome (Eckardt et al. 1998; Antzelevitch 2005). The first mechanism is the frequent development of early afterdepolarizations (EADs) under conditions of prolonged action potential duration. EADs are depolarizations that occur during phase 2 and 3 of the cardiac action potential. These depolarizations can give rise to premature action potentials or even series

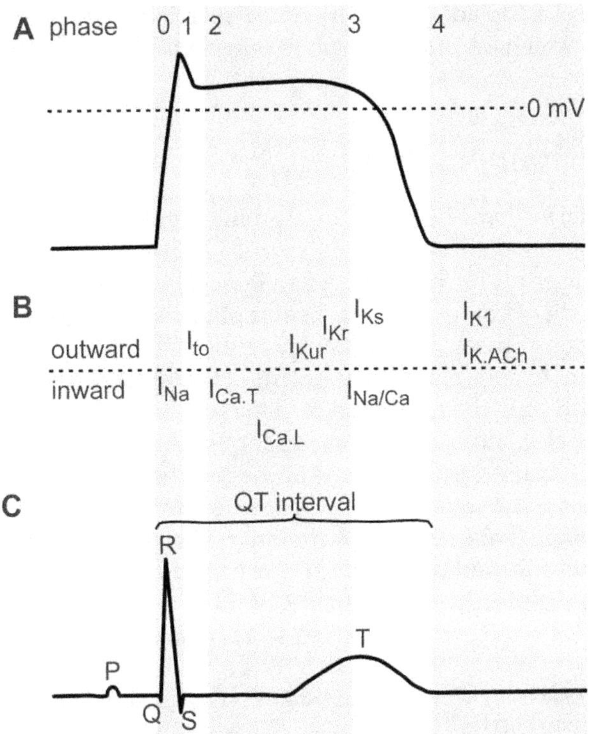

Figure 1. The Cardiac Action Potential, the Underlying Ionic Currents and the Correlation with the Surface ECG. Panel (A) shows the schematic representation of a ventricular action potential. The different phases 0-4 are annotated. Panel (B) represents the temporal pattern of the occurrence of the most important ionic currents that shape the action potential. Net outward currents are placed above, net inward currents below the dotted line. Panel (C) shows a representation of a surface ECG in temporal correlation with the ventricular action potential. The names of the peaks (P, Q, R, S, T) as well as the definition of the QT interval are shown. Abbreviations: I_{Na}, sodium current. I_{to}, transient outward current. $I_{Ca.T}$, low-voltage-activated, transient calcium current. $I_{Ca.L}$, high-voltage-activated, long-lasting calcium current. I_{Kur}, ultra-rapid component of the delayed rectifying potassium current (I_K). I_{Ks}, slow component of the of I_K. I_{Kr}, rapid component of I_K. $I_{Na/Ca}$, net inward current through the electrogenic sodium/calcium exchanger. I_{K1}, inward rectifier potassium current. I_{KAch}, acetylcholine-sensitive potassium current. Based on Ackerman and Clapham, 2004.

of action potentials, which are then called triggered activity. The second mechanism that contributes to the generation of Torsades de pointes is the so-called dispersion of repolarization. This terminus means that across the myocardium, the action potential duration is longer in the mid-myocardial cells (M cells) than in the sub-epicardial or sub-endocardial myocytes. The M cells are particularly sensitive to a prolongation of the action potential duration in response to physiological stimuli or drugs. Thus, in patients affected by the long-QT syndrome, conditions that lead to a prolongation of the QT interval also lead to an increased dispersion of repolarization between M cells and adjacent cells. This dispersion of repolarization is the substrate on which EADs can trigger re-entrant tachycardias (like Torsades de pointes).

Congenital long-QT-syndrome

The congenital forms of the long-QT syndrome have been classified according to different affected chromosomal loci into (at the time of the preparation of this manuscript) at least 13 types (Table 1). Generally speaking, the congenital long-QT syndrome is a disease of cardiac ion channels involved in the shaping of the cardiac action potential. The genes mutated in the different forms encode either components of the ion channels themselves or proteins involved in the regulation or in the sub-cellular trafficking of these ion channels (see Table 1). For several forms like the long-QT syndrome type 1 (LQT1, caused by mutations of the ion channel sub-unit KCNQ1), the mechanisms of several disease-causing mutations have been more precisely defined. For example, the mutation M520R impairs the trafficking of the channel sub-unit by impairing the transport from the endoplasmic reticulum to the plasma membrane (Schmitt et al. 2007). Accordingly, the phenotype of affected heterozygous patients is caused by haploinsufficiency,

Table 1. Genetically-defined long-QT syndromes. Data compiled from Lehnart et al. 2007, Wu et al. 2008 and Yang et al. 2010.

Syndrome	Gene	Gene Product (Synonyms)	Function of Gene Product
LQT1	*KCNQ1*	K_v7.1α (KvLQT1)	α-subunit of channel responsible for I_{Ks}
LQT2	*KCNH2*	K_v11.1 (HERG)	α-subunit of channel responsible for I_{Kr}
LQT3	*SCN5A*	Na_v1.5α	α-subunit of channel responsible for I_{Na}
LQT4	*ANK2*	Ankyrin-B	adaptor protein involved in membrane targeting of sodium/potassium-ATPase and sodium/calcium-exchanger
LQT5	*KCNE1*	MinK β	β-subunit of channel responsible for I_{Ks}
LQT6	*KCNE2*	MiRP1 β	β-subunit of channel responsible for I_{Kr}
LQT7	*KCNJ2*	Kir2.1 α	α-subunit of channel responsible for I_{K1}
LQT8	*CACNA1C*	$Ca_v1.2\ a_{1c}$	α-subunit of channel reponsible for $I_{Ca,L}$
LQT9	*CAV3*	Caveolin-3	membrane protein presumably binding to sodium channels
LQT10	*SCN4B*	Na_v1.5 β4	β-subunit of channel responsible for I_{Na}
LQT11	*AKAP9*	Yotiao (AKAP9)	A-kinase anchoring protein mediating regulation of I_{Ks} by protein kinase A
LQT12	*SNTA1*	syntrophin-α_1	cytoskeletal protein interacting with channel responsible for I_{Na}
LQT13	*KCNJ5*	Kir3.4 (GIRK4)	subunit of channel responsible for I_{KACh}

since only the channel sub-units transcribed from the non-mutated allele reach the plasma membrane. In contrast, the mutation P320H, in which the pore region of the channel is mutated, exerts a dominant-negative effect on co-expressed wild type channel sub-units (Thomas et al. 2010). This presumably happens because mutated channel sub-units interact with wild type sub-units to form channels that are infunctional because of a disruption of the normal architecture of the pore region.

Acquired long-QT-syndrome: a major problem in drug development

More common than the monogenetically caused congenital long-QT syndromes is the so-called "acquired" long-QT syndrome. Several conditions, such as heart failure, electrolyte disturbances, thyroid disorders or, most importantly, treatment with specific drugs, lead to a prolonged QT interval in the ECG, indicative of prolonged single-myocyte action potentials. Similar to congenital long-QT syndrome, this condition leads to an increased susceptibility to potentially fatal ventricular arrhythmias. Although the precise pathophysiology of acquired long-QT syndrome is far from being completely understood, it has been demonstrated that most drugs known to cause acquired long-QT syndrome are inhibitors of the HERG potassium channel, which is responsible for a repolarizing potassium current occurring during the repolarization phase of the cardiac action potential, called the rapid component of the delayed rectifier potassium current (I_{kr}) (Sanguinetti et al. 1995).

Drug-induced QT interval prolongation is a major and increasingly recognized problem of drug safety, which is reflected by the observation that it was the most important cause for restrictions of use or the withdrawal of drugs from the US market in the recent time (Lasser et al. 2002). It is recognized that a drug inducing QT interval prolongation in some or all patients might increase the mortality risk of these patients by increasing the likelihood of fatal ventricular arrhythmias. However small this risk may be, it can be only tolerated if the drug is needed to treat a serious condition and if no safer alternative exists. Intriguingly, the problem of drug-induced QT interval prolongation is not limited to drugs or drug candidates intended for cardiac use, but it is also a highly relevant problem of substances intended for non-cardiac use. For the companies involved in drug development, this poses the risk of tremendous financial losses, since the investments during the early phases of drug development are lost if the drug candidate turns out to lead to acquired long-QT syndrome. Therefore, and also as a requirement imposed by the regulatory agencies of many countries, several pre-clinical test systems are regularly used before the drug candidates are used in clinical trials (Giorgi et al. 2010). Since most of these tests so far rely on surrogate parameters like the extent of I_{Kr} blockade induced by the drug candidate, iPSC-based test systems that allow the evaluation of the drug effect on human cells would be highly relevant tools for drug development.

The Need for Models of Long-QT Syndrome

As outlined above, both the congenital long-QT syndrome and the acquired long-QT syndrome are clinically relevant disease entities. A better understanding of the pathophysiology of these disorders might provide a basis for the development of new specific therapies (in the case of the congenital long-QT syndrome) as well as a means to circumvent the problem of acquired long-QT syndrome in the development of new drugs. However, since the available methods to directly study the disease in affected human individuals are limited, disease models are fundamental for such kind of research. However, the iPSC-based disease models discussed in this article are only a complement to the already existing pharmacologically- and genetically-defined animal models.

Pharmacological models

Freshly isolated cardiomyocytes from small animals, e.g., neonatal rat cardiomyocytes or adult rabbit cardiomyocytes, have been traditionally used as standard tools by cellular electrophysiologists. The identification of the proteins encoded by the genes mutated in congenital long-QT syndromes (see Table 1), together with the availability of specific pharmacological blockers of the involved ion channels, has led to the development of pharmacological models of long-QT syndromes. For example, the suppression of the I_{Kr} current in isolated rabbit ventricular cardiomyocytes by use of the channel blocker E-4031 leads to a prolongation of the action potential plateau phase and to the occurrence of early afterdepolarizations (Studenik et al. 2001). However, the findings of such experiments cannot always be easily translated to genetic diseases in humans. The properties of an ion channel blocked by a drug (which might bind in a voltage-dependent manner) can very well differ from those of the same ion channel altered by a disease-causing mutation. Moreover, the specificity of pharmacological channel blockers is often questionable.

Genetically defined models

Modification of the mouse genome, either by a transgenic approach or by gene targeting, has proven efficient to generate several model systems for the long-QT syndrome and to confirm the relation between specific ion channel-encoding genes and the corresponding cardiac ionic currents (Salama and London 2007). However, differences in the physiology between murine and human cardiomyocytes sometimes hamper the interpretation of results from these studies.

For example, several mouse lines in which the KCNQ1 (alias KVLQT) gene has been targeted have been developed to model the long-QT syndrome type 1 (LQT1), which is caused by mutations in this gene. KCNQ1 encodes the alpha sub-unit of an ion channel that constitutes the slow component of the delayed rectifier potassium current (I_{Ks}). The functional ion channel is composed of four KCNQ1 sub-units and at least one regulatory KCNE1 sub-unit. A homozygous targeted disruption of the murine KCNQ1 gene (Lee et al. 2000) led to deafness and anatomical disruption of the cochlear and vestibular organs, but not to electrocardiographic abnormalities. In contrast, a similar KCNQ1 knockout mouse line generated by targeting the same exon (Casimiro et al. 2001) displayed a phenotype characterized not only by deafness and a disruption of the inner ear organs, but also by abnormal T- and P-waves and a prolonged QT interval in the ECG measured *in vivo*. However, this electrophysiological phenotype could not be reproduced when isolated hearts from these mice were investigated. A different approach to a mouse model for LQT1 was to generate a transgenic mouse line characterized by overexpression of a dominant-negative KCNQ1 sub-unit (the human KCNQ1 isoform 2) under the control of a cardiac-specific α-myosin heavy chain promoter (Demolombe et al. 2001). The KCNQ1 isoform 2 is an alternatively spliced and thus truncated KCNQ1 gene product that has been demonstrated to exert a dominant-negative effect on KCNQ1-mediated currents (Demolombe et al. 1998). Transgenic mouse lines overexpressing this dominant-negative KCNQ1 sub-unit in the heart displayed a cardiac phenotype characterized by a markedly prolonged QT interval, but also other features not typically present in human LQT1 patients, such as sinus node dysfunction and atrio-ventricular block (Demolombe et al. 2001).

Why do we need human cell-based models?

Although the above-mentioned models, especially the genetic mouse models, have provided invaluable insights into the pathophysiology of long-QT syndrome, several problems preclude the generalization of these results to human physiology. Obviously, mice and men differ in several aspects, but these differences become especially evident upon looking at disorders of cardiac repolarization, like the long-QT syndrome. The heart rate of a mouse is about ten times faster than that of a human, requiring the cardiac action potential to be much shorter. Accordingly, major differences exist in the shape of the cardiac action potential and in the underlying ionic currents between murine and human cardiomyocytes (London 2001; Nerbonne et al. 2001). The cardiac action potential of a mouse is triangular and lacks the plateau phase that is characteristic of a human ventricular action potential. The major repolarizing currents in the mouse are not the slow and fast components of the delayed rectifier potassium current (I_{kr} and I_{ks}) as in the human, but the transient outward potassium currents and the rapidly activating, slowly inactivating delayed rectifier currents ($I_{to,f}$, $I_{to,s}$, $I_{K,slow1}$ and $I_{K,slow2}$). Finally, fatal ventricular arrhythmias are not a common phenomenon in mice. Taken together, these interspecies differences make disease models based on human cells a desirable goal.

Induced Pluripotent Stem Cell-based Models for Long-QT Syndrome: Current State of the Art

Models developed so far

At the time of the preparation of this manuscript, four iPSC-based model systems for the long-QT syndrome have been published: while our group has focused on LQT1 (Moretti et al. 2010b), other groups were successful in modeling LQT2 (Itzhaki et al. 2011; Matsa et al. 2011) as well as the Timothy syndrome (a disease characterized by long-QT syndrome (LQT8), syndactyly, immune deficiency and autism) (Yazawa et al. 2011). The approaches used by the different groups were similar in many aspects. In the following section describing practical aspects of these models, we will, thus, focus on our LQT1 model (Moretti et al. 2010b) and mention relevant distinct aspects of the work of the other groups where appropriate.

Generation of patient-specific and control iPSC lines

In the seminal paper describing for the first time the reprogramming of somatic cells to iPSC (Takahashi and Yamanaka 2006), murine embryonic fibroblasts and fibroblasts isolated from the tail tips of adult mice were used as the source of cells for reprogramming. In the first report of the generation of human iPSC from the same group (Takahashi et al. 2007), fibroblasts isolated from skin were used. Accordingly, when this technology was adopted by other groups and exploited to generate patient-specific iPSC lines, most investigators chose to stick to skin fibroblasts and acquired skin biopsies from the patients or probands to isolate those cells. A simple isolation procedure consists of mincing the skin biopsy, placing the minced tissue on a culture dish, and waiting for fibroblasts to migrate out of these explants. These fibroblasts are then harvested and passaged before they are ready for reprogramming.

A skin biopsy is a small, not very invasive procedure that can be performed under local anesthesia. However, it would be even easier to recruit patients or probands for reprogramming if an even less invasive procedure would suffice to harvest the cells needed for reprogramming. The reprogramming of keratinocytes from plucked hair follicles (Novak et al. 2010) or the reprogramming of T-lymphocytes isolated from blood (Brown et al. 2010) are potential alternatives to achieve this goal.

The classical "cocktail" of reprogramming factors—Oct3/4, Sox2, c-Myc and Klf-4—as well as the original method to deliver these factors into the cells—by use of retroviral gene transfer—were soon modified by several groups (reviewed recently in Sidhu 2011). These factor-free reprogramming approaches are extremely valuable for future applications in human cell therapy, where it would be essential to use cells that neither overexpress the known oncogene, c-Myc, nor have retroviruses randomly integrated into their genome, potentially leading to the activation of endogenous oncogenes. However, to use the cells for *in vitro* disease modeling, such considerations are not necessary, and, thus, most groups still rely on retroviral transfer of the classical four reprogramming factors.

To analyze the phenotype of patient-specific iPSC-derived cardiomyocytes, the observations in the patient-specific cells must be compared with observations in control cardiomyocytes, which should be iPSC-derived cells generated by the same differentiation protocol. However, the ideal source for these control iPSC is still a matter of debate. Up to date, the usual practice is to generate control iPSC lines from healthy probands unrelated to the patient who are unaffected by the disease under study (Moretti et al. 2010b; Itzhaki et al. 2011; Yazawa et al. 2011). However, this approach bears the risk that differences between patient and control cardiomyocytes might arise from confounding genetic factors unrelated to the disease. One way to circumvent this problem would be to rely on not just one control iPSC line but on a panel of control lines derived from genetically diverse subjects. The major drawback of this approach is the increased labor needed to generate and maintain the cell lines and to perform multiple control experiments with cells derived from the different lines, which increases cost and may not be feasible for the majority of laboratories. Another way to limit the genetic variance between patient and control cells is the derivation of control iPSC lines from a close relative of the patient who is not affected by the disease. Finally, when a monogenetic disease is investigated, the ultimate control iPSC line could be constructed by correcting the disease-causing mutation in the patient iPSC by a gene-targeting approach. Furthermore, an experiment designed this way could unequivocally prove that the mutation under consideration—and nothing else—is responsible for the phenotypic differences between patient and control cells. While the huge effort in terms of money, labor, and time needed for gene repair in human iPSC has prevented these types of controls from becoming a standard at this point, advances in gene targeting of human iPSC may make this kind of control more feasible in the future.

Differentiation of the iPSC towards a cardiomyocyte phenotype

Several protocols have been developed to direct the differentiation of iPSC towards cardiomyocytes. The yield of cardiomyocytes achieved by the different protocols varies significantly. For the purpose of modeling arrhythmogenic heart disease, the readout often consists of physiological measurements in single cells, such as recordings of the membrane potential using the whole cell patch clamp technique or calcium measurements using fluorescent indicator dyes. For this type of measurement, it is not necessary for the differentiation protocol to lead to the generation of a pure preparation of cardiomyocytes, as long as the target cells can be identified in advance. For cardiomyocytes, the identification can be based simply on contractions evoked by electrical pacing. Thus, the yield of cardiomyocytes that a differentiation protocol can provide is not as important as its robustness and reproducibility for disease modeling purposes.

The differentiation protocols used for the first iPSC-based models for arrhythmogenic heart disease (Moretti et al. 2010b; Matsa et al. 2011; Itzhaki et al. 2011; Yazawa et al. 2011) are all based on the differentiation of the cells in "embryoid bodies" (EBs), which form spontaneously when undifferentiated iPSCs aggregate under appropriate culture conditions (Keller 1995). In these three-dimensional structures, spontaneous differentiation into lineages representing all three germ layers starts, eventually leading to differentiation of some cells to cardiomyocytes. To promote cardiac differentiation, supplements such as ascorbic acid (Takahashi et al. 2003) or Wnt3a (Tran et al. 2009) among many others have been added to the cell culture medium. The areas of the EBs in which the cells differentiate into cardiomyocytes can be easily identified by spontaneous contractions in these areas that appear usually after 10–20 days of differentiation. These spontaneously contracting areas can be dissected manually from the rest of the EBs and cultured further to allow additional maturation of the cells. To perform experiments that require single cells, the cells can be dissociated by collagenase digestion. It should be taken into consideration that cardiomyocytes without pacemaker activity could be missed by this approach. However, by applying external field stimulation, cardiomyocytes that contract in response to pacing can be selected for further experiments.

In Fig. 2A, immunofluorescence stainings of human iPSC-derived cardiomyocytes are shown. The cardiomyocytes can be identified by the expression of the cardiac isoform of troponin T (cTnT) in ordered sarcomeric structures. Moreover, a sub-population of these cells also stains

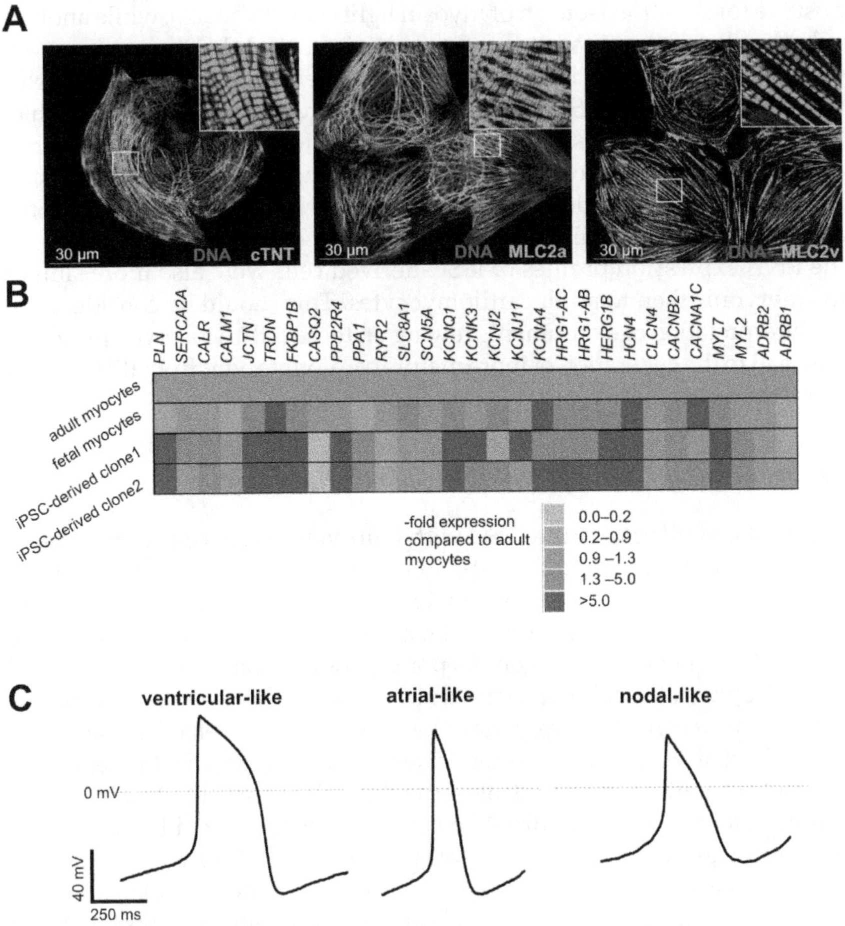

Figure 2. Cardiomyocytes Derived from Human Induced Pluripotent Stem Cells. Panel (A) shows immunofluorescence images of representative iPSC-derived cardiomyocytes stained for cardiac troponin T (cTnT) and the atrial (MLC2a) or the ventricular (MLC2v) isoform of myosin light chain 2 as indicated. The insets show higher magnifications of the areas indicted by the white squares. Panel (B) shows the result of a gene expression analysis using quantitative RT-PCR with primers specific for the indicated cardiomyocyte-specific genes in adult cardiomyocytes, fetal cardiomyocytes and cardiomyocytes generated from two different clones of iPSC. The expression levels are reported relative to the expression in adult myocytes. Panel (C) shows representative action potential recordings of ventricular-like, atrial-like and nodal-like iPSC-derived myocytes as indicated (upper panel) along with representative gels loaded with DNA samples generated by single-cell-PCR using specific primers for the indicated transcripts (lower panel). HCN4, hyperpolarization-activated, cyclic nucleotide-gated channel 4.

Color image of this figure appears in the color plate section at the end of the book.

positive for the atrial isoform of myosin light chain (MLC2a), while another sub-population expresses the ventricular isoform MLC2b (see Fig. 2A). Co-expression of both isoforms was also observed in a fraction of cells. Morphologically, the iPSC-derived cells are more similar to embryonic than to adult cardiomyocytes. To more closely investigate the degree of maturity of these cells, quantitative RT-PCR was performed to assess the expression levels of a panel of cardiomyocyte-specific markers in isolated embryonic and adult as well as in iPSC-derived cardiomyocytes. As shown in Fig. 2B, the RNA expression profiles of iPSC-derived cells were also more similar to embryonic than to adult cardiomyocytes. This should be considered in the interpretation of experiments relying on these cells. Moreover, protocols that lead to the generation of more mature cardiomyocytes from iPSC would be extremely valuable.

Different subgroups of iPSC-derived cardiomyocytes

Due to the similarity of the iPSC with embryonic stem cells (ESC), many of the above mentioned differentiation protocols have been adapted from protocols originally developed for ESC. Experiments with ESC-derived cardiomyocytes have shown that the cardiomyocytes generated by these protocols are not one homogenous population, but can be classified based on electrophysiological characteristics as well as on the expression of specific markers as ventricular-like, atrial-like and nodal-like cells (Maltsev et al. 1993; He et al. 2003). This was replicated in the cardiomyocytes generated from LQT1 patient- and control-derived iPSC (Moretti et al. 2010b). The action potential of ventricular-like cells was characterized by a negative maximum diastolic potential, a large "overshoot" (positive membrane potential reached by the initial depolarization) and a marked plateau phase (Fig. 2C). Single-cell RT-PCR analysis of cells that exhibited ventricular-like action potentials demonstrated that these cells expressed the ventricular isoform of myosin light chain 2 (MLC2v; see Fig. 2C). Atrial-like cells produced action potentials with a less-pronounced overshoot and without a marked plateau phase and expressed the atrial isoform of myosin light chain 2 (MLC2a). Action potentials of nodal-like cells differed from ventricular-like and atrial-like cells by displaying a less-negative maximum diastolic potential, a slower upstroke velocity and smaller amplitude. Cells displaying nodal-like action potentials were characterized in single-cell RT-PCR by the expression of the hyperpolarization-activated, cyclic nucleotide-gated channel 4 (HCN4; see Fig. 2C). The characteristic features of the distinct types of action potentials were reflected in objective parameters like the APD90/APD50 ratio (the duration from the beginning of the action potential until the repolarization is completed by 90%, divided by the duration until 50% repolarization), which were calculated for representative action potentials of each cell. Based on these values, the cells were characterized as ventricular-like, atrial-like or nodal-like, and further analysis was carried out separately for these different cell populations.

Recapitulation of the disease phenotype in iPSC-derived cardiomyocytes

To model LQT1 with patient-specific iPSC (Moretti et al. 2010b), dermal fibroblasts from two patients (father and son) who were clinically apparent with a prolonged QT interval and proven by genomic sequencing to carry a LQT1-associated mutation of the KCNQ1 locus (R190Q) were reprogrammed to generate patient-specific R190Q-iPSC lines. Control iPSC lines were derived from fibroblasts of an unrelated healthy proband without history of cardiac disease. When action potentials of ventricular-like myocytes were elicited by electrical pacing at 1 Hz in R190Q and control cells, it was obvious that the R190Q myocytes displayed prolonged action potentials as compared to control cells (Fig. 3A). The same observation was made for spontaneously occurring action potentials in unstipulated ventricular cells. When the control cells were paced at different rates, it was observed in control cells that increasing the pacing rate led to a decrease of the action potential duration (APD, see Fig. 3A). This is consistent with normal cardiac physiology, where the QT interval (which reflects the action potential duration of ventricular myocytes) decreases with increasing heart rates. In the R190Q myocytes, the action potentials were already prolonged at a slow pacing rate (1 Hz) and the decrease of the APD at increased pacing rates was blunted (see Fig. 3A). Similarly, catecholamine stimulation led to a much lesser reduction of the APD in R190Q cells than in control myocytes. This points out that in the R190Q myocytes, the APD is not only prolonged under basal conditions, but the normal regulation mechanisms which lead to a shortening of the action potential at situations of increased catecholamine stimulation or heart rate (which are the situations in which LQT1 patients typically develop Torsades de pointes) are malfunctional. This was corroborated by an experiment in which spontaneously beating myocytes were subjected to stimulation with the catecholminergic agonist isoproterenol (Fig. 3B). In control cells, this led to an increased beating rate, but concomitantly to a decreased APD, resulting in a reduction of the ratio between the APD90 and the beat-to-beat interval. In the R190Q cells, however, the increased beating rate could not be compensated by a shortening of the action potentials, indicated by an increase in the APD90/beat-to-beat interval ratio. Moreover, under conditions of catecholamine stimulation, the cells frequently displayed EAD. The β receptor antagonist propranolol (reflecting a class of medications that is typically beneficial for patients affected by LQT1) reversed both effects in the R190Q cells (see Fig. 3B).

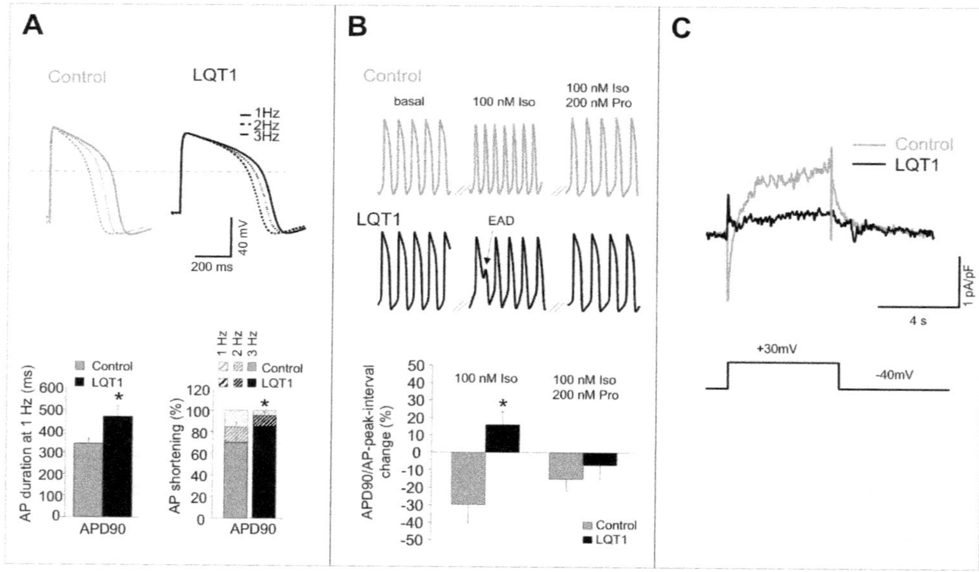

Figure 3. Recapitulation of Key Features of Long-QT Syndrome Type 1 in Cardiomyocytes Derived from Patient-Specific iPSC. Panel (A) shows representative tracings of action potentials (AP) recorded from control and KCNQ1-R190Q (LQT1) myocytes at three different pacing rates (1, 2, and 3 Hz) as indicated. The left bar graph shows statistics for the absolute value of APD90 (the duration from the beginning of the action potential until repolarization is accomplished by 90%) at 1 Hz pacing in control and LQT1 myocytes. The right bar graph shows the relative shortening of the APD90 upon increasing the pacing rate from 1 Hz to 2 or 3 Hz (as indicated) in control and LQT1 myocytes. Panel (B) shows representative membrane potential recordings of spontaneously beating control and LQT1 myocytes before and after incubation with 100 nM isoproterenol (Iso), in the presence or in the absence of 200 nM propranolol (Pro). In the tracing from the LQT1 cell, an early afterdepolarization (EAD) is indicated. The bar graph shows statistics for the ratio of the APD90 divided by the interval between two action potentials under isoproterenol stimulation in the presence and in the absence of propranolol. Panel (C) shows representative recordings of I_{Ks} current (the slow component of the delayed rectifying potassium current) in control and LQT1 cells. The used voltage protocol is shown below the current recordings. Asterisks indicate statistically significant differences ($p<0.05$). Adapted from data published in Moretti *et al.*, 2010b.

The KCQ1 gene mutated in the LQT1 patients encodes the α-sub-unit of the ion channel responsible for the I_{Ks} current. We therefore measured I_{Ks} in the R190Q myocytes and found that I_{Ks} was reduced to about 25% as compared to control myocytes (Fig. 3C). This reduction by more than 50% indicated that the mutation might exhibit a dominant-negative effect in the cells heterozygous for the R190Q mutation. Indeed, we could demonstrate that the R190Q mutation exerts a dominant-negative effect by forming multimers with wild type sub-units and interfering with their trafficking to the plasma membrane (Moretti et al. 2010b).

Other Promising Targets for Patient-specific Stem Cell-based Disease Modeling

As stated above, the widely used rodent models have specific drawbacks when used to investigate disorders of the heart rhythm. Thus, besides the long-QT syndrome, the following inherited rhythmological disease entities appear to be valuable targets for human iPSC-based disease modeling.

Short-QT syndrome

A rare familial syndrome with autosomal dominant inheritance, characterized by a shortened QT interval, episodes of syncope, susceptibility to atrial fibrillation and to sudden cardiac arrest has been termed short-QT syndrome (SQTS) (Gaita et al. 2003; Giustetto et al. 2006). The disease has been associated with gain-of-function mutations in KCNH2 (encoding HERG; SQTS1), KCNQ1 (SQTS2), and KCNJ2 (SQTS3), as well as loss-of-function mutations in genes encoding the l-type calcium channels CACNA1C and CACNB2 (Giustetto et al. 2006; Antzelevitch et al. 2007). Considering these mutations, an electrophysiological situation opposite to the one present in long-QT syndrome appears the probable phenotype on a single cell level. However, little is known about the pathophysiology that leads to the susceptibility to life threatening arrhythmias in affected individuals, particularly as healthy individuals with relatively short QT intervals may not have a comparable risk for sudden cardiac arrest (Gallagher et al. 2006). Induced pluripotent stem cell technology seems ideal to study the molecular mechanisms of different mutations and test the electrophysiological properties of affected human cardiomyocytes *in vitro*.

Brugada syndrome

Brugada syndrome is defined by specific ECG patterns and typical clinical features and is a cause of sudden cardiac death in persons with structurally normal hearts (Wilde et al. 2002; Antzelevitch et al. 2005). Sudden unexpected nocturnal death syndrome (SUNDS), described in males in Southeast Asia, is considered to be the same syndrome (Vatta et al. 2002). Mutations in the gene SCN5A, coding for sub-units of a cardiac sodium channel, have been found in patients with Brugada syndrome, a gene that is also affected in LQT3 (Antzelevitch et al. 2005). Mutations in other cardiac ion transport genes have been described as well (Weiss et al. 2002; Watanabe et al. 2008). Although there is an autosomal dominant expression pattern, penetrance is highly variable and the typical ECG pattern can be found more often in men than in women for reasons not completely understood (Benito et al. 2008). Furthermore, the known mutations are only present in a fraction of patients with Brugada syndrome and the mutation status is not sufficient to predict the risk of sudden cardiac arrest in affected individuals (Antzelevitch et al. 2005). In conclusion, patients with Brugada syndrome appear to be a genetically heterogeneous population and risk stratification of individuals has been proven difficult (Priori et al. 2002; Probst et al. 2010).

Studying SCN5A mutation-positive Brugada syndrome using iPSC models in a similar fashion as outlined above for other channelopathies may be an important step towards understanding the pathogenesis of the disease, the mechanisms of arrhythmia in affected patients, and potential treatment options. Furthermore it might help identifying asymptomatic patients at risk for sudden cardiac death. Modeling of mutation-negative Brugada syndrome in iPSC appears a big challenge, but it may also offer novel possibilities. For example, the electrophysiological features of iPSC-derived human cardiomyocytes from mutation-negative patients with Brugada syndrome could be examined in order to test ion channel function, action potential properties, and thus define a functional phenotype regardless of the genotype. This cellular phenotype in turn might then help to improve risk stratification, identify exogenic factors that promote arrhythmia, and identify or develop drugs that can prevent sudden cardiac arrest in affected patients.

Catecholaminergic polymorphic ventricular tachycardia

Another inheritable disorder characterized by cardiac arrhythmias has been termed catecholaminergic polymorphic ventricular tachycardia (CPVT; reviewed recently in Priori and Chen 2011). The typical clinical presentation is the development of ventricular tachycardias in situations of increased catecholamine secretion, e.g., physical exercise or emotional stress. These tachycardias often present as so-called bidirectional ventricular tachycardias, in which two different morphologies of the QRS complex alternate from beat to beat, suggesting two alternating arrhythmogenic foci. In contrast to the disorders mentioned above, in which the primary pathomechanism is the disruption of the normal function of plasma membrane ion channels, CPVT is a disorder of intracellular calcium cycling. One of the key functions of the cardiac action potential is to regulate the entry of calcium ions into the cytoplasm, which then act as a trigger to activate calcium release from the intracellular calcium stores, constituted mainly by the sarcoplasmic reticulum (SR). The resulting increase in the cytoplasmic calcium concentration is a major regulator of cardiac contraction. This system of calcium storage and release, which results in coordinated oscillations of the cytoplasmic calcium concentration at every heartbeat, is tightly regulated. The CPVT-causing mutations known to date are either dominant mutations in the gene encoding the cardiac ryanodine receptor (RyR2), which is the calcium release channel in the membrane of the SR, or recessive mutations in calsequestrin, which is a calcium-buffering protein located in the SR lumen. A unifying concept of the pathophysiology of CPVT is that in patients, overactive mutated ryanodine receptors or a lowered SR calcium buffering capacity due to mutated calsequestrin lead to a lowered threshold for calcium release from the SR under conditions of increased SR calcium loading which arise during catecholaminergic stress. This in turn leads to increased calcium release from the SR resulting in arrhythmogenesis (see Priori and Chen 2011).

We have recently generated iPSC lines from a patient suffering from CPVT affected by a novel RyR2 mutation (S406L). In Fig. 4, preliminary results of an analysis of ryanodine receptor function in cardiomyocytes generated from these patient-specific iPSC are shown. Since RyR2 forms a calcium release channel that is activated by cytoplasmic calcium, the calcium release from one RyR2 channel may activate neighboring RyR2 channels. Indeed, RyR2-mediated calcium release occurs in so-called calcium "sparks" that originate from the coordinated calcium release of groups of about 100 RyR2 channels called calcium release units (CRU) (Berridge et al. 2000). These single calcium sparks can be visualized by high-speed confocal microscopy using fluorescent calcium indicators. The properties of the recorded calcium sparks are directly correlated to the gating properties of the RyR2 channels. Figure 4 shows typical examples of calcium sparks recorded from CPVT and control myocytes under basal conditions and after stimulation with the catecholaminergic agonist isoproterenol. In the control cells, the low basal spark frequency increases upon isoproterenol application. In the CPVT cells, the spark frequency is already higher at basal conditions and increases even further upon stimulation with isoproterenol. The size (calculated as full width at half maximum intensity, FWHM) and the duration (calculated as decay time constant Tau) of the sparks was also increased in CPVT myocytes under isoproterenol stimulation as compared to control myocytes (see Fig. 4), indicating altered, "overactive" RyR2 gating

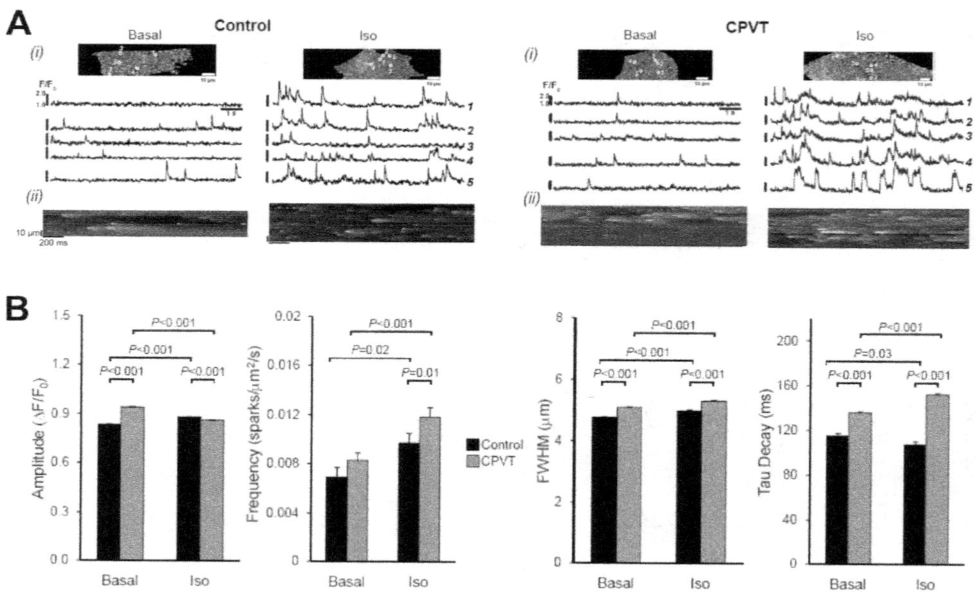

Figure 4. Enhanced Basal and Isoproterenol-Induced Calcium Spark Activity in Cardiomyocytes Derived from CPVT Patient-Specific iPSC. Panel (A) shows representative results of calcium spark imaging in fluo-4-AM-loaded control and CPVT myocytes in the absence (Basal) or in the presence (Iso) of 1 μM isoproterenol. Part (i) shows pseudo-colored images (upper row) together with typical Ca^{2+} traces corresponding to each of the 5 individual regions of interest marked in the top images, imaged at 105 images/s (lower row). Part (*ii*) shows line-scan images of Ca^{2+} sparks at a higher temporal resolution (1000 lines/s). Panel (B) shows the statistics of Ca^{2+} spark characteristics from control (black) and CPVT (red) myocytes in absence or presence of 1 μM isoproterenol. FWHM, full width at half maximum. Between 38 and 142 cells from 3 iPSC lines were analyzed in each group. Data are means ± s.e.m. from 4 independent experiments. *P* values from two-tailed *t* test are given. Experiments were performed in collaboration with Peter Lipp at the University of Homburg/Saar, Germany.

Color image of this figure appears in the color plate section at the end of the book.

properties. These data indicate that basic features of the disease can be recapitulated in the iPSC-based disease model for CPVT. This model might thus provide a platform to test, in a human cell-based system, potential drugs for CPVT that target the overactive mutant RyR2 channels (Kobayashi et al. 2009; Watanabe et al. 2009).

Disease Modeling with Patient-specific Stem Cells: Possible Applications

A new instrument in the tool cabinet of the basic scientist

Up to date, the majority of reports of using iPSC to model cardiac diseases have been, at least to a large extent, proof-of-principal studies that did not reveal fundamentally new aspects of the underlying pathophysiology. This is not surprising given the novelty of the field. When a new methodology is introduced, it is wise to first apply it to an area of the research landscape that has been already mapped well using conventional methodology. This facilitates the interpretation of results given by the new methodology and also allows the correlation of the new model system with the already known aspects of the world it is supposed to represent. However, as stated above, the use of patient-specific human cells bears the possibility of gaining insights that could not be gained by using the conventional rodent-based models. As the iPSC methodology becomes increasingly standardized and, thus, better accessible to basic scientists working in different fields all over the world, it will be exciting to see how the inclusion of this new instrument into their tool cabinets will lead not only to more refined maps of well-known territories, but hopefully also to the discovery of unknown shores that were inaccessible before the advent of iPSC technology.

Standardized tests for drug safety

Given the impact of QT interval prolongation on the development of new drugs, human cell-based systems to screen for QT interval prolongation would be a desirable goal. It is easy to envision a test system consisting of human iPSC-derived cardiomyocytes whose action potentials are monitored (e.g., by whole-cell patch clamp recordings) before and after exposure to a candidate drug. However, when it comes to the question which human iPSC lines should be used for such a test system, several different strategies are feasible. Since a modulation of HERG channel activity is a frequent mechanism of drug-induced QT interval prolongation, iPSCs derived from a

LQT2 patient carrying a HERG mutation could be used to generate the test cardiomyocytes (Itzhaki et al. 2011). However, drug-induced QT interval prolongation frequently occurs in patients who do not carry mutations in their HERG (alias KCNH2) locus (Yang et al. 2002). Thus, another strategy would be to use a panel of several iPSC lines derived from a random sample of (at best genetically diverse) subjects. Finally, since a genetic predisposition seems to play a role for the susceptibility to drug-induced QT interval prolongation (although the involved genes are not known) (Kannankeril et al. 2005), a panel of iPSC lines derived from patients who have already reacted to different drugs with a prolonged QT interval might be used to generate the cardiomyocytes that constitute the test system.

Patient-specific therapy

Treatment options are limited especially in arrhythmogenic diseases and the effect of a therapy often remains uncertain. By using iPSC-derived cardiomyocytes of a patient with an arrhythmogenic disorder, the effect of different drug candidates on the individual target cell could be assessed. Thus, one could predict the potential hazardous and beneficial effects for the individual patient and distinguish patients who benefit from a certain therapy from those who are at risk for developing side effects. Moreover, these cells could also be used in developing new drugs and open the possibility of tailor-made medication. By studying the electrophysiological properties of the individual target cell and measuring effects of drugs and small molecules directly in the living cell, new treatment options might be discovered and drugs might be modified to fit the patient's specific disease. So far, such experiments are mainly performed in non-human cell systems or *in vivo* models. Recently, new small molecules which shorten the QT interval were identified using high-throughput screening in a zebrafish long-QT model (Peal et al. 2011). The results from these studies are partially limited by the electrophysiological differences between zebrafish and human cardiomyocytes. Newly developed models of human long-QT syndromes (Moretti et al. 2010b; Itzhaki et al. 2011; Yazawa et al. 2011; Matsa et al. 2011) offer a platform for similar screening experiments without this limitation.

Disease Modeling with Patient-specific Stem Cells: Future Challenges

Increasing the throughput

To date, the generation of cardiomyocytes from human iPSCs is a laborious task. The amount of qualified manual cell culture work and the cost of the cell culture reagents make iPSC-derived cardiomyocytes a very precious material. Accordingly, labor intensive and time consuming techniques such as whole-cell patch clamp recordings or single-cell-RT-PCR have been applied for the analysis of the physiology of these cells (Moretti et al. 2010b). However, one important goal of current research in the iPSC field is to increase the efficiency of the cell culture and differentiation protocols (Fujiwara et al. 2011; Shafa et al. 2011), which will eventually make patient-specific cardiomyocytes a more common good. At this point, it will be helpful to also increase the throughput of the methodology used to study these cells. One step into this direction that has already been taken is the use of multielectrode arrays (Itzhaki et al. 2011), which allow the recording of field action potentials from larger numbers of myocytes. Other medium- to high throughput analysis methods like automated patch clamping (Jones et al. 2009) will likely aid iPSC-based disease modeling in the future.

Genetic marking of specific cell types

The heterogeneity of the iPSC-derived cardiomyocytes, which can be classified at least as ventricular-like, atrial-like and nodal-like cardiomyocytes (see "Different subgroups of iPSC-derived cardiomyocytes"), might become a problem if the cells are analyzed using a methodology that cannot discriminate between those cell types and if the parameter of interest differs between the different cell types. Genetic marking techniques might be helpful to identify specific sub-populations of cardiomyocytes. Bearing this in mind we have designed an adenoviral construct in which the red fluorescent protein dsRed is expressed under the control of a myosin light chain 2v (mlc-2v) promoter (Fig. 5A). Since transcription from this promoter is limited to ventricular cardiomyocytes, expression of this construct should lead to dsRed expression only in those cells. Indeed, when iPSC-derived cardiomyocytes were infected with this adenovirus, red fluorescence was found only in a sub-population of cardiomyocytes, which co-stained for the ventricular isoform of myosin light chain 2 (Fig. 5B, C). Using such a marking strategy, the red fluorescent cells could then be selected for single-cell measurements like electrophysiological recordings or fluorometric calcium measurements. Alternatively, the cells of interest could be purified by fluorescence-activated cell sorting for subsequent analysis.

Addressing disorders with more complex genetics

As stated above, the initial reports of iPSC-based disease modeling have concentrated on monogenetic disorders. However, this is not a necessity

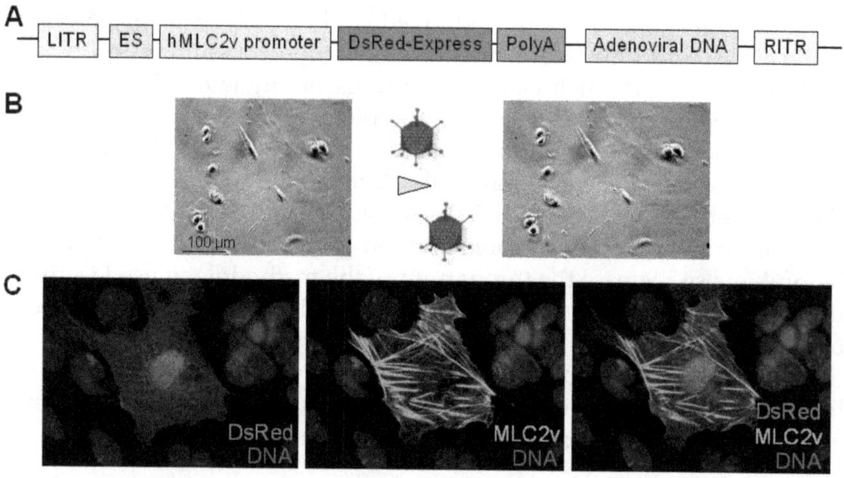

Figure 5. Genetic Marking of Ventricular Cardiomyocytes. Panel (A) shows a schematic representation of the adenoviral vector construct used for genetic marking. Panel (B) shows a phase contrast image (left image) of iPSC-derived cardiomyocytes infected with this virus and the same image merged with the red fluorescence of DsRed (right image). Panel (C) shows a closer magnification of a DsRed-expressing cardiomyocyte (DsRed fluorescence in red) merged with nuclear staining (blue) and immunofluorescent staining for the ventricular isoform of myosin light chain 2v (MLC2v, green) as indicated.

Color image of this figure appears in the color plate section at the end of the book.

for iPSC-based disease models. The basic concept of patient-specific stem cells is founded on the principle that the iPSC lines and their differentiated progeny are genetically identical to the patient. Thus, they provide a platform that might also be used to model complex multifactorial diseases like atherosclerosis, diabetes mellitus, or arterial hypertension, which are responsible for most of the cardiovascular morbidity in industrialized countries. In these diseases, genetic factors act together with lifestyle and environmental factors to precipitate the disease phenotype in single patients. By using iPSC technology, it would be possible to investigate the genetic factors in patient-specific cells while keeping the environmental factors constant in patient and control cells.

Another problem that could be addressed by iPSC-based disease modeling is incomplete penetrance, meaning that not all individuals who carry a specific disease-causing mutation develop the disease phenotype to the same extent (Zlotogora 2003). The causes for incomplete penetrance can be environmental factors as well as additional genetic factors. Somatic cells from several members of a family affected by a disease with incomplete penetrance could be used to generate iPSC lines. These iPSC could be then differentiated to the cell type affected by the disease and the phenotype of the cells could be correlated to the phenotype of the respective patients. By using this strategy, the genetic factors that contribute to incomplete penetrance could be investigated without confounding by environmental factors.

From cardiac cells to cardiac tissue

Arrhythmogenic diseases are valuable targets for iPSC-based disease modeling because key features of these diseases like altered action potentials, altered calcium homeostasis or the generation of afterpotentials can be studied in single cells. However, other important aspects of cardiac arrhythmias like the generation of re-entrant arrhythmias are not cell-autonomous and thus not accessible to experiments that rely only on isolated cardiomyocytes. Modern tissue engineering techniques (Tiburcy et al. 2011; Tulloch et al. 2011) might be adopted to generate artificial human myocardium from patient-specific iPSC lines. Such engineered heart tissue could provide a platform to study the underlying pathophysiology of emergent phenomena like Torsades de pointes or bi-directional ventricular tachycardia in networks of cells genetically identical to patients affected by diseases like long-QT syndrome or CPVT.

Conclusion

The development of disease models based on patient-specific iPSCs, as outlined above in detail for the long-QT syndrome, is a new technology bearing the potential to spawn a wealth of new applications in translational medicine (Fig. 6). Hopefully, the patients affected by the diseases of interest are not only important as donors of cells, but will eventually profit from the results of such research. A better understanding of the pathophysiology gained by such experiments might lead to the development of new therapies. Induced pluripotent stem cell-based drug screening systems might both lead to the discovery of new drugs and help to circumvent problems of unwanted side effects during drug development. Finally, this technology even offers the potential of tailoring a patient-specific therapy to the precise needs of a single patient.

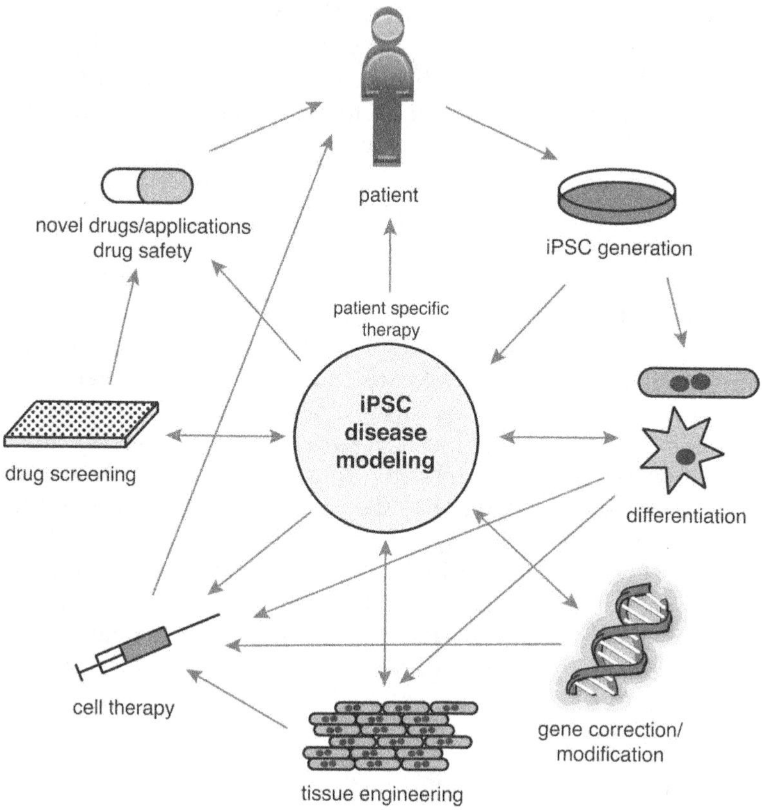

Figure 6. Disease Modeling at the Center of Potential iPSC Future Applications. Generation of patient-specific iPSC and differentiation into diseased cell types is the prerequisite for *in vitro* disease modeling by iPSC. Genetic modifications and gene repair will improve disease models and might find application in future cell therapies. While defects on a single cell and subcellular level have been successfully modeled using iPSC, engineered tissue might be a key of future disease models addressing defects involving cell-cell interactions, potentially involving different cell types. iPSC-based disease models might be a platform for development of new drugs and a new tool to improve drug safety. Finally, in some cases iPSC-based disease models will help to find specific therapies for single patients, particularly in multigenetic diseases or diseases with unknown pathophysiology.

Color image of this figure appears in the color plate section at the end of the book.

References

Ackerman, M. J. and Clapham, D. E. Excitability and Conduction. *In:* K. R. Chen [ed.]. 2004. Molecular basis of cardiovascular disease: a companion to Braunwald's Heart Disease. 2nd Edition. Saunders an imprint of Elsevier, Philadelphia, USA.

Antzelevitch, C. 2005. Role of transmural dispersion of repolarization in the genesis of drug-induced Torsades de pointes. Heart Rhythm 2 (2 Suppl.), S9–15.

Antzelevitch, C., Brugada, P., Borggrefe, M., Brugada, J., Brugada, R., Corrado, D., Gussak, I., LeMarec, H., Nademanee, K., Perez Riera, A. R., Shimizu, W., Schulze-Bahr, E., Tan,

H., and Wilde, A. 2005. Brugada syndrome: report of the second consensus conference: endorsed by the Heart Rhythm Society and the European Heart Rhythm Association. Circulation 111(5), 659.

Antzelevitch, C., Pollevick, G. D., Cordeiro, J. M., Casis, O., Sanguinetti, M. C., Aizawa, Y., Guerchicoff, A., Pfeiffer, R., Oliva, A., Wollnik, B., Gelber, P., Bonaros, E. P. Jr., Burashnikov, E., Wu, Y., Sargent, J. D., Schickel, S., Oberheiden, R., Bhatia, A., Hsu, L.F., Haïssaguerre, M., Schimpf, R., Borggrefe, M., and Wolpert, C. 2007. Loss-of-function mutations in the cardiac calcium channel underlie a new clinical entity characterized by ST-segment elevation, short QT intervals, and sudden cardiac death. Circulation 115(4), 442.

Benito, B., Sarkozy, A., Mont, L., Henkens, S., Berruezo, A., Tamborero, D., Arzamendi, D., Berne, P., Brugada, R., Brugada, P., and Brugada, J. 2008. Gender differences in clinical manifestations of Brugada syndrome. J. Am. Coll. Cardiol. 52(19), 1567.

Berridge, M. J., Lipp, P., and Bootman, M.D. 2000. The versatility and universality of calcium signalling. Nature Rev. Mol. Cell. Biol. 1, 11–21.

Brown, M. E., Rondon, E., Rajesh, D., Mack, A., Lewis, R., Feng, X., Zitur, L. J., Learish, R. D., and Nuwaysir, E. F. 2010. Derivation of induced pluripotent stem cells from human peripheral blood T lymphocytes. PLoS One 5(6): e11373.

Casimiro, M. C., Knollmann, B. C., Ebert, S. N., Vary, Jr. J. C., Greene, A. E., Franz, M. R., Grinberg, A., Huang, S. P., and Pfeifer, K. 2001. Targeted disruption of the Kcnq1 gene produces a mouse model of Jervell and Lange-Nielsen syndrome. Proc. Natl. Acad. Sci. USA 98(5), 2526–2531.

Demolombe, S., Baró, I., Péréon, Y., Bliek, J., Mohammad-Panah, R., Pollard, H., Morid, S., Mannens, M., Wilde, A., Barhanin, J., Charpentier, F., and Escande, D. 1998. A dominant negative isoform of the long-QT syndrome 1 gene product. J. Biol. Chem. 237(12), 6837–6843.

Demolombe, S., Lande, G., Charpentier, F., van Roon, M. A., van den Hoff, M. J. B., Toumaniantz, G., Baro, I., Guihard, G., Le Berre, N., Corbier, A., de Bakker, J., Opthof, T., Wilde, A., Moorman, A. F. M., and Escande, D. 2001. Transgenic mice overexpressing human KvLQT1 dominant-negative isoform. Part I: phenotypic characterisation. Cardiovasc. Res. 50, 314–327.

Dessertenne, F. 1966. La tachycardie ventriculaire à deux foyers opposés variables. Arch. Mal. Coeur 59, 263–272.

Ebert, A. D., Yu, J., Rose, Jr. F. F., Mattis, V. B., Lorson, C. L., Thomson, J. A., and Svendsen, C. N. 2009. Induced pluripotent stem cells from a spinal muscular atrophy patient. Nature 457, 277–280.

Eckardt, L., Haverkamp, W., Borggrefe, M., and Breithardt, G. 1998. Experimental models of Torsade de pointes. Cardiovascular Research 39, 178–193.

Fujiwara, M., Yan, P., Otsuji, T. G., Narazaki, G., Uosaki, H., Fukushima, H., Kuwahara, K., Harada, M., Matsuda, H., Matsuoka, S., Okita, K., Takahashi, K., Nakagawa, M., Ikeda, T., Sakata, R., Mummery, C. L., Nakatsuji, N., Yamanaka, S., Nakao, K. and Yamashita, J. K. 2011. Induction and enhancement of cardiac cell differentiation from mouse and human induced pluripotent stem cells with cyclosporin-A. PLoS One 6(2: e16734).

Gaita, F., Giustetto, C., Bianchi, F., Wolpert, C., Schimpf, R., Riccardi, R., Grossi, S., Richiardi, E. and Borggrefe, M. 2003. Short QT Syndrome: a familial cause of sudden death. Circulation 108(8), 965–70.

Gallagher, M. M., Magliano, G., Yap, Y. G, Padula, M., Morgia, V., Postorino, C., Di Liberato, F., Leo, R., Borzi, M., and Romeo, F. 2006. Distribution and prognostic significance of QT intervals in the lowest half centile in 12,012 apparently healthy persons. Am. J. Cardiol. 98(7), 933.

Giorgi, M. A., Bolanos, R., Gonzalez, C. D., and Di Girolamo, G. 2010. QT interval prolongation: preclinical and clinical testing arrhythmogenesis in drugs and regulatory implications. Curr. Drug Safety 5, 54–57.

Giustetto, C., Di Monte, F., Wolpert, C., Borggrefe, M., Schimpf, R., Sbragia, P., Leone, G., Maury, P., Anttonen, O., Haissaguerre, M., and Gaita, F. 2006. Short QT syndrome: clinical findings and diagnostic-therapeutic implications. Eur. Heart 27(20), 2440.

He, J-Q., Ma, Y., Lee, Y., Thomson, J. A., and Kamp, T. J. 2003. Human embryonic stem cells develop into multiple types of cardiac myocytes. Action potential characterization. Circ. Res. 93: 32–39.

Itzhaki, I., Maizels, L., Huber, I., Zwi-Dantsis, L., Caspi, O., Winterstern, A., Feldman, O., Gepstein, A., Arbel, G., Hammerman, H., Boulos, M., and Gepstein, L. 2011. Modelling the long-QT syndrome with induced pluripotent stem cells. Nature 471(7337), 225–229.

Jones, K. A., Garbati, N., Zhang, H., and Large, C. H. 2009. Automated patch clamping using the QPatch. Methods Mol. Biol. 565, 209–223.

Kannankeril, P. J., Roden, D. M., Norris, K. J., Whalen, S. P., George, Jr., A. L., and Murray, K. T. 2005. Genetic susceptibility to acquired long QT syndrome: pharmacologic challenge in first-degree relatives. Heart Rhythm 2(2), 134–140.

Keller, G. M. 1995. *In vitro* differentiation of embryonic stem cells. Curr. Opin. Cell Biol. 7(6), 862–869.

Kobayashi, S., Yano, M., Suetomi, T., Ono, M., Tateishi, H., Mochizuki, M., Xu, X., Uchinoumi, H., Okuda, S., Yamamoto, T., Koseki, N., Kyushiki, H., Ikemoto, N., and Matsuzaki, M. 2009. Dantrolene, a therapeutic agent for malignant hyperthermia, markedly improves the function of failing cardiomyocytes by stabilizing interdomain interactions within the ryanodine receptor. J. Am. Coll. Card. 53(21): 1993–2005.

Lasser, K. E., Allen, P. D., Woolhandler, S. J., Himmelstein, D. U., Wolfe, and Bor, D. H. 2002. Timing of new black box warnings and withdrawals for prescription medications. JAMA 287, 2215–2220.

Lee, M. P., Ravenel, J. D., Hu, R-J., Lustig, L. R., Tomaselli, G., Berger, R. D., Brandenburg, S. A., Litzi, T. J., Bunton, T. E., Limb, C., Francis, H., Gorelikow, M., Gu, H., Washington, K., Argani, P., Goldenring, J. R., Coffey, R. J., and Feinberg, A. P. 2000. Targeted disruption of the Kvlqt1 gene causes deafness and gastric hyperplasia in mice. J. Clin. Invest 106, 1447–1455.

Lee, G., Papapetrou, E. P., Kim, H., Chambers, S. M., Tomishima, M. J., Fasano, C. A., Ganat, Y. M., Menon. J., Shimizu, F., Viale, A., Tabar, V., Sadelain, M., and Studer, L. 2009. Modelling pathogenesis and treatment of familial dysautonomia using patient-specific iPSCs. Nature 461: 402–406.

Lehnart, S. E., Ackerman, M. J., Benson, W., Brugada, R., Clancy, C. E., Donahue, J. K., George, Jr., A. L., Grant, A. O., Groft, S. C., January, C. T., Lathrop, D. A., Lederer, J., Makielski, J. C., Mohler, P. J., Moss, A., Nerbonne, J. M., Olson, T. M., Przywara, D. A., Towbin, J. A., Wang, L-H., and Marks, A.R. 2007. Inherited arrhythmias. A national heart, lung, and blood institute and office of rare diseases workshop consensus report about the diagnosis, phenotyping, molecular mechanisms, and therapeutic approaches for primary cardiomyopathies of gene mutations affecting ion channel function. Circulation 116, 2325–2345.

London, B. 2001. Cardiac arrhythmias: from (transgenic) mice to men. J. Cardiovasc. Electrophysiol. 12(9), 1089–1091.

Maltsev, V. A., Rohwedel, J., Hescheler, J., and Wobus, A. M. 1993. Embryonic stem cells differentiate in vitro into cardiomyocytes representing sinusnodal, atrial and ventricular cell types. Mech. Dev. 44(1), 41–50.

Matsa, E., Rajamohan, D., Dick, E., Young, L., Mellor, I., Stainforth, A., and Denning, C. 2011. Drug evaluation in cardiomyocytes derived from human induced pluripotent stem cells carrying a long-QT syndrome type 2 mutation. Eur. Heart J. 32(8), 952–962.

Moretti, A., Bellin, M., Jung, C. B., Thies, T. M., Takashima, Y., Bernshausen, A., Schiemann, M., Fischer, S., Moosmang, S., Smith, A. G., Lam, J. T., and Laugwitz, K. L. 2010a. Mouse and human induced pluripotent stem cells as a source for multipotent Isl1+ cardiovascular progenitors. FASEB J. 24, 700–711.

Moretti, A., Bellin, M., Welling, A., Jung, C. B., Lam, J. T., Bott-Flügel, L., Dorn, T., Goedel, A., Höhnke, C., Hofmann, F., Seyfarth, M., Sinnecker, D., Schömig, A., and Laugwitz, K-L. 2010b. Patient-specific induced pluripotent stem-cell models for long-QT syndrome. N. Engl. J. Med. 363(15), 1397–1409.

Nerbonne, J. M., Nichols, C. G., Schwarz, T. L., and Escande, D. 2001. Genetic manipulation of cardiac K+ channel function in mice. What we have learned, and where do we go from here? Circ. Res. 89, 944–956.

Novak, A., Shtrichman, R., Germanguz, I., Segev, H., Zeevi-Levin, N., Fishman, B., Mandel, Y. E., Barad, L., Domev, H., Kotton, D., Mostoslavsky, G., Binah, O., and Itskovitz-Eldor, J. 2010. Enhanced reprogramming and cardiac differentiation of human keratinocytes derived from plucked hair follicles, using a single excisable lentivirus. Cell. Reprogram 12(6), 665–678.

Peal, D. S., Mills, R. W., Lynch, S. N., Mosley, J. M., Lim, E., Ellinor, P. T., January, C. T., Peterson, R. T., and Milan, D. J. 2011. Novel Chemical Suppressors of Long QT Syndrome Identified by an *in vivo* Functional Screen. Circulation, published online before print, DOI 10.1161/CIRCULATIONAHA.110.003731.

Priori, S. G., Napolitano, C., Gasparini, M., Pappone, C., Della Bella, P., Giordano, U., Bloise, R., Giustetto, C., De Nardis, R., Grillo, M., Ronchetti, E., Faggiano, G., and Nastoli, J. 2002. Natural history of Brugada syndrome: insights for risk stratification and management. Circulation 105(11), 1342.

Priori, S. and Chen, S. R. W. 2011. Inherited dysfunction of sarcoplasmic reticulum Ca2+ handling and arrhythmogenesis. Circ. Res. 108, 871–883.

Probst, V., Veltmann, C., Eckardt, L., Meregalli, P. G., Gaita, F., Tan, H. L., Babuty, D., Sacher, F., Giustetto, C., Schulze-Bahr, E., Borggrefe, M., Haissaguerre, M., Mabo, P., Le Marec, H., Wolpert, C., and Wilde, A. A. 2010. Long-term prognosis of patients diagnosed with Brugada syndrome: Results from the FINGER Brugada Syndrome Registry. Circulation 121(5), 635.

Roden, D. M. 2004. Drug-induced prolongation of the QT interval. N. Engl. J. Med. 350, 1013–1022.

Salama, G. and London, B. 2007. Mouse models of long-QT syndrome. J. Physiol. 578.1, 43–53.

Sanguinetti, M. C., Jiang, C., Curran, M. E., and Keating, M. T. 1995. A mechanistic link between an inherited and an acquired cardiac arrhythmia: HERG encodes the IKr potassium channel. Cell 81, 299–307.

Schmitt, N., Calloe, K., Nielsen, N. H., Buschmann, M., Speckmann, E. J., Schulze-Bahr, E., and Schwarz, M. 2007. The novel C-terminal KCNQ1 mutation M520R alters protein trafficking. Biochem. Biophys. Res. Commun. 358(1), 304–310.

Schwartz, P. J., Stramba-Badiale, M., Crotti, L., Pedrazzini, M., Besana, A., Bosi, G., Gabbarini, F., Goulene, K., Insolia, R., Mannarino, S., Mosca, F., Nespoli, L., Rimini, A., Rosati, E., Salice, P., and Spazzolini, C. 2009. Prevalence of the congenital long-QT syndrome. Circulation 120, 1761–1767.

Shafa, M., Sjonesen, K., Yamashita, A., Liu, S., Michalak, M., Kallos, M. S., and Rancourt, D. E. 2011. Expansion and long-term maintenance of induced pluripotent stem cells in stirred suspension bioreactors. J. Tissue Eng. Regen. Med. doi: 10.1002/term.450. [Epub ahead of print].

Sidhu, K. S. 2011. New approaches for the generation of induced pluripotent stem cells. Expert Opin. Biol. Ther. 11(5), 569–579.

Studenik, C. R., Zhou, Z., and January, C. T. 2001. Differences in action potentials and early afterdepolarization properties in LQT2 and LQT3 models of long-QT syndrome. Br. J. Pharmacol. 132, 85–92.

Takahashi, T., Lord, B., Schulze, P. C., Fryer, R. M., Sarang, S. S., Gullans, S. R., and Lee, R. T. 2003. Ascorbic acid enhances differentiation of embryonic stem cells into cardiac myocytes. Circulation 107, 1912–1916.

Takahashi, K. and Yamanaka, S. 2006. Induction of pluripotent stem cells from mouse embryonic and adult fibroblast cultures by defined factors. Cell 126(4), 663–676.

Takahashi, K., Tanabe, K., Ohnuki, M., Narita, M., Ichisaka, T., Tomoda, K., and Yamanaka, S. 2007. Induction of pluripotent stem cells from adult human fibroblasts by defined factors. Cell 131(5), 861–872.

Thomas, D., Khali, M., Alter, M., Schweizer, P. A., Karle, C. A., Wimmer, A. B., Licka, M., Katus, H. A., Koenen, M., Ulmer, H. E., and Zehelein, J. 2010. Biophysical characterization of KCNQ1 P320 mutations linked to long QT syndrome 1. J. Mol. Cell. Cardiol. 48(1), 230–237.

Tiburcy, M., Didié, M., Boy, O., Christalla, P., Doeker, S., Naito, H., Karikkineth, B. C., El-Armouche, A., Grimm, M., Nose, M., Eschenhagen, T., Zieseniss, A., Katschinski, D., Hamdani, N., Linke, W. A., Yin, X., Mayr, M., and Zimmermann, W-H. 2011. Circ. Res. Published online before print, DOI: 10.1161/CIRCRESAHA.111.251843.

Tulloch, N. L., Muskheli, V., Razumova, M. V., Korte, F. S., Regnier, M., Hauch, K. D., Pabon, L., Reinecke, H., and Murry, C. E. 2011. Circ. Res. 109, 47–59.

Tran, T.H., Wang, X., Browne, C., Zang, Y., Schinke, M., Izumo, S., and Burcin, M. 2009. Wnt3a-induced mesoderm formation and cardiomyogenesis in human embryonic stem cells. Stem Cells 27, 1869–1878.

Vatta, M., Dumaine, R., Varghese, G., Richard, T. A., Shimizu, W., Aihara, N., Nademanee, K., Brugada, R., Brugada, J., Veerakul, G., Li, H., Bowles, N. E., Brugada, P., Antzelevitch, C., and Towbin, J. A. 2002. Genetic and biophysical basis of sudden unexplained nocturnal death syndrome (SUNDS), a disease allelic to Brugada syndrome. Hum. Mol. Genet. 11(3), 337.

Watanabe, H., Koopmann, T. T., Le Scouarnec, S., Yang, T., Ingram, C. R., Schott, J. J., Demolombe, S., Probst, V., Anselme, F., Escande, D., Wiesfeld, A. C., Pfeufer, A., Kääb, S., Wichmann, H. E., Hasdemir, C., Aizawa, Y., Wilde, A. A., Roden, D. M., and Bezzina, C. R. 2008. Sodium channel beta1 sub-unit mutations associated with Brugada syndrome and cardiac conduction disease in humans. J. Clin. Invest 118(6), 226.

Watanabe, H., Chopra, N., Laver, D., Hwang, H. Y., Davies, S. S., Roach, D. E., Duff, H. J., Roden, D. M., Wilde, A. A. M., and Knollmann, B. C. 2009. Flecainide prevents catecholaminergig polymorphic ventricular tachycardia in mice and humans. Nature Med. 15(4), 380–383.

Weiss, R., Barmada, M. M., Nguyen, T., Seibel, J. S., Cavlovich, D., Kornblit, C. A., Angelilli, A., Villanueva, F., McNamara, D. M., and London, B. 2002. Clinical and molecular heterogeneity in the Brugada syndrome: a novel gene locus on chromosome 3. Circulation 105(6), 707.

Wilde, A. A., Antzelevitch, C., Borggrefe, M., Brugada, J., Brugada, R., Brugada, P., Corrado, D., Hauer, R. N., Kass, R. S., Nademanee, K., Priori, S. G., and Towbin, J. A. Study Group on the Molecular Basis of Arrhythmias of the European Society of Cardiology. 2002. Proposed diagnostic criteria for the Brugada syndrome. Eur. Heart J. 23(21), 1648.

Wu, G., Ai, T., Kim, J. J., Mohapatra, B., Xi, Y., Li, Z., Abbasi, S., Purevjav, E., Samani, K., Ackerman, M. J., Qi, M., Moss, A. J., Shimizu, W., Towbin, J. A., Cheng, J., and Vatta, M. 2008. □-1-syntrophin mutation and the long-QT syndrome: a disease of sodium channel disruption. Circ. Arrhythm. Electrophysiol. 1(3), 193–201.

Yang, P., Kanki, H., Drolet, B., Yang, T., Wei, J,. Viswanathan, P. C., Hohnloser, S. H., Shimizu, W., Schwartz, P. J., Stanton, M., Murray, K. T., Norris, K., George, Jr., A. L., and Roden, D. M. 2002. Allelic variants in long-QT disease genes in patients with drug-associated torsades de pointes. Circulation 105(16), 1943–1948.

Yang, Y., Yang, Y., Liang, B., Liu, J., Li, J., Grunnet, M., Olesen, S-P., Rasmussen, H. B., Ellinor, P. T., Gao, L., Lin, X., Li, L., Wang, L., Xiao, J., Liu, Y., Liu, Y., Zhang, S., Liang, D., Peng, L., Jespersen, T., and Chen, Y-H. 2010. Identification of a Kir3.4 mutation in congenital long QT syndrome. Am. J. Hum. Genet. 86, 872–880.

Yazawa, M., Hsue, B., Jia, X., Pasca, A., Bernstein, J. A., Hallmayer, J., and Dolmetsch, R. E. 2011. Using induced pluripotent stem cells to investigate cardiac phenotypes in Timothy syndrome. Nature 471(7337), 230–234.

Ye, L., Chang, J. C., Lin, C., Sun, X., Yu J., and Kan, Y. W. 2009. Induced pluripotent stem cells offer new approach to therapy in thalassemia and sickle cell anemia and option in prenatal diagnosis in genetic diseases. Proc. Natl. Acad. Sci. USA 106(24), 9826–9830.

Zlotogora, J. 2003. Penetrance and expressivity in the molecular age. Genet. Med. 5(5), 347–352.

CHAPTER 17

In vitro Generation of Efficacious Platelets from iPS Cells

Naoya Takayama[a] and Koji Eto[b,*]

Introduction

Blood transfusion is a vital part of the treatment offered patients with fatal anemia and thrombocytopenia. However, the declining population of blood donors and the increasing risk of viral infection make an alternative transfusion system, one that is not dependent on donor blood, highly desirable. In this context, a number of different alternative sources of megakaryocytes have been suggested, including hematopoietic stem cells (HSCs) from bone marrow (Debili et al. 1995), cord blood (Matsunaga et al. 2006; Schipper et al. 2003) or adipose tissue (Matsubara et al. 2009). But it has proven difficult to generate megakaryocytes and platelets on a large scale, *in vitro*, probably due to the limited *ex vivo* expansion capability of HSCs and hematopoietic multipotent progenitor cells (HMPCs). Human embryonic stem cells (hESCs) are pluripotent cells that can proliferate almost infinitely *in vitro* (Thomson et al. 1998), and thus represent a potential source of platelets for transfusion and for studying the ontogeny of hematopoiesis (Ma et al. 2008; Takayama et al. 2008; Vodyanik et al. 2005; Wang et al. 2004; Yokoyama et al. 2009). It is well known, however, that repeated transfusion induces the production of antibodies against allogenic human leukocyte antigen (HLA) on the transfused platelets (Schiffer 2001), which renders the

Clinical Application Department, Center for iPS Cell Research and Application, Kyoto University, 606-8507, 53 Kawahara-cho, Shogoin, Sakyo-ku, Kyoto 606-8507, Japan.
[a]Email: naoya.takayama@cira.kyoto-u.ac.jp
[b]Email: kojieto@cira.kyoto-u.ac.jp
*Corresponding author

patient unresponsive to platelet transfusion therapy. To avoid such immune reactions following multiple allogeneic platelet transfusions, it is desirable to use HLA-matched or autologous platelets. In this context, we propose that the use of HLA-matched human induced pluripotent stem cells (hiPSCs) may be an effective strategy for generating functional platelets for patients requiring repeated transfusion (Fig. 1). Furthermore, derivation of iPS cells from patients with congenital platelet abnormalities should contribute to a better understanding of the pathogenesis of these abnormalities, as well as to the development of gene correction therapies.

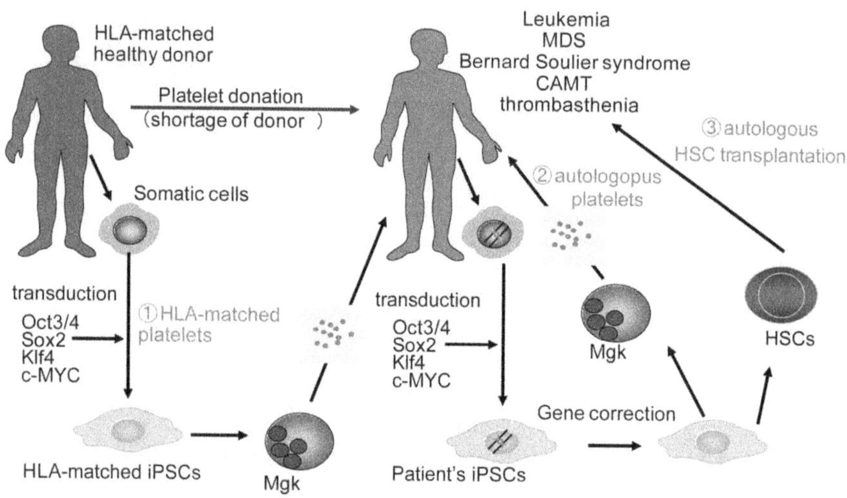

Figure 1. Alternative treatment of thrombocytopenia using hiPSCs. iPS cell technology enables us to consider innovative therapy models in the field of hematology. iPS cells are a potential source for blood transfusion. Validated and selected iPS clones from an HLA-identical donor can be useful for stable generation of platelets. Alternatively, iPS cell technology may enable "tailor-made gene therapy" using normalized self-iPS-derived stem cells after gene correction.

Color image of this figure appears in the color plate section at the end of the book.

Generation of Multipotent Hematopoietic Progenitors from Human ESCs and iPSCs via a Unique Sac-like Structure, the "ES/iPS-sac"

The culture methods for maintaining the pluripotency of mouse ESCs are now well established. In addition, differentiation protocols are being developing to generate a variety of cells and tissues, and several groups have reported producing platelets and/or their precursor cells, megakaryocytes, *in vitro* (Eto et al. 2002; Fujimoto et al. 2003; Nishikii et al. 2008). The first report of megakaryocyte generation from hESCs was presented by Gaur et al. (Gaur et al. 2006). The hESCs were differentiated on OP9 cells, a bone marrow stromal cell line derived from mice lacking macrophage colony stimulating factor (Nakano et al. 1994). OP9 cells are useful feeder cells and support differentiation of both mouse ESCs and hESCs into hematopoietic cells *in vitro*. Gaur's group observed that ESC-derived megakaryocytes began to express specific surface antigens, including the lineage markers CD41a (integrin αIIb) and CD42b (GPIbα), after 15–17 days of co-culture in the presence of thrombopoietin (TPO); however, no platelets were detected in this system (Gaur et al. 2006).

By testing various combinations of feeder cells, cytokines and culture durations, we eventually succeeded in inducing relatively mature, platelet-generating megakaryocytes from hESCs *in vitro* (Takayama et al. 2008). When we co-cultured 10T1/2 cells, a fetal fibroblast cell line derived from a C3H mouse strain (Hiroyama et al. 2006), in the presence of vascular endothelial growth factor (VEGF), unique sac-like structures derived from the hESCs (ES-sacs) appeared on day 14 of culture. These ES-sacs, particularly their external layer, expressed several endothelial markers, including CD31, CD34 and VEGF-receptor 2. They also contained CD34+/CD43+/CD45+ hematopoietic progenitors that exhibited hematopoietic colony forming potential in semisolid culture and differentiated into most kinds of mature cells, including megakaryocytes capable of releasing platelets, erythrocytes, T-cells, B-cells, NK cells, macrophages and granulocytes, under specific culture conditions. Early hematopoietic cells reportedly originate from hemangioblasts (Choi et al. 1998), which are a common precursor of vascular endothelial cells and hematopoietic cells. ES-sacs may mimic the early hematopoietic organ, within which hematopoiesis and vasculogenesis occur simultaneously, and may also contain hematopoietic niche circumstances for HSCs and/or HMPCs. We later confirmed that various types of hiPSCs can be differentiated into mature megakaryocytes that release platelets *in vitro* (Takayama et al. 2010). All of the different hiPSC clones that we examined generated sac-like structures (iPS-sacs) that led, to one degree or another, to production of functional platelets *in vitro*. We have since confirmed that the ES/iPS-sac method is a stable culture system for generating hematopoietic cells *in vitro*.

Generation of HLA-identical Platelets from Human iPSCs

During megakaryocyte maturation, internal membrane systems, granules and organelles are assembled in bulk. High ploidy megakaryocytes form an extensive internal demarcation membrane, which is continuous with the plasma membrane and serves primarily as a membrane reservoir for the formation of precursors of cytoplasmic extensions called proplatelets.

The open canalicular system (OCS) for granule release is formed prior to the initiation of proplatelet assembly, and when platelets are stimulated, the granules are released through the OCS (Patel et al. 2005).

Co-culture for an additional 7 to 11 days in the presence of 100 ng/mL TPO, 50 ng/mL stem cell factor (SCF) and 25 U/mL heparin promotes differentiation of many hematopoietic progenitors within ES/iPS-sacs into polyploid megakaryocytes forming proplatelets (Fig. 2) (Takayama et al. 2008; Takayama et al. 2010). Fifty percent to 60% of cells consistently express CD41a, CD42a and CD42b, which are essential and specific cell surface molecules on megakaryocytes and platelets (den Dekker et al. 2003; Tomer 2004). When the surface markers of hESC/iPSC-derived platelet-like particles were examined using flow cytometry with the same forward- and side-scatter gates used for fresh adult human platelets, $CD41a^+/CD42b^+$ particles were detected in the culture supernatants. Electron microscopic examination of the cytosolic structures of hESC/iPSC-derived platelets revealed the morphology to be similar to human platelets, with normal microtubules, granules and OCS. These platelets also displayed integrin $\alpha IIb\beta 3$ activation and spreading in response to the physiological platelet agonists ADP or thrombin, indicating their functionality *in vitro*. Using hiPSC specimens, we confirmed that human iPSCs generated from dermal fibroblasts through artificial reprogramming could be differentiated *in vitro* into functional platelets that could serve to supply HLA-identical platelets or, with use of disease-specific iPSCs, could serve as a platform for the study of abnormal thrombopoiesis.

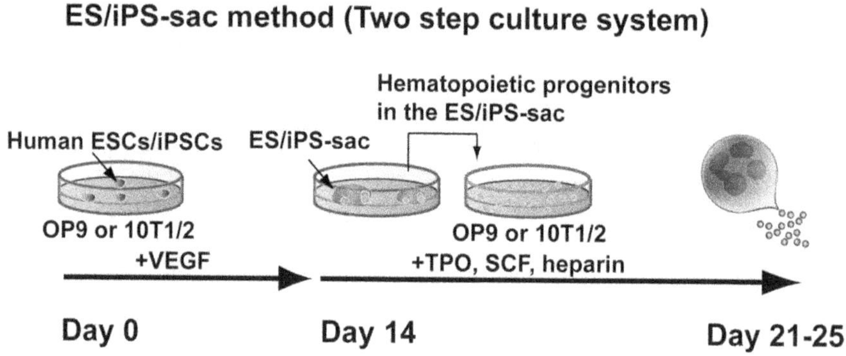

Figure 2. Human ESC/iPSC-derived megakaryocytes and platelets generated via ""sac"-like structures (ES/iPS-sacs). Schematic diagram of the in vitro differentiation protocol for hESC/hiPSC-derived megakaryocytes and platelets. (Step 1) hESC/iPSC are differentiated to Sac-like structure which contains multipotent hematopoietic progenitors on day 14 of culture. (Step 2) Hematopoietic progenitors within sac are transferred to new feeder cells and further cultivated. Mature megakaryocytes yielding platelets are generated on day 21-25. VEGF, vascular endothelial growth factor; TPO, thrombopoietin; SCF, stem cell factor.

Color image of this figure appears in the color plate section at the end of the book.

Role of *c-MYC* in iPSCs and Platelet Generation

It has already been reported that *c-Myc* plays an essential role in both embryonic and adult hematopoiesis, though its effects on megakaryopoiesis and thrombopoiesis in various mouse models remain unclear unclear (Chanprasert et al. 2006; Guo et al. 2009; Thompson et al. 1996a; Thompson et al. 1996b). For example, two studies of inducible *c-Myc* overexpression under the control of megakaryocyte-specific differentiation revealed that *c-Myc* plays a positive role in the proliferation of megakaryocyte progenitors (Thompson et al. 1996a; Thompson et al. 1996b). Moreover, *c-Myc* is reportedly essential for the TPO-c-mpl axis in megakaryopoiesis (Chanprasert et al. 2006). On the other hand, recent studies using *c-Myc*-deficient mice showed that the absence of the gene actually led to a cell autonomous increase in the number of low ploidy megakaryocytes and platelets (Guo et al. 2009). The reasons for this apparent discrepancy are so far unclear. Likewise, the effects of *c-MYC* on human megakaryocytes have not yet been completely understood. That said, when we compared several human iPS clones established with or without *c-MYC*, we noticed a novel activity of *c-MYC* in human thrombopoiesis. Upon comparing the efficiency of megakaryocyte and platelet generation from several iPS clones established using four (*OCT3/4*, *SOX2*, *KLF4* and *c-MYC*) or three (without *c-MYC*) reprogramming factors with retroviral vectors, we detected considerable heterogeneity in the production of iPS-Sacs and CD34$^+$ hematopoietic progenitors showing multilineage colony forming potential, even from iPSCs derived from the same source. This is consistent with earlier observations with hESCs (Osafune et al. 2008). Nonetheless, when equal numbers of hematopoietic progenitors from ES/iPS-Sacs were seeded onto fresh culture dishes in the presence of TPO, SCF and heparin, the numbers of CD42b (GPIbα; von Willebrand factor receptor)$^+$ megakaryocytes obtained with four-factor (including *c-MYC*) iPSC clones were higher than obtained with three-factor (without *c-MYC*) iPSC clones or an ESC clone. Quantitative PCR analyses revealed that the four-factor clones reactivated expression of exogenous *c-MYC* at the megakaryocyte stage, but that the clone yielding the largest numbers of platelets showed a reduction in total (exogenous plus endogenous) *c-MYC* expression during the megakaryocyte maturation phase. This suggests that although *c-MYC* is beneficial for megakaryocyte proliferation, its sustained expression inhibits platelet release. To test that idea, we used retroviral vectors to separately introduce each reprogramming factor into hematopoietic progenitors derived from hESCs. As expected, only overexpression of *c-MYC* recapitulated the time course of the enhanced megakaryopoiesis. In addition, Flow cytometry revealed that most of the cells in specimens overexpressing *c-MYC* were CD41a$^+$CD42b$^+$, but cells in the other specimens were not, though only mononuclear and lower ploidy cells were present. The numbers of platelets per megakaryocyte, one indicator of megakaryocyte maturation, were quite low in specimens overexpressing *c-MYC*, and the released "platelet"-like particles displayed a lack of CD42b expression, suggesting the megakaryocytes from specimens overexpressing *c-MYC* were immature. Collectively, these findings suggest that excessive expression of *c-MYC* in hESCs may promote lineage commitment into megakaryopoiesis without maturation. Moreover, overexpression of *c-MYC* in hematopoietic cells to levels that are inhibitory for platelet generation activates expression of the senescence/apoptosis-inducing *p14 (ARF)* and *p16 (INK4A)* genes, leading to senescence and apoptosis without maturation. Thus a narrow window of *c-MYC* expression is required for efficient platelet development.

To further confirm whether an increase and subsequent decline in *c-MYC* is critical for megakaryopoiesis leading to an efficient platelet yield, we prepared a Sendai viral vector (SeV) harboring the four reprogramming genes, which enabled RNA viral transduction during the generation of hiPSCs without integration of DNA into the chromosome (Nishimura et al. 2007). Thereafter, a doxycycline (DOX)-inducible *c-MYC* overexpression system in a lentiviral vector was applied to the SeV-based hiPSCs. This approach enabled us to confirm that transient upregulation at the level of megakaryocyte progenitors and subsequent decline in *c-MYC* expression increased the total numbers of mature megakaryocytes, proplatelets and $CD42b^+$ platelets (Takayama et al. 2010).

Preservation of Platelet Function *in vitro* Using an ADAM-17 Inhibitor

The functional platelet paradigm in hemostasis and thrombosis is the initiation of platelet adhesion to the extracellular matrix (Ruggeri 2002). One key event in that process is the interaction between glycoprotein (GP) Ibα (the platelet receptor, CD42b) and von Willebrand factor (vWF) present in the extracellular matrix (Ruggeri 2002). Simultaneously, platelets can interact with surface-bound collagen via the platelet receptors GPVI and integrin α2β1 (Nieswandt and Watson 2003). The net result is activation of integrin αIIbβ3, turning it into a competent fibrinogen receptor, which leads to the formation of platelet aggregates (Ruggeri 2002). Notably, aged platelets easily shed GPV and the extracellular domain of GPIbα, which contains the binding sites for vWF and thrombin (Canobbio et al. 2004; Rabie et al. 2005). This phenomenon makes it difficult to preserve platelet concentrates for more than seven days. The process involves the action of a disintegrin and metalloproteinase (ADAM) 17 (also referred to as tumor necrosis factor α converting enzyme, TACE) (Bergmeier et al. 2004) and leads to a reduction in platelet function (Bergmeier et al. 2003; Bergmeier et al. 2004). We observed that, in culture, mouse ESC-derived platelets shed the extracellular domains of GPIbα and GPV, as well as GPVI, possibly through the action of ADAM10, which reduced αIIbβ3 activation and actin polymerization and impaired thrombus formation. However, administration of the metalloproteinase inhibitor GM6001 (EMD Millipore) during differentiation increased the expression of GPIbα and GPVI and improved both thrombogenesis *in vitro* and post-transfusion recovery *in vivo* (Nishikii et al. 2008). Administration of GM6001 also inhibited shedding of the extracellular domains of human GPIbα from hiPSC-derived platelets (Takayama et al. 2010). These data suggest that regulation of metalloproteinases in culture could be useful for obtaining high quality and efficacious hiPSC-derived platelets for transfusion.

In vivo Functionality of Human iPSC-derived Platelets

To assess the ability of iPSC-platelets to form thrombi *in vivo*, we recently developed a novel *in vivo* imaging system that employs high spatio-temporal resolution confocal laser microscopy to visualize the behavior of individual platelets upon initiation of adhesion to an injured vessel wall and during the subsequent steps in thrombus formation under flow within vessels (Takizawa et al. 2010). With our system, administration of hematoporphyrin and subsequent laser exposure causes vascular injury, which induces formation of a microthrombus.

iPSC-platelets stained with tetramethylrhodamine ethyl ester, a membrane-permeant, cationic, red-orange fluorescent dye that is readily sequestered by active mitochondria, were transfused into NOG mice, where red staining then indicated living iPSC-platelets (specimens with non-functional mitochondria may be incorporated). Following laser-induced vessel injury, we clearly observed that iPSC-platelets initially adhered to the injured vessel wall, coordinating with host platelets, and that this process ultimately led to thrombus formation and vessel occlusion, which suggests that the *in vivo* functionality of iPSC-derived platelets is intact (Takayama et al. 2010).

Perspectives

The rarity of megakaryocytes within bone marrow limits the use of bone marrow specimens for studies of megakaryopoiesis and platelet generation. We have therefore endeavored to establish a protocol for obtaining megakaryocytes capable of generating platelets from mouse ESCs (Eto et al. 2002; Nishikii et al. 2008) or hESCs (Takayama et al. 2008) *in vitro*. These methods have enabled us to obtain relatively large numbers of megakaryocytes and platelets from both ESCs and iPSCs for the study of thrombopoiesis. Furthermore, megakaryocytes derived from patient-specific hiPSCs represent a powerful tool for investigating unresolved aspects of the mechanisms underlying thrombocytopenia and also for screening novel therapeutic agents for the treatment of diseases affecting platelet production and/or function.

We have shown that iPSC-derived platelets are functional *in vivo* and could represent a novel solution for the shortage of donated platelets. However, the following questions remain. 1) How do we obtain very large numbers of cells? 2) How do we ensure a sufficient level of cell quality? 3) How do we overcome the short shelf life of platelets? To address these questions, we are now attempting to use genetic screening to establish an immortalized megakaryocyte progenitor cell line derived from hiPSCs. Once immortalized, the cells could be cryopreserved for long periods. This is a safe strategy because platelets do not have a nucleus and can be irradiated before transfusion to avoid leukemia. Additional developments, such as a device to add shear stress to promote platelet release from megakaryocytes would be expected to increase the efficiency of platelet generation, further advancing the process toward the ultimate aim of use in a clinical setting.

Acknowledgments

The authors thank Dr. S. Nishimura for developing a novel *in vivo* imaging system to visualize single platelet behavior. The Sendai viral vector for integration-free hiPSC generation was a kind gift from Dr. M. Nakanishi. Drs. T. Hiroyama, Y. Nakamura and H. Nakauchi provided valuable discussion. The other authors declare no competing financial interests.

References

Bergmeier, W., Burger, P. C., Piffath, C. L., Hoffmeister, K. M., Hartwig, J. H., Nieswandt, B., and Wagner, D. D. 2003. Metalloproteinase inhibitors improve the recovery and hemostatic function of *in vitro*-aged or -injured mouse platelets. Blood 102, 4229–4235.

Bergmeier, W., Piffath, C. L., Cheng, G., Dole, V. S., Zhang, Y., von Andrian, U. H., and Wagner, D. D. 2004. Tumor necrosis factor-alpha-converting enzyme (ADAM17) mediates GPIbalpha shedding from platelets *in vitro* and *in vivo*. Circ. Res. 95, 677–683.

Canobbio, I., Balduini, C., and Torti, M. 2004. Signalling through the platelet glycoprotein Ib-V-IX complex. Cell Signal 16, 1329–1344.

Chanprasert, S., Geddis, A. E., Barroga, C., Fox, N. E., and Kaushansky, K. 2006. Thrombopoietin (TPO) induces c-myc expression through a PI3K- and MAPK-dependent pathway that is not mediated by Akt, PKCzeta or mTOR in TPO-dependent cell lines and primary megakaryocytes. Cell Signal 18, 1212–1218.

Choi, K., Kennedy, M., Kazarov, A., Papadimitriou, J. C., and Keller, G. 1998. A common precursor for hematopoietic and endothelial cells. Development 125, 725–732.

Debili, N., Wendling, F., Katz, A., Guichard, J., Breton-Gorius, J., Hunt, P., and Vainchenker, W. 1995. The Mpl-ligand or thrombopoietin or megakaryocyte growth and differentiative factor has both direct proliferative and differentiative activities on human megakaryocyte progenitors. Blood 86, 2516–2525.

den Dekker, E., van Abel, M., van der Vuurst, H., van Eys, G. J., Akkerman, J. W., and Heemskerk, J. W. 2003. Cell-to-cell variability in the differentiation program of human megakaryocytes. Biochim. Biophys. Acta. 1643, 85–94.

Eto, K., Murphy, R., Kerrigan, S. W., Bertoni, A., Stuhlmann, H., Nakano, T., Leavitt, A. D., and Shattil, S. J. 2002. Megakaryocytes derived from embryonic stem cells implicate CalDAG-GEFI in integrin signaling. Proc. Natl. Acad. Sci. USA 99, 12819–12824.

Fujimoto, T. T., Kohata, S., Suzuki, H., Miyazaki, H., and Fujimura, K. 2003. Production of functional platelets by differentiated embryonic stem (ES) cells *in vitro*. Blood 102, 4044–4051.

Gaur, M., Kamata, T., Wang, S., Moran, B., Shattil, S. J., and Leavitt, A. D. 2006. Megakaryocytes derived from human embryonic stem cells: a genetically tractable system to study megakaryocytopoiesis and integrin function. J. Thromb. Haemost. 4, 436–442.

Guo, Y., Niu, C., Breslin, P., Tang, M., Zhang, S., Wei, W., Kini, A. R., Paner, G. P., Alkan, S., Morris, S.W. et al. 2009. c-Myc-mediated control of cell fate in megakaryocyte-erythrocyte progenitors. Blood 114, 2097–2106.

Hiroyama, T., Miharada, K., Aoki, N., Fujioka, T., Sudo, K., Danjo, I., Nagasawa, T., and Nakamura, Y. 2006. Long-lasting *in vitro* hematopoiesis derived from primate embryonic stem cells. Exp. Hematol. 34, 760–769.

Ma, F., Ebihara, Y., Umeda, K., Sakai, H., Hanada, S., Zhang, H., Zaike, Y., Tsuchida, E., Nakahata, T., Nakauchi, H. et al. 2008. Generation of functional erythrocytes from human embryonic stem cell-derived definitive hematopoiesis. Proc. Natl. Acad. Sci. USA 105, 13087–13092.

Matsubara, Y., Saito, E., Suzuki, H., Watanabe, N., Murata, M., and Ikeda, Y. 2009. Generation of megakaryocytes and platelets from human subcutaneous adipose tissues. Biochem. Biophys. Res. Commun. 378, 716–720.

Matsunaga, T., Tanaka, I., Kobune, M., Kawano, Y., Tanaka, M., Kuribayashi, K., Iyama, S., Sato, T., Sato, Y., Takimoto, R. et al. 2006. *Ex vivo* large-scale generation of human platelets from cord blood CD34+ cells. Stem Cells 24, 2877–2887.

Nakano, T., Kodama, H., and Honjo, T. 1994. Generation of lymphohematopoietic cells from embryonic stem cells in culture. Science 265, 1098–1101.

Nieswandt, B. and Watson, S. P. 2003. Platelet-collagen interaction: is GPVI the central receptor? Blood 102, 449–461.

Nishikii, H., Eto, K., Tamura, N., Hattori, K., Heissig, B., Kanaji, T., Sawaguchi, A., Goto, S., Ware, J., and Nakauchi, H. 2008. Metalloproteinase regulation improves *in vitro* generation of efficacious platelets from mouse embryonic stem cells. J. Exp. Med. 205, 1917–1927.

Nishimura, K., Segawa, H., Goto, T., Morishita, M., Masago, A., Takahashi, H., Ohmiya, Y., Sakaguchi, T., Asada, M., Imamura, T. et al. 2007. Persistent and stable gene expression by a cytoplasmic RNA replicon based on a noncytopathic variant Sendai virus. J Biol Chem 282, 27383–27391.

Osafune, K., Caron, L., Borowiak, M., Martinez, R. J., Fitz-Gerald, C. S., Sato, Y., Cowan, C. A., Chien, K. R., and Melton, D. A. 2008. Marked differences in differentiation propensity among human embryonic stem cell lines. Nat Biotechnol 26, 313–315.

Patel, S. R., Hartwig, J. H., and Italiano, J. E., Jr. 2005. The biogenesis of platelets from megakaryocyte proplatelets. J. Clin. Invest. 115, 3348–3354.

Rabie, T., Strehl, A., Ludwig, A., and Nieswandt, B. 2005. Evidence for a role of ADAM17 (TACE) in the regulation of platelet glycoprotein V. J. Biol. Chem. 280, 14462–14468.

Ruggeri, Z. M. 2002. Platelets in atherothrombosis. Nat. Med. 8, 1227–1234.

Schiffer, C. A. 2001. Diagnosis and management of refractoriness to platelet transfusion. Blood Rev. 15, 175–180.

Schipper, L. F., Brand, A., Reniers, N., Melief, C. J., Willemze, R., and Fibbe, W. E. 2003. Differential maturation of megakaryocyte progenitor cells from cord blood and mobilized peripheral blood. Exp. Hematol. 31, 324–330.

Takayama, N., Nishikii, H., Usui, J., Tsukui, H., Sawaguchi, A., Hiroyama, T., Eto, K., and Nakauchi, H. 2008. Generation of functional platelets from human embryonic stem cells *in vitro* via ES-sacs, VEGF-promoted structures that concentrate hematopoietic progenitors. Blood 111, 5298–5306.

Takayama, N., Nishimura, S., Nakamura, S., Shimizu, T., Ohnishi, R., Endo, H., Yamaguchi, T., Otsu, M., Nishimura, K., Nakanishi, M. et al. 2010. Transient activation of c-MYC expression is critical for efficient platelet generation from human induced pluripotent stem cells. J. Exp. Med. 207, 2817–2830.

Takizawa, H., Nishimura, S., Takayama, N., Oda, A., Nishikii, H., Morita, Y., Kakinuma, S., Yamazaki, S., Okamura, S., Tamura, N. et al. 2010. Lnk regulates integrin alphaIIbbeta3 outside-in signaling in mouse platelets, leading to stabilization of thrombus development *in vivo*. J. Clin. Invest. 120, 179–190.

Thompson, A., Zhang, Y., Kamen, D., Jackson, C. W., Cardiff, R. D., and Ravid, K. 1996a. Deregulated expression of c-myc in megakaryocytes of transgenic mice increases megakaryopoiesis and decreases polyploidization. J Biol Chem 271, 22976–22982.

Thompson, A., Zhao, Z., Ladd, D., Zimmet, J., and Ravid, K. 1996b. A new transgenic mouse model for the study of cell cycle control in megakaryocytes. Stem Cells 14 Suppl 1, 181–187.

Thomson, J. A., Itskovitz-Eldor, J., Shapiro, S. S., Waknitz, M. A., Swiergiel, J. J., Marshall, V. S., and Jones, J. M. 1998. Embryonic stem cell lines derived from human blastocysts. Science 282, 1145–1147.

Tomer, A. 2004. Human marrow megakaryocyte differentiation: multiparameter correlative analysis identifies von Willebrand factor as a sensitive and distinctive marker for early (2N and 4N) megakaryocytes. Blood 104, 2722–2727.

Vodyanik, M. A., Bork, J. A., Thomson, J. A., and Slukvin, II. 2005. Human embryonic stem cell-derived CD34+ cells: efficient production in the coculture with OP9 stromal cells and analysis of lymphohematopoietic potential. Blood 105, 617–626.

Wang, L., Li, L., Shojaei, F., Levac, K., Cerdan, C., Menendez, P., Martin, T., Rouleau, A., and Bhatia, M. 2004. Endothelial and hematopoietic cell fate of human embryonic stem cells originates from primitive endothelium with hemangioblastic properties. Immunity 21, 31–41.

Yokoyama, Y., Suzuki, T., Sakata-Yanagimoto, M., Kumano, K., Higashi, K., Takato, T., Kurokawa, M., Ogawa, S., and Chiba, S. 2009. Derivation of functional mature neutrophils from human embryonic stem cells. Blood 113, 6584–6592.

CHAPTER 18

Neuron Regeneration Using iPS Cells

Satoshi Nori,[1,2,a] Osahiko Tsuji,[2,b] Kyoko Miura,[1,c] Yoshiaki Toyama,[2,d] Masaya Nakamura[2,e] and Hideyuki Okano[1,f,*]

Introduction

The adult mammalian central nervous system (CNS), especially the spinal cord, has been considered a representative example of an organ in which regeneration is difficult. However, in 1992, Reynolds, and Weiss identified cells with the potential to generate new neurons or neural stem cells, which they named neurospheres (Reynolds and Weiss 1992). Subsequently, methods were established to culture mammalian neural stem/progenitor cells (NS/PCs) from animals, and humans (Davis and Temple 1994; Kilpatrick and Bartlett 1993; Palmer et al. 1995). Since the report that the transplantation of fetal spinal cord tissue leads to functional recovery from spinal cord injury (SCI) in the rat (Bregman et al. 1993), NS/PCs have been considered an effective cell source for transplantation therapy in SCI. Recently, stem cell based approaches such as NS/PC transplantation have been further identified as promising therapies for various CNS disorders (Bjorklund and Lindvall 2000; Gage 2000; Lindvall and Kokaia 2006).

[1]Department of Physiology, School of Medicine, Keio University, 35 Shinanomachi, Shinjuku, Tokyo 160-8582, Japan.
[a]Email: satoshi_nori@2003.jukuin.keio.ac.jp
[c]Email: miura@z8.keio.jp
[f]Email: hidokano@a2.keio.jp
[2]Department of Orthopedic Surgery, School of Medicine, Keio University, 35 Shinanomachi, Shinjuku, Tokyo 160-8582, Japan.
[b]Email: osahiko@2003.jukuin.keio.ac.jp
[d]Email: toyama@z6.keio.jp
[e]Email: masa@a8.keio.jp
*Corresponding author

Our group has been studying stem cell based therapy for repair of the damaged CNS, and particularly repair following SCI. Previous studies report functional recovery after transplantation of NS/PCs into the injured spinal cord of rodents, and non-human primates (Cummings et al. 2005; Hofstetter et al. 2005; Iwanami et al. 2005; Ogawa et al. 2002; Okada et al. 2005; Salazar et al. 2010). Furthermore, recent studies reveal that embryonic stem cells (ESCs) can generate neural cells including NS/PCs (Kumagai et al. 2009; Okada et al. 2008; Okada et al. 2004), and oligodendrocyte precursor cells (OPCs) (Keirstead et al. 2005; Sharp et al. 2010). Therefore, human ESC-based therapies are moving out of the laboratory, and into clinical treatments for SCI (Keirstead et al. 2005; Strauss 2010). However, there has been a long-standing debate on the ethical issues of human ESC research, and its clinical application.

Recently, induced pluripotent stem cells (iPSCs) that possess ESC-like pluripotency, and proliferative capacity have been produced by transducing several genes into somatic cells (Okita et al. 2007; Takahashi et al. 2007; Takahashi and Yamanaka 2006). If performing cell transplantation therapy using iPSC-derived NS/PCs were possible, the ethical problems of using human fetal tissue could be avoided. In this chapter, we outline our series of studies on the *in vitro* culture methods for inducing iPSC-derived NS/PCs, discuss their safety at transplantation, and describe current research on their transplantation into SCI models *in vivo*.

iPSC-derived NS/PCs, and Safety Issues

In vitro models of neural development using mouse ESC-derived NS/PCs

Neural stem/progenitor cells (NS/PCs) are defined as having self-renewal potential, and pluripotency. However, their ability to differentiate, and proliferate is strictly governed by their time, and place of birth, and not all NS/PCs have precisely the same properties (Okano and Temple 2009). NS/PCs are already present around embryonic day 5 (E5.5), and can be cultured in the presence of leukemia inhibitory factor (LIF) (Hitoshi et al. 2004). Soon thereafter, and still at a relatively early stage of embryonic development (E8.5–12.5), NS/PCs can be cultured *in vitro* in the presence of fibroblast growth factor-2 (FGF-2). From this stage until the late stage of embryonic development, the radial glia located around the cerebral ventricles possess the properties of NS/PCs, such as self-renewal by symmetric cell division, and produce neurons by asymmetric cell division (Miyata et al. 2001; Noctor et al. 2001). In the late stage of embryonic development, NS/PCs proliferate *in vitro* in response to epidermal growth factor (EGF) as well as FGF-2. However, the NS/PCs that emerge from this stage onward are no longer able to produce early-born projection neurons such as forebrain-type cholinergic neurons, dopaminergic neurons, or motor neurons, which are born only during the early stage of development. Finally, from the late stage of embryonic development through to the neonatal period, and into adulthood, NS/PCs in the brain are mainly present around the cerebral ventricles, and produce glial cells (astrocytes, and oligodendrocytes) as well as neurons (Okano and Temple 2009; Temple 2001). Thus, the differentiation, and proliferation capacity of NS/PCs is strictly controlled by the spatial, and temporal specificities that cells are endowed with at the time, and place where they are born.

In previous reports from our group (Naka et al. 2008; Okada et al. 2008), we used mouse ESCs derived from the inner cell mass to induce NS/PCs with high plasticity that are present at a relatively early stage of development. Thus, we succeeded in constructing a neural development model culture system *in vitro* that mimics neural development, and reflects its temporal, and spatial specificity. This culture system eliminates the requirement for LIF to maintain the ESCs in their undifferentiated state; also, by using floating cultures it allows the formation of embryoid bodies (EBs), which contain cells derived from the three germ layers. EBs contain relatively early stage NS/PCs, and by growing EBs in floating cultures containing serum-free neural stem cell culture medium supplemented with FGF-2, it is possible to selectively culture NS/PCs in the form of neurospheres, in which progenitor cells rather than stem cells are dominantly present. Moreover, inhibiting bone morphogenic protein (BMP) signaling during EB formation promotes differentiation into neuroepithelium, while adding a low concentration of Noggin, which plays an important role in the formation of the forebrain, or retinoic acid, which is important in neural induction and the development of the hindbrain and anterior spinal cord, increases the proportion of NS/PCs in EBs and the efficiency of neurosphere formation.

Primary neurospheres formed in this manner can be subcultured as secondary, and tertiary neurospheres. Intriguingly, although primary neurospheres produce neurons almost exclusively, secondary, and tertiary neurospheres produce glial cells (astrocytes, and oligodendrocytes) as well as neurons. The ability of neurospheres to be subcultured repeatedly, and to produce three CNS cell lineages demonstrates that mouse ESC-derived neurospheres contain NS/PCs, as defined by their potential for self-renewal, and pluripotency. Moreover, the changes in differentiation potential associated with subculturing closely reflect the temporal changes seen in the development of the nervous system *in vivo*, where only neurons are born at the initial stage of development, and glial cells first begin to be born during the middle stage of development. Furthermore, by adding Noggin or changing the concentration of retinoic acid during EB formation, NS/PCs with anterior-posterior axis region-specific properties can be induced.

Similarly, dorsal-ventral region-specific control has been achieved by adding the ventralization factor Sonic hedgehog (SHH) or the dorsalization factor BMP4 or Wnt3a to the culture during the formation of primary neurospheres (Okada et al. 2008). Based on these findings, adding the right factor at the right time during culture appears to permit the region-specific control of NS/PC properties.

Assessment of the Safety of Mouse iPSC-derived NS/PCs

iPSCs are somatic cells that have been reprogrammed by transducing such genes as *Oct4*, *Sox2*, *Klf4*, and *c-Myc* into mouse or human fibroblasts (Okita et al. 2007; Takahashi et al. 2007; Takahashi, and Yamanaka 2006). They are pluripotent stem cells that have the same proliferative ability, and differentiation potential as ESCs. It appears possible to generate iPSCs from somatic cells, a technology that is expected to resolve the ethical problems that hinder the use of ESCs.

iPSCs can be induced into neurospheres by the same culture system used in ESCs. We recently demonstrated that the responsiveness of iPSCs to neural differentiation, and their safety after transplantation vary greatly according to differences in somatic cell origin at the time they were generated (Miura et al. 2009). We induced differentiation of neurospheres using 36 mouse iPSC clones that were previously generated in our laboratory, and then evaluated their differentiation capacity *in vitro*, and safety after transplantation into the striatum of the brains of non-obese, diabetic-severe combined immunodeficient (NOD-SCID) mice. Almost all of the iPSC lines analyzed were capable of differentiating into neurospheres. However, detailed flow cytometry analysis revealed large differences in the percentages of *Nanog* gene-driven, enhanced green fluorescence, protein-positive, undifferentiated cells that remained in the neurospheres according to the type of somatic cells from which the iPSCs had been derived.

A mouse embryonic fibroblast (MEF)-derived iPSC clone demonstrated responsiveness to induction of differentiation that was comparable to that of ESCs, and hardly any undifferentiated cells remained in the neurospheres. Teratoma formation in mice after transplantation of neurospheres derived from this MEF-iPSC clone was infrequent, and minor, comparable to the teratoma formation obtained with mice transplanted with ESC-derived neurospheres. Moreover, no teratoma formation was observed during a 16-week observation period in a group transplanted with two iPSC clones that had been produced from adult gastric epithelial cells (Stm). By contrast, an iPSC clone derived from adult tail-tip fibroblasts (TTFs) demonstrated statistically significant resistance to differentiation, and many undifferentiated cells remained in the neurospheres after inducing differentiation. Significantly larger teratomas were observed in the group of mice that had been transplanted with these TTF-iPSC-derived neurospheres, and many of the mice soon became debilitated or died. The responsiveness to induction of differentiation, and the teratoma-forming propensity of an adult hepatocyte (Hep)-derived iPSC clone were intermediate between those of the MEF-iPSC, and TTF-iPSC clones. Use of a *c-Myc* transgene had no effects on the teratoma-forming propensity of iPSC-derived NS/PCs.

Incomplete reprogramming, that is, the persistence of the gene expression patterns of the original somatic cells from which they were derived, appears to underlie the variations in the differentiation capacity of iPSCs. In connection with this, according to a recent report, early-passage iPSCs obtained by reprogramming adult mouse tissue with transcription factors results in iPSCs with traces of the DNA methylation pattern of the original somatic cell tissue. Such iPSCs demonstrate a tendency to differentiate along cell lineages that resemble those of the donor cells, and their cell fate choices are limited. This sort of donor tissue "epigenetic memory" is reported to contribute largely to the properties of iPSCs (Kim et al. 2010; Sakurada 2010).

Transplantation of "safe" Mouse iPSC Clone-derived NS/PCs for the Treatment of SCI

We previously assessed the safety of each iPSC clone by transplanting neurospheres into the brains of NOD-SCID mice (Miura et al. 2009). We then used a mouse iPSC clone whose safety had been confirmed (clone 38C2) to study the effectiveness of treating SCI by transplanting 38C2-derived neurospheres into the lesion site in a mouse model (Tsuji et al. 2010). Transplantation of 38C2-iPSC-derived secondary neurospheres (38C2-SNSs) was performed during the sub-acute stage at day 9 after SCI. We used luciferase luminescence to bio-image the grafted cells, and found that approximately 20% of the transplanted 38C2-SNSs survived within the injured spinal cord, and differentiated into neurons (31.4% ± 1.1%), astrocytes (49.3% ± 4.5%), and oligodendrocytes (14.4% ± 3.0%), suggesting that almost all grafted 38C2-SNSs differentiated into definable CNS cell types. Severe atrophic change, and demyelination occurred in the injured spinal cord after the contusive injury, but these changes were significantly diminished in the group treated with 38C2-SNSs. Quantification of LFB-positive areas at the lesion epicenter 42 d after injury revealed that 38C2-SNS-grafted mice demonstrated a significantly larger myelinated area than PBS-treated control mice. Detailed observation indicated that graft (38C2-SNS)-derived oligodendrocytes contributed to remyelination. Moreover, in the 38C2-SNS-grafted group, serotonergic (5-hydroxytryptamine (5HT)-positive) fiber innervation of the distal cord was promoted, and supported by grafted 38C2-SNS-derived glial fibrillary acidic protein (GFAP)-positive astrocytes compared to the control group. Interestingly, these graft-derived astrocytes exhibited a bipolar morphology with long processes, extending along the axis of the spinal cord caudal to the lesion epicenter, and in close association with the 5HT-positive host serotonergic fibers, which are known to be associated with the locomotor function of hind limbs (Kaneko et al. 2006).

Quantitative analysis of the serotonergic innervation of the distal cord revealed a significant difference between 38C2-SNS-, and PBS-treated mice. Notably, serotonergic innervation of the distal cord was significantly enhanced by the grafted 38C2-SNSs 6 wk after SCI (Tsuji et al. 2010). Furthermore, evaluation of hind limb motor function on the Basso mouse scale (BMS) revealed significantly better recovery of function in the group treated with iPSC-derived neurospheres than in the control groups (PBS-treated, and fibroblast-transplanted groups) (Tsuji et al. 2010). Namely, 42 d after injury, the 38C2-SNS-grafted animals could lift their trunks (BMS: 4.3 ± 0.28), and had significantly higher BMS scores than PBS-treated control animals (BMS: 2.8 ± 0.2), which were unable to support their body weight with their hind limbs. From a clinical perspective, the recovery of weight-supported plantar steps in the 38C2-SNS-grafted group was noteworthy. Collectively, the observed functional recovery was apparently attributable to both cell-autonomous, and non-cell-autonomous effects, such as the promotion of axonal regrowth in the raphespinal tract supported by graft-derived bipolar astrocytes, and remyelination by the transplanted cells, in addition to the mitigation of the atrophic change, and demyelination mentioned above.

Next, considering the clinical setting, we conducted a similar transplantation experiment using adult-tissue (TTF)-derived iPSCs as a model. There were six TTF-derived clones among the 36 mouse iPSC clones (Miura et al. 2009), but only one, clone 335D1, was eventually confirmed to be safe in our teratoma assay. We used clone 335D1 along with two "unsafe" clones (256H13, 256H18) in a transplantation experiment, in which the clones were induced to form neurospheres, and transplanted into the mouse SCI model as described for clone 38C2. We found that, although functional recovery was achieved after transplantation of neurospheres from all of the clones (335D1, 256H13, 256H18) into the injured spinal cord, the improvement was eventually lost at 6 weeks after injury in the mice with transplants of the "unsafe"-clone-derived neurospheres. Furthermore, most of these mice then died suddenly. Subsequent histological analysis revealed huge teratomas within the spinal cords of the animals that had been treated with neurospheres derived from the "unsafe" clones. Moreover, these grafted "unsafe" cells exhibited Nanog-, and Oct4-immunoreactivity, suggesting that the grafted, and surviving "unsafe"-clone-derived cells remained in the undifferentiated state, and failed to differentiate into definable CNS cell types of neurons, astrocytes, and oligodendrocytes, instead resulting in teratoma formation within the injured cord. By contrast, when the "safe" 335D1 clone was used, no tumor formation was observed in any of the mice, as expected, and their functional recovery was statistically significant in comparison with the control group, and equivalent to that obtained with ESC-derived neurospheres.

These findings demonstrated that adult-tissue-derived iPSC clones were generally not as safe as clones derived from embryonic tissue. However, extremely wide variation was observed among the clones with respect to their safety after transplantation. If their safety is rigorously assessed in advance, iPSCs may a useful source of cells for the treatment of SCI (Tsuji et al. 2010).

Grafted Human iPSC-derived NS/PCs Promote Motor Functional Recovery After SCI in Mice

Human iPSCs (hiPSCs), which do not entail ethical concerns, may become a preferred cell source for regenerative medicine in human patients. Therefore, we next investigated the therapeutic potential of transplanting hiPSC-derived neurospheres (hiPSC-NSs) into spinal cord injured NOD-SCID mice (Nori et al. 2011). For this aim, we used a hiPSC clone (201B7) established by transducing four reprogramming factors (*Oct4*, *Sox2*, *Klf4*, and *c-Myc*) into adult human fibroblasts (Takahashi et al. 2007). hiPSC-derived primary neurospheres were dissociated, and infected with lentivirus expressing Venus fluorescent protein under control of the EF promoter (pCSII-EF-Venus). Before transplantation, the primary neurospheres were passaged into secondary, and tertiary neurospheres. Contusive SCI was induced at the Th10 level in NOD-SCID mice, and 5×10^5 Venus-positive hiPSC-neurospheres (NSs) or PBS (control group) was injected into the lesion epicenter 9 d after injury (Nori et al. 2011). Subsequently, histological analyses were performed 56 d after SCI.

We found that the grafted hiPSC-NSs survived, and migrated into the host spinal cord. The transplanted hiPSC-NSs differentiated into neuronal nuclei (NeuN)-positive, and β-tubulin isotype III (βIII tubulin)-positive neurons, GFAP-positive astrocytes, and adenomatous polyposis coli CC-1 (APC)-positive oligodendrocytes. βIII tubulin-positive/Venus-positive neurons comprised 49.1% ± 2.0% of Venus-positive cells, and mature NeuN-positive/Venus-positive neurons comprised 22.9% ± 1.0% of Venus-positive cells. Thus, approximately 50% of the grafted hiPSC-NSs differentiated into neurons, about half of which were mature neurons. GFAP-positive/Venus-positive astrocytes comprised 17.0% ± 1.2% of cells, but APC-positive/Venus-positive oligodendrocytes were rare (3.0% ± 0.4%). Nestin-positive/

Venus-positive NS/PCs made up 10.7% ± 2.2% of the total. Since 22.9% of hiPSC-NSs differentiated into mature neurons, and 15.8% ± 2% of Venus-positive cells were glutamic acid decarboxylase 67 (GAD67) positive, 69% (15.8%/22.9% = 69.0%) of the hiPSC-NS-derived mature neurons were GABAergic. We also found small numbers of Venus-positive/tyrosine hydroxylase-positive dopaminergic neurons, and choline acetyltransferase-positive cholinergic neurons.

Interestingly, graft-derived human neurons were found to form synapses with host mouse neurons. Grafted βIII tubulin-positive/human nuclear protein (HNu)-positive cells in parenchymal location were contacted by synaptic boutons of the host origin neurons, suggesting that the host axons innervated hiPSC-derived neurons. Immuno-electron microscopic examination revealed graft-derived (Venus-positive) presynaptic, and postsynaptic structures as well as synapse formation between host neurons at the injured site. We observed that most hiPSC-derived mature neurons were GABAergic, a neurotransmitter type that plays important roles in the spinal cord by controlling the levels of motoneuron output, and sensory input by modulating primary afferent transmitter release as well as direct postsynaptic inhibition of motoneurons (Malcangio and Bowery 1996, Schneider and Fyffe 1992). Furthermore, hypofunction of spinal GABAergic inhibition is involved in pathological pain states that develop due to SCI (Kim et al. 2010). Thus, we propose that synapse formation by donor-derived GABAergic neurons may be important for motor coordination within the spinal neuronal network, as well as for the suppression of SCI-induced spasticity (Iwanami et al. 2005; Watanabe et al. 2004), and pain (Kim et al. 2010).

Next, we examined the non-cell autonomous angiogenic effects of hiPSC-NS transplantation (Nori et al. 2011). We observed significantly more platelet endothelial cell adhesion molecule-1-positive blood vessels at the lesion epicenter in the hiPSC-NS group compared to the control group. To clarify the underlying angiogenic signals, we examined the expression of the angiogenic growth factor vascular endothelial growth factor (VEGF) in the grafted spinal cord. Quantitative analyses revealed a significantly larger VEGF-positive area at the lesion epicenter in the hiPSC-NS group than the control group. Furthermore, GFAP-positive/Venus-positive hiPSC-derived astrocytes as well as GFAP-positive/Venus-negative host astrocytes were found to express VEGF. This observation was consistent with the results of RT-PCR. Notably, the mouse-specific *Vegf* expression level was higher in the hiPSC-NS-grafted mice than in PBS-injected mice.

In association with angiogenesis, and its positive effect on tissue sparing, we examined atrophic changes in the injured spinal cord by hematoxylin-eosin staining. Compared to the hiPSC-NS group, atrophic changes of the injured spinal cord were prominent in the control group.

Quantitative analysis revealed significant differences in the transverse area of the spinal cord between the hiPSC-NS, and control group, suggesting that hiPSC-NS transplantation prevented atrophy of the spinal cord. Luxol fast blue (LFB) staining revealed a greater preservation of myelinated areas in the hiPSC-NS group compared to the control group. Angiogenesis after SCI promotes endogenous repair, and supports axonal outgrowth (Beattie et al. 1997; Kim et al. 2009). Here, we observed that hiPSC-NS transplantation enhanced angiogenesis, and tissue sparing after SCI. By contrast, astrocytes are known to express angiogenic growth factors such as VEGF under hypoxic conditions (Kumagai et al. 2009; Mense et al. 2006; Yoshida et al. 2002). Consistent with this finding, hiPSC-derived astrocytes as well as host astrocytes expressed VEGF, suggesting that the transplantation of hiPSC-NSs promoted VEGF expression in both host-, and graft-derived astrocytes.

We also evaluated the effects of hiPSC-NS transplantation on axonal regrowth after SCI through immunohistochemical analyses of 200-kDa neurofilament (NF-H), 5HT, and growth-associated protein 43 (GAP43) in the injured spinal cord. We observed significantly higher numbers of NF-H-positive neuronal fibers in the hiPSC-NS group than in the control group. In all mice, 5HT-positive fibers were observed at the lumbar intumescence. Although no additional increase in the areas of 5HT-positive fibers was observed in the control group, the hiPSC-NS group demonstrated enhanced innervation of the distal cord by 5HT-positive fibers. Quantitative analyses of the serotonergic innervation of the distal cord revealed significant differences in the hiPSC-NS group compared to the control group. GAP43-positive axons, which represent regrowing axons (Kobayashi et al. 1997), were also detected in the distal cords of all mice. Significantly more GAP43-positive fibers were detected in the hiPSC-NS group in the caudal to the lesion epicenter. These results suggest that the transplantation of hiPSC-NSs promoted axonal regrowth in the injured spinal cord.

Notably, we also observed that hiPSC-NS-derived astrocytes closely associated with NF-H-positive fibers, and 5-HT-positive fibers, suggesting the possibility that graft-derived astrocytes exerted non-cell autonomous effects for facilitating the regrowth or regeneration of host axons. Thus, we examined the expression of graft-, and host-derived neurotrophic factors, including nerve growth factor (NGF), brain-derived neurotrophic factor (BDNF), and hepatocyte growth factor (HGF), by RT-PCR analysis. Significant expression of these factors, which are associated with the axonal growth, and survival of existing neurons, was observed in both the grafted hiPSC-NSs, and the host mouse tissues. Graft-derived astrocytes are reported to promote the regrowth of NF-H-positive, and 5HT-positive fibers by offering a growth-permissive substrate (Hofstetter et al. 2002; Kumagai et al. 2009; Tsuji et al. 2010). Furthermore, neurotrophic factors such as NGF, BDNF, and HGF play critical roles in axonal growth, and the survival of existing neurons (Brock et al. 2010; Jakeman et al. 1998; Jones et al. 2001; Kitamura et al. 2007; Tuszynski 1997; Tuszynski et al. 1996). Consistently, we observed that hiPSC-NS transplantation promoted axonal regrowth in the distal spinal cord.

Since the treatment induced axonal regrowth, we also examined the pain afferents entering lamina III of the spinal cord, which are closely associated with NS/PC-transplantation-induced allodynia (Hofstetter et al. 2005). Thus, we investigated the changes in expression of calcitonin gene-related peptide (CGRP), which is involved in peripheral, and spinal pain mechanisms (Hofstetter et al. 2005; Krenz et al. 1999; Krenz and Weaver 1998). We quantified the areas of CGRP-positive fibers in lamina III 4 mm rostral, and 4 mm caudal to the lesion epicenter. No significant differences in the area of CGRP-positive fibers of lamina III were observed between the hiPSC-NS, and control groups. Thus, our data suggested that hiPSC-NS transplantation did not induce abnormal innervation by pain afferents, in contrast to a previous study in which NS/PCs derived from adult rat spinal cord were transplanted into a rat SCI model (Hofstetter et al. 2005).

We evaluated motor functional recovery by three methods. BMS indicated that the hiPSC-NS group exhibited significantly better functional recovery than the control group at 21 d, and thereafter. The rotarod test revealed significantly longer running time in the hiPSC-NS group than the control group at 56 d after SCI. Furthermore, we examined treadmill gait analysis using the DigiGait system, which enabled the objective evaluation of motor function (Li et al. 2005; Springer et al. 2010). The profile of stride length at 8 cm/s speed clearly demonstrated significantly better recovery of motor function in the hiPSC-NS-grafted mice compared with control mice. To electrophysiologically evaluate functional recovery, motor-evoked potential (MEP) was measured in the mice in all groups. The waves were detected in most of the hiPSC-NS group (14 of 17 mice), whereas they were not detected in the control group (0 of 15 mice).

The greatest potential drawback of hiPSC-based therapies is their potential for tumorigenicity. Therefore, we performed a long-term safety study of the grafted hiPSC-NSs by extending the follow-up period to 112 d after SCI. Functional recovery was maintained in the hiPSC-NS group until 112 d after SCI. Consistently, the hiPSC-NS-grafted mice demonstrated no tumor formation, and the grafted cells exhibited normal neural differentiation, and maturation. We determined the proportion of cells immunopositive for each cell type-specific marker among Venus-positive-grafted cells at 112 d after SCI. Grafted hiPSC-NS differentiated into NeuN-positive cells (40.2% ± 2.8%), βIII tubulin-positive cells (50.7% ± 2.0%), GFAP-positive cells (18.1% ± 2.2%), APC-positive cells (8.9% ± 1.6%), and Nestin-positive cells (7.5% ± 1.0%) at 112 d after SCI, whereas the cells differentiated into NeuN-positive cells (22.9% ± 1.0%), βIII tubulin-positive cells (49.1% ± 2.0%), GFAP-positive cells (17.0% ± 1.2%), APC-positive cells (3.0% ± 0.4%), and Nestin-positive cells (10.7% ± 2.2%) at 56 d after SCI. Note that significantly higher percentages of NeuN-positive mature neurons, and APC-positive oligodendrocytes were observed at 112 d compared to those at 56 d after SCI. By contrast, Nestin-positive NS/PCs slightly decreased at 112 d compared to those at 56 d after SCI.

We also examined the number of proliferating grafted cells positive for Ki-67. Ki-67-positive/HNu-positive cells significantly decreased over time from 1.1% ± 0.2% at 56 d to 0.7% ± 0.1% at 112 d after SCI. Ki-67-positive/HNu-positive cells were dispersed throughout the graft area without evidence of clustering at specific sites. Consistent with our findings, previous studies of human fetal NS/PCs reported some Nestin-positive/HNu-positive cells that were negative for the proliferative markers Ki-67, and proliferating cell nuclear antigen in the striatum of NOD-SCID mice at six months after transplantation (Ogawa et al. 2009); 28–34% Nestin-positive/HNu-positive, and 2–4% Ki-67-positive/HNu-positive cells in the injured spinal cord of NOD-SCID mice at four months after transplantation (Salazar et al. 2010); and 11–14% Nestin-positive/HNu-positive, and 3–5% Ki-67-positive/HNu-positive cells in the intact spinal cord of nude rats at six months after transplantation (Yan et al. 2007). Despite the presence of Nestin-positive-, and Ki-67-positive-grafted cells, tumor was not observed in any of these studies, even after long-term observation. Collectively, the increase in the number of mature neural cells over time, the lack of tumors, and low Nestin, and Ki-67-positive cell numbers support the potential that this approach will be safe in human therapy.

Future Tasks, and Prospects

We summarize here, the future tasks, and prospects of iPSC-based cell therapy with a particular focus on safety issues. As discussed in the preceding sections, and our recent reports (Nori et al. 2011; Tsuji et al. 2010; Okano et al. 2011), iPSCs have great potential as a cell transplantation source for the treatment of SCI. However, in initial reports, reprogramming factors were transduced for iPSC establishment using retroviruses (Takahashi et al. 2007), which are often integrated in the vicinity of gene promoters where they risk altering the expression pattern of nearby endogenous genes, and causing tumor formation. For this reason, increasing efforts have recently been directed toward the generation of insertionless or insertion-free iPSCs using chemical compounds (Efe and Ding 2011; Ichida et al. 2009; Zhu et al. 2010), adenovirus vectors (Stadtfeld et al. 2008), transposons (Kaji et al. 2009; Woltjen et al. 2009), plasmids (Okita et al. 2008), recombinant proteins 1 episomal vectors (Okita et al. 2010; Okita et al. 2011), Sendai virus vectors (Ban et al. 2011; Fusaki et al. 2009; Jin et al. 2003; Seki et al. 2010), and modified RNA (Warren et al. 2010; reviewed by Okano et al. 2011). The use of iPSCs produced without exogenous gene insertion appears to be preferable for cell transplantation therapy.

Efforts are also underway to modify the transgenes, and/or chemical compounds employed in an attempt to improve both the quality of iPSCs, and the efficiency of iPSC generation. These strategies include the development of somatic inhibitors for histone deacetylase (Huangfu et al. 2008), and protein kinases (Silva et al. 2008), and the replacement of *c-Myc* with *L-Myc* (Nakagawa et al. 2010). Reprogramming of somatic cells is also being explored using mature microRNAs (Anokye-Danso et al. 2011; Miyoshi et al. 2011), and the maternal transcription factor *Glis1* (Maekawa et al. 2011). However, detailed comparisons of the properties of iPSC clones generated by these various methods are absolutely necessary to determine whether their pluripotency, and *in vitro* differentiation capacity are comparable to those of iPSCs produced with retroviruses. Furthermore, issues of genomic instability, including copy number variations, and whole chromosome aneuploidies, are very important in considering the safety issues of iPSC-based therapy, irrespective of the methods of iPSCs establishment.

Recently, clinical trials of human stem cell-based therapy for SCI were launched using human NS/PCs (Cummings et al. 2005; Salazar et al. 2010) or human ESC-derived OPCs (Strauss 2010). In this paradigm, OPC-mediated restoration of myelination, and trophic effects are the most likely mechanisms for the resulting benefits. However, the use of human ESCs (hESCs) may be problematic due to associated immune responses, and ethical viewpoints. The studies presented here suggest that grafted hiPSC-NSs may be more beneficial than hESCs because hiPSC-NSs gave rise to GABAergic neurons, which can help to suppress SCI-induced spasticity, and pain (Iwanami et al. 2005; Kim et al. 2010; Watanabe et al. 2004). In addition, as described above, we found that hiPSC-NSs differentiated into astrocytes, and oligodendrocytes, which can exert positive effects through both cell-autonomous, and non-cell-autonomous (trophic) mechanisms (Nori et al. 2011). Nevertheless, our results are only a first step toward clinical applications (Okano et al. 2011). The safety, and effectiveness of hiPSC-derived NS/PCs should be more intensively investigated, for example, using non-human primate SCI models (Iwanami et al. 2005).

A recent study indicated that the immunogenicity of mouse iPSCs was stronger than that of mouse ESCs upon their transplantation into C57BL/6 mice (Zhao et al. 2011). However, no tumorigenesis or immunogenicity was observed after transplantation of pre-differentiated NS/PCs from appropriately pre-evaluated mouse iPSCs into the injured mouse spinal cord (Tsuji et al. 2010). Furthermore, the iPSC-derived NS/PCs differentiated into normal tri-lineage neural cells in the injured spinal cord in a similar manner to ESC-derived NS/PCs (Kumagai et al. 2009), and fetal CNS-derived NS/PCs (Okada et al. 2005; Okano et al. 2011).

Previous research supports the effectiveness of iPSC-derived NS/PC therapy for SCI when high-quality, appropriate iPSC clones are used. With a view to clinical application, the quality of both iPSCs, and iPSC-derived NS/PCs should be evaluated from the perspectives of their differentiation ability *in vitro,* and *in vivo,* genetic, and epigenetic status, and tumorigenicity.

Acknowledgements

The original investigation of iPSCs was performed in collaboration with Professor Shinya Yamanaka (Center for iPS Cell Research & Application, Kyoto University, Japan). This work was supported by Grants-in-Aid for Scientific Research from JSPS, and the Ministry of Education, Culture, Sports, Science, and Technology of Japan (MEXT), the Project for Realization of Regenerative Medicine, and Support for Core Institutes for iPS Cell Research from the MEXT, JST-CIRM collaborative program; the Kanrinmaru Project from Keio University; Research Fellowships for Young Scientists from the Japan Society for the Promotion of Science; Keio Gijuku Academic Development Funds; and by a grant-in-aid for the Global COE program from MEXT to Keio University.

References

Anokye-Danso, F., Trivedi, C. M., Juhr, D., Gupta, M., Cui, Z., Tian, Y., Zhang, Y. Yang, W., Gruber, P. J., Epstein, J. A., and Morrisey, E. E. 2011. Highly efficient miRNA-mediated reprogramming of mouse, and human somatic cells to pluripotency. Cell Stem Cell 8, 376–388.

Ban, H., Nishishita, N., Fusaki, N., Tabata, T. , Saeki, K., Shikamura, M., Takada, N., Inoue, M. , Hasegawa, M. , Kawamata, S., and Nishikawa, S. 2011. Efficient generation of transgene-free human induced pluripotent stem cells (iPSCs) by temperature-sensitive Sendai virus vectors. Proc. Natl. Acad. Sci. USA 108, 14234–14239.

Beattie, M. S., Bresnahan, J. C., Komon, J. , Tovar, C. A., Van Meter, M., Anderson, D. K., Faden, A. I., Hsu, C. Y., Noble, L. J., Salzman, S., and Young, W. 1997. Endogenous repair after spinal cord contusion injuries in the rat. Exp. Neurol. 148, 453–463.

Bjorklund, A. and Lindvall, O. 2000. Cell replacement therapies for central nervous system disorders. Nat. Neurosci. 3, 537–544.

Bregman, B. S., Kunkel-Bagden, E., Reier, P. J., Dai, H. N., McAtee, M., and Gao, D. 1993. Recovery of function after spinal cord injury: mechanisms underlying transplant-mediated recovery of function differ after spinal cord injury in newborn, and adult rats. Exp. Neurol. 123, 3–16.

Brock, J. H., Rosenzweig, E. S., Blesch, A., Moseanko, R., Havton, L. A., Edgerton, V. R., and Tuszynski, M. H. 2010. Local, and remote growth factor effects after primate spinal cord injury. J. Neurosci. 30, 9728–9737.

Cummings, B. J., Uchida, N., Tamaki, S. J., Salazar, D. L., Hooshmand, M., Summers, R., Gage, F. H., and Anderson, A. J. 2005. Human neural stem cells differentiate, and promote locomotor recovery in spinal cord-injured mice. Proc. Natl. Acad. Sci. USA 102, 14069–14074.

Davis, A. A. and Temple, S. 1994. A self-renewing multipotential stem cell in embryonic rat cerebral cortex. Nature 372, 263–266.

Efe, J. A. and Ding, S. 2011. The evolving biology of small molecules: controlling cell fate, and identity. Philos. Trans. R. Soc. Lond. B. Biol. Sci. 366, 2208–2221.

Fusaki, N., Ban, H., Nishiyama, A., Saeki, K., and Hasegawa, M. 2009. Efficient induction of transgene-free human pluripotent stem cells using a vector based on Sendai virus, an RNA virus that does not integrate into the host genome. Proc. Jpn. Acad. Ser. B. Phys. Biol. Sci. 85, 348–362.

Gage, F. H. 2000. Mammalian neural stem cells. Science. 287, 1433–1438.

Hitoshi, S., Seaberg, R. M., Koscik, C., Alexson, T., Kusunoki, S., Kanazawa, I., Tsuji S., and van der Kooy, D. 2004. Primitive neural stem cells from the mammalian epiblast differentiate to definitive neural stem cells under the control of Notch signaling. Genes Dev. 18, 1806–1811.

Hofstetter, C.P., Holmstrom, N. A., Lilja, J. A., Schweinhardt, P., Hao, J., Spenger, C., Wiesenfeld-Hallin, Z., Kurpad, S. N., Frisen, J., and Olson, L. 2005. Allodynia limits the usefulness of intraspinal neural stem cell grafts; directed differentiation improves outcome. Nat. Neurosci. 8, 346–353.

Hofstetter, C. P., Schwarz, E. J., Hess, D., Widenfalk, J., El Manira, A., Prockop, D. J., and Olson, L. 2002. Marrow stromal cells form guiding strands in the injured spinal cord, and promote recovery. Proc. Natl. Acad. Sci. USA 99, 2199–2204.

Huangfu, D., Osafune, K., Maehr, R., Guo, W., Eijkelenboom, A., Chen, S., Muhlestein, W., and Melton, D. A. 2008. Induction of pluripotent stem cells from primary human fibroblasts with only Oct4, and Sox2. Nat. Biotechnol. 26, 1269–1275.

Ichida, J. K., Blanchard, J., Lam, K., Son, E. Y., Chung, J. E., Egli, D., x. Loh, D., Carter, A. C., Di Giorgio, F. P., Koszka, K., Huangfu, D., Akutsu, H., Liu, D. R., Rubin, L. L., and Eggan, K. 2009. A small-molecule inhibitor of tgf-Beta signaling replaces sox2 in reprogramming by inducing nanog. Cell Stem Cell. 5, 491–503.

Iwanami, A., Kaneko, S., Nakamura, M., Kanemura, Y., Mori, H., Kobayashi, S., Yamasaki, M., Momoshima, S., Ishii, H., Ando, K., Tanioka, Y., Tamaoki, N., Nomura, T., Toyama, Y., and Okano, H. 2005. Transplantation of human neural stem cells for spinal cord injury in primates. J. Neurosci. Res. 80, 182–190.

Jakeman, L. B., Wei, P., Guan, Z., and Stokes, B. T. 1998. Brain-derived neurotrophic factor stimulates hindlimb stepping, and sprouting of cholinergic fibers after spinal cord injury. Exp. Neurol. 154, 170–184.

Jin, C. H., Kusuhara, K., Yonemitsu, Y., Nomura, A., Okano, S., Takeshita, H., Hasegawa, M., Sueishi, K., and Hara, T. 2003. Recombinant Sendai virus provides a highly efficient gene transfer into human cord blood-derived hematopoietic stem cells. Gene Ther. 10, 272–277.

Jones, L. L., Oudega, M., Bunge, M. B., and Tuszynski, M. B. 2001. Neurotrophic factors, cellular bridges, and gene therapy for spinal cord injury. J. Physiol. 533, 83–89.

Kaji, K., Norrby, K., Paca, A., Mileikovsky, M., Mohseni, P., and Woltjen, K. 2009. Virus-free induction of pluripotency, and subsequent excision of reprogramming factors. Nature 458, 771–775.

Kaneko, S., Iwanami, A., Nakamura, M., Kishino, A., Kikuchi, K., Shibata, S., Okano, H. J., Ikegami, T., Moriya, A., Konishi, O., Nakayama, C., Kumagai, K., Kimura, T., Sato, Y., Goshima, Y., Taniguchi, M., Ito, M., He, Z., Toyama, Y., and Okano, H. 2006. A selective Sema3A inhibitor enhances regenerative responses, and functional recovery of the injured spinal cord. Nat. Med. 12, 1380–1389.

Keirstead, H. S., Nistor, G., Bernal, G., Totoiu, M., Cloutier, F., Sharp, K., and Steward, O. 2005. Human embryonic stem cell-derived oligodendrocyte progenitor cell transplants remyelinate, and restore locomotion after spinal cord injury. J. Neurosci. 25, 4694–4705.

Kilpatrick, T. J. and Bartlett, P. F. 1993. Cloning, and growth of multipotential neural precursors, requirements for proliferation, and differentiation. Neuron. 10, 255–265.

Kim, D., Kim, C. H., Moon, J. I., Chung, Y. G., Chang, M. Y., Han, B. S., Ko, S., Yang, E., Cha, K. Y., Lanza, R., and Kim, K. S. 2009. Generation of human induced pluripotent stem cells by direct delivery of reprogramming proteins. Cell Stem Cell. 4, 472–476.

Kim, D. S., Jung, S. J., Nam, T. S., Jeon, Y. H., Lee, D. R., Lee, J. S., Leem, J. W., and Kim, D. W. 2010. Transplantation of gabaergic neurons from ES cells attenuates tactile hypersensitivityfollowing spinal cord injury. Stem Cells.

Kim, H. M., Hwang, D. H., Lee, J. E., Kim, S. U., and Kim, B. G. 2009. *Ex vivo* VEGF delivery by neural stem cells enhances proliferation of glial progenitors, angiogenesis, and tissue sparing after spinal cord injury. PLoS One. 4, e4987.

Kim, K., Doi, A., Wen, B., Ng, K., Zhao, R., Cahan, P. , Kim, J., Aryee, M. J., Ji, H., Ehrlich L. I., Yabuuchi, A., Takeuchi, A., Cunniff, K. C., Hongguang, H., McKinney-Freeman, S., Naveiras, O., Yoon, T. J., Irizarry, R. A., Jung, N., Seita, J., Hanna, J., Murakami, P., Jaenisch, R., Weissleder, R., Orkin, S. H., Weissman, I. L., Feinberg, A. P., and Daley, G. Q. 2010. Epigenetic memory in induced pluripotent stem cells. Nature 467, 285–290.

Kitamura, K., Iwanami, A., Nakamura, M., Yamane, J., Watanabe, K., Suzuki, Y., Miyazawa, D., Shibata, S., Funakoshi, H., Miyatake, S., Coffin, R.S., Nakamura, T., Toyama, Y., and Okano, H. 2007. Hepatocyte growth factor promotes endogenous repair, and functional recovery after spinal cord injury. J. Neurosci. Res. 85, 2332–2342.

Kobayashi, N. R., Fan, D. P., Giehl, K. M., Bedard, A. M., Wiegand, S. J., and Tetzlaff, W. 1997. BDNF, and NT-4/5 prevent atrophy of rat rubrospinal neurons after cervical axotomy, stimulate GAP-43, and Talpha1-tubulin mRNA expression, and promote axonal regeneration. J. Neurosci. 17, 9583–9595.

Krenz, N. R., Meakin, S. O., Krassioukov, A. V., and Weaver, L. C. 1999. Neutralizing intraspinal nerve growth factor blocks autonomic dysreflexia caused by spinal cord injury. J. Neurosci. 19, 7405–7414.

Krenz, N. R. and Weaver, L. C. 1998. Sprouting of primary afferent fibers after spinal cord transection in the rat. Neuroscience 85, 443–458.

Kumagai, G., Okada, Y., Yamane, J., Nagoshi, N., Kitamura, K., Mukaino, M., Tsuji, O., Fujiyoshi, K., Katoh, H., Okada, S., Shibata, S., Matsuzaki, Y., Toh, S., Toyama, Y., Nakamura M., and Okano, H. 2009. Roles of ES cell-derived gliogenic neural stem/progenitor cells in functional recovery after spinal cord injury. PLoS One 4, e7706.

Li, S., Kim, J. E., Budel, S., Hampton, T. G., and Strittmatter, S. M. 2005. Transgenic inhibition of Nogo-66 receptor function allows axonal sprouting, and improved locomotion after spinal injury. Mol. Cell Neurosci. 29, 26–39.

Lindvall, O. and Kokaia, Z. 2006. Stem cells for the treatment of neurological disorders. Nature 441, 1094–1096.

Maekawa, M., Yamaguchi, K., Nakamura, T., Shibukawa, R., Kodanaka, I., Ichisaka, T., Kawamura, Y., Mochizuki, H., Goshima, N., and Yamanaka, S. 2011. Direct reprogramming of somatic cells is promoted by maternal transcription factor Glis1. Nature 474, 225–229.

Malcangio, M. and Bowery, N. G. 1996. GABA, and its receptors in the spinal cord. Trends Pharmacol. Sci. 17, 457–462.

Mense, S. M., Sengupta, A., Zhou, M., Lan, C., Bentsman, G., D. J. Volsky, and Zhang, L. 2006. Gene expression profiling reveals the profound upregulation of hypoxia-responsive genes in primary human astrocytes. Physiol. Genomics 25, 435–449.

Miura, K., Okada, Y., Aoi, T., Okada, A., Takahashi, K., Okita, K., Nakagawa, M., Koyanagi, M., Tanabe, K., Ohnuki, M., Ogawa, D., Ikeda, E., Okano, H., and Yamanaka, S. 2009. Variation in the safety of induced pluripotent stem cell lines. Nat. Biotechnol. 27, 743–745.

Miyata, T., Kawaguchi, A., Okano, H., and Ogawa, M. 2001. Asymmetric inheritance of radial glial fibers by cortical neurons. Neuron. 31, 727–741.

Miyoshi, N., Ishii, H., Nagano, H., Haraguchi, N., Dewi, D. L., Kano, Y., Nishikawa, S., Tanemura, M., Mimori, K., Tanaka, F., Saito, T., Nishimura, J., Takemasa, I., Mizushima,

T., Ikeda, M., Yamamoto, H., Sekimoto, M., Doki, Y., and Mori, M. 2011. Reprogramming of mouse, and human cells to pluripotency using mature microRNAs. Cell Stem Cell. 8, 633–638.

Naka, H., Nakamura, S., Shimazaki, T., and Okano, H. 2008. Requirement for COUP-TFI, and II in the temporal specification of neural stem cells in CNS development. Nat. Neurosci. 11, 1014–1023.

Nakagawa, M., Takizawa, N., Narita, M., Ichisaka, T., and Yamanaka, S. 2010. Promotion of direct reprogramming by transformation-deficient Myc. Proc. Natl. Acad. Sci. USA 107, 14152–14157.

Noctor, S. C., Flint, A. C., Weissman, T. A., Dammerman, R. S., and Kriegstein A. R. 2001. Neurons derived from radial glial cells establish radial units in neocortex. Nature 409, 714–720.

Nori, S., Okada, Y., Yasuda, A., Tsuji, O., Takahashi, Y., Kobayashi, Y., Fujiyoshi, K., Koike, M., Uchiyama, Y., Ikeda, E., Toyama, Y., Yamanaka, S., Nakamura, M., and Okano, H. 2011. Grafted human-induced pluripotent stem-cell-derived neurospheres promote motor functional recovery after spinal cord injury in mice. Proc. Natl. Acad. Sci. USA 108, 16825–16830.

Ogawa, D., Okada, Y., Nakamura, M., Kanemura, Y., Okano, H. J., Matsuzaki, Y., Shimazaki, T., Ito, M., Ikeda, E., Tamiya, T., Nagao, S., and Okano, H. 2009. Evaluation of human fetal neural stem/progenitor cells as a source for cell replacement therapy for neurological disorders: properties, and tumorigenicity after long-term *in vitro* maintenance. J. Neurosci. Res. 87, 307–317.

Ogawa, Y., Sawamoto, K., Miyata, T., Miyao, S., Watanabe, M., Nakamura, M., Bregman, B.S., Koike, M., Uchiyama, Y., Toyama, Y., and Okano, H. 2002. Transplantation of *in vitro*-expanded fetal neural progenitor cells results in neurogenesis, and functional recovery after spinal cord contusion injury in adult rats. J. Neurosci. Res. 69, 925–933.

Okada, S., Ishii K., Yamane J., Iwanami A., Ikegami T., Katoh H., Iwamoto Y., Nakamura M., Miyoshi, H., Okano, H. J., Contag, C. H., Toyama, Y., and Okano, H. 2005. *In vivo* imaging of engrafted neural stem cells, its application in evaluating the optimal timing of transplantation for spinal cord injury. FASEB J. 19, 1839–1841.

Okada, Y., Matsumoto, A., Shimazaki, T., Enoki, R., Koizumi, A., Ishii, S., Itoyama, Y., Sobue, G., and Okano, H. 2008. Spatiotemporal recapitulation of central nervous system development by murine embryonic stem cell-derived neural stem/progenitor cells. Stem Cells 26, 3086–3098.

Okada, Y., Shimazaki, T., Sobue, G., and Okano, H. 2004. Retinoic-acid-concentration-dependent acquisition of neural cell identity during *in vitro* differentiation of mouse embryonic stem cells. Dev. Biol. 275, 124–142.

Okano, H. and Temple, S. 2009. Cell types to order: temporal specification of CNS stem cells. Curr Opin. Neurobiol. 19, 112–119.

Okano, H., Nakamura, M., Yoshida, K., Okada, Y., Tsuji, O., Nori, S., Ikeda, E., Yamanaka, S., and Miura, K. 2011 Steps toward safe cell therapy using induced pluripotent stem cells. Circ. Res. in press.

Okita, K., Hong, H., Takahashi, K., and Yamanaka, S. 2010. Generation of mouse-induced pluripotent stem cells with plasmid vectors. Nat. Protoc. 5, 418–428.

Okita, K., Ichisaka, T., and Yamanaka, S. 2007. Generation of germline-competent induced pluripotent stem cells. Nature 448, 313–317.

Okita, K., Matsumura, Y., Sato, Y., Okada, A., Morizane, A., Okamoto, S., Hong, H., Nakagawa, M., Tanabe, K., Tezuka, K., Shibata, T., Kunisada, T., Takahashi, M., Takahashi, J., Saji, H., and Yamanaka, S. 2011. A more efficient method to generate integration-free human iPS cells. Nat. Methods 8, 409–412.

Okita, K., Nakagawa, M., Hyenjong, H., Ichisaka, T., and Yamanaka, S.. 2008. Generation of mouse induced pluripotent stem cells without viral vectors. Science 322, 949–953.

Palmer, T. D., Ray, J., and Gage, F. H. 1995. FGF-2-responsive neuronal progenitors reside in proliferative, and quiescent regions of the adult rodent brain. Mol. Cell Neurosci. 6, 474–486.

Reynolds, B. A. and Weiss, S. 1992. Generation of neurons, and astrocytes from isolated cells of the adult mammalian central nervous system. Science 255, 1707–1710.

Sakurada, K. 2010. Environmental epigenetic modifications, and reprogramming-recalcitrant genes. Stem Cell Res. 4, 157–164.

Salazar, D. L., Uchida, N., Hamers, F. P., Cummings, B. J., and, Anderson, A. J. 2010. Human neural stem cells differentiate, and promote locomotor recovery in an early chronic spinal cord injury NOD-scid mouse model. PLoS One 5, e12272.

Schneider, S. P. and Fyffe, R. E. 1992. Involvement of GABA, and glycine in recurrent inhibition of spinal motoneurons. J. Neurophysiol. 68, 397–406.

Seki, T., Yuasa, S., Oda, M., Egashira, T., Yae, K., Kusumoto, D., Nakata, H., Tohyama, S., Hashimoto, H., Kodaira, M., Okada, Y., Seimiya, H., Fusaki, N., Hasegawa, M., and Fukuda, K. 2010. Generation of induced pluripotent stem cells from human terminally differentiated circulating T cells. Cell Stem Cell. 7, 11–14.

Sharp, J., Frame, J., Siegenthaler, M., Nistor, G., and Keirstead, H. S. 2010. Human embryonic stem cell-derived oligodendrocyte progenitor cell transplants improve recovery after cervical spinal cord injury. Stem Cells 28, 152–163.

Silva, J., Barrandon, O., Nichols, J., Kawaguchi, J., Theunissen, T. W., and Smith, A. 2008. Promotion of reprogramming to ground state pluripotency by signal inhibition. PLoS Biol. 6, e253.

Springer, J. E., Rao, R. R., Lim, H. R., Cho, S. I., Moon, G. J., Lee, H. Y., Park, E. J., Noh, J. S., and Gwag, B. J. 2010. The functional, and neuroprotective actions of Neu2000, a dual-acting pharmacological agent, in the treatment of acute spinal cord injury. J. Neurotrauma. 27, 139–149.

Stadtfeld, M., Nagaya, M., Utikal, J., Weir, G., and Hochedlinger, K. 2008. Induced pluripotent stem cells generated without viral integration. Science 322, 945–949.

Strauss, S. 2010. Geron trial resumes, but standards for stem cell trials remain elusive. Nat. Biotechnol. 28, 989–990.

Takahashi, K., Tanabe, K., Ohnuki, M., Narita, M., Ichisaka, T., Tomoda, K., and Yamanaka, S. 2007. Induction of pluripotent stem cells from adult human fibroblasts by defined factors. Cell. 131, 861–872.

Takahashi, K. and Yamanaka, S. 2006. Induction of pluripotent stem cells from mouse embryonic, and adult fibroblast cultures by defined factors. Cell. 126, 663–676.

Temple, S. 2001. The development of neural stem cells. Nature 414, 112–117.

Tsuji, O., Miura, K., Okada, Y., Fujiyoshi, K., Mukaino, M., Nagoshi, N., Kitamura, K., Kumagai, G., Nishino, M., Tomisato, S., Higashi, H., Nagai, T., Katoh, H., Kohda, K., Matsuzaki Y., Yuzaki, M., Ikeda, E., Toyama, Y., Nakamura, M., Yamanaka, S., and Okano, H. 2010. Therapeutic potential of appropriately evaluated safe-induced pluripotent stem cells for spinal cord injury. Proc. Natl. Acad. Sci. USA 107, 12704–12709.

Tuszynski, M. H. 1997. Gene therapy for nervous system disease. Ann. N. Y. Acad. Sci. 835, 1–11.

Tuszynski, M. H., Gabriel, K., Gage, F. H., Suhr, S., Meyer, S., and Rosetti, A. 1996. Nerve growth factor delivery by gene transfer induces differential outgrowth of sensory, motor, and noradrenergic neurites after adult spinal cord injury. Exp. Neurol. 137, 157–173.

Warren, L., Manos, P. D., Ahfeldt, T., Loh, Y. H., Li, H., Lau, F., Ebina, W., Mandal, P. K., Smith, A. Meissner, Z. D., Daley, G. Q., Brack, A. S., Collins, J. J., Cowan, C., Schlaeger, T. M., and Rossi, D. J. 2010. Highly efficient reprogramming to pluripotency, and directed differentiation of human cells with synthetic modified mRNA. Cell Stem Cell. 7, 618–630.

Watanabe, K., Nakamura, M., Iwanami, A., Fujita, Y., Kanemura, Y., Toyama, Y., and Okano, H. 2004. Comparison between fetal spinal-cord-, and forebrain-derived neural stem/progenitor cells as a source of transplantation for spinal cord injury. Dev. Neurosci. 26, 275–287.

Woltjen, K., Michael, I. P., Mohseni, P., Desai, R., Mileikovsky, M., Hamalainen, R., Cowling, R., Wang, W., Liu, P., Gertsenstein, M., Kaji, K., Sung, H. K., and Nagy, A. 2009. piggyBac transposition reprograms fibroblasts to induced pluripotent stem cells. Nature 458, 766–770.

Yan, J., Xu, L., Welsh, A. M., Hatfield, G., Hazel, T., Johe, K., and Koliatsos, V. E. 2007. Extensive neuronal differentiation of human neural stem cell grafts in adult rat spinal cord. PLoS Med. 4, e39.

Yoshida, H., Imaizumi, T., Tanji, K., Matsumiya, T., Sakaki, H., Kimura, D., Cui, X. F., Kumagai, M., Tamo, W., Shibata, T., Hatakeyama, M., Sato, Y., and Satoh, K. 2002. Platelet-activating factor enhances the expression of vascular endothelial growth factor in normal human astrocytes. Brain Res. 944, 65–72.

Zhao, T., Zhang, Z. N., Rong, Z., and Xu, Y. 2011. Immunogenicity of induced pluripotent stem cells. Nature. 474, 212–215.

Zhou, H., Wu, S., Joo, J. Y., Zhu, S., Han, D. W., Lin, T., Trauger, S., Bien, G., Yao, S., Zhu, Y., Siuzdak, G., Scholer, H. R., Duan, L., and Ding, S. 2009. Generation of induced pluripotent stem cells using recombinant proteins. Cell Stem Cell. 4, 381–384.

Zhu, S., Li, W., Zhou, H., Wei, W., Ambasudhan, R., Lin, T., Kim, J., Zhang, K., and Ding, S. 2010. Reprogramming of human primary somatic cells by OCT4, and chemical compounds. Cell Stem Cell. 7, 651–655.

Index

M

N

O

P

R

S

T

V

W

Color Plate Section

Chapter 3

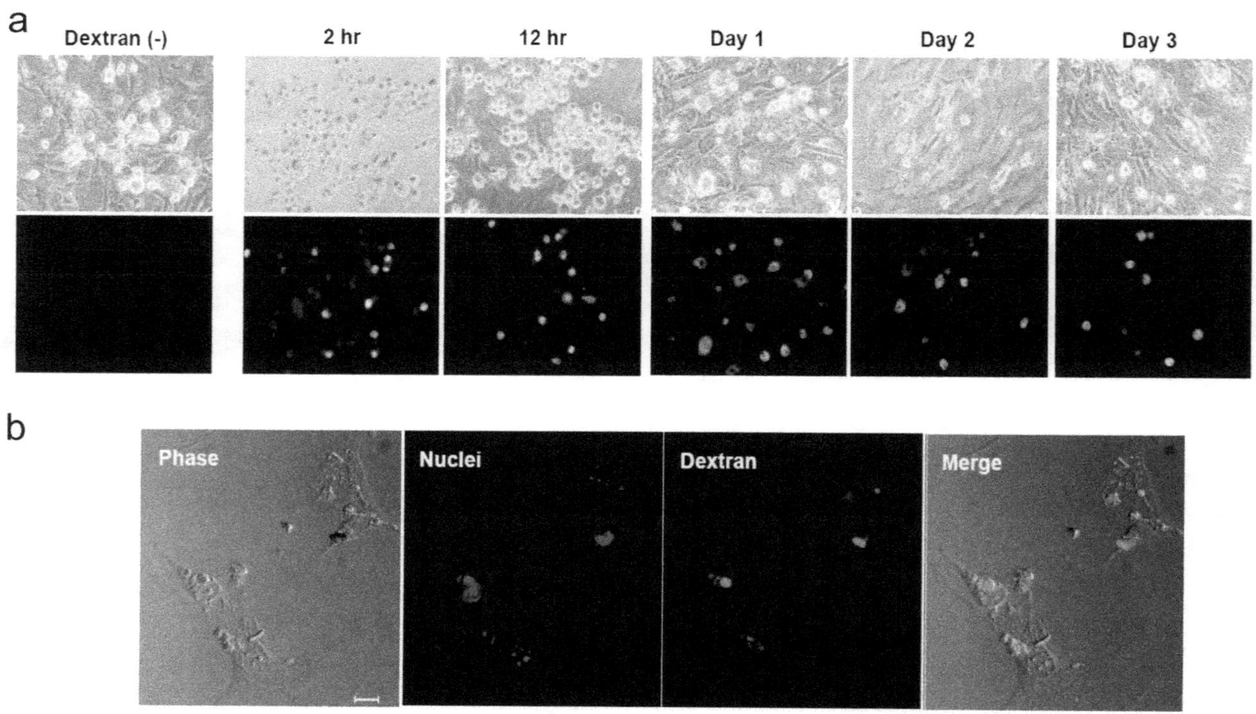

Figure 1. Generation of protein-iPS cells from cardiac fibroblasts by protein delivery. (a) Fibroblasts were reversibly permeabilized, incubated with 70 kDa Texas Red–conjugated dextran and resealed. Serial fluorescent images. (b) Confocal images at day 7.

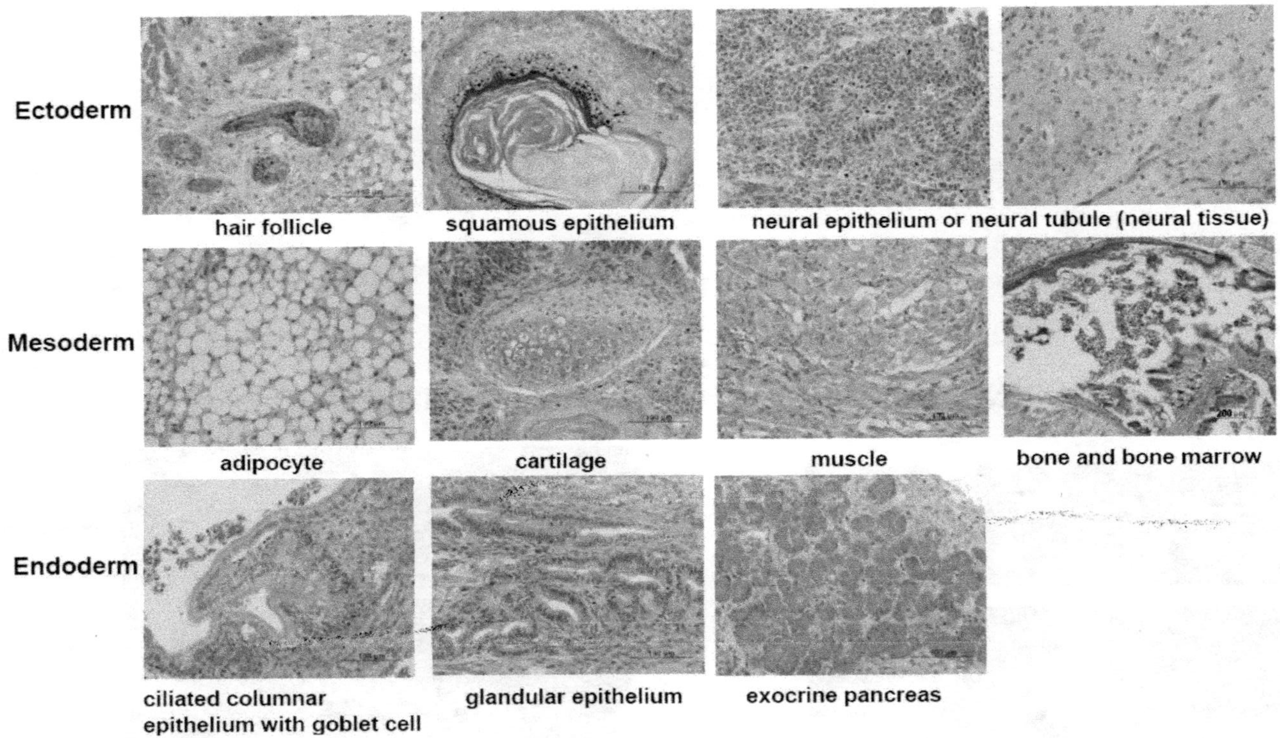

Figure 2. *In vivo* differentiation potentials of the protein-iPS cells. Hematoxylin and eosin (H&E) staining of a tumor from cFB-protein-iPS cells showing a well-differentiated teratoma.

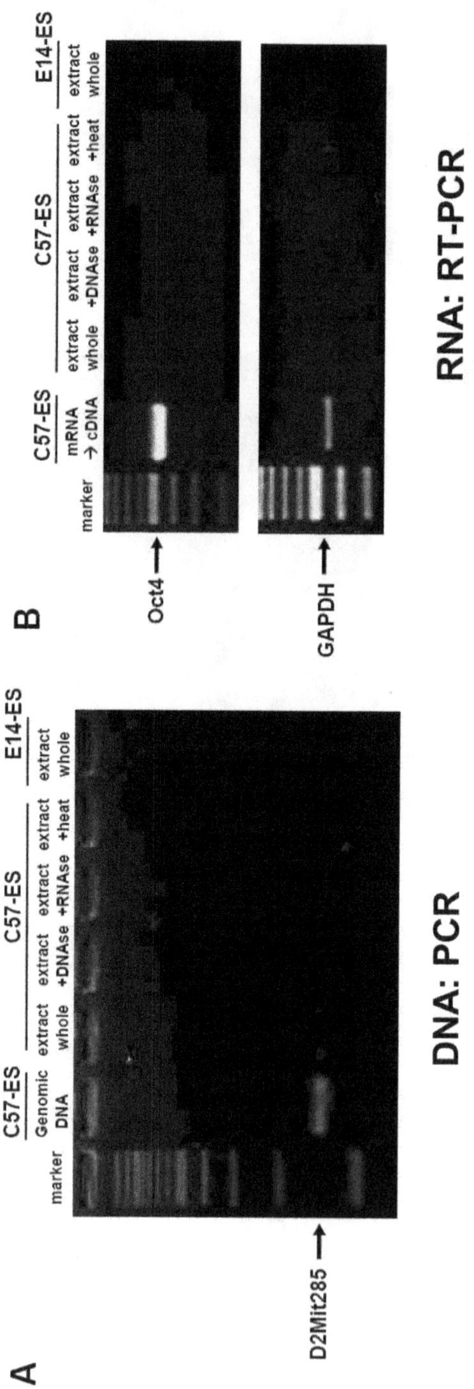
A
B
marker
C57-ES
Genomic DNA
extract whole
extract +DNAse
extract +RNAse
extract +heat
E14-ES
extract whole
D2Mit285
DNA: PCR
mRNA → cDNA
Oct4
GAPDH
RNA: RT-PCR

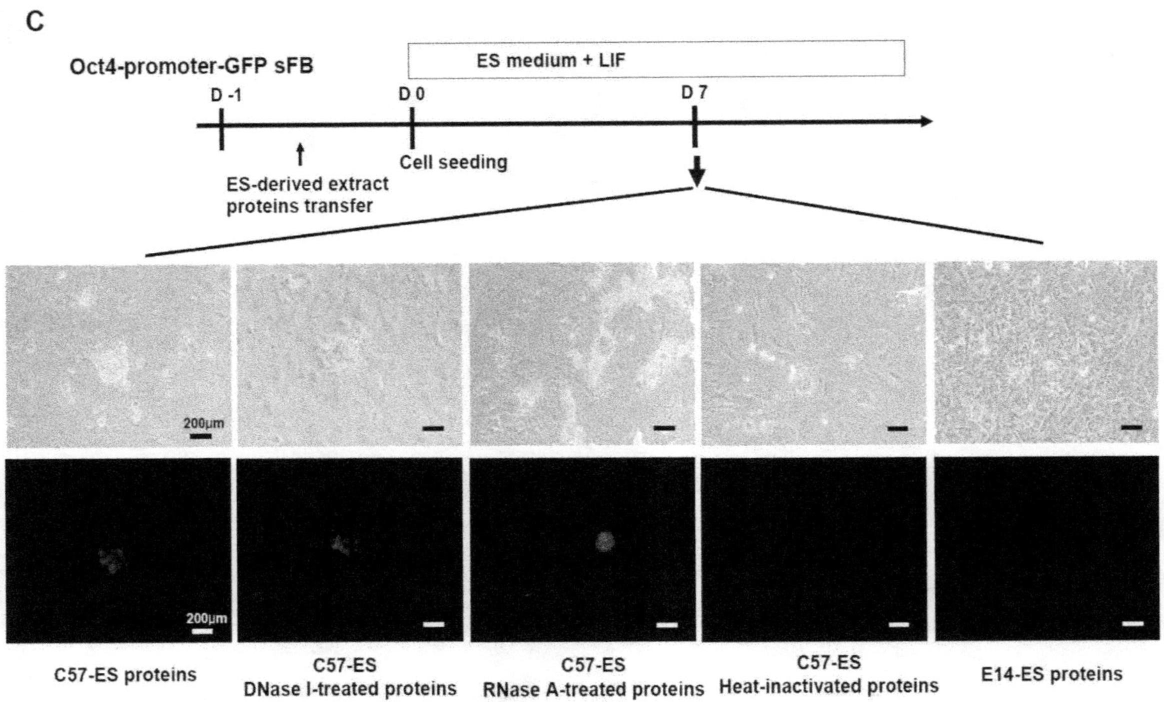

Figure 3. Effector of reprogramming. Protein-based reprogramming is not mediated by DNAs/RNAs from donor ES cell. (A-B) The absence of contaminating DNA or RNA in the extract proteins was confirmed by PCR and RT-PCR. (C) Colony formation of Oct4-promoter–driven GFP reporter cells (skin fibroblast) at seven days after ES derived extract proteins transfer.

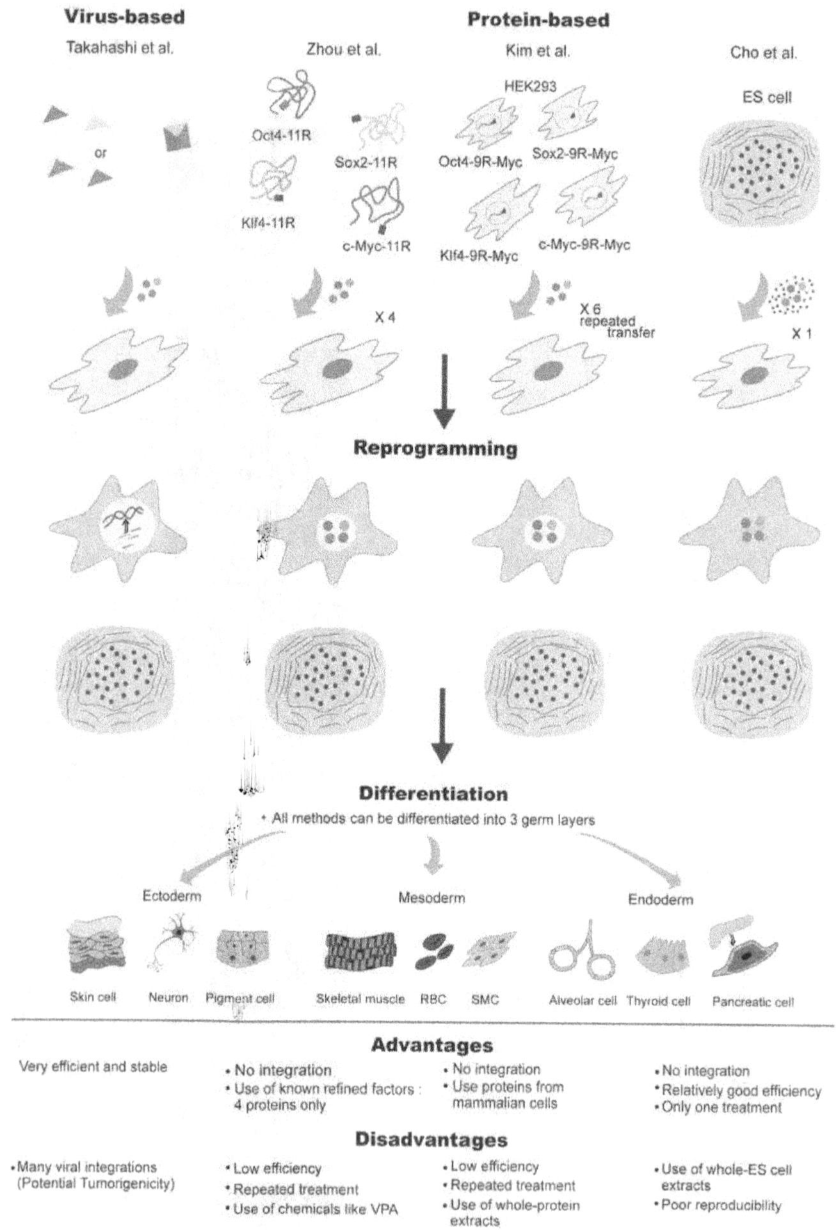

Figure 4. Methods of genetic and non-genetic derived iPS. A flow diagram summarizing the methods to generate iPS cells, with their advantages and disadvantages shown below.

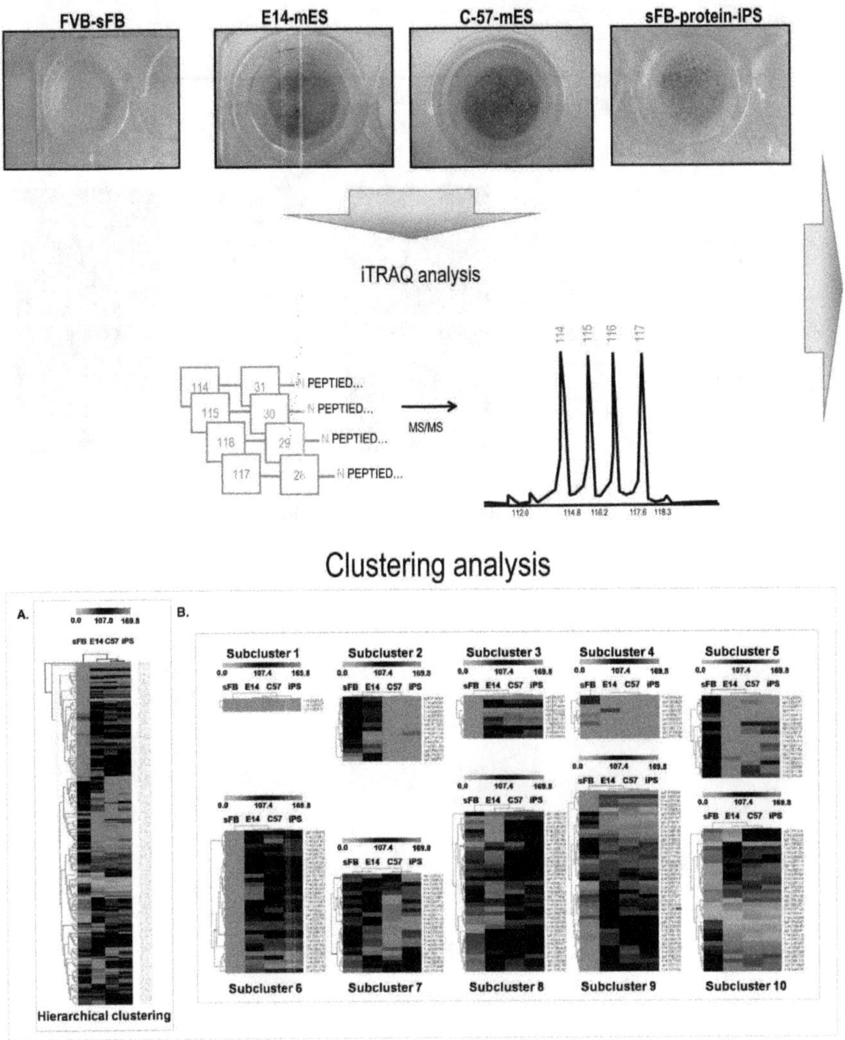

Figure 5. Analysis of differential proteomes of induced pluripotent stem cells by protein-based reprogramming of fibroblasts.

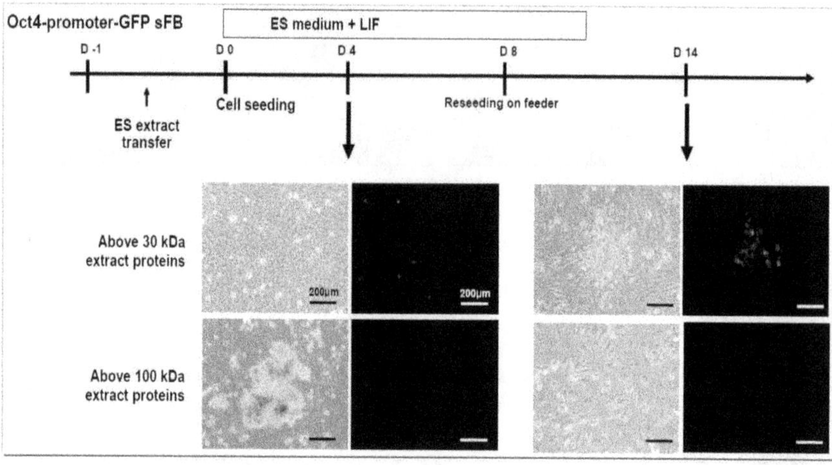

Figure 6. The size of speculated effector of reprogramming. Extract proteins above 30 kDa induced reprogramming, whereas proteins above 100kDa did not.

Chapter 4

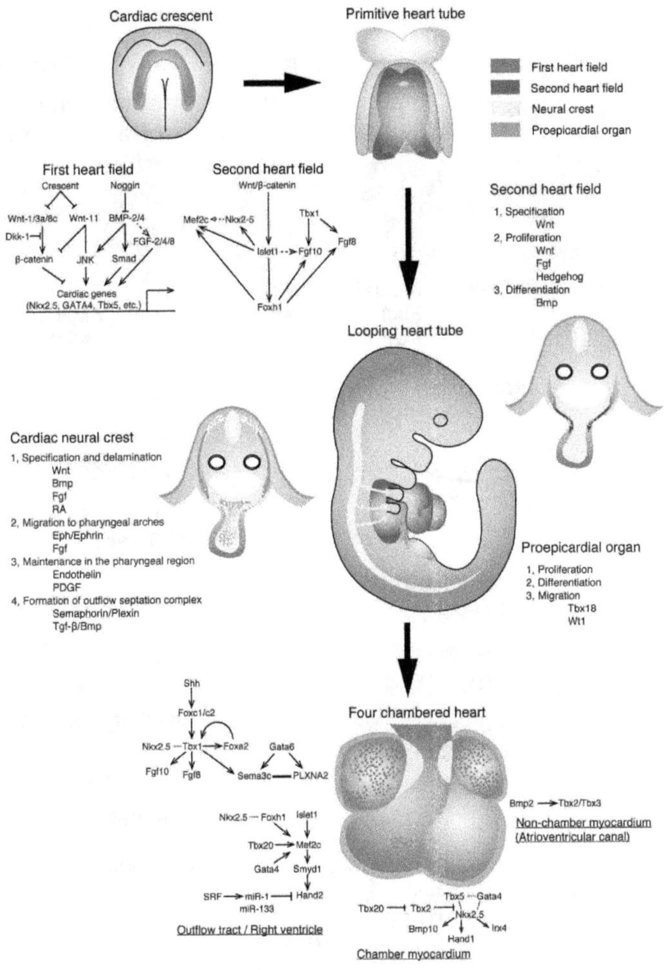

Figure 1. Developmental origins of the heart. Frontal views of embryos at cardiac crescent and primitive heart tube stage, and a lateral view of embryo at looping heart stage with transverse sections, and the four-chambered heart are shown. In the cardiac crescent, the first heart field (FHF) cells form a crescent shape in the anterior region of embryo with the second heart field (SHF) cells medial to it. In the primitive heart tube, the SHF cells lie dorsal to the heart tube in the splanchnic mesoderm of pharyngeal region. In the looping heart tube, the SHF cells migrate into the anterior and posterior ends of the tube (blue arrows). Subsequently, cardiac neural crest cells also migrate into the outflow tract from the dorsal neural tube (yellow arrows) to remodel the outflow tract and the aortic arch system. In the four-chambered heart, the FHF cells form the left ventricle and atria (red) whereas the SHF cells form the outflow tract, the right ventricle, and part of atria (blue). Cardiac neural crest cells contribute to the outflow tract cushions (yellow) that eventually form the aorto-pulmonary septum. Epicardium is derived from proepicardial organ (green). Signaling and transcriptional network for the development of the cardiac progenitors is shown (See detail in the text).

Chapter 5

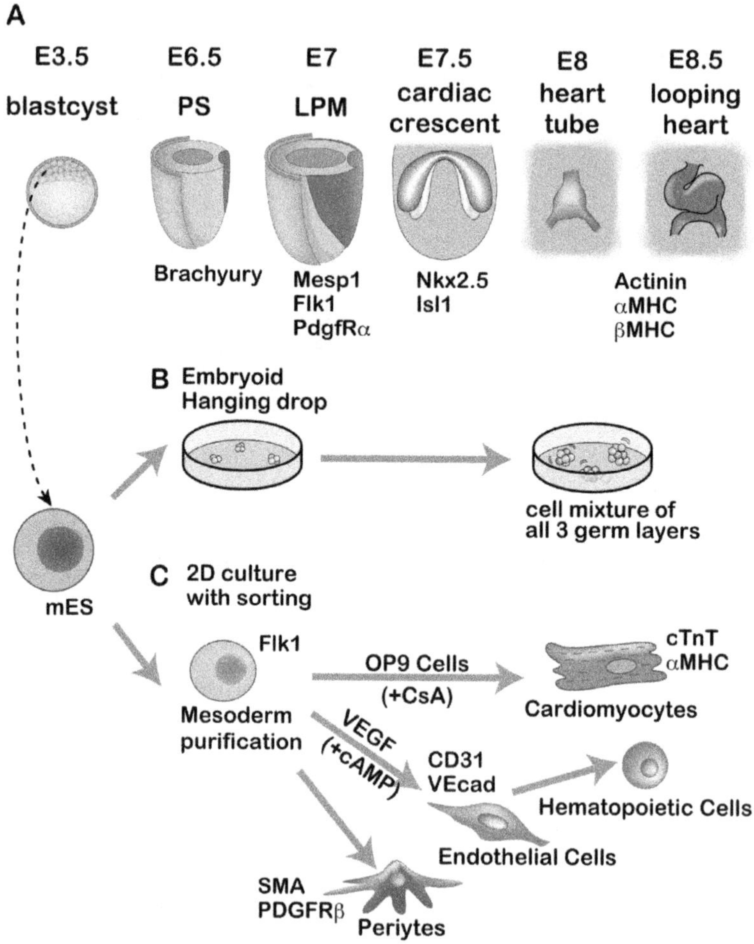

Figure 1. Heart development *in vivo* and cardiovascular differentiation *in vitro*. (A) Heart development in mouse. At gastrulation stage, mesendodem cells express Brachyury. Part of mesoderm cells migrate lateral anterior to form lateral plate mesoderm expressing *Mesp1*, Flk1 and PdgfRa. These cells further migrate to form cardiac crescent expressing *Nkx2.5* and/or *Isl1*, then fuse midline to form heart tube and start beating. Linear heart tube loops to form four-chambered heart. Abbreviations; PS: primitive streak, LPM: lateral plate mesoderm. (B-C) Embryonic stem cell (ESC) differentiation. ESCs are generated from inter cell mass (ICM) of blastocysts. (B) Conventional differentiation method is called hanging drop or embryoid body methods, in which ESC differentiation is induced as small aggregates. These aggregates spontaneously differentiate into cells containing all three germ layers. (C) Two-dimensional culture and sequential cell sorting allow dissecting differentiation processes in detail. As the first step, Flk1-positive mesoderm cells are sorted by FACS or MACS. Cardiomyocytes are induced by culturing Flk1-positive cells on OP9 feeder cells with supplementation of cyclosporin A. Endothelial cells and mural cells are induced on collagen IV or gelatin-coated dishes with VEGF treatment (modified from Uosaki and Yamashita 2011a).

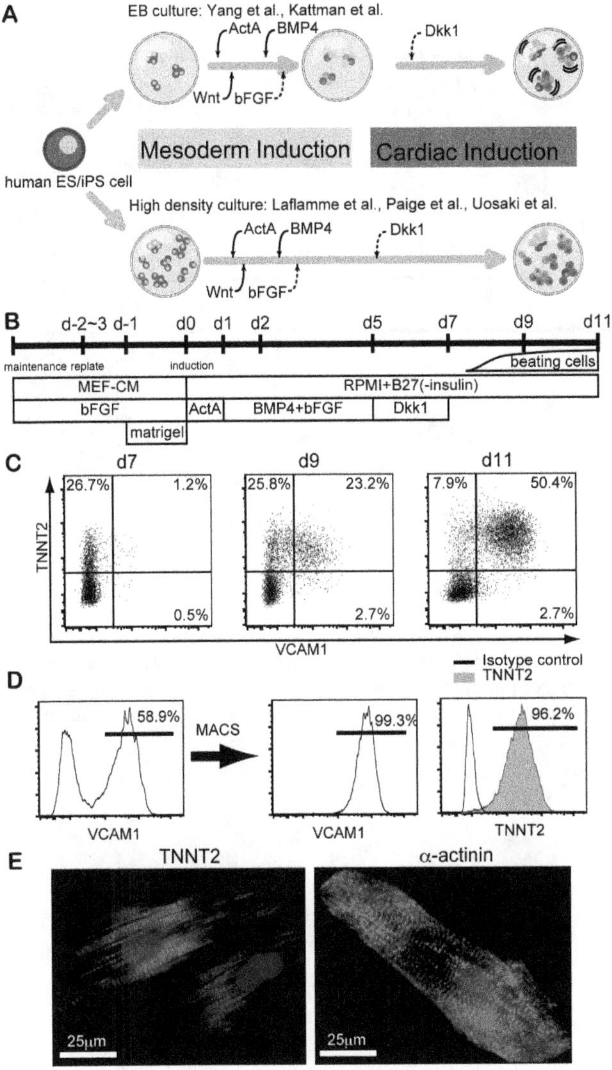

Figure 2. Directed differentiation and cardiomyocyte specific cell surface marker. (A) Schemes of two established directed differentiation methods —EB culture and high density culture methods. In both of them, cells are treated with Activin A and BMP4 to induce differentiation of mesoderm. Wnt3a, bFGF, or Dkk1 are occasionally added to enhance differentiation to mesoderm or cardiomyocytes, respectively. (B) Differentiation scheme of modified directed differentiation. (C) TNNT2 (cTnT)-positive cells start to appear at day 7 after Activin A treatment. Approximately 80% of TNNT2-positive cells become VCAM1-positive cells at day 11 after induction. (D) Approximately 95% of cells were positive for cTnT after purification of VCAM1-positive cells by MACS. (E) Purified cells showed clear sarcomeric structure (modified from Uosaki et al. PLoS One 2011).

Chapter 7

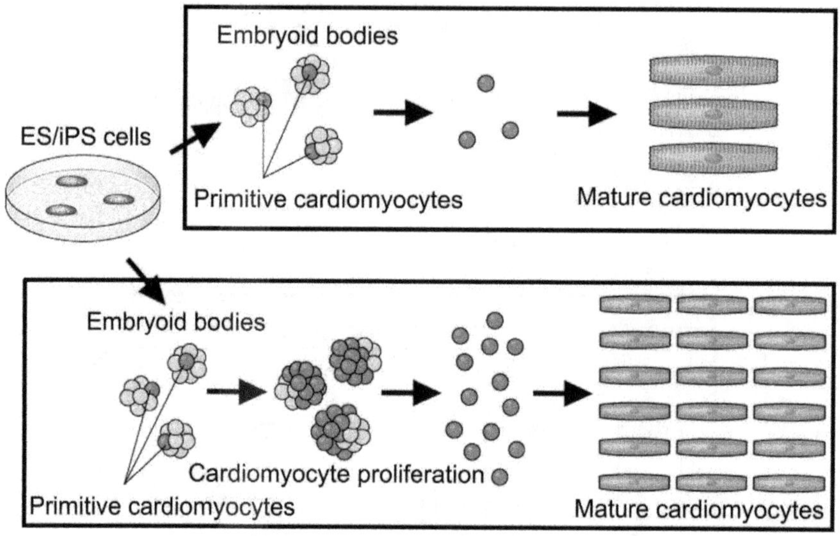

Figure 1. The effects of factors on cardiomyocyte proliferation. Embryonic stem (ES)/induced pluripotent stem (iPS) cells can differentiate into many cell types, including cardiomyocytes. Conventionally, primitive cardiomyocytes emerge spontaneously and differentiate into mature cardiomyocytes (upper panel). Cardiomyocyte proliferation promoting factors can promote the proliferation of primitive cardiomyocytes, boosting the yield of mature cardiomyocytes (lower panel).

Chapter 8

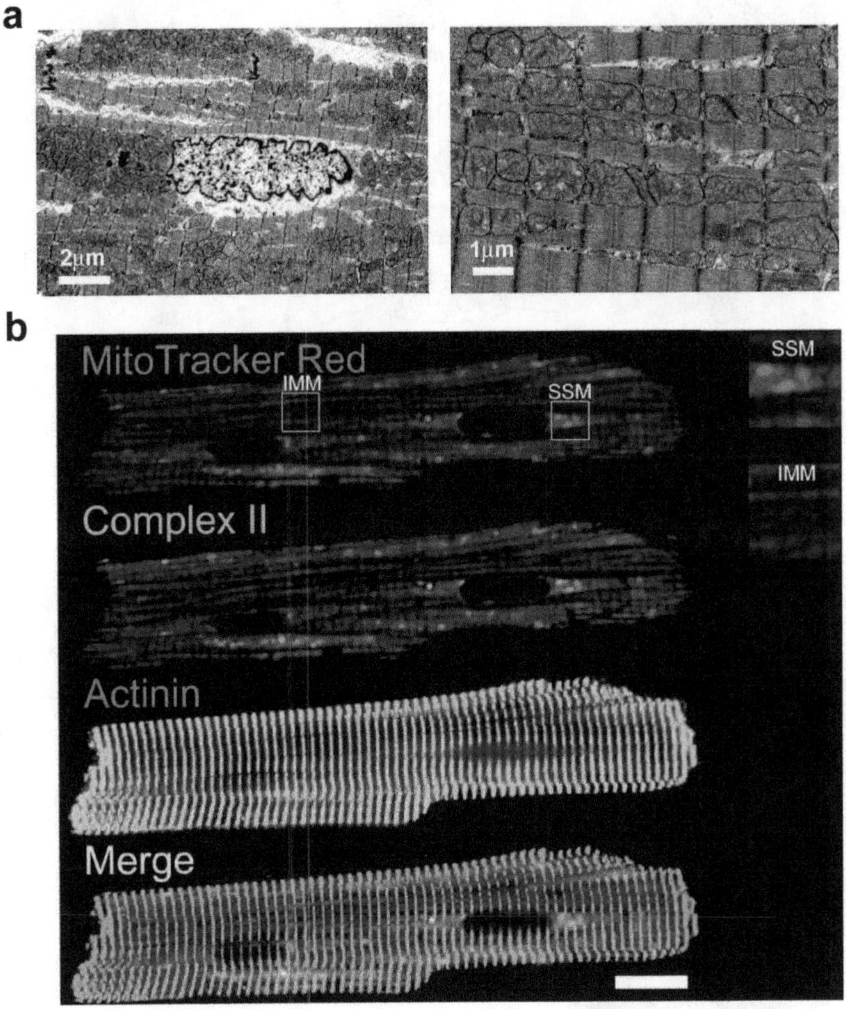

Figure 1. Mitochondrial abundance in cardiomyocytes. (a) Ultrastructure of adult human cardiomyocytes. Note the large numbers of mitochondria in the cardiomyocytes. (b) Fluorescent staining of cardiac mitochondria by MitoTracker Red, immunohistochemical staining for mitochondrial electron transport chain complex II and sarcomeric actinin, and a merged image, as indicated. Cardiomyocytes contains two distinguishable mitochondria: subsarcomeric (SSM) and intermyofibrillar mitochondria (IFM). Scale bar, 20 μm.

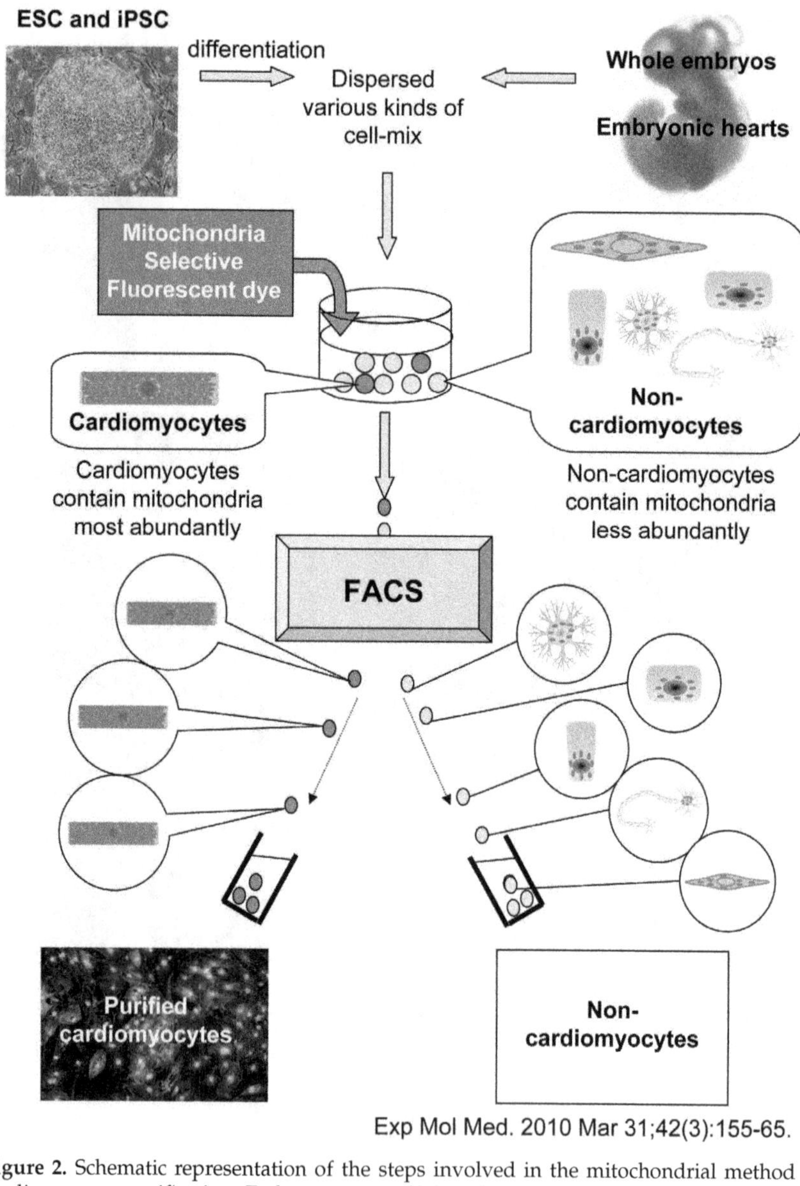

Figure 2. Schematic representation of the steps involved in the mitochondrial method of cardiomyocyte purification. Embryonic stem (ES) and induced pluripotent stem (iPS) cell derivatives or primary cells are dispersed and stained with tetramethylrhodamine methyl ester (TMRM) before being separated by fluorescence-activated cell sorting (FACS) (Fig. reproduced with permission from Hattori et al. 2010; Hattori and Fukuda 2010).

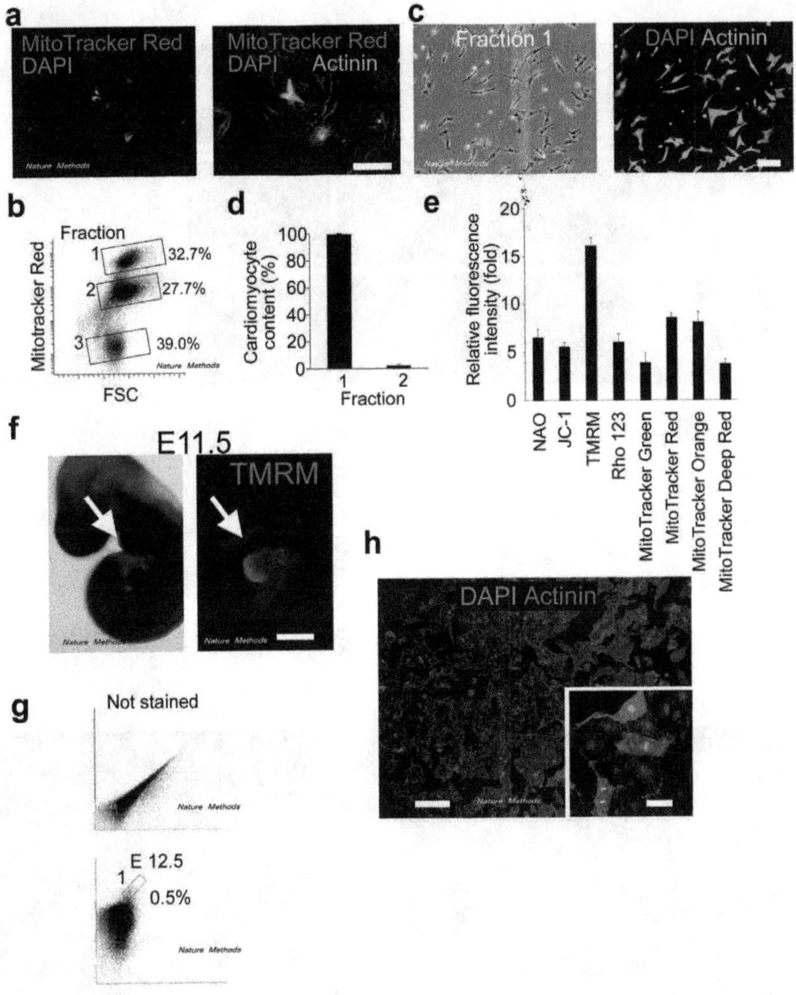

Figure 3. Characterization of mitochondria-selective dyes for primary cardiomyocyte purification. (a) Primary cultured neonatal rat heart cells prestained with MitoTracker Red (red) were immunostained for α-actinin (green). 4′,6′-Diamidino-2-phenylindole (DAPI; blue) was used to indicate nuclear staining. (b) Fluorescence-activated cell sorting (FACS) analysis of neonatal rat heart-derived cells stained with MitoTracker Red resulted in cells being divided into three fractions (1, 2, and 3). (c) Cells in Fraction 1 were sorted further (left) and stained immunofluorescently for α-actinin (green) and DAPI (blue), as shown on the right. (d) The cardiomyocyte content in Fractions 1 and 2 is shown. (e) Representative FACS charts of various mitochondrial indicators. NAO, nonyl acridine orange. (f) Tetramethylrhodamine methyl ester (TMRM)-stained living whole embryos [embryonic day (E) 11.5]. (g) FACS analysis of cells derived from whole embryos treated with (lower) or without (upper) TMRM. (h) Immunofluorescent staining for α-actinin in cells obtained from Fraction 1 in (g, lower panel). Inset shows proliferating cardiomyocytes. Scale bars, 100 μm (a,c,e,h); 20 μm (h, inset) (Figs. reproduced with permission from Hattori et al. 2010 (Hattori et al. 2010)).

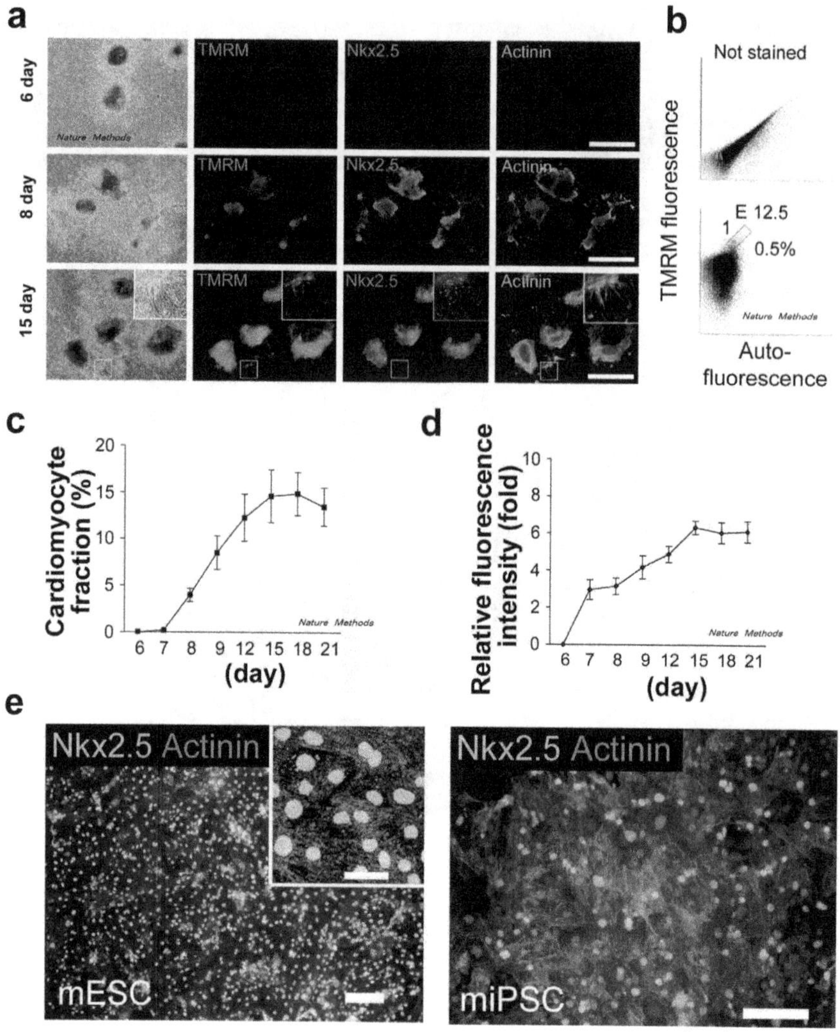

Figure 4. Purification of mouse pluripotent stem (PS) cell-derived cardiomyocytes. (a) Representative photographs of tetramethylrhodamine methyl ester (TMRM) fluorescence, as well as Nkx2.5 and α-actinin immunofluorescence, in differentiated mouse embryonic stem (ES) cells (on Days 6, 8, and 15). (b) Two examples of the fluorescence-activated cell sorting (FACS) analysis of embryoid body (EB)-derived cells 6–25 days after EB formation. (c) Time course of the relative fluorescence intensity of cells in Fraction 1 versus Fraction 2. (d) Time course of the percentage of cells in Fraction 1 (cardiomyocytes) 6–21 days after EB formation. (e) Immunofluorescent staining for Nkx2.5 (green) and α-actinin (red) in the sorted cells (Fraction 1) from EB derived from mESCs and induced PSCs (miPSC). Scale bars, 1 mm (a); 100 μm (e); 30 μm (e, inset) (Figs. reproduced with permission from Hattori et al. 2010 (Hattori et al. 2010)).

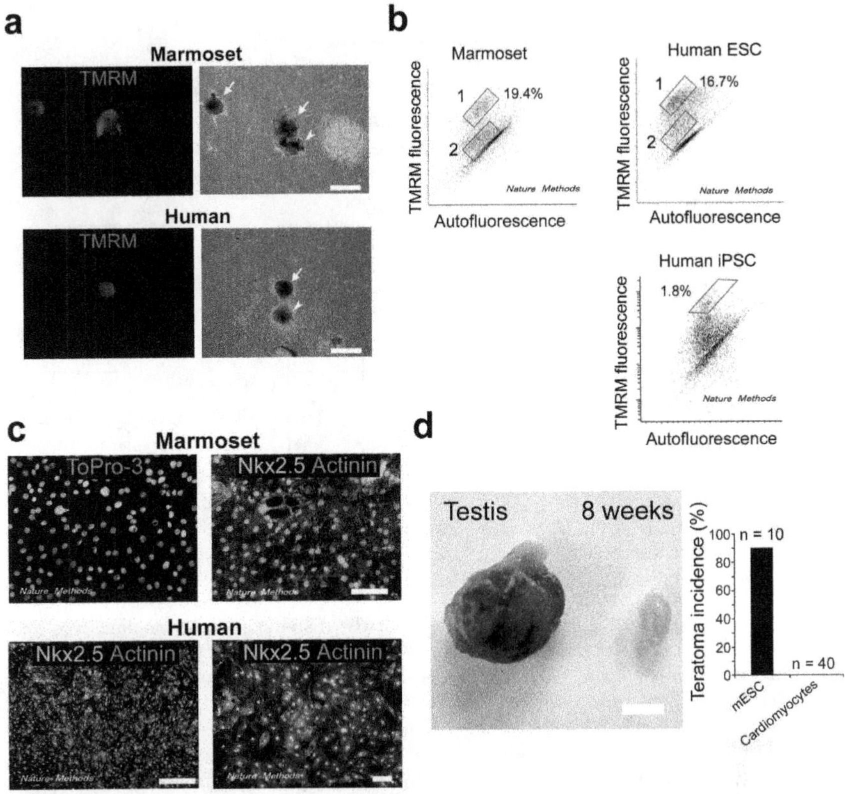

Figure 5. Purification of marmoset embryonic stem (ES) cell- and human pluripotent stem (PS) cell-derived cardiomyocytes. (a) Marmoset and human embryoid bodies (EB) treated with tetramethylrhodamine methyl ester (TMRM). The panels on the left show TMRM fluorescence, whereas the panels on the right show phase contrast photography of the same field. Arrows and arrowheads indicate spontaneously beating and non-beating areas, respectively. (b) Two-dimensional fluorescence-activated cell sorting (FACS) separation of TMRM-stained common marmoset ESC (cmESC) and hESC- and iPSC-derivatives. (c) The upper panels show immunofluorescent staining for α-actinin (red) and Nkx2.5 (green) in cells from cmESC derivatives. ToPro-3 was used to highlight nuclear staining. The lower panels show immunofluorescent staining for α-actinin (red) and Nkx2.5 (green) in cells from hESC and iPSC derivatives. (d) Transplantation of 250 undifferentiated ESC into testes resulted in teratoma formation, whereas the transplantation of 1.9×10^5 purified, sorted ESC-derived cardiomyocytes did not. The mean incidence of teratoma formation after transplantation of undifferentiated ESC and purified and sorted cardiomyocytes is summarized in the graph. Scale bars, 500 μm (a); 100 μm (c); 5 mm (d) (Figs. reproduced with permission from Hattori et al. 2010 (Hattori et al. 2010)).

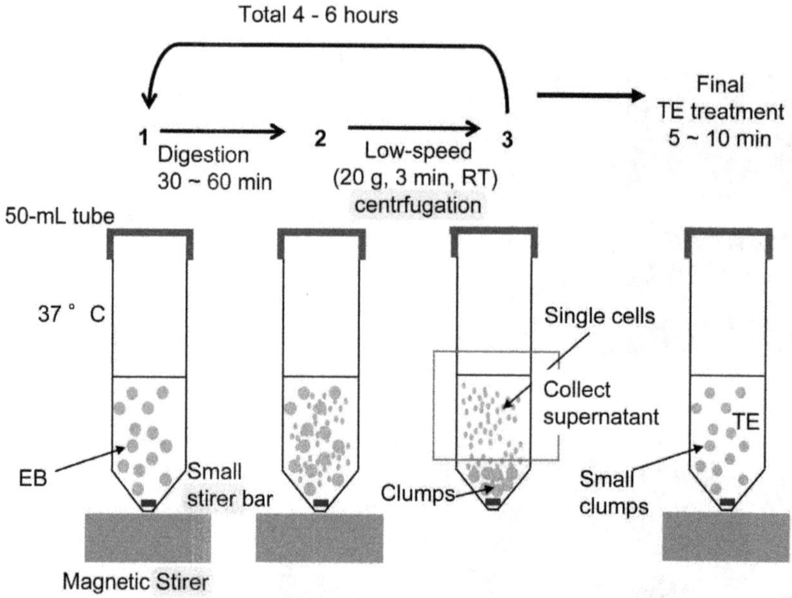

Figure 7. Step 1 of the mitochondrial method for the dispersion of hearts, embryos and embryoid bodies (EB). After dispersal, the EB are stained with 0.1% collagenase type 3 (Worthington), 0.25% Trypsin (DIFCO), and 10–20 nM tetramethylrhodamine methyl ester (TMRM) in Ads buffer [composition (in mM): NaCl 116; HEPES 20; NaH_2PO_4 12.5; glucose 5.6; KCl 5.4; $MgSO_4$ 0.8, pH 7.35] with stirring. Note that we use a 50-mL tube and small magnetic stirrer, with the tubes set up in a cell incubator. Digestion (30–60 min) is repeated several times, with the total digestion time ranging between 4 and 6 hours (this is determined after inspection of the tubes after each single bout of digestion). Finally, cell clumps consisting of a few cells are treated with 0.25% trypsin and 0.02% EDTA (TE) for 5–10 min, with stirring, to obtain single cells. All supernatants and TE-treated cells are combined and applied to fluorescence-activated cell sorting (FACS) sorting.

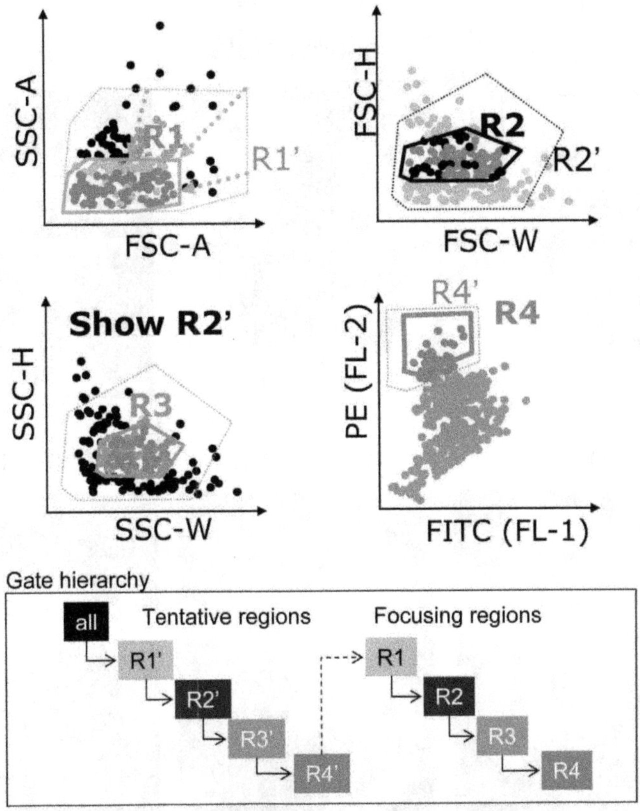

Figure 8. Step 2 of the mitochondrial method: gate settings. First, all data are displayed on a two-dimensional dot plot of forward scatter (FSC) area versus side scatter (SSC) area (FSC-A and SSC-A, respectively). Tentative gates are set for all cells (region 1: R1′), and then the events included in R1′ are dotted in a plot of FSC-height (H) and FSC-width (W). Another tentative gate is set (R2′) to eliminate doublets as per the instructions supplied by the manufacturer. Similarly, events included in R2′ are dotted in a plot of SSC-H and SSC-W and another tentative gate (R3′) is set to eliminate doublets. The events included in R3′ are dotted in a plot of phycoerythrin (PE) area and fluorescein isothiocyanate (FITC) area. The population of cardiomyocytes is displayed in the uppermost section for PE-A and furthest away from the diagonal line. Another tentative gate (R4′) is set for cardiomyocytes. The events included in R4′ are colored and displayed in the above four dot plots. Then, all regions are focused towards the population gated by R4′, meaning to make R1, 2 and 3. Close attention should be paid to the dot plot of PE-A and FITC-A; more distinguishable populations may be evident on this plot. Finally, the gate settings for R4 can be determined. Higher gating according to the PE-A axis results in a higher purity but lower yield; in turn, lower gating results in a lower purity but higher yield.

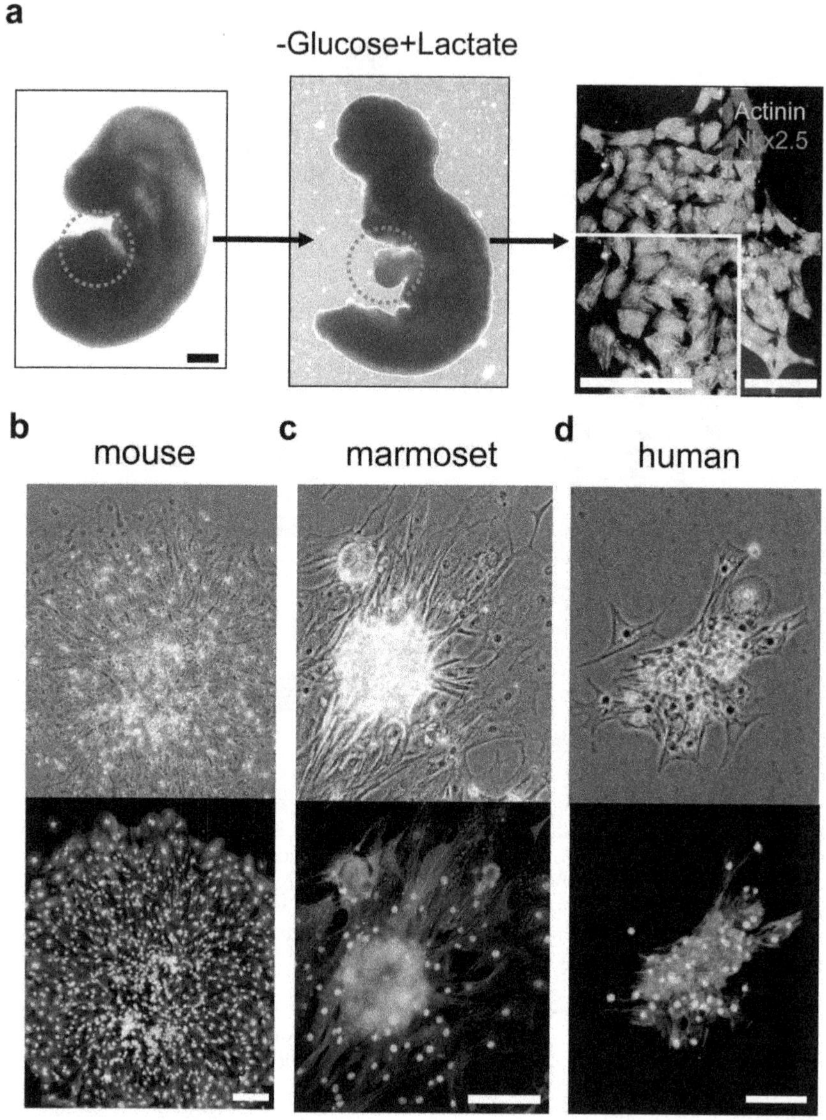

Figure 9. Purification of cardiomyocytes from various sources using the lactate method. Cardiomyocytes can be purified from rat embryonic day (E) 11.5 whole embryos. The E11.5 rat embryo is cultured with glucose-depleted, lactate-containing Dulbecco's minimal essential medium (DMEM) for 1 week. Following culture, all cells comprising the embryo appear to be dead, except for the heart tube, which is still beating. The surviving cells are recovered and seeded onto fibronectin-coated plates. The attached cells are fixed with 4% paraformaldehyde and stained immunohistochemically for actinin (green) and Nkx2.5 (red). (b) Mouse, marmoset and human embryonic stem cell-derived embryoid bodies were cultured in glucose-depleted, lactate-containing DMEM and the surviving cells transferred to fibronectin-coated plates before being stained for actinin (red) and Nkx2.5 (green). Scale bars, 100 μm.

Chapter 11

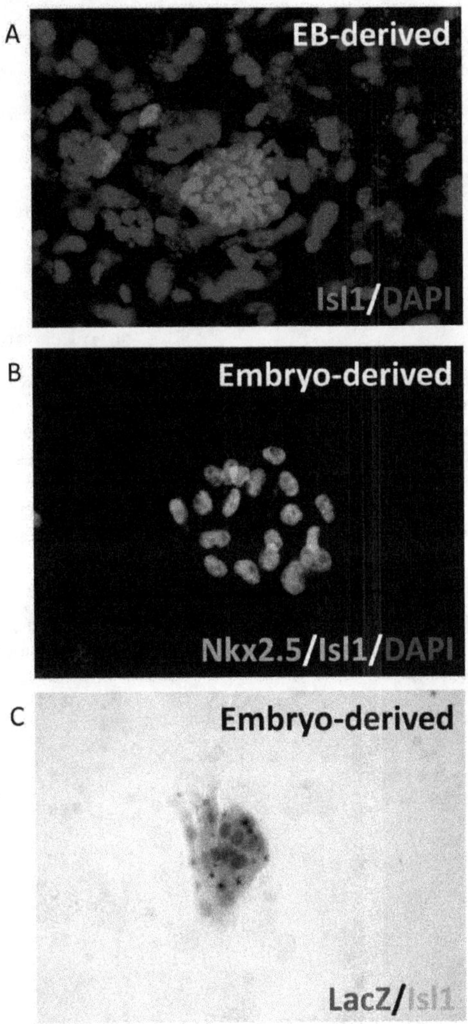

Figure 3. Isl1⁺ cardiac progenitors isolated on cardiac mesencymal feeder cells. (A) ES-derived Isl1+ cardiac progenitor colony on cardiac mesenchymal feeder layer. (B) Mouse embryo-derived Isl1+ cardiac progenitor colony on cardiac mesenchymal feeder layer. Cells in the colony express both Isl1 and Nkx2.5. (C) A colony of Isl1+ cardiac cells derived from posterior SHF at later stage. Progenitors were isolated from *Sln-Cre; R26R* mouse embryo at E10.5 and grown on feeder layer. LacZ staining indicates that these progenitors are derived from the posterior sublineage. Note that these Isl1+ cells display myofibrils.

Chapter 12

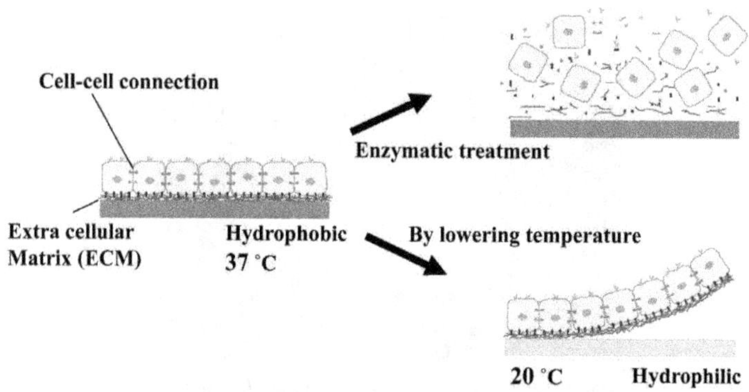

Figure 1. Cell harvest from temperature-responsive culture dishes. At 37°C, the surface of the temperature-responsive culture dishes is hydrophobic, so cells can attach and proliferate onto them. By lowering the temperature below 32°C, the surface reversibly changes to the hydrophilic state, which leads to the detachment of cells from the dishes. Typical cell harvest using enzymatic digestion results in the disruption of both adhesive proteins and membrane receptors. When using temperature-responsive culture dishes, cell sheets with preserved cell to cell connection and extracellular matrixes and intact membrane receptors are harvested.

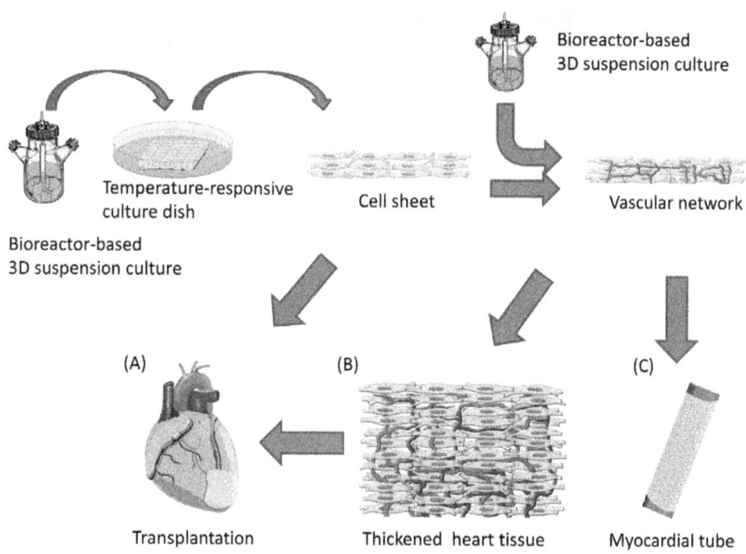

Figure 2. Future views of myocardial tissue reconstruction using cell sheet-based tissue engineering. Cell sheets are created using stem cell-derived cardiomyocytes and fibroblasts after the expansion and purification in the bioreactor system. (A) Transplantation of the layered constructs to repair the injured heart is one of the directions. (B) Thickened heart tissue as disease heart models and for reconstructing the broad heart injury. (C) Myocardial tubes as independent cardiac assist device.

Chapter 13

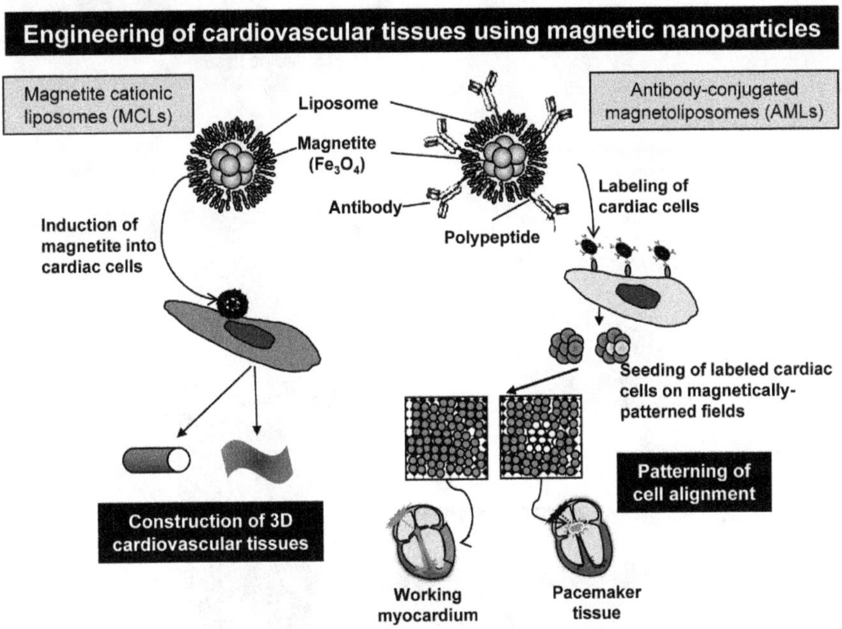

Figure 1. Schematic explanation of tissue engineering of cardiovascular tissues using magnetic nanoparticles and magnetic force. Magnetite cationic liposomes (MCLs) and antibody-conjugated magnetoliposomes (AMLs). MCLs are mainly used in multi-layering of cardiac tissues such as myocardial and blood vascular tissues. On the other hand, AMLs are suitable to label target cells specifically.

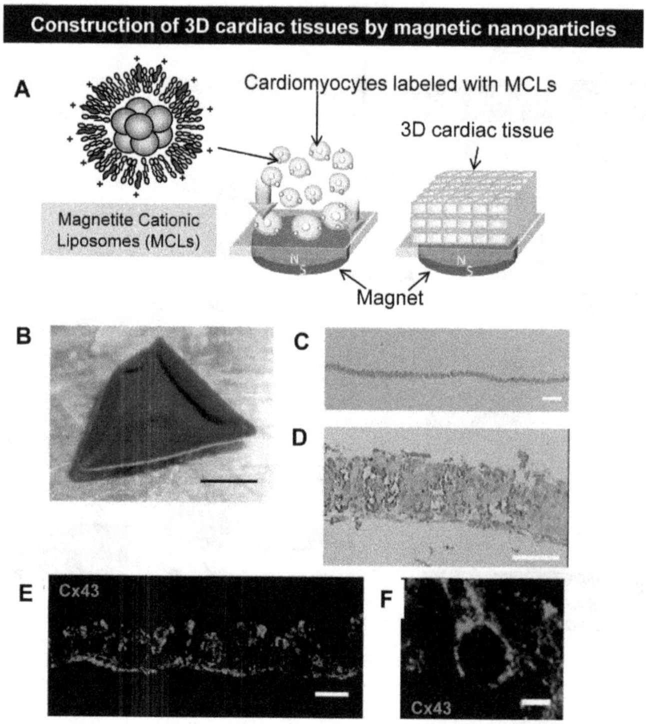

Figure 2. (A) Schemes to explain multilayering technique using magnetite cationic liposomes (MCLs) and magnetic force. To label CMs magnetically, MCLs were added to the culture medium of CMs at the concentration of 100 pg/cell and incubated for 4 hr. Then, the magnetically labeled CMs were recovered and used to construct the CM sheets. The CM sheets were fixed in 10% formalin solution and embedded in paraffin. (B) Bright-field photograph of CM sheet constructed by Mag-TE; scale bar, 5 mm. (C) Cross-sections of CM sheet; scale bar, 500 mm (Hematoxylin and eosin stain). (D) Magnified view of C (scale bar, 100 mm). (E) Connexin43expression in engineered myocardial tissues (Immunofluorescence stained with anti- Cx43 antibody; scale bar, 100 mm. (F) Magnified view of (E) (Cited Shimizu et al. Biotechnol. Bioeng. 2007) {Shimizu, 2006 2351 /id}

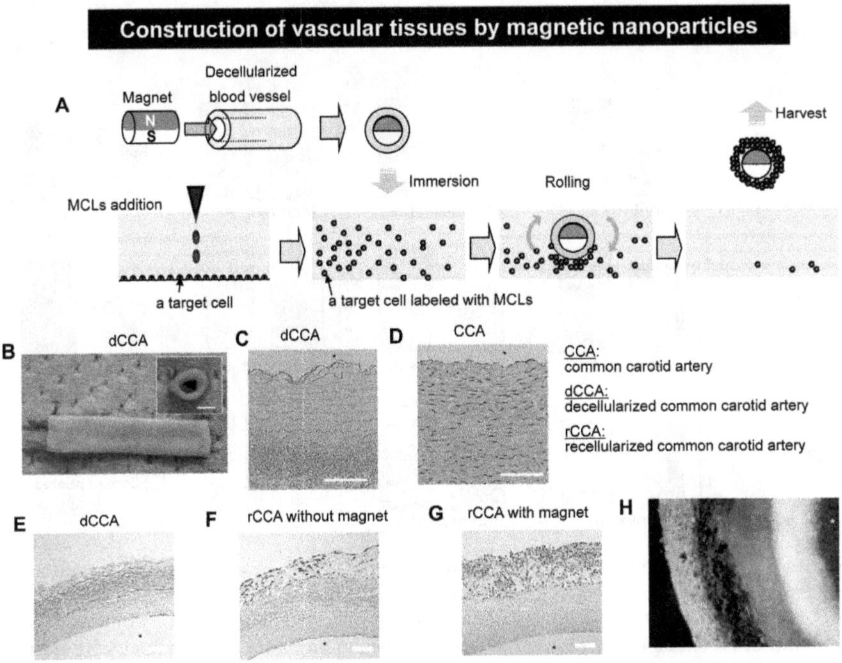

Figure 3. (A) A schematic illustration of Mag-seeding onto tubular scaffolds. MCLs were added to the medium to label target cells magnetically. Decellularized blood vessel into which a magnet was inserted was immersed in the cell suspension and rolled in the suspension to seed cell evenly. (B) A gross view of decellularized common carotid arteries (dCCA). The dCCAs were cut into 20 mm for subsequent experiments. Scale bar in the right upper box, 2 mm. (C, D) Cross-sectional view of dCCA (C) and CCA (D) with HE stain (E, F, G). Cross section of dCCA seeded with 3T3s with (F, G) or without (E) magnet. * Lumen. (H) Fluorescence microscopy of the cross section of dCCA seeded with Smooth muscle cells (SMCs) and Normal Human Dermal Fibroblasts (NHDFs). SMCs and NHDFs were probed with blue and green dye, respectively. * Lumen (Cited from Shimizu et al. J. Biosci. Bioeng. 2007) {Shimizu, 2007 2755 /id}.

Chapter 14

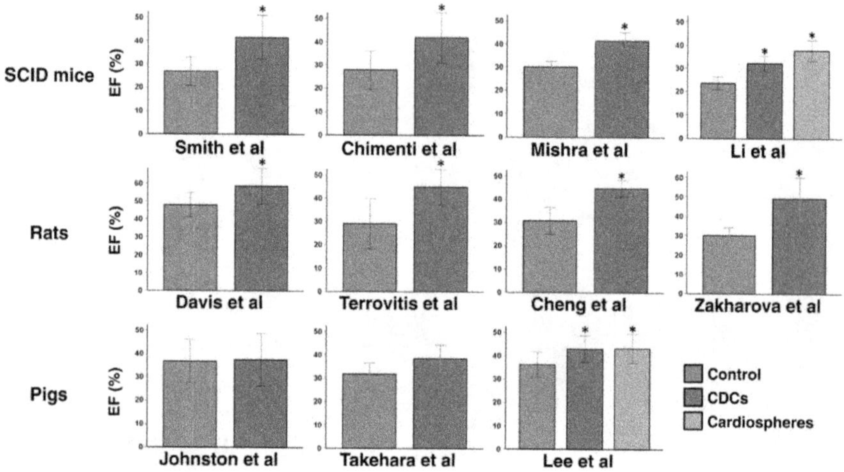

Figure 2. CDCs and CSps improve cardiac function post-MI in mice, rats and pigs, compared with sham-injected controls (*$P < 0.05$ compared with controls; in the study by Zakharova et al.,(Zakharova et al. 2010) CDCs were transplanted as a sheet, while in the study by Takehara et al. (Takehara et al. 2008) CDCs were injected with a basic fibroblast growth factor incorporating hydrogel). Reproduced from reference (Malliaras and Marbán 2011).

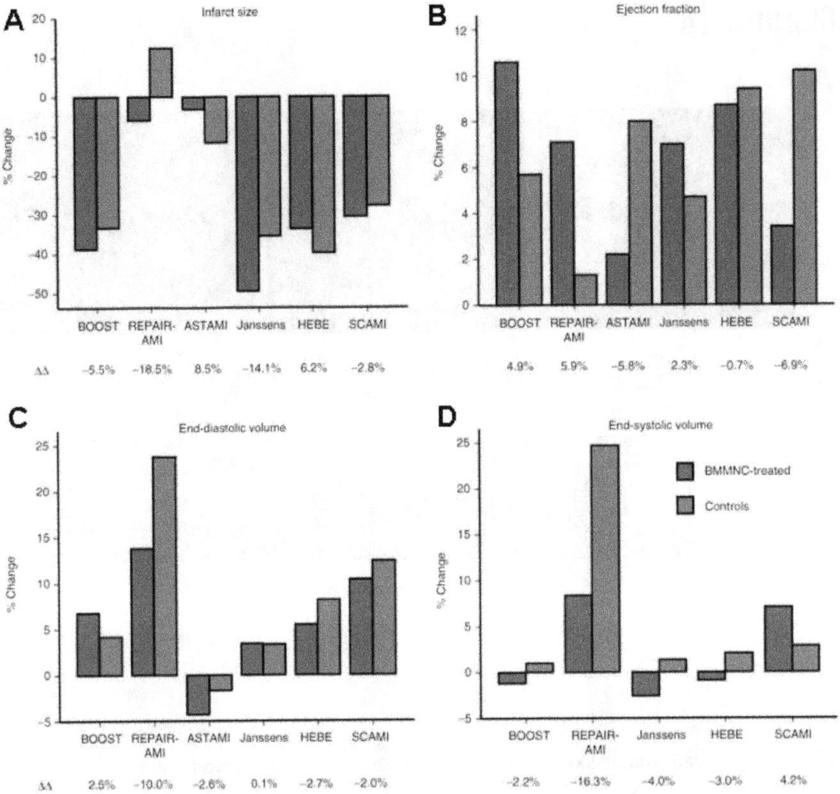

Figure 4. BMMNC-therapy treatment effects on cardiac structure and function as measured with magnetic resonance imaging (MRI). A synopsis of the cell therapy treatment effects on infarct size (A), left-ventricular ejection fraction (B), end-diastolic volume (C), and end-systolic volume (D) as measured with cardiac MRI in the large randomized controlled intracoronary BMMNC trials is provided. BOOST (Meyer et al. 2006), REPAIR-AMI (Dill et al. 2009), ASTAMI (Lunde et al. 2006), HEBE (Hirsch et al. 2011), SCAMI (Wöhrle et al. 2010), and the study by Janssens et al. (Janssens et al. 2006) are included in the analysis. The relative treatment effect normalized to baseline (% change) is presented for both treated patients and controls; the difference in relative treatment effects between treated patients and controls (ΔΔ) is also provided. BMMNC, bone marrow mononuclear cell. Reproduced from reference (Malliaras et al. 2011).

Chapter 16

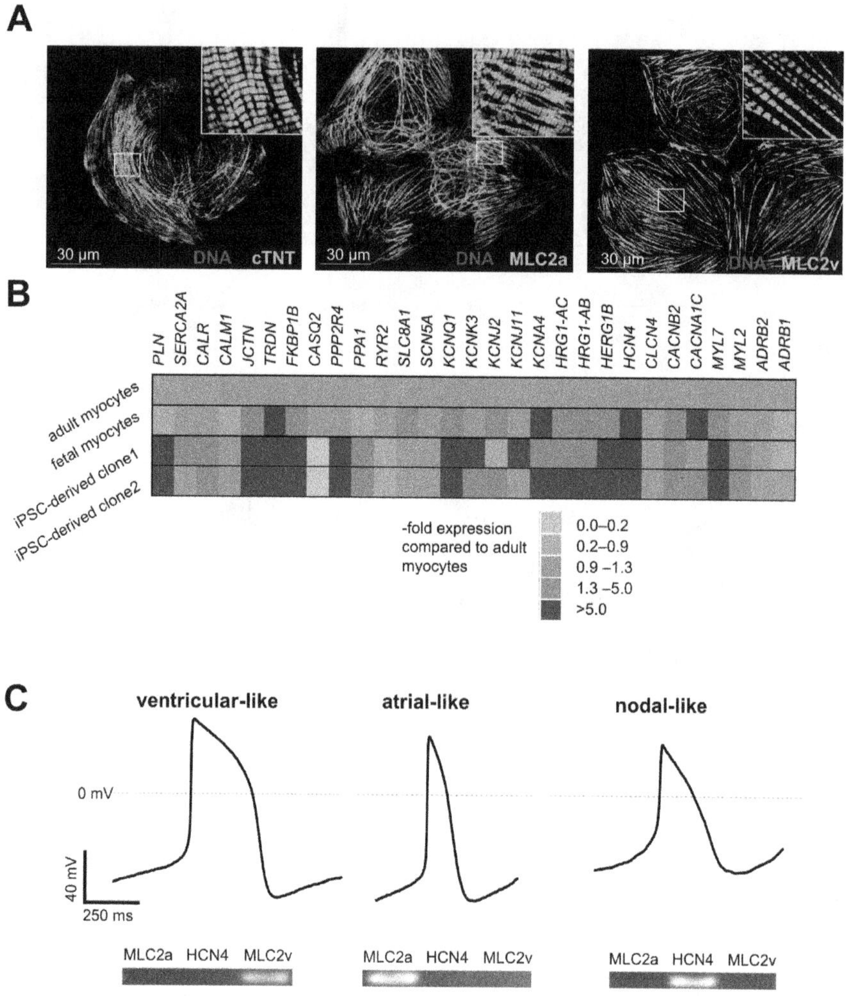

Figure 2. Cardiomyocytes Derived from Human Induced Pluripotent Stem Cells. Panel (A) shows immunofluorescence images of representative iPSC-derived cardiomyocytes stained for cardiac troponin T (cTnT) and the atrial (MLC2a) or the ventricular (MLC2v) isoform of myosin light chain 2 as indicated. The insets show higher magnifications of the areas indicted by the white squares. Panel (B) shows the result of a gene expression analysis using quantitative RT-PCR with primers specific for the indicated cardiomyocyte-specific genes in adult cardiomyocytes, fetal cardiomyocytes and cardiomyocytes generated from two different clones of iPSC. The expression levels are reported relative to the expression in adult myocytes. Panel (C) shows representative action potential recordings of ventricular-like, atrial-like and nodal-like iPSC-derived myocytes as indicated (upper panel) along with representative gels loaded with DNA samples generated by single-cell-PCR using specific primers for the indicated transcripts (lower panel). HCN4, hyperpolarization-activated, cyclic nucleotide-gated channel 4.

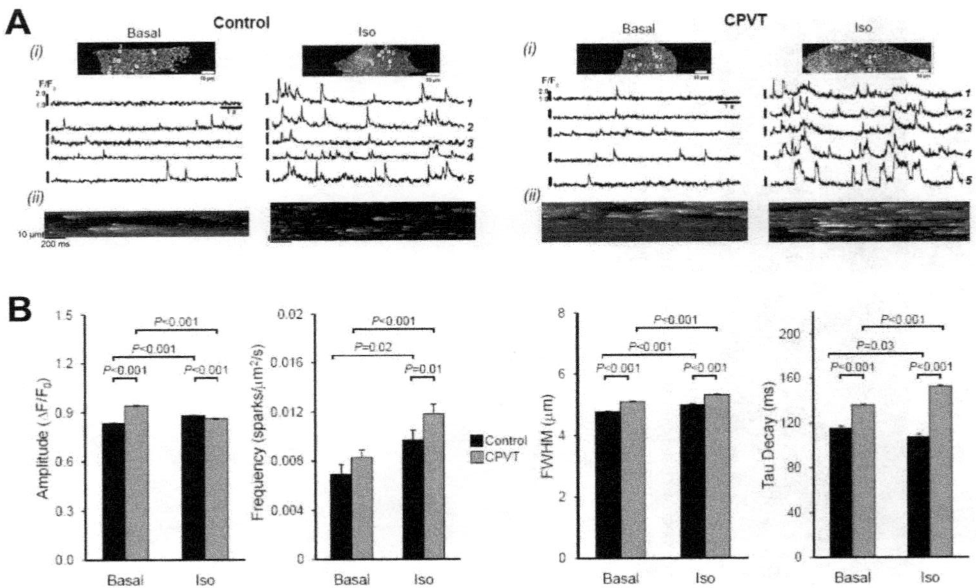

Figure 4. Enhanced Basal and Isoproterenol-Induced Calcium Spark Activity in Cardiomyocytes Derived from CPVT Patient-Specific iPSC. Panel (A) shows representative results of calcium spark imaging in fluo-4-AM-loaded control and CPVT myocytes in the absence (Basal) or in the presence (Iso) of 1 μM isoproterenol. Part (i) shows pseudo-colored images (upper row) together with typical Ca^{2+} traces corresponding to each of the 5 individual regions of interest marked in the top images, imaged at 105 images/s (lower row). Part (*ii*) shows line-scan images of Ca^{2+} sparks at a higher temporal resolution (1000 lines/s). Panel (B) shows the statistics of Ca^{2+} spark characteristics from control (black) and CPVT (red) myocytes in absence or presence of 1 μM isoproterenol. FWHM, full width at half maximum. Between 38 and 142 cells from 3 iPSC lines were analyzed in each group. Data are means ± s.e.m. from 4 independent experiments. *P* values from two-tailed *t* test are given. Experiments were performed in collaboration with Peter Lipp at the University of Homburg/Saar, Germany.

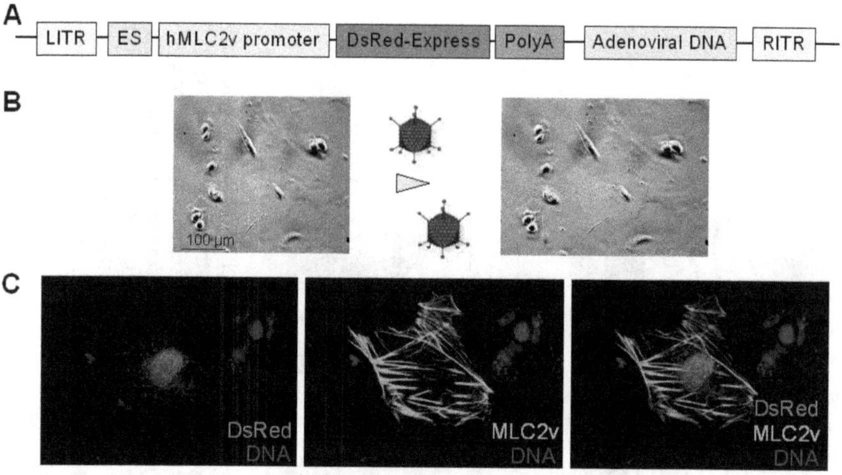

Figure 5. Genetic Marking of Ventricular Cardiomyocytes. Panel (A) shows a schematic representation of the adenoviral vector construct used for genetic marking. Panel (B) shows a phase contrast image (left image) of iPSC-derived cardiomyocytes infected with this virus and the same image merged with the red fluorescence of DsRed (right image). Panel (C) shows a closer magnification of a DsRed-expressing cardiomyocyte (DsRed fluorescence in red) merged with nuclear staining (blue) and immunofluorescent staining for the ventricular isoform of myosin light chain 2v (MLC2v, green) as indicated.

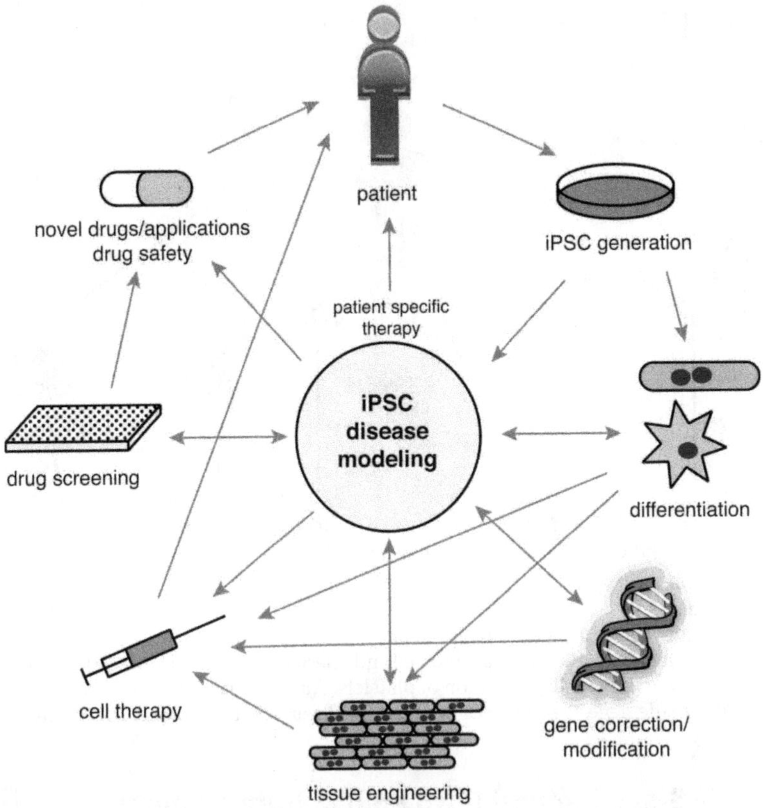

Figure 6. Disease Modeling at the Center of Potential iPSC Future Applications. Generation of patient-specific iPSC and differentiation into diseased cell types is the prerequisite for *in vitro* disease modeling by iPSC. Genetic modifications and gene repair will improve disease models and might find application in future cell therapies. While defects on a single cell and subcellular level have been successfully modeled using iPSC, engineered tissue might be a key of future disease models addressing defects involving cell-cell interactions, potentially involving different cell types. iPSC-based disease models might be a platform for development of new drugs and a new tool to improve drug safety. Finally, in some cases iPSC-based disease models will help to find specific therapies for single patients, particularly in multigenetic diseases or diseases with unknown pathophysiology.

Chapter 17

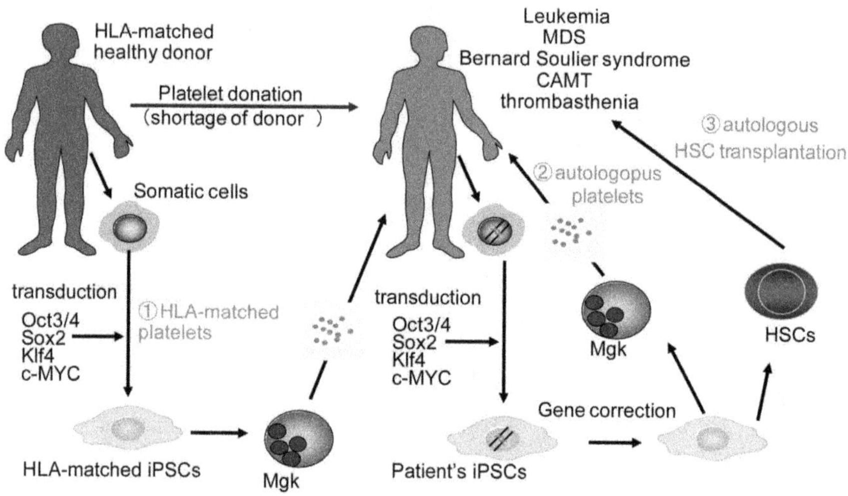

Figure 1. Alternative treatment of thrombocytopenia using hiPSCs. iPS cell technology enables us to consider innovative therapy models in the field of hematology. iPS cells are a potential source for blood transfusion. Validated and selected iPS clones from an HLA-identical donor can be useful for stable generation of platelets. Alternatively, iPS cell technology may enable "tailor-made gene therapy" using normalized self-iPS-derived stem cells after gene correction.

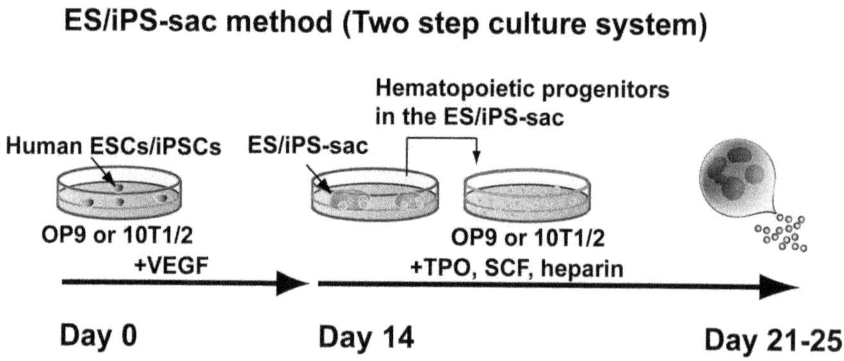

Figure 2. Human ESC/iPSC-derived megakaryocytes and platelets generated via ""sac"-like structures (ES/iPS-sacs). Schematic diagram of the in vitro differentiation protocol for hESC/hiPSC-derived megakaryocytes and platelets. (Step 1) hESC/iPSC are differentiated to Sac-like structure which contains multipotent hematopoietic progenitors on day 14 of culture.
(Step 2) Hematopoietic progenitors within sac are transferred to new feeder cells and further cultivated. Mature megakaryocytes yielding platelets are generated on day 21-25. VEGF, vascular endothelial growth factor; TPO, thrombopoietin; SCF, stem cell factor.

For Product Safety Concerns and Information please contact our EU representative GPSR@taylorandfrancis.com
Taylor & Francis Verlag GmbH, Kaufingerstraße 24, 80331 München, Germany

www.ingramcontent.com/pod-product-compliance
Lightning Source LLC
Chambersburg PA
CBHW060747150926
51501CB00003B/115
9781466578395